CONTENTS

PART 2

MACHINE TOOL DESIGN

v

Contents

METAL CUTTING AND GRINDING

AUTOMATION

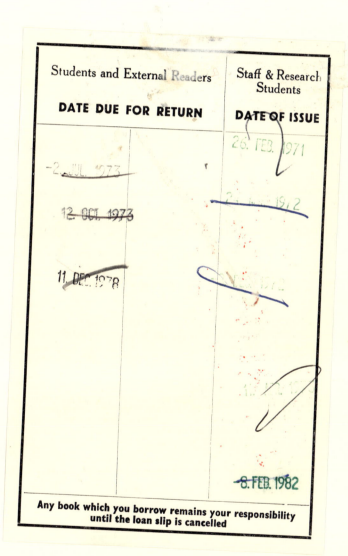

ADVANCES IN
MACHINE TOOL DESIGN
AND RESEARCH
1968

PART 2

ADVANCES IN

MACHINE TOOL DESIGN AND RESEARCH

1968

(in two parts)

PROCEEDINGS OF THE

9TH INTERNATIONAL M.T.D.R. CONFERENCE

THE UNIVERSITY OF BIRMINGHAM

SEPTEMBER 1968

PART 2

EDITED BY

S. A. TOBIAS

AND

F. KOENIGSBERGER

PERGAMON PRESS

OXFORD · LONDON · EDINBURGH · NEW YORK

TORONTO · SYDNEY · PARIS · BRAUNSCHWEIG

Pergamon Press Ltd., Headington Hill Hall, Oxford
4 & 5 Fitzroy Square, London W.1

Pergamon Press (Scotland) Ltd., 2 & 3 Teviot Place, Edinburgh 1

Pergamon Press Inc., Maxwell House, Fairview Park, Elmsford, New York 10523

Pergamon of Canada Ltd., 207 Queen's Quay West, Toronto 1

Pergamon Press (Aust.) Pty. Ltd., 19a Boundary Street, Rushcutters Bay,
N.S.W. 2011, Australia

Pergamon Press S.A.R.L., 24 rue des Écoles, Paris 5^e

Vieweg & Sohn GmbH, Burgplatz 1, Braunschweig

First edition 1969

Library of Congress Catalog Card No. 74–79867

Printed in Great Britain by Page Bros. (Norwich) Ltd., Norwich

08 013369 X

PART 1

HIGH ENERGY RATE FORMING

Contents

DEVELOPMENT OF DIGITAL COMPUTER PROGRAMS FOR THE ANALYSIS OF MACHINE TOOL STRUCTURES AND ELEMENTS

H. Opitz and W. Döpper

Technological University of Aachen, Germany

1. INTRODUCTION

During the last few years the demand on power and workpiece accuracy of a machine tool has very much increased. This is based on the new technologies of aerospace industries and the increasing applicability of numerically controlled machines. The experience in this case showed that both requirements, power and accuracy, cannot always be fulfilled concerning the dynamic stability of machine tools. The stability border is responsible for the workpiece accuracy and in some cases it does not allow us to use all the installed power of a machine.

The behavior of machine tools built in series can still be corrected by analyzing a prototype, but this method is impracticable for special machine tools. Here it is clearly necessary to precalculate the static and dynamic behavior of machine tools already in the state of design. Rough estimations as normally carried out are completely insufficient and even great experience is not sufficient if new projects have to be designed without a basis on which to work.

Methods to calculate the behavior of machine tools with accuracy are often unknown or developed insufficiently at present. If they are known the effort of computation is such that it is impossible to calculate by hand. Common analog computers are usually too small, although differential equations and systems of those equations describing the dynamic behavior of machine tools can be solved better on analog computers than on digital computers. Apart from the bigger capacity of digital computers it is also advisable to use them in this case because these methods, once they are programmed, can be used again and again and even by people who cannot program any statement at all.

Owing to these facts the aim of our work is now quite clear: digital computer programs will be set and be placed at any designer's disposal. These programs are based on methods which in several cases have to be developed or refined before programming. With these programs the designer can already get information on the static and dynamic behavior of his machine tool in the design state without calculating by himself, i.e. practical aids in form of digital computer programs have to be developed for the designer.

2. COMPUTER PROGRAM AS AID FOR DESIGNING

As shown in the block diagram of Fig. 1, a design order has to be analyzed first in order to find practical starting points for this work on one hand and reasonable limitations on the other hand.

The diagram shows that the design work is delimited to the sales department and also to the production. According to the formulation of the design order a first rough sketch can be made which has to be calculated. The computation results have to be compared with the demands which are also given in the order formulation. If they are fulfilled the workshop drawing has to be prepared, otherwise the design has to be changed, improved and computed again until the demands are fulfilled.

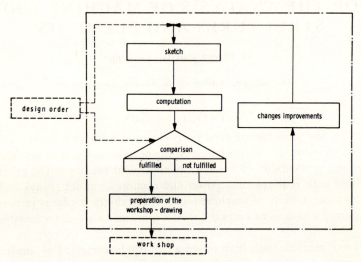

FIG. 1. General block diagram for designing.

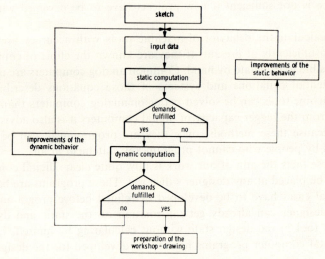

FIG. 2. Block diagram for a static and dynamic loaded construction.

Looking at the computation in more detail it turns out that for statically and dynamically loaded machine tool elements a static and a dynamic calculation has to be done (Fig. 2).

In the block diagram of Fig. 2 a double loop occurs, showing that the influence of static and dynamic improvements can be contrary. Furthermore, it is shown in this figure that changes and improvements of the design only require changed input data for the computation. If the method of calculation is right, the accuracy of the results therefore depends on how carefully the input data were prepared. Three important categories of data can be indicated, as shown in Fig. 3:

1. Data which can be calculated exactly on the basis of drawings, called geometrical data.
2. Data which can be determined on the basis of the geometrical information obtained under 1.
3. Data obtained experimentally and from experience.

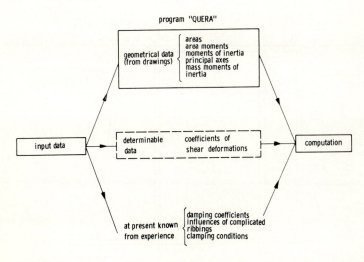

FIG. 3. Classification of the input data for the computation of a machine tool element.

It is very complicated to calculate all the data under 1 and some of them under 2 by hand. For this reason a program called QUERA has been developed in order to calculate those data also by means of a computer.

The handling of this program will be briefly described to get an impression of how much effort is required to use a program, especially if the user is not familiar with the application of digital computers.

In Fig. 4 an arbitrary cross-section of a machine tool column is given. To calculate all cross-section values which are of interest, the designer has to declare an arbitrary chosen perpendicular system of coordinates. The cross-section itself is assumed to be a polygon and the points have to be numbered in an ascending way while the material should always be on the left-hand side.

This requirement is given by Gauss' Integral Law which is the basis of this method. Finally, the coordinates of all points have to be noted by the user.

In the table of Fig. 5 all input data of a cross-section are shown as they are put out by the computer for checking purposes.

All computed cross-section values are shown in Fig. 6 as they are given by the computer.

Computer programs of this type can be handled even by untrained people. The computing time is only a few seconds on a CD 6400 where one second amounts to DM 1,– in Aachen.

3. COMPUTATION OF MACHINE TOOL ELEMENTS AND MACHINE TOOLS

From literature some different methods are known for calculating the tooth-deflection of a gear. The theoretical work published by C. Weber and Banaschek solves the problem most satisfactorily (Fig. 7). Here, the computed total deflection in direction of the force P_n, consists of two components: the deformation of the tooth and the share of the wheel-body. However, an exact calculation of the deflection w depending on the position of the

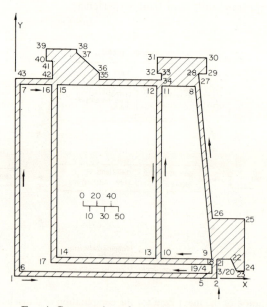

FIG. 4. Cross-section of a machine tool column.

INPUT DATA

NUMBER OF POINTS		43
LENGTH OF THE ELEMENT		1.380000000000000 E + 02
SPECIFIC MASS		8.099999999999987 E - 10
SPECIFIC DENSITY		7.850000000000001 E - 06

COORDINATES OF THE POINTS

POINT NR.	X - AXIS	Y-AXIS
1	0.	0.
2	2.810000 00000000 E + 02	0.
3	2.810000 00000000 E + 02	1.000000000000000 E + 01
4	2.730000 00000000 E + 02	1.000000000000000 E + 01
5	2.730000 00000000 E + 02	5.000000000000000 E + 00
6	5.500000 00000000 E + 00	5.000000000000000 E + 00
7	5.500000 00000000 E + 00	2.530000000000000 E + 02
8	2.430000 00000000 E + 02	2.530000000000000 E + 02
9	2.715000 00000000 E + 02	2.520000000000005 E + 01
10	2.005000 00000000 E + 02	2.520000000000005 E + 01
11	2.005000 00000000 E + 02	2.521999999999998 E + 02
12	1.955000 00000000 E + 02	2.521999999999998 E + 02
13	1.955000 00000000 E + 02	2.520000000000005 E + 01
14	5.850000 00000000 E + 01	2.520000000000005 E + 01
15	-5.850000 00000000 E + 01	2.525000000000000 E + 02
16	5.350000 00000000 E + 01	2.521999999999998 E + 02
17	5.350000 00000000 E + 01	2.020000000000005 E + 01
18	2.725000 00000000 E + 02	2.020000000000005 E + 01
19	2.735000 00000000 E + 02	1.000000000000000 E + 01
20	2.815000 00000000 E + 02	1.000000000000000 E + 01

FIG. 5. Input information (for program QUERA).

force acting point requires considerable effort due to the solution of the integrals I_1 and I_2 in Fig. 7 as no explicit solutions are known. The solution of these integrals would be possible by the aid of a numerical method programmed for a digital computer. A comparison between the costs for a conventional and a computer aided computation shows quite well the economy of the programmed solution: the costs of the conventional method would be about

RESULTS

SHEAR AREA	1.646210000000126 E + 04
NEUTRAL X - AXIS	1.657594231598659 E + 02
NEUTRAL Y - AXIS	1.526712716887050 E + 02
X - AXIS AREA MOMENT	2.513289741666824 E + 06
Y - AXIS AREA MOMENT	2.728748200000241 E + 06
X - AXIS INERTIA MOMENT	5.739666355889435 E + 08
Y - AXIS INERTIA MOMENT	6.178074763687782 E + 08
NEUTRAL X - AXIS INERTIA MOMENT	1.902594946064949 E + 08
NEUTRAL Y - AXIS INERTIA MOMENT	1.654917487882195 E + 08
COUNTER CLOCKWISE ANGLE OF ROTATION OF PRINCIPLE AXES	3.721662272419167 E + 01
MOMENT OF INERTIA ABOUT PRINCIPLE X - AXIS	2.239131314025421 E + 08
MOMENT OF INERTIA ABOUT PRINCIPLE Y - AXIS	1.318381119921713 E + 08
POLAR MASS MOMENT OF INERTIA	1.988293699333053 E + 01
TRANSVERSE MOMENT ABOUT THE X - AXIS	1.207914765633603 E + 01
TRANSVERSE MOMENT ABOUT THE Y - AXIS	1.069487834255261 E + 01
TRANSVERSE MOMENT ABOUT THE PRINCIPLE X - AXIS	1.396004941686709 E + 01
TRANSVERSE MOMENT ABOUT THE PRINCIPLE Y - AXIS	8.813976582021496 E + 00
WEIGHT OF ELEMENT	8.916696465000598 E + 00

FIG. 6. Results (of program QUERA).

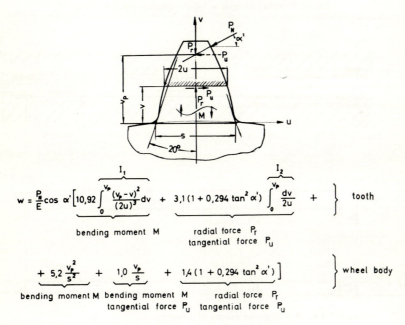

$$w = \frac{P_N}{E}\cos \alpha' \left[\underbrace{10,92 \int_0^{v_p} \frac{(v_p - v)^2}{(2u)^3} dv}_{\substack{\text{bending moment } M}} + \underbrace{3,1(1 + 0,294 \tan^2 \alpha') \int_0^{v_p} \frac{dv}{2u}}_{\substack{\text{radial force } P_r \\ \text{tangential force } P_u}} + \right.$$

$$\left. + \underbrace{5,2 \frac{v_p^2}{s^2}}_{\substack{\text{bending moment } M}} + \underbrace{1,0 \frac{v_p}{s}}_{\substack{\text{bending moment } M}} + \underbrace{1,4(1 + 0,294 \tan^2 \alpha')}_{\substack{\text{radial force } P_r \\ \text{tangential force } P_u}} \right]$$

FIG. 7. Deformations of a single tooth.

DM 200,– as estimated by a well-trained engineer. On the other hand, the computer needs about 2 sec. Including the time for data preparation, the costs would be approximately DM 20,–.

Furthermore, a set of programs has been developed for the computation of machine tool elements and machine tools which is given in the table of Fig. 8. The programs SPINDEL 1 and SPINDEL 2 have already been running for more than 250 hours and can be regarded as fully proved. The handling of these programs is shown in Figs. 9 and 10. It can be seen that a given spindle system has to be abstracted into a mathematical model. The information

PROGRAM	COMPUTES
SPINDEL 1	Static deformations of spindle systems.
SPINDEL 2	Natural frequencies and mode shapes of spindle systems.
MESTA 1	Static deformations of single pinned beam systems. (One column machines)
MESTA 2	Static deformations of multiple pinned beam systems. (Portalframes)
DYNA	Natural frequencies and mode shapes of single pinned beam systems for an arbitrary chosen degree of freedom. (One column machines)
DYNA 3	Natural frequencies, mode shapes and response locus of single pinned beam system for an arbitrary chosen degree of freedom. (One column machines)
DYNA 4/5 (in state of test)	Natural frequencies, mode shapes and response locus of multiple pinned beam systems for an arbitrary chosen degree of freedom. (Portalframes)

Fig. 8. Existing programs for computing machine tools.

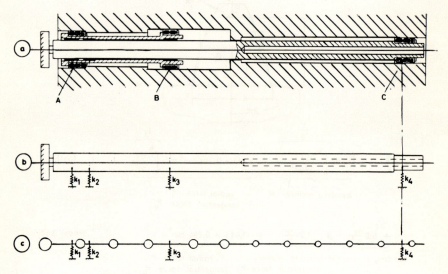

Fig. 9. Computer model of a boring spindle system.

at all declared points of the system has to be listed in a given form. After that, it is transferred onto cards and put into the computer. The accuracy to be achieved computing the static deformation can be seen from Fig. 11 showing the results of a lathe spindle system.

Figure 12 shows the measured and computed undamped natural frequencies of a spindle system of a boring machine. The computed values are compared with the measured results of five machine tools of the same type for different overhangs of the spindle system. The highest and lowest natural frequency of any of the five machines is plotted for each overhang. Furthermore, the average value of all measurements is indicated. The figure shows that nearly all computed results were found within the variation of the five machines.

Another program which has been developed is MESTA 1, which is based on a mathematical method different from that of the spindle programs. The program allows the influence of joints to be included at the foundation and between the elements of the machine.

Card No.	12	3456	7890	1234	567890	1	234567	890123	456789	012345
	P Nr.	da (mm)	di (mm)	lk (mm)	Power (Kp)		K (Kp/mm)	Weight (Kp)	K Kpm/Gr.rad	M (mkp)
1	01	0120								
2	02	0120	0000	0050	000 035	1	100 000	035 000		
3	03	0120		0080						
4	04	0130		0040	000 000	0	100 000	008 000		
5	05	0130		0040						
6	06	0190		0100						
7	07	0190		0050				067 000		
8	08	0190		0150						
9	09	0130		0090				019 000		
10	10	0130		0090						
11	11	0180		0045				018 000		
12	12	0180		0045						
13	13	0130		0090				008 500		
14	14	0130		0045						
15	15	0160		0060				019 000		
16	16	0160		0060						
17	17	0130		0125				025 000		
18	18	0130		0125						
19	19	0130	0060	0200				033 000		
20	20	0130	0060	0450	075 000			040 000		
21	21	0090	0060	0060				005 000		
22	22	0090	0060	0090						
23	23									
24	24									
25	25									
26	26									
27	27									
28	28									
29	29									
30	30									
31	99	Final Card								

Fig. 10. Form for data preparation (spindle computation).

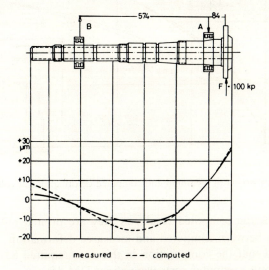

—·— measured   --- computed

Fig. 11. Comparison of measured and computed spindle deformations.

In Fig. 13 the measured and computed results of two simple column models are compared. The computation includes the shear-force deformation. Having done some computations on simple models with different ribbings more complicated ones were taken into consideration. Figure 14 shows the computation of a column with a complicated cross-section which has already been shown above to explain the program QUERA.

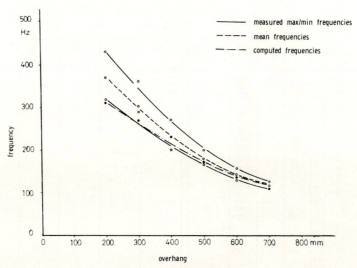

FIG. 12. Comparison of measured and computed natural frequencies of a spindle system.

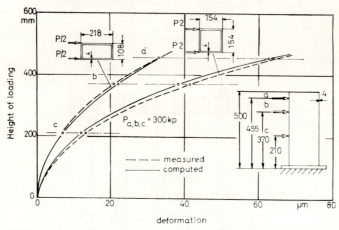

FIG. 13. Comparison of measured and computed deformations of column models.

The figure contains the computed results as they are printed out by the computer as well as a comparison between computed and measured deformation. Regarding the cost, it should be mentioned that the computation in this case required 85 sec on the CD 6400, in other words, the pure computer costs were about DM 85,–.

SCHEME OF LOADING

NR	PX	PY	PZ	MX	MY	MZ
1	0.	0.	0.	0.	0.	0.
2	0.	0.	0.	0.	0.	0.
3	0.	0.	0.	0.	0.	0.
4	0.	0.	0.	0.	0.	0.
5	0.	0.	0.	0.	0.	0.
6	0.	1.00000000E+00	0.	0.	0.	0.
7	0.	0.	0.	0.	0.	0.
8	0.	0.	0.	0.	0.	0.
9	0.	0.	0.	0.	0.	0.

DEFORMATION

NR	SX	SY	SZ	PHIX	PHIY	PHIZ
1	0.	2.87149803E-06	0.	0.	0.	1.55664124E-08
2	0.	6.70762955E-06	0.	0.	0.	2.48311263E-08
3	0.	9.84207133E-06	0.	0.	0.	3.01128992E-08
4	0.	1.89381289E-05	0.	0.	0.	5.63433402E-08
5	0.	2.92808977E-05	0.	0.	0.	6.03255976E-08
6	0.	3.46573639E-05	0.	0.	0.	6.09075445E-08
7	0.	3.88599845E-05	0.	0.	0.	6.09075445E-08
8	0.	4.29407900E-05	0.	0.	0.	6.09075445E-08
9	0.	4.33671428E-05	0.	0.	0.	6.09075445E-08

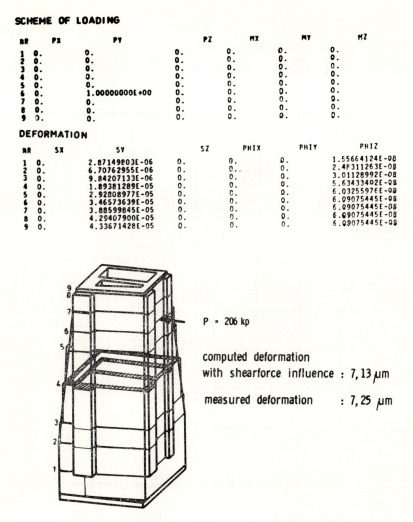

P = 206 kp

computed deformation
with shearforce influence : 7,13 μm

measured deformation : 7,25 μm

FIG. 14. Comparison of measured and computed deformations of a column.

However, it must be pointed out that the experience gained with the programs listed in Fig. 8 is not sufficient to guarantee the accuracy obtained in this example. Important parameters influencing the accuracy of computation are the different types of joints. A generally valid and simple method to determine the stiffness of joints is not yet available. It is, however, no problem to include them into the computation if those parameters are known.

To demonstrate the applicability of the DYNA programs all computations which are possible by means of these programs are put together in Fig. 15 using the above-mentioned column as an example.

The difference between the measured and computed undamped natural frequencies and response loci could be kept within the limit 15%. All input data which had to be prepared for the MESTA and DYNA program have been precalculated by the QUERA program, except the damping coefficients, which were assumed from experience.

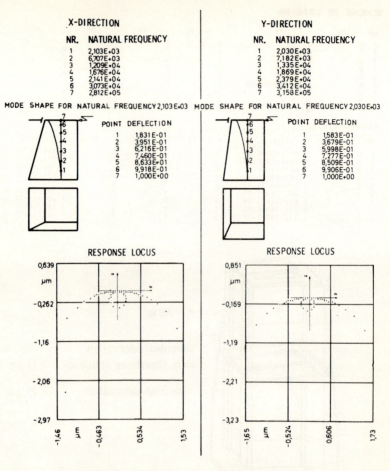

FIG. 15. Results of the dynamic computation for a column.

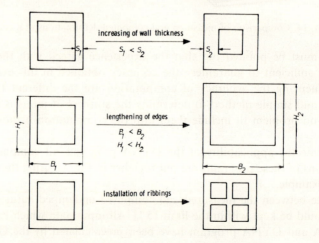

FIG. 16. Possibilities to improve the static flexibility.

4. PROSPECT ON FURTHER DEVELOPMENT

Looking at Fig. 2 it appears that the static and dynamic behavior of machine tools and machine tool elements can be predicted with sufficient accuracy in the near future by means of digital computer programs. Thus, these programs are suitable to reduce the effort of design and to diminish the risk of a new development.

Another important development becomes conceivable in Fig. 2. The task of the designer is represented by the block changes and improvements. The specifications of the design order can be met by means of a simple comparison.

If the known improvements are tabulated and programmed successfully the comparison may be done automatically by the program. In this case the possibility of automatic design becomes feasible.

It is not the purpose of this paper to discuss whether or not automatic designing will ever be possible in a general and complete way. On the other hand, there is no doubt that this possibility would be very helpful in cases where a standard part has to be varied continually in a limited range owing to different loadings or other requirements. First applications of this method are already known from American press builders.

To investigate this possibility in general a static test program AKM 1 was set up and applied first to the design of a machine tool column. The input information for this program consists of a rough geometrical description of the column in different heights and the required stiffness in two directions at an arbitrary chosen point.

The program improves the given construction gradually until the desired stiffness is reached and also prints out the drawings of the cross-sections at some discrete heights.

The ways of changing and improving the static behavior of the column, which this program is able to do, are the increasing of wall thickness, lengthening of edges and building in ribbings (Fig. 16).

Apart from increasing the thickness of the outside walls every improvement can be made in two directions. Thus, seven possibilities altogether are available.

How the program works can easily be seen from the following simple example. The diagrams shown here are all original outputs of the computer.

In Fig. 17 the rectangular cross-section of a box beam or a column is given. This first

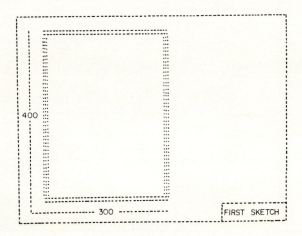

Fig. 17. First sketch.

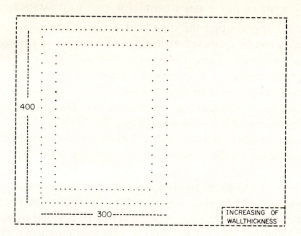

Fig. 18. Increasing of walls.

The required stiffness for the system was fulfilled as follows :

	Required stiffness (MICROMETER/KP)	Fulfilled stiffness (MICROMETER/KP)
in Y – direction	2.349999999999999E-05	2.163558066436890E-05
in Z – direction	2.349999999999999E-05	2.163558066436890E-05

Indication of the cross-sections of all elements of the system in rising stages

Fig. 19. Data for the first possible cross-section.

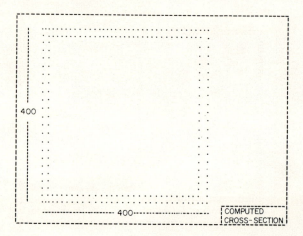

Fig. 20. First possible cross-section.

sketch is automatically checked whether or not it already fulfils the demanded requirements. In this special case, equal stiffness for both directions was demanded. Thus, a quadratic cross-section can be expected. As shown in Fig. 18, the computer first increased the wall thickness. By this time the requirements were not yet fulfilled. The computer therefore increased the length of the smaller edge in several discrete steps; the appropriate stages are not shown here. Finally, a cross-section was produced which fulfilled the requirements. This is indicated by the data of Fig. 19; the cross-section itself is shown in Fig. 20, a quadratic cross-section as expected.

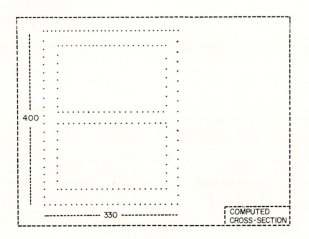

FIG. 21. Second possible cross-section.

However, this is only one of the results produced by the computer. In Fig. 21 another result is shown which also fulfils the requirements but in addition it includes some special limitations concerning the maximum outside contours. The required stiffness in this case was achieved by building a ribbing into a rectangular. Finally, at this point it turns out that if the computer can bring out different design results it would appear feasible to optimize these results with the aid of the computer.

5. CONCLUSIONS

It can be pointed out that a set of programs has already been developed which can be regarded as a practical aid for designers.

The first results of these programs are very encouraging. However, one must remember that the accuracy of the results depends on the accuracy of the input data, which have to be carefully prepared.

REFERENCES

1. H. FLESSNER, Ein Beitrag zur Ermittlung von Querschnittswerten, *Bauingenieur*, Heft **4**, 1962.
2. W. DÖPPER, Möglichkeiten zur Berechnung von Werkzeugmaschinen mit Digitalrechner-Programmen, *Industrieanzeiger*, No. 17, 1968
3. H. OPITZ, Chatter Behavior of Heavy Machine Tools, Quarterly Technical Reports AF 61 (052)-916, I, II, III, IV, VI.

COMPUTER AIDED DESIGN OF A PLANING MACHINE STRUCTURE

S. TAYLOR

Department of Mechanical Engineering, University of Birmingham

1. INTRODUCTION

The machine tool industry stands to benefit considerably, both in improved designs and in reductions of design time, cost and development time, by adopting the digital computing techniques which are currently becoming available. Moreover, computing is increasing in importance because the cost of computing continues to fall and is likely to continue decreasing whereas traditional design procedures are progressively becoming more expensive.

The major cost and work involved in a computer application are those of writing, testing, developing and verifying the program. This is comparable with the work involved in designing, testing and developing a machine tool or other product. Programs for major computations are best prepared by experts in the appropriate fields, the cost of each program being spread over many applications in the same way that the design costs to a machine tool are spread among the purchasers. Programs which are applicable to a wide range of design calculations have been prepared by M.T.I.R.A., P.E.R.A., universities and other organisations and are available to industry or at least to supporting organisations. Modern computer languages are sufficiently simple for companies to write their own programs for the many simple calculations which are useful in a design office.

The design of machine tool structures is a typical and important example of a computer design procedure. A computer technique for analysing a wide variety of machine tool structures is described below and the use of the analyses for design purposes is discussed. This technique is now available for use by industrial designers for whose applications the system is intended. Special care has been taken to ensure that the use of the programs is within the capabilities of machine tool designers and that computer experts are not required. The designer is only asked to learn an outline of computing and a formal numerical method of describing structures.

Industrial use of the programs is illustrated by computations used in the design of a planing machine. The computations were carried out by two men from the company concerned. They had no previous knowledge of computing but had full control over the work and made their own choices of alternative designs and other parameters.

2. ANALYSIS OF MACHINE TOOL STRUCTURES

A lumped parameter flexibility method for analysing machine tool structures has been developed by Taylor and Tobias.[1, 2, 3] As quite extensive calculations are involved, a digital computer is essential. However, the cost and work involved in analysing these structures by computer are well within the resources of most machine tool companies.

A machine tool structure is analysed in two stages. The first[1] depends on setting up a lumped parameter representation of the structure as shown in Fig. 1. The model consists of a number of lumped masses connected together by elastic beams and represents the general arrangement of the machine tool. This technique permits a complex structure to be described in convenient numerical terms and in a suitable form for digital computation. The configuration of the structure is defined by the three dimensional coordinates of the lumped masses, the mass and its distribution determined the magnitude of the lumped masses and the flexibility of the structure is represented by the beams which have bending

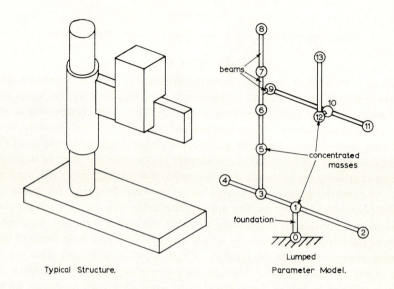

FIG. 1. Radial drill and lumped parameter model.

torsion, shear and axial flexibilities. Each elastic beam may be sub-divided into any number of small elements as necessary to describe the structure in sufficient detail. Each element gives rise to its own data which conveniently defines its mass in addition to its flexibility. The preparation of the element data is the predominant part of the computing work. The program for the first stage computes the static flexibilities at any of the lumped masses and the specified number of the lowest natural frequencies of vibration, each with its associated mode shape.

The second stage, the computation of responses to excitation and of dynamic stability,[2] uses the modes of vibration already computed and assumed values for the damping are introduced. The exciting forces acting on the structure and the relevant positions and directions of the displacements have to be specified. The response locus for these conditions is then computed. If the forces and displacements are those associated with a cutting operation, the resulting locus is an operative response locus from which the dynamic stability of the machine tool may be computed. Thus the vibration responses and the chatter resistance of the machine tool are entirely computed.

The computing techniques used are still basically those previously described[1, 2] but many detailed improvements have since been introduced to give an appreciable improvement in the efficiency of the system. These modifications and the present scope of the programs are reviewed in the Appendix.

3. DESIGN OF MACHINE TOOL STRUCTURES

This analytical procedure forms the basis of a design technique for machine tool structures. Since it can be used for predicting the behaviour of a machine tool under working conditions from information obtained at a very early stage in the design process, the optimum design can be approached very rapidly and before any manufacture has commenced.

The strength of computer techniques lies in their convenience for comparing alternative designs. If each possible alternative design for a machine tool is analysed in turn, the best design may be selected with due regard to cost and other factors. Comparisons are based on the assumption of identical beams, masses, joints, tolerances and other factors unless these are deliberately modified in some of the alternative designs. For example, the stiffness of a column depends on the stiffness of the connections at its base as well as on its structural stiffness. However, columns of different construction can be compared quite accurately using the same nominal estimate of the stiffness of the base joint because the consequent unknown error will be the same for each column. Similarly, stability computations rely on arbitrary assumptions concerning cutting conditions and material properties. Hence the computed dynamic stabilities for various designs are not precise but give a reliable comparison of the designs if several typical conditions are considered.

Once a structure has been analysed, modifications may very rapidly be introduced into the data and further computations quickly yield the required comparative information. The comparison of designs has been illustrated by the analysis of several designs for the column and arm of an openside planing machine.[3] Only the first stage of the computations was available at the time but useful deductions were made. A comparison of responses and dynamic stability for three designs of a horizontal milling machine were made in Ref. 2.

In contrast, model analysis cannot detect small changes since the effects of small modifications are masked by different construction and tolerance errors as well as experimental errors. Prototypes also have different tolerances and other effects which make similar machines difficult to compare. Hence computations are very good in comparative terms even though they may be poor in absolute terms.

4. PLANER-MILLING MACHINE STRUCTURE

The practical application of this design technique is illustrated by the design of a planer-milling machine.* The computations were carried out by two engineers from R.T.M.† under the guidance of the author. These engineers decided all the details of the computations and of the alternative designs themselves in the light of the company's needs. The men had no previous knowledge of computing but rapidly learnt how to apply the technique. They would not need any further guidance or training to use the technique in the design of any other machine tool. Hence this application is typical of the envisaged industrial use of the design technique.

4.1 Lumped Parameter Representation

The planer-milling machine is shown in Fig. 2 with the leading dimensions in mm. This is the configuration analysed and reported in Ref. 4. Figure 3 is the lumped parameter representation of the structure consisting of masses numbered 0 to 20 distributed over the

* Sant Eustacchio.
† Istituto per le Ricerche di Tecnologia Meccanica, Vico Canavese, Torino, Italy.

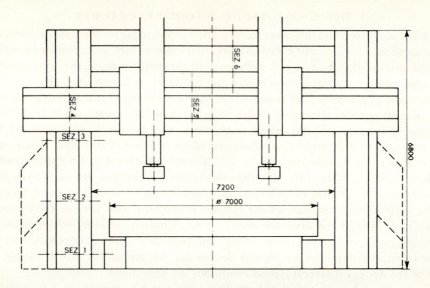

Fig. 2. Planer-milling machine.

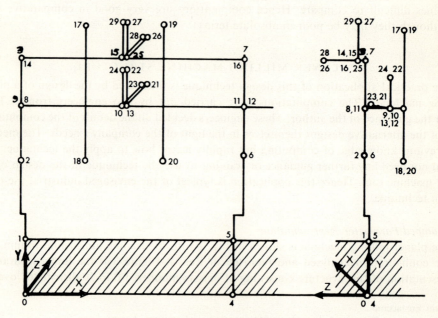

Fig. 3. Lumped parameter model of structure.

columns, the cross-rail, the top-rail and the tool boxes, these being connected together by elastic beams. Masses numbered 21 to 29 are temporary masses used in a device which allows computations to be made on structures which have closed loops. These disappear early in the computations so that there are effectively 20 masses excluding number 0 which is a fixed mass. Mass 4 is also a fixed mass and the beams connecting mass 0 with 1 and 4 with 5 were intended to represent the flexibility of the foundation, shown shaded. However, all the computations eventually were made for rigid foundations so that masses 1 and 5 are also fixed. The beams connecting mass 1 with 2 and 5 with 6 should preferably have been drawn as straight lines to correspond with the computer interpretation of the data.

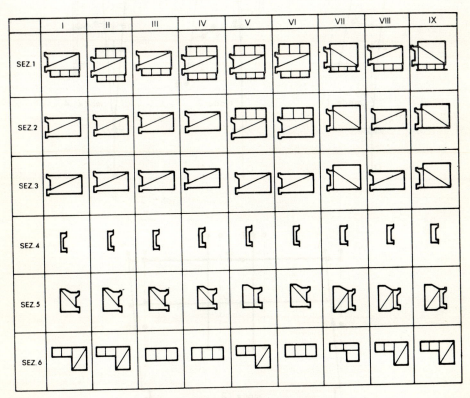

FIG. 4. Comparison of cross-sections.

Data was prepared for the original structure, I, by listing the coordinates of the masses and certain other details and then by preparing the data for each element. Each element may be used to represent bending in two planes, shear in two planes, torsion and axial deflection as well as mass distribution between the lumped masses and rotary inertia. This structure required sub-division into 37 elements to give sufficient detail.

A further eleven designs, numbered II to XII, were then analysed by modifying the data for structure I. The designs differed in the six cross-sections marked SEZ 1 to SEZ 6 in Fig. 2. The cross-sections for structures I to IX are shown in Fig. 4 which shows the various methods and combinations of methods proposed for stiffening the structure. Structure X is the same as I with the columns reinforced at the back and front, structure XI shows the effect of a very light top-rail on I and structure XII is the extreme case of I without a top-rail.

4.2 *Static Flexibilities and Modes of Vibration*

The static flexibilities were computed at mass 18, which is the position of a cutting tool. Those at mass 20 are the same because all the structures are symmetrical. Table 1 shows these flexibilities for each of the 12 structures, each flexibility being for both the force and the displacement parallel to the same coordinate axis. The natural frequencies of vibration of the

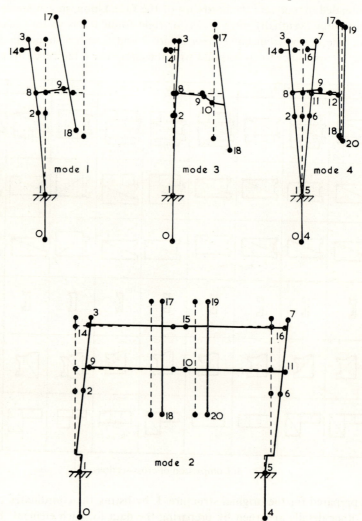

FIG. 5. Mode shapes.

first four modes, those with the lowest frequencies, are also given in Table 1. The deflected shape of each mode for structure I is shown in Fig. 5, the mode shapes for the other structures being similar. Broken lines are used for the undeflected structure and full lines for the mode shapes. Mode 1 consists of a rocking motion in the *z* direction for which the column stiffness* is the predominant factor. Mode 2 is a rocking motion in the *x* direction and the

---

* Stiffness is the reciprocal of flexibility.

column stiffness is again important. Torsional and bending deflections of the cross-rail are predominant in mode 3. In each of the modes 1 to 3 the two columns move together in phase, but in mode 4 the columns move in opposite directions, out of phase, in the z direction. None of the modes have appreciable movement perpendicular to the plane of its diagram, Fig. 5. Modes 3 and 4, Table 1, had close natural frequencies and are therefore likely to occur together in practice and were not always completely separated in the computations. In all, eight modes were computed for each structure but those described are the most important.

Examination of Table 1 shows that under static conditions structure V is the stiffest in the x direction and structure IX in both y and z directions. Structures II, V, VI and X had high frequencies for modes 1, 2 and 4 and may therefore be expected to have the stiffest

TABLE 1. FLEXIBILITIES AND FREQUENCIES

| Structure | Static flexibilities at mass 18 (10^{-6} mm/kgf) | | | Natural frequencies (Hz) | | | |
	x direction	y direction	z direction	Mode 1 Rocking z	Mode 2 Rocking x	Mode 3 Cross-rail	Mode 4 Torsion
I	26·1	24·6	76·1	12·9	14·2	22·7	23·7
II	22·2	24·1	75·4	13·8	17·1	22·8	26·0
III	28·3	24·6	76·2	13·1	13·1	22·6	22·0
IV	23·6	24·1	75·6	13·9	15·7	22·7	23·5
V	19·8	23·6	74·7	14·3	19·0	23·0	27·2
VI	20·7	23·6	74·9	14·5	17·6	22·9	24·7
VII	21·1	20·3	57·0	13·5	15·6	22·1	24·8
VIII	20·4	20·2	56·9	9·5	15·7*	20·3	14·0*
IX	23·7	19·7	53·8	12·9	14·0	26·7	23·9
X	24·0	23·3	75·0	14·7	15·0	23·1	26·6
XI	26·8	24·7	76·2	13·3	14·4	22·8	23·6
XII	29·6	26·1	76·7	14·9	15·2	22·4	23·5

* Mode shapes slightly different from other structures.

columns under dynamic conditions since these modes depend largely on the columns. Structure IX had the highest frequency for mode 3 which depends on the cross-rail stiffness. However, natural frequencies and mode shapes do not give a complete picture of the dynamic behaviour of a structure.

4.3 Response to Excitation and Dynamic Stability

The vibration characteristics of the structure under working conditions were compared by computing the dynamic stability of each structure for a selection of machining operations on the assumption that each structure had the same damping. The nature of the cutting constants was not defined but was assumed to be the same in each case.

The different conditions considered, columns 1 to 9 of Table 2, are various directions of the cutting or exciting force and of the relevant displacement. All are for excitation at the cutting tool at mass 18 and for displacement of the same point. Columns 1 to 4 represent planing operations where vertical displacements in the y direction, Fig. 3, are of interest. The exciting forces in columns 1, 3 and 4 are in the yz plane and at angles of 60°, 50° and 70° respectively to the y axis. The force used in column 2 is similar to that of column 1 except that the force is at 10° to the yz plane, measured towards the x axis. Columns 5 and 6 are for milling operations with displacements in the x direction and forces in the xz plane. The angles between the x axis and the force are respectively $-60°$ and $+60°$ measured towards the z axis. Columns 7, 8 and 9 are for both the exciting force and the displacement in the same direction and parallel respectively to the x, y and z axes. These last three are not true cutting conditions but they do give a realistic comparison of the dynamic stiffnesses of the structures when the force and displacement coincide. Since the structures are symmetrical, corresponding conditions at mass 20 would lead to identical results.

Figure 6 is the response locus computed for structure I with the exciting conditions as defined for column 1 of Table 2. The amplitude of vibration has been resolved into components in phase and out of phase with the exciting forces, the out of phase component

TABLE 2. DYNAMIC STABILITY COMPARISON

Structure	Planing				Milling		Comparison of dynamic stiffness		
	1	2	3	4	5	6	7	8	9
I	6458	6696	9843	4642	8572	7923	4142	6959	1822
II	6281	6522	10,188	4582	11,473	10,469	5477	6147	1872
III	6309	6612	10,055	4580	9450	5795	3576	5880	1852
IV	6287	6535	10,041	4591	11,474	9776	5299	5929	1853
V	6278	6568	10,537	4578	14,977	12,593	6842	6092	1878
VI	6322	6619	10,640	4588	15,281	12,439	6855	5920	1870
VII	5730	5933	8990	4307	8783	9122	4472	5179	1989
VIII	10,526	10,908	14,388	8059	12,771	13,427	6546	4779	2446
IX	10,221	10,269	10,887	6872	8544	8327	4194	5620	2181
X	6468	6819	11,468	4597	10,742	8128	4634	5350	1831
XI	6311	6585	9747	4521	8273	7484	3933	5275	1825
XII	7203	7683	9658	4644	7719	5577	3246	3128	1738

leading by 90°. The locus shows the variation of amplitude with the frequency of the exciting forces, the frequencies being marked along the curve. The first loop is associated with mode 1 at 12·9 Hz and the large upper loop is associated with either or both of modes 3 and 4. Mode 2 does not produce a large amplitude in this case. The loop at 37 Hz arises from one of the smaller modes not described.

Dynamic instability or machine tool chatter can occur when the width of cut exceeds a certain defined value.[5] The magnitude of this limiting width of cut depends on the cutting constants, material properties and other factors but the contribution of the structure is always related to the reciprocal of $-2x$ where x is the largest negative ordinate of the operative response locus, as shown in Fig. 6 for this particular case. Here $x = -77 \times 10^{-6}$ mm/kgf, hence the entry in Table 2 is 6458 and the chatter frequency is 22·0 Hz. Therefore Table 2 compares the contribution of each structure to dynamic instability for various cutting operations, the largest coefficients representing the best structures, but does not give absolute values of the width of cut.

Table 2 shows that structure VIII was the best for all the planing operations and for one of the milling operations, structure VI being the best for the other milling operation, column

5. Structure VII was the worst structure for use as a planing machine and XII was the worst milling machine. The comparison of dynamic stiffnesses parallel to the x, y and z axes, columns 7, 8 and 9, show that structure VIII is clearly the stiffest in the z direction, nearly the best in the x direction and poor in the y direction. Also, the static flexibility of each

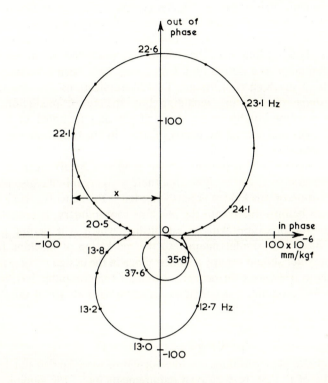

FIG. 6. Typical response locus.

machine in the z direction is much greater than in the other directions, Table 1. It seems, therefore that both statically and dynamically, the greatest improvements are to be made by increasing the structural stiffness in the z direction, that is, in a direction perpendicular to the frame.

4.4 *Cost and Time*

As a guide to the cost and time involved in an analysis of the kind described, the work was carried out by two engineers who had first to learn the necessary techniques. In four weeks, two planer-milling machine structures were analysed, about 60% of the time being spent on the analysis described. The computations for the one structure, as described, required about 100 minutes* of computing time on a KDF9 computer which costs £180 per hour to hire. The time and cost can vary appreciably with the complexity of the computations.

---

* The latest versions of the programs would have reduced this to about 70 minutes, without affecting the results in any way, in addition to reducing the time for preparing the data. See Appendix.

5. CONCLUSION

Table 2 shows that structure VIII is the best under dynamic conditions and Table 1 shows that it is good but not the best under static conditions. Hence the conclusion may be drawn that it is the best overall design. Such a conclusion does not make allowance for the relative costs of the structures, the relative importance of the conditions considered, the mounting of equipment, drives and so on, the company's design policy or aesthetic considerations. The final choice of design must be made by the company taking into account all these factors as well as the information obtained from the computations.

The design of a planer-milling machine structure, as described, is but one example of the use of a digital computer as a design tool. Computers have been successfully applied to a variety of aspects of machine tool design. Typical applications are structural design for practically any shape of structure including box structures, predictions of structural behaviour under working conditions, analysis of spindles mounted in flexible bearings, computations of areas and second moments of area, hydrostatic bearings, cutting forces, gears and other calculations.[6]

Machine tool designers can, at present, obtain programs by arrangement with Research Organisations, Universities or certain firms, but there is no organised register of the programs available. Information of this kind is urgently needed if the benefits arising from computer aided design are to be introduced into the machine tool industry. Eventually, no doubt, a marketing system for programs will be established which will permit programs to be bought, sold or hired in a similar manner to the marketing of machine tools and other products. The computer should occupy a similar place in the design process to that occupied by machine tools in a production process, although the computer has not yet taken its inevitable place. The economic considerations concerning both are of the same type.

ACKNOWLEDGEMENTS

The analyses of the planer-milling machine structure were carried out by Ing. E. Duze and Ing. G. Balbo of R.T.M. to their own requirements under the guidance of the author. The co-operation of Ing. R. Graziosi, Director of R.T.M., is also gratefully acknowledged and permission to publish the work is appreciated. Figures 2, 3 and 4 are reproduced from Ref. 4.

REFERENCES

1. S. TAYLOR and S. A. TOBIAS. Lumped constants method for the prediction of the vibration characteristics of machine tool structures, *Advances in Machine Tool Design and Research*, 1964, Pergamon.
2. S. TAYLOR. The design of machine tool structures using a digital computer, *Advances in Machine Tool Design and Research*, 1966, Pergamon.
3. S. TAYLOR. A computer analysis of an openside planing machine, *Advances in Machine Tool Design and Research*, 1965, Pergamon.
4. E. DUSE and G. BALBO. Struttura di una macchina utensile a portale, R.T.M., Italy, *Bollettino* No. 3, Dec. 1967.
5. G. SWEENEY and S. A. TOBIAS. An algebraic method for the determination of the dynamic stability of machine tools, *Int. Res. in Prod. Eng.*, 1963.
6. S. TAYLOR and S. A. TOBIAS. Computer methods for the structural analysis of machine tools, presented at C.I.R.P. General Assembly, Sept. 1968.
7. J. P. GURNEY and S. A. TOBIAS. A graphical analysis of regenerative machine tool instability, *Trans. Amer. Soc. Mech. Engrs.* B., Feb. 1962.

APPENDIX
DETAILS OF COMPUTER PROGRAMS

The computations are in two stages. The first is the computation of the static flexibilities and the natural frequencies of vibration, each with its mode shape, and the second stage is the computation of response to excitation and of dynamic stability data. Each stage has a separate computer program.

Flexibilities and Modes

Details of the original program and an outline of the mathematical procedure were given in Ref. 1. This program allowed a maximum of 19 lumped masses, any number of elements and any reasonable number of modes to be computed. Closed loops could not be computed but a technique for multiple bases was described. Mercury Autocode and five-hole punched paper tape were used.

The program was later extended[2] to include a technique for analysing structures containing closed loops (redundancies) which, in addition, superseded the rather inefficient technique for multiple bases. The method used was to cut each closed loop, to compute the flexibility of the resulting statically determinate structure and then to compute the forces necessary to close the gaps and re-align the structure. From this data the flexibility matrix for the redundant structure was deduced. Additional masses were needed at the cuts so it was necessary to improve the program to cater for up to 32 masses. The vibration computations for natural frequencies were not affected except that the additional masses were removed before starting this part of the computation. More than 26 masses was uneconomical for the vibration analysis. A system for checking the data on input and for printing out detectable errors was added to the program.

Subsequently a more powerful version of Mercury Autocode called k-code became available, permitting a number of improvements to be made. Computer space as well as computing time were saved by storing only half of the symmetrical flexibility matrix and by storing only the non-zero terms in the mass matrix which contains mostly zeros. Consequently 37 masses could be used in the redundancy computations but the vibration analysis remained at 26 because the final matrix is unsymmetrical so all the terms are required. The program permitted 20 modes to be computed, 40 redundant forces to be used and up to 200 elements to be used. These limits do not impose any practical limitations for machine tool design purposes. This version of the program was used for the planer-milling machine analyses.

More recently, the program has been rewritten in Fortran using punched cards. Both the data preparation and the program itself have been completely reorganised to take advantage of the more powerful computer code, thus reducing the quantity of data appreciably and improving the efficiency of the computations. The consequent reduction in data and the ease with which it can be modified result in a useful reduction in the time taken to prepare the data and modify it for alternative designs. The reduction in computer costs is about 30% compared with the previous program. Up to 50 masses may now be used in redundancy computations and 35 in vibration computations.

Thus both the power and the economics of the program have improved appreciably since the original program was reported.

Responses and Stabilities

This stage of the computation technique was described in Ref. 2 which included an outline of the mathematical technique used for response loci. The stability computations were for regenerative machine tool instability by the method described in Ref. 7 but without allowance for the penetration rate effect. Mercury Autocode and five-hole punched paper tape were used.

When k-code became available, the program was rewritten using this code with numerical improvements. The method of Ref. 5 was adopted for the stability computations. This is a more general theory of instability which extends the first method to partial regenerative chatter where each cut only partially overlaps the previous cut. In addition the penetration rate effect was included in the program.

The program has recently been changed to Fortran using punched cards so that it is compatible with that for flexibilities and modes of vibration.

In a typical application the computing cost associated with the second stage is likely to be less than 20% of the total but could be quite large if many stability charts were to be computed.

COMPUTER ANALYSIS OF MACHINE TOOL STRUCTURES BY THE FINITE ELEMENT METHOD

A. C. STEPHEN

Coventry Gauge and Tool Co. Ltd., formerly University of Birmingham

S. TAYLOR

Department of Mechanical Engineering, University of Birmingham

1. INTRODUCTION

The analysis of large complex structures has until recently been the preserve of the aircraft industry since only in airframe design could the high cost of optimisation be justified by the resulting increase in the payload/total weight ratio. However, the availability of high speed digital computers coupled with the sophisticated requirements of present day mechanical and civil engineering design has brought about the application of the techniques of Argyris,[1] Clough[2] and others to more "down to earth" problems. Of particular note is the work of Zienkiewicz[3] on concrete structures, including arch dams.

Previous papers have dealt with the application of computers to the static and dynamic analysis of machine tool structures using a matrix force (or flexibility) method[4, 5, 6] and the Myklestad method.[7] This paper introduces a third technique, the matrix displacement (or stiffness) method in the form of the Finite Element Technique. This is a much more powerful method which enables the designer to analyse plate and beam assemblies such as ribbed boxes without the necessity of reducing them to equivalent beams as was done in Refs. 4, 5, 6 and 7.

2. ANALYTICAL TECHNIQUES

In all large-scale methods of linear structural analysis the first step is to subdivide the real structure into simple elements such as beams and polygonal plates, these elements being interconnected at specified nodal or station points only. Relationships between the forces and displacements at these nodal points are then derived in the form of element matrices,[3] the manipulation of these element matrices giving the required information about the whole structure. Matrix methods are used as these are ideally suited to digital computation and allow programmes to be written which are capable of analysing a wide range of structures.[8]

(a) Flexibility Method

The flexibility method in which the structural deformations are expressed as a function of the applied forces, as in equation (1)*,

$$\{\delta\} = [F] \{p\} \tag{1}$$

has two serious disadvantages when applied to large highly redundant structures such as grinding machines. Firstly, the formation of a redundant flexibility matrix can be extremely time consuming as it is necessary to first eliminate each redundancy by cutting the structure

---

* See page 759 for nomenclature.

at suitable points. This is illustrated in Fig. 1 which shows a double portal frame in which two cuts must be made to reduce it to a determinate problem. The flexibility matrix for the resulting statically determinate structure is then formed and the final redundant matrix found by applying suitable forces to close the cuts. The preliminary matrix will be much larger than the final as each time a cut is made six degrees of freedom are added to the structure, thus increasing the order of the structural matrix by six. As the time taken for many matrix operations is approximately proportional to the cube of the order, this is a significant factor. The second disadvantage is that a flexibility matrix is generally fully

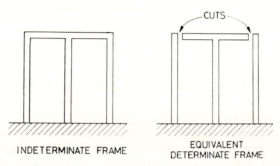

FIG. 1. Removal of redundancies by cutting structure.

populated, i.e. has no zero terms, so that for a structure with say 50 nodes and 300 degrees of freedom the resulting 300 by 300 matrix would require the storage of 90,000 numbers. This requires a computer with a large fast random access store if the cost of the analysis is to be kept within reasonable limits. This storage requirement can be reduced to one quarter by considering only three degrees of freedom per node[3] but this advantage can be offset by the extra six degrees which are still required for each redundancy.

(b) Stiffness Method

With a displacement or stiffness formulation in which the applied force is given as a function of the nodal displacements as in equation (2),

$$\{P\} = [S]\,\{\delta\} \tag{2}$$

neither of the two disadvantages mentioned above exists since a redundant stiffness matrix for a complete structure is formed by simply adding the small individual element stiffness matrices into the appropriate locations. This process is known as Assembly. Also the resulting matrix is sparsely populated i.e. contains many zero terms since a displacement given to one mode can only produce forces at nodes to which it is directly connected by an element. By careful numbering of the nodes in the structure all the non-zero terms can be grouped around the diagonal to form a band matrix as shown in Fig. 2. Only the diagonal area need be stored in the computer thus reducing the storage requirements to roughly 10% of that for the equivalent flexibility matrix.

Having obtained the stiffness matrix it is necessary to extract the desired information about the structure. Usually the applied loading will be known and the displacements required. Given a flexibility matrix the displacements for any number of loads can be found by simple matrix multiplication. For the stiffness method no such direct method exists and it is necessary either to invert the stiffness matrix to form the corresponding flexibility matrix, which requires much computation and raises the storage problem mentioned above, or to solve the set of simultaneous linear equations represented by equation (2). The choice depends on the number of separate loading cases which have to be considered. If, as is normally the case, only 20 to 30 different loadings are applied then solution of the equations is best as it can be carried out without destroying the band form. If, however, the number of load cases is large then inversion of the matrix may be preferable. A further advantage of inversion is that it need only be done once and the flexibility matrix stored for any number of subsequent load cases to be analysed.

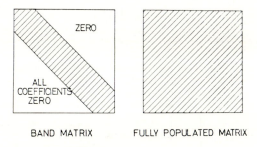

BAND MATRIX   FULLY POPULATED MATRIX

FIG. 2. Comparison of normal and band matrices.

3. FINITE ELEMENT TECHNIQUE

(a) *Static*

In the finite element technique the displacement method is used almost exclusively because the idealisation of structures into assemblies of plate and beam elements as shown in Fig. 3, invariably results in a high degree of redundancy. This is illustrated by Fig. 4, which shows a simple space frame with 12 nodes which requires 15 cuts to be made to remove all the redundancies. The figure also shows the relative sizes and populations of the stiffness matrix and the equivalent statically determinate flexibility matrix.

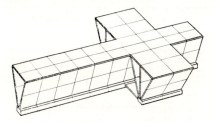

FIG. 3. Typical finite element idealisation.

A selection from the various types of elements which can be employed as building blocks in the finite element technique is shown in Fig. 5. Beam elements are the most accurate as they give exactly the same deflections and stresses as would be obtained using simple beam theory. The most useful of the straight sided elements is the triangle since it is capable of representing double curvature but it is not as accurate as the equivalent rectangular element. The simple elements are easy to form in the computer as they may be obtained explicitly, i.e. as algebraic expressions. The elements with curved boundaries and many nodes, although

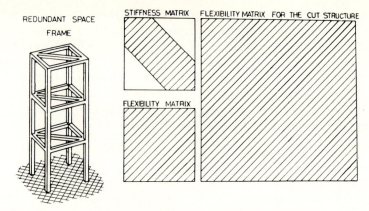

Fig. 4. Redundant structure with relative sizes of matrices.

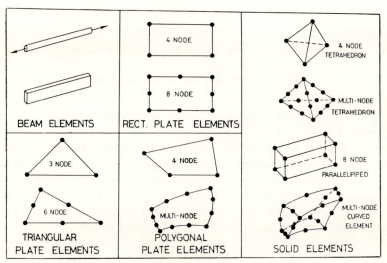

Fig. 5. Types of finite elements.

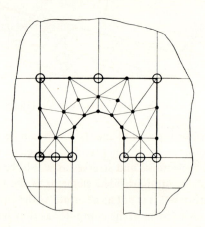

Fig. 6. Typical super-element.

better able to represent complex shapes and stress patterns, require numerical integration and thus use more computer time.

The introduction of cutouts into panels gives rise to large stress gradients which the simpler elements are not capable of representing. It thus becomes necessary to formulate complex elements to handle such cases or to use a large number of simple elements. Unless the element is to be used many times the former course is not an economic proposition whilst the latter results in an increased bandwidth which is also undesirable. A method for avoiding both of these difficulties is to combine a number of simple elements into a super-element and to suppress some of the extra nodes by a suitable matrix transformation. Figure 6 shows such a super-element in which only the nodes marked with a circle were required in the final analysis, the others being eliminated. This method can be extended to form complete sections of the structure as super-elements. This technique is incorporated in the program specified in Appendix 1.

The accuracy of finite element analysis depends on how closely the elements themselves are capable of representing the structural stresses and deformations. Ideally there should be continuity of displacements, bending moments and shear forces across element boundaries. The complexity of the element stress and deformation patterns is related to the total number of degrees of freedom associated with the element nodes, thus simple elements can only represent simple patterns. This places restrictions on the actual number of continuity requirements which can be satisfied and it is usual to satisfy only the displacement criterion as it can be shown that this results in the computed deflections always being less than true values, with the error decreasing as the number of elements increases. For three dimensional work the number of degrees of freedom associated with each node is normally three (the mutually perpendicular displacements u, v and w) or six (three displacements plus three rotations), since the use of any other number makes the structural matrix difficult to assemble. As mentioned above using three degrees per node reduces the storage required but this is offset by the increase in the number of nodes necessary for a given accuracy. Three degree elements have also the disadvantage that bending moments at nodes cannot be represented thus restricting their application to pin ended beams and thin plates and shells.

(b) Dynamic

The prediction of vibration characteristics is theoretically simple but presents certain computational difficulties. By assuming simple harmonic motion, a matrix can be formed which accurately represents the mass or inertia of the structure. This mass matrix is banded in a similar manner to that of the stiffness matrix and expresses the nodal inertia forces as a function of the nodal displacements. If the response of the structure to an undamped sinusoidal load is required the problem is formulated as equation (3),

$$[S] \{\delta\} - \omega^2[M] \{p\} = \{p\} \tag{3}$$

in which the matrix $[S - \omega^2 M]$ may be formed and the resulting set of simultaneous equations solved as for the static case. Usually, however, the natural frequencies are required, in which case no external loading is applied and equation (3) reduces to the eigenvalue problem of equation (4) in which the unknown is ω.

$$[[S] - \omega^2[M]] \{\delta\} = 0 \tag{4}$$

The simplest method of extracting eigenvalues is the iteration method as used by Taylor.[3]

Unfortunately, in the case of a stiffness matrix, an inversion is required if the process is to give the important lower natural frequencies. Another technique currently in use is to solve equation (5) by assuming values of ω and hunting for zero values of the determinant

$$|S - \omega^2 M| = 0 \qquad (5)$$

In practice changes in the sign of the determinant are looked for as the absolute values can be very large. If the approximate natural frequencies are known this method is quick but it is very easy to miss frequencies by using too large an increment between tries. Using a small interval will obviously help but as the time taken to calculate the determinant is almost as long as that taken to solve the equivalent set of equations, it becomes very expensive to use this method. A further disadvantage is that a matrix having a near zero determinant is ill-conditioned with consequent loss of accuracy during computation.

At the expense of the accuracy of the higher natural frequencies, the amount of calculation can be reduced by expressing the deflections of selected secondary nodes as a function of the remaining primary nodes. The secondary nodes can then be removed from the structural matrices by suitable transformations. The technique used by Taylor[3] for introducing any number of small elements between adjacent mass points achieves this effect by linearly interpolating between the station points.

(c) Extensions of the Finite Element Method

The finite element method can be extended to calculate stresses and also to handle non-linear phenomena.

Stresses are calculated by simply forming an element stress matrix which relates the stresses to the deformation at the nodes. Because the finite element method does not normally require continuity of stresses across element boundaries, the stress patterns produced must be smoothed before being used. This point is illustrated by Fig. 7 which shows the stresses predicted for a circular hold in a plate twice the hole diameter. The use of multi-node elements will reduce these discontinuities as will the introduction of a finer element mesh.

Non-linear structures may be analysed by an incremental linear approach.[1] Both geometrical non-linearity (large deflection) and non-linear stress–strain relationships are amenable to this technique whereby the load is applied in small increments over which the

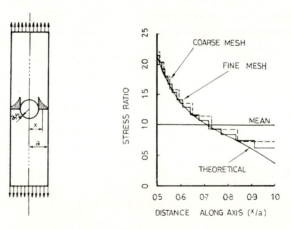

Fig. 7. Calculation of stress concentration factor.

structure is assumed to behave linearly. Between increments the structural matrices are modified as required to allow for the non-linearity. This is, however, expensive since each increment is equivalent to a static analysis. Anisotropic materials which exhibit directional stress characteristics, such as wood, can also be included by formulating suitable element matrices.

Elements can also be formulated to deal with axisymmetric and solid three dimensional stresses and deformations. Typical elements are shown in Fig. 5.

Limitations on Type of Structure

The size of structure which can be analysed depends on whether or not the program to be used has a maximum allowable nodal bandwidth for the structural matrix. If unlimited bandwidth is allowed then any idealisation can be used and the only problems are computer size and cost. If, however, a limit exists, although hundreds of nodes can still be used the actual structure must be long and thin. Limited bandwidth programs are more easily written as all the assembly and solution can be carried out in the fast access core store. Immediately the computer backing store is used, as is necessary with unlimited bandwidth, the time taken increases enormously and efficient programming becomes difficult, especially if the backing store is on magnetic tape.

This is becoming less important as modern computers have more core store, better backing stores (drums and discs) and have time sharing facilities so that if one program is waiting for a number from the backing store, a second program is entered.

Limited bandwidth is particularly important if the structure has cutouts as the stress gradients introduced require many nodes for adequate representation with consequent

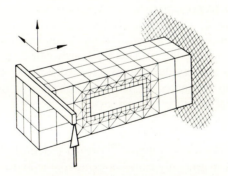

FIG. 8. Idealisation used for box beam with hole.

local increase in bandwidth. Figure 8 shows an idealisation used in an investigation into the effects of cutouts on a column. Without a cutout the coarse mesh was adequate but simulation of the hole by simply removing three elements did not reduce the stiffness by the correct amount. It was necessary to refine the mesh around the hole increasing the section size from 24 to 43 nodes and exceeding the program limit of 33 nodes. The fine mesh was introduced by employing the super-element technique described above.

If the structure exhibits symmetry about one or more axes the response of the whole can be found by considering only part of it and applying suitable constraints along the axes of symmetry. The grinding machine sub-frame shown in Fig. 3 was analysed by this method as the program bandwidth was exceeded on the projecting wings.

4. SOURCES OF ERROR

(a) Finite Element Method

The accuracy with which the structure is idealised depends on the number, type and distribution of the elements. Unfortunately the ideal element which is best for all occasions, has not been found and it is necessary for the analyst to know the properties and limitations of the various kinds when making a choice. The only general rules are that long narrow elements are to be avoided and that in regions of high stress gradients many nodes are required.

A further possible source of error lies in the process used to solve the large set of equations which result from the finite element method. However, static structural matrices are of the type known as positive definite symmetric and are stable. One of the simplest and fastest methods is Gaussian Elimination[9] without row interchange followed by back substitution. This method has been used on the KDF9 to solve sets of up to 2000 equations with band-widths up to 200. On checking by substituting the answers into the original equations nodal unbalanced forces of the order 10^{-10} were found. The more advanced Choleski Triangular Decomposition Method was also used as it is a highly stable method[10] but this did not result in a significant increase in accuracy.

(b) Data

From the program user's point of view one of the most important aspects of the finite element method is how the structure is described to the computer. The simplest programs require as data the coordinates of each nodal point, details of any constraints and loadings, the types of elements and the nodes to which they are to be connected. For a structure with 300 nodes and 400 elements approximately 3500 numbers would be required. Whilst the computer can be programmed to reject obvious errors such as negative section constants, "non-rectangular" rectangles, etc., it is extremely difficult to check that the structure actually analysed was that intended. If many similar structures are to be processed the physical labour of punching the data tape or cards can be avoided by writing a short preprogram to convert a few key dimensions into data for the main program. In practice this is rarely the case and it is necessary to write sophisticated input routines in the main program which will accept specifications for groups of elements.

(c) Computer

The high reliability of the modern digital computer practically eliminates machine error and even when it occurs it generally causes a catastrophic end to the program. However, it should never be assumed that the computer is always correct and checking routines should be written into the program to detect the rare occasions when non-catastrophic error has occurred. In finite element analysis most of the arithmetic operations are carried out by the solution routine which is the most easily checked by simply substituting the answer into the original equations.

5. COST OF ANALYSIS

Finite element programs can be written to run on practically any computer, but the increased time taken if a small machine is used produces a disproportionate increase in the cost and a machine of KDF9 size is desirable (see Appendix 2). It is impossible to general-ise on the cost of an analysis as it depends on the computer, the type of program, the type of structure and the amount of information required. Exact costs are given below for specific analyses on the KDF9 using the programs described in Appendix 1.

The analysis of the box beam shown in Fig. 8 with 140 nodes and 720 degrees of freedom took approximately 8 min for four load cases giving a computer cost of £24. The preparation of the input data took about 4 hours. The analysis of the large box beams shown in Fig. 9

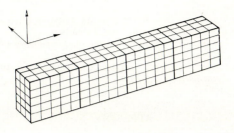

FIG. 9. Large box beam.

took 29 minutes on the computer, costing £87. Because of the high degree of regularity of the structure the data was prepared by a small program which took about an hour to write and ran for 4 min. This gives a total computing cost of £99.

6. CONCLUSIONS

The Finite Element Method has been presented and shown to be a valuable addition to the analytical techniques available to machine tool structural designers. At the present time the method is best used for static analysis but with the advent of larger and faster computers dynamic analysis will become an economic proposition.

In the interim the Finite Element method could be used to form a flexibility matrix for selected station points on the structure, an approximate lumped mass matrix formed and the techniques employed by Taylor[4] used to predict the dynamic response,

ACKNOWLEDGEMENT

The authors wish to thank the Coventry Gauge and Tool Co. Ltd. who suggested and sponsored this project, which was carried out at the Department of Mechanical Engineering, University of Birmingham, under the supervision of Professor S. A. Tobias.

NOMENCLATURE

$[F]$ Flexibility matrix, ft/lbf.
$[S]$ Stiffness matrix, lbf/ft.
$[M]$ Mass (or inertia) matrix, lbf sec^2/ft.
$\{p\}$ Load vector, lbf.
$\{\delta\}$ Displacement vector, ft.
ω Frequency, rad/sec.

REFERENCES

1. J. H. ARGYRIS, S. KELSEY and H. KAMEL. *Matrix Methods of Structural Analysis. A precis of recent developments.* AGARDograph 72, Ed. de Veubeke, Pergamon Press, 1964.
2. R. W. CLOUGH. The finite element method in structural mechanics, Chapter 7 of *Stress Analysis*, Ed. Zienkiewicz and Holister, Wiley, 1965.

3. O. C. Zienkiewicz. *The Finite Element Method in Structural and Continuum Mechanics*, McGraw-Hill, 1967.
4. S. Taylor and S. A. Tobias. Lumped constants method for the prediction of the vibration characteristics of machine tool structures, *Machine Tool Design and Research*, 1964, Pergamon.
5. S. Taylor. A computer analysis of an open side planning machine, *Machine Tool Design and Research*, 1965, Pergamon.
6. S. Taylor. The design of machine tool structures using a digital computer, *Machine Tool Design and Research*, 1966, Pergamon.
7. B. C. Cuppan and J. G. Bollinger. Simulation of a machine tool drive and structure on an analogue computer, *Machine Tool Design and Research*, 1966, Pergamon.
8. J. M. Prentice and F. A. Leckie. *Mechanical Vibrations: An Introduction to Matrix Methods*, Longmans, 1963.
9. L. Fox. *An Introduction to Numerical Linear Algebra*, Clarendon Press, 1964.
10. Wilkinson. *Numerische Mathematik*, Vol. 7, 1965.

APPENDIX I

Specification of Finite Element Programs Available at the University of Birmingham, Department of Mechanical Engineering, Machine Tool Group

---

Author—A. C. Stephen
Language—I.C.I. K Autocode + KDF9 Machine Code
Input—5 Hole Paper Tape

Programme No.	50300	60100
Degrees of freedom per node	6	3
Maximum section* size (nodes)	33	66
Maximum no. of sections	60	60
Elements (assembly time, sec)	Beam (0·15)	Pin end beam (0·1)
	3 Node triangle (0·5)	3 Node triangle, Trim 3 (0·15)
	4 Node rectangle (0·3)	6 Node triangle, Trim 6 (0·6)
	Super-elements	4 Node rectangle (0·2)
Deflections	Yes	Yes
Stresses	No	Yes
Time to assemble structural matrix	Multiply total element time by two	
Time to solve equations	Proportional to (No. of Sections) (Section size)^2 Maximum section takes approx. 40 sec.	

---

* Program is band limited so that three sections cannot meet at any node.

APPENDIX II

Specification of KDF9 Computer

Manufacturer	—English Electric Leo Marconi
Designation	—KDF9
Storage	—32,768 words of core store on 6 μsec access, word length 48 bits.
	8 magnetic tape units with a transfer rate of 5000 words per second, each tape holds approximately 1,200,000 words
	4,000,000 word disc with a transfer rate of 15,000 words/sec on 1 sec random access (maximum)
Arithmetic Unit	—Integer addition 1 μsec integer multiplication 6 μsec
	Floating point addition 6 μsec, multiplication 16 μsec
Input	—Paper tape, cards
Output	—Paper tape, cards and line printer (600 1/m alpha numeric)

SOME FACTORS INFLUENCING THE DESIGN OF BOX-TYPE MACHINE TOOL STRUCTURAL ELEMENTS

R. H. THORNLEY and P. KUMAR

U.M.I.S.T.

SUMMARY

This paper compares the characteristics of plain box-type structures with those of various forms of buttressed box-type structures when loaded both statically and dynamically under torsional and bending forces. In lightweight beam structures of box-type sections deformation of the cross-section plays a large part in the overall deflection. This is particularly the case when the beam-type structure is subjected to torsional forces. General theories assume that there is no longitudinal extension when the beam is subjected to torsional forces. Although this is satisfactory for solid beams, torsion in thin-walled beams produces some longitudinal extension. This results in the warping of the cross-section and a reduction of the overall torsional stiffness of the beam. A theoretical analysis backed by experimental evidence is put forward for the determination of the static torsional stiffness under different end conditions.

INTRODUCTION

The overall performance of a machine tool depends primarily upon the static and dynamic behaviour of the machine structure. This in turn is dependent upon the individual elements and their joints which make up the structure.

The structure has to resist the deformations caused by the weight of the moving parts of the machine and the cutting forces and possible inertia forces under dynamic conditions.

With the ever-increasing demands for higher productivity, precision working and good surface quality in machining, the design of the basic structure is receiving more attention. The deficiency of a scientific approach to the design of structures is to some extent due to the present lack of factual data on the design of basic structural elements which make up the machine tool structure.

The design of the machine tool structure differs from that of other structures in that the allowable limits of deflections are very small and the areas involved relatively large, so that the stresses set up in the members are far below the permissible stress of the material. Thus the governing factors in the design of machine tool structures are, static stiffness to reduce deflection under heavy loads applied to the structure and dynamic stiffness to prevent resonance vibrations under high working frequencies.

The most common element used in the design of machine tool structures is the thin-walled box-type beam, as in uprights and cross-slides of planing machines, vertical and horizontal boring machines, milling machines and several other machines. Simple calculations show that the box-type structural elements are for many purposes far superior to elements of other cross-sectional designs.[1-3] Considerations other than those of the structural design often lead the designer into compromise with respect to shape and dimensions of the structural elements. These compromises result in various additional features such as ribbings, cut-aways, dividing walls, buttressings, etc. As a result, a design which

763

started off as a simple thin wall rectangular structure, the characteristics of which could be calculated easily, often emerges as a very complex type structure. With the improved methods and knowledge of model techniques the study of the effects of these varying additional features has been made necessary.[4,5] This paper aims at showing the designer how various forms of buttressing affect the static and dynamic characteristics of thin-walled box section beams. It was mentioned earlier that the joints play a part in the performance of the structure. The joint stiffness characteristics are dependent upon the pre-load on the joint and the manner in which the force is transmitted from one element to another.[6,7] This in turn has an effect upon the warping characteristics of the different elements. Tests have therefore been carried out to determine the effect of different end conditions upon the torsional stiffness and warping of the cross-section of thin wall box-type structures.

SPECIMENS

As a basis for comparison three different buttressing methods for three different depth-to-length ratios were compared with beams without buttressing. In all this gave a total of twelve beams studied and, by applying the rules of model analysis to the dimensions of the beams, covered the normal range of machine tools where buttressing is used in the design of various elements. The beams were of welded steel structure and had a constant overall length of 36 in. The cross-sectional dimensions are given in Fig. 1 and photographs of the actual beams in Fig. 2.

EXPERIMENTAL INVESTIGATIONS

In the design of structural elements the torsional and bending characteristics are of interest to the designer. The following tests were therefore carried out to determine both the static

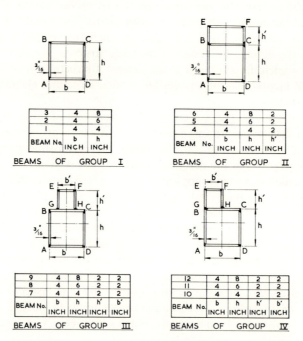

FIG. 1. Cross-sectional dimensions and grouping of beams.

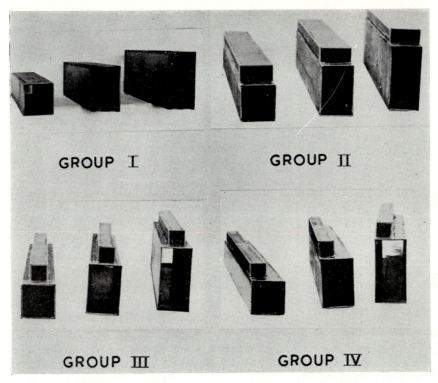

FIG. 2. Beam models.

and dynamic characteristics of the various beam configurations:
  (i) Static torsion.
  (ii) Static bending:
      (a) about axis XX;
      (b) about axis YY (Fig. 1).
  (iii) Dynamic torsion.
  (iv) Dynamic bending.
  In the static tests angle of twist, deflection of the beam and warping of the sections were the main parameters considered, whereas in the dynamic tests natural frequencies and damping were considered.
  The test rigs and methods of testing were the same as those reported in ref. 8.

DISCUSSION OF RESULTS

Under Static Torsional Loads

  Table 1 gives the results of experimental and theoretical values. It will be observed that the experimental values of the angle of twist are 10–20% higher than the theoretically calculated values. This is attributed to the fact that in spite of the efforts made to obtain rigid end conditions, there was still some distortion of the section. (This point is considered in more detail later in the paper.)

TABLE 1

BEAM No.	CROSS-SECTIONAL SHAPE	WEIGHT OF BEAM W lb	TORQUE T lb in	ANGLE OF ROTATION θ rad x 10⁻⁴		STIFFNESS K_t lb in/rad x 10⁶		$\frac{K_t}{W}$	
				THEORETICAL	EXPERIMENTAL	THEORETICAL	EXPERIMENTAL	THEORETICAL	EXPERIMENTAL
1		30·3	3600	8·50	9·5	4·24	3·79	13·99	12·5
2		37·8	3600	4·72	5·7	7·63	6·31	20·2	16·7
3		45·4	3600	3·18	3·55	11·31	10·12	25·0	22·38
4		45·4	3600	4·72	5·6	7·63	6·43	16·8	14·15
5		53·0	3600	3·18	3·8	11·31	9·47	21·4	17·88
6		60·5	3600	2·38	2·65	15·11	13·58	25·0	22·42
7		41·6	3600	6·80	7·90	5·29	4·56	12·6	10·97
8		49·1	3600	4·08	4·45	8·83	8·09	17·97	16·45
9		56·7	3600	2·94	3·4	12·22	10·59	21·6	18·65
10		41·6	3600	6·80	8·2	5·29	4·39	12·6	10·52
11		49·1	3600	4·08	4·35	8·83	8·28	17·97	16·83
12		56·7	3600	2·94	3·45	12·22	10·41	21·6	18·38

The following conclusions are made from the results:

(a) The stiffness of the beams increases directly with the square of area bounded by the outer periphery of the beam cross-section and inversely with the length of the periphery of the beam cross-section. Therefore, there is an increase in stiffness and stiffness to weight ratio when the dimensions of the cross-section are increased.

(b) The torsional stiffness for a square section is a maximum among rectangular sections for the same perimeter and thickness of wall. (c) When the beams are buttressed similar to the beams of group II there is an increase in the stiffness-to-weight ratio. In fact the stiffness of these box section beams is equal to that of beams having the same outer dimensions. The contribution of the middle longitudinal strip to the stiffness of the beam is negligible.

(d) When the beams are buttressed as those of groups III and IV there is a decrease in the stiffness-to-weight ratio.

Under Static Bending Loads

From the graphs of load/deflection the results in Table 2 have been compiled. As a basis

TABLE 2

BEAM No.	CROSS-SECTIONAL SHAPE	WEIGHT OF BEAM W lb.	LOAD P ton f.	DEFLECTION in x 10⁻³		STIFFNESS K lb/in x 10⁶		$\frac{K}{W}$ x 10⁴	
				THEORETICAL	EXPERIMENTAL	THEORETICAL	EXPERIMENTAL	THEORETICAL	EXPERIMENTAL
1		30·3	4	35·8	36·5	·25	·245	·82	·809
2		37·8	4	15·3	14·85	·585	·605	1·54	1·6
3		45·4	4	8·1	7·64	1·105	1·172	2·43	2·58
4		45·4	4	13·8	14·3	·649	·626	1·43	1·38
5		53·0	4	7·1	7·4	1·26	1·21	2·37	2·28
6		60·5	4	4·3	4·1	2·08	2·18	3·43	3·6
7		41·6	4	16·3	17·1	·549	·525	1·32	1·26
8		49·1	4	8·3	8·5	1·079	1·053	2·19	2·14
9		56·7	4	4·9	5·1	1·829	1·76	3·22	3·1
10		41·6	4	16·3	18·0	·549	·499	1·32	1·2
11		49·1	4	8·3	8·2	1·079	1·091	2·19	1·22
12		56·7	4	4·9	5·1	1·829	1·76	3·22	3·1

of comparison for the various sections the stiffness-to-weight ratio was used. From the results the following comments can be made:

(a) As the depth of the beam section is increased, the stiffness and stiffness-to-weight ratio values are increased.

(b) When the beam is buttressed with another box, the stiffness and stiffness-to-weight ratio values are increased with respect to the basic beam. This is because when the beam is buttressed the depth of the beam is increased.

(c) Positioning of buttressing from centre to one side, viz. beams group III and IV, has little effect upon the stiffness-to-weight ratio values.

(d) For the beams of group II the longitudinal cross-spar formed by buttressing of two sections reduces the stiffness-to-weight ratio values for the same outer dimension. Similarly, for the beams of groups III and IV, the stiffness-to-weight ratio values are very much less compared with beams of rectangular cross-section of the same depth and width. Therefore, the simple rectangular box-type structure with the same overall dimensions is better than buttressed beams with regard to its static bending stiffness.

Dynamic Torsional Characteristics

The experimental results and theoretical values of the first torsional natural frequency and damping ratio are given in Table 3.

TABLE 3

BEAM No.	CROSS-SECTIONAL SHAPE	POLAR MOMENT OF INERTIA I_p	TORSIONAL MOMENT OF INERTIA J_d in^4	FIRST NATURAL FREQUENCY		DAMPING RATIO × 10^4
				EXPERI-MENTAL	THEORE-TICAL	
1		16·0	12·0	342	317	42·6
2		31·5	21·6	322	304	21·9
3		54·0	32·0	290	283	13·8
4		30·8	21·6	329	306	29·1
5		58·25	32·0	321	271	15·2
6		92·7	42·9	275	249	10·4
7		26·5	15·0	326	275	27·2
8		48·63	24·5	320	260	14·1
9		80·58	34·7	272	240	10·1
10		27·3	15·0	319	270	19·4
11		49·5	24·5	309	256	10·2
12		81·48	34·7	267	236	9·7

The following conclusions can be drawn from these results:

(a) For the beams of group I it is clear that when the beam has a square cross-section, the first torsional natural frequency and damping ratio are a maximum. When the (h/b) ratio is increased, both these values decrease (see also ref. 9).

A similar pattern is followed in the other groups of beams.

(b) When the rectangular cells are welded together, the natural frequency and damping ratio decreases with respect to the basic structure.

(c) Beams 2 and 4 have the same outer dimensions, but beam 4 has an additional longitudinal strip. The natural frequencies of these beams are almost the same, i.e. longitudinal strip does not increase the natural frequency appreciably. However, the damping ratio of beam 4 is higher due to the addition of more rubbing faces. The same phenomenon is displayed by beams 3 and 5.

(d) Buttressing of beams as shown in groups III and IV produces lower natural frequencies and damping ratios than their equivalent beams in group II. Also beams in group IV have lower values than those of group III.

Dynamic Bending Characteristics

For the sake of convenience the discussion of these results is divided into two parts—vibration of the beam as a whole and vibration of the plates constituting the beam.

Vibration of Beam as a Whole

Table 4 shows the various natural frequencies of vibration. The first mode and second mode are mainly due to the elastic supports, while the third mode represents the vibration of the beam.[8] The natural frequencies of the different beams follow a similar pattern to

TABLE 4

BEAM No.	CROSS-SECTIONAL SHAPE	FIRST MODE		SECOND MODE		THIRD MODE	
		FREQUENCY c/sec.		FREQUENCY c/sec.		FREQUENCY c/sec.	
		EXPERI-MENTAL	THEORE-TICAL	EXPERI-MENTAL	THEORE-TICAL	EXPERI-MENTAL	THEORE-TICAL
1		397	350	470	410	592	560
2		368	325	404	376	722	752
3		339	315	419	371	930	965
4		399	356	475	412	769	759
5		331	335	435	404	900	930
6		261	302	382	352	1045	1110
7		422	369	480	417	702	725
8		313	321	446	408	853	890
9		275	310	396	358	960	1045
10		419	369	491	417	751	725
11		328	321	459	408	863	890
12		270	310	410	358	970	1045

that of the stiffness for static tests. The third mode is taken as the basis of the analysis and is termed as the first natural frequency. The following conclusions can be made:

(a) Increasing the depth of the beam results in higher natural frequencies.

(b) For all types of buttressings considered the natural frequency of the beams increases in a manner similar to the stiffness-to-mass ratio obtained in the static tests.

(c) When comparing the beams whose overall dimensions are the same, but in one there is a longitudinal cross-strip as in beams 4 and 5, the natural frequencies remain almost the same. This suggests that the longitudinal strip has little effect upon dynamic beinding stiffness as was the case with static bending.

Vibration of Plates

Different modes of vibrations of the plates of the beams are given in Table 5. It is seen that as the depth of beam increases a greater number of modal shapes of the plates fall within the frequency range measured.

Due to the lack of suitable instrumentation at the time of carrying out the tests no attempt was made in these investigations to determine the damping ratio of the beams under bending vibrations.

TABLE 5

BEAM No.	CROSS SECTIONAL SHAPE	FREQUENCIES OF PLATE VIBRATIONS c/sec.			
		MODES OF SIDE PLATE			MODES OF BOTTOM AND TOP PLATE
1		–	–	–	–
2		686	–	–	–
3		462	563	825	870
4		–	–	–	–
5		752	–	–	–
6		471	570	791	940
7		–	–	–	–
8		693	–	–	–
9		500	569	761	910
10		–	–	–	–
11		691	–	–	–
12		522	585	782	930

EFFECT OF END CONDITIONS

Whilst carrying out tests to determine the torsional characteristics of various box-type elements the importance of the end fixings was realized. In order to investigate this point further a series of tests were developed and carried out.

For this series of tests a beam of outer dimensions $4 \times 4 \times \frac{3}{16}$ in wall thickness and 36 in long was rigidly secured and prevented from rotating at one end whilst a torque was applied through various fixings at the opposite end. At the fixed end the contour of the beam was prevented from deforming throughout the test whilst at the free end contour deformation was allowed to occur according to the different methods of torque application.

The tests were carried out with the beam mounted in the chuck of a large lathe as shown in Fig. 3, the torque being applied by a lever mechanism and transmitted by a spindle and chuck arrangement which replaced the tailstock barrel. The headstock chuck was theoretically prevented from rotating by a stop block inserted between the chuck and the bed. A dial indicator, however, was situated as shown in the figures to measure the small amount of rotation due to deformation of the members, and the final readings corrected accordingly.

The variation in end fixings at the points of torque application are shown in Fig. 4a. In case A the end contour was prevented from distorting by the packing plates. For the later tests these packing plates were removed and also some of the screws as shown, until finally only two screws transmitted the full torque to the beam. By positioning the screws as shown it was possible to transmit the full torque whilst at the same time allowing the contour to deform freely.

As can be observed in Fig. 4b, as the end conditions are varied from A to D the angle of rotation of the beam increases compared with that of a beam with rigid ends. In case D, where the torque was applied through two screws only, the angle of rotation was almost 10 times the value of that obtained when the end was prevented from distorting. This shows that the distortion of contour under torsional loads is of high importance and may be a deciding factor in the design of thin-walled box-type structures.

In order to consider further the effect of contour deformation upon the torsional stiffness the following investigation was carried out.

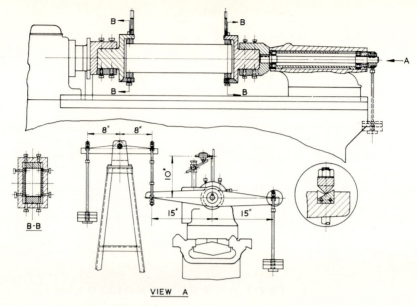

VIEW   A

FIG. 3. General arrangement of static torsion test rig.

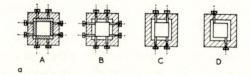

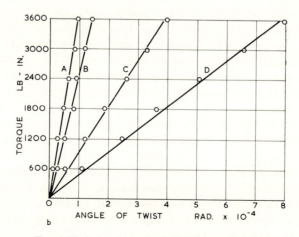

FIG. 4. Variation in method of torque application.

THEORETICAL ANALYSIS

Theory of pure torsion is based on the assumption that there is no longitudinal extension in the beam. This is applicable in the case of solid beams; thin-walled beams, however undergo longitudinal extension as a result of torsion. Consequently, longitudinal norma stresses proportional to these strains are created, which lead to an internal equilibrium of the longitudinal forces in each cross-section. These complementary longitudinal normal stresses, which arise as a result of relative warping of the section and which are not examined in the theory of pure torsion, can attain very large values in thin-walled beams with closed flexible cross-section.

The present theoretical analysis is based on similar lines as suggested by Kaminskaya.[10,11] The following assumptions are made:

1. Shear and normal stresses on any cross-section are distributed uniformly over the thickness of the wall.

2. The actual rigidity of the walls to bending in the plane of least rigidity and to torsion is small enough to be neglected.

Considering a beam with one end rigidly fixed and acted on by a torque Pb at the other end (see Fig. 5a) when the beam deforms the angle of rotation of horizontal and vertical walls will not be the same.

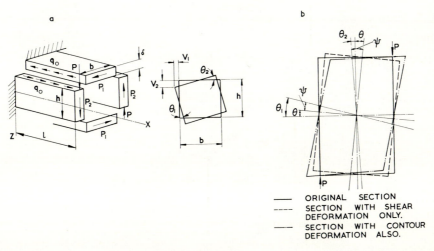

———— ORIGINAL SECTION
----- SECTION WITH SHEAR DEFORMATION ONLY.
—·— SECTION WITH CONTOUR DEFORMATION ALSO.

FIG. 5. (a) Diagram of loading of individual walls of a box beam in torsion. (b) Deformation of section in torsion.

Considering the equilibrium of each wall, there is a tangential stress acting in the transverse section of the walls which results in transverse forces P_1 and P_2 and normal stress σ_x acting on the walls. At the longitudinal edges of the walls, tangential stress $q = q(x)$ is acting.

The shear in a wall is given by

$$\gamma = \frac{q}{G\delta} = \frac{\partial u}{\partial z} + \frac{\partial v}{\partial x} \tag{1}$$

where u and v are the deflections in the direction of x and z axes respectively.

For small angle of rotation (Fig. 5b)

$$v_1 = \theta_1 \cdot \frac{h}{2}; \quad v_2 = \theta_2 \cdot \frac{b}{2} \tag{2}$$

where θ_1 and θ_2 are the angle of rotation of the vertical and horizontal walls respectively. Thus for the horizontal wall,

$$\frac{qz}{G\delta} = u_1 + \theta \frac{h}{2} z$$

or

$$u_1 = \frac{qz}{G\delta} - \theta_1 \frac{h}{2} z \tag{3}$$

Similarly, for the vertical wall,

$$u_2 = \frac{qz}{G\delta} - \theta_2 \frac{b}{2} z \tag{4}$$

assuming that θ_1 and θ_2 are constant along z.

Now, since at a point on the corner of the outline

$$u_1 \left(\frac{b}{2} \right) = u_2 \left(-\frac{h}{2} \right)$$

we have

$$q = \frac{hb \, (\dot{\theta}_1 + \dot{\theta}_2) \, G\delta}{2(h + b)} \tag{5}$$

therefore

$$\sigma_{1x} = E \frac{\partial u_1}{\partial x} = \frac{Ezh \, (b\ddot{\theta}_2 - h\ddot{\theta}_1)}{2(h + b)} \tag{6}$$

$$\sigma_{2x} = E \frac{\partial u_2}{\partial x} = \frac{Ezb \, (h\ddot{\theta}_1 - b\ddot{\theta}_2)}{2(h + b)} \tag{7}$$

$$\tau_1 = \int \frac{\partial \sigma_{1x}}{\partial x} \, dz + \frac{q}{z}$$

$$= \frac{Eh \, (b\dddot{\theta}_2 - h\dddot{\theta}_1) \, z^2}{4(h + b)} + \frac{hb \, (\dot{\theta}_1 + \dot{\theta}_2) \, G\delta}{2(h + b)} \tag{8}$$

$$\tau_2 = \int \frac{\partial \sigma_{2x}}{\partial x} \, dz + \frac{q}{z}$$

$$= \frac{Eb \, (h\dddot{\theta}_1 - b\dddot{\theta}_2) \, z^2}{4(h + b)} + \frac{hb \, (\dot{\theta}_1 + \dot{\theta}_2) \, G\delta}{2(h + b)} \tag{9}$$

The expression for transverse bending moments caused by the deformations of the walls out of their planes can be written as:

$$M_1 = \frac{E\delta^3 z}{b(h + b)} \, (\theta_1 - \theta_2) \tag{10}$$

$$M_2 = -\frac{E\delta^3 z}{h(h + b)} \, (\theta_1 - \theta_2) \tag{11}$$

$_1$ and θ_2 can be determined by the variation method, without constructing a differential quation. Introducing the auxiliary functions α and β such that

$$\left. \begin{array}{l} \theta_1 = ba - \beta \\[2mm] \theta_2 = ha + \beta \end{array} \right\} \tag{12}$$

and

equations (6), (7), (8), (9), (10) and (11) can be written in terms of a and β

$$\sigma_{1x} = E \frac{hz}{2} \ddot{\beta} \tag{13}$$

$$\sigma_{2x} = - E \frac{hz}{2} \ddot{\beta} \tag{14}$$

$$\tau_1 = \frac{Ehz^2}{4} \dddot{\beta} + G \frac{hb}{2} \dot{a} \tag{15}$$

$$\tau_2 = \frac{Ebz^2}{4} \dddot{\beta} + G \frac{hb}{2} \dot{a} \tag{16}$$

$$M_1 = \frac{E\delta^3 z}{h(h+b)} \{ (b-h) a - 2\beta \} \tag{17}$$

$$M_2 = - \frac{E\delta^3 z}{b(h+b)} \{ (b-h) a - 2\beta \} \tag{18}$$

The general solution will be of the form

$$\left. \begin{array}{l} a = A_1 + B_1 x \\[2mm] \beta = A_2 + B_2 x + C_2 x^2 + D_2 x^3 \end{array} \right\} \tag{19}$$

and

The boundary conditions are:

(i) At $x = 0$; $\theta_1 = \theta_2 = 0$

or

$$a = \beta = 0$$
$$\therefore A_1 = A_2 = 0.$$

(ii) At $x = 0$; $\dfrac{\partial u}{\partial z} = 0$

$$\left(\frac{\partial u_1}{\partial z} \right)_{x=0} = \frac{q}{G\delta} - \theta_1 \frac{h}{2}$$

$$= \frac{h}{2} \beta$$

$$\therefore \beta_2 = 0.$$

(iii) Torque $T = P_1 h + P_2 b$ at $x = l$

$$\therefore T = \delta \int_{b-/2}^{b/2} \tau_1 h dz + \delta \int_{-h/2}^{h/2} \tau_2 b dz$$

For approximation, from eq. (1), the shearing force is given by (q/δ). Therefore,

$$T = G\delta h^2 b^2 \dot{a}.$$

or

$$\beta_1 = \frac{T}{G\delta h^2 b^2}$$

(iv) The constants C_2 and D_2 will be determined from the theorem of minimum potential energy. The potential energy of the system is given by

$$u = \frac{\delta}{2E} \int_0^1 \left\{ 2 \int_{-b/2}^{+b/2} \left(\sigma_{1x}^2 + \frac{E}{G} \tau_1^2 \right) dz + 2 \int_{-h/2}^{+h/2} \left(\sigma_{2x}^2 + \frac{E}{G} T_2^2 \right) dz \right\} dx$$

$$+ \frac{12}{2E\delta^3} \int_0^1 \left\{ 2 \int_{-b/2}^{+b/2} M_1^2 \, dz + 2 \int_{-h/2}^{+h/2} M_2^2 \, dz \right\} dx - T\theta_2(l) \qquad (20)$$

Substituting the corresponding values and applying the theorem of minimum potential energy, i.e.

$$\frac{\partial u}{\partial C_2} = 0 = \frac{\partial u}{\partial D_2}$$

the constants C_2 and D_2 can be found.

$$C_2 = \frac{12\epsilon}{(1 + t)t} \times \frac{(3 + 9r - 0.048s + 0.046sr)}{(3 + 0.97s + 0.003s^2)} \times \frac{T}{E\delta b^2}$$

and

$$D_2 = \frac{12}{(1 + t)t} \times \frac{(S/15 - 1 - 2.5r - Sr/30)}{(3 + 0.97s + 0.003s^2)} \times \frac{T}{E\delta b^2}$$

where

$$t = \frac{h}{b},$$

$$\epsilon = \frac{l}{h}$$

$$s = \frac{48\,\epsilon^4 t^4}{(1 + t)^2} \times \frac{\delta^2}{h^2}$$

and

$$r = \frac{(1 - t)}{(1 + t)} \times \frac{\delta^2}{h^2} \times \epsilon^2 t^2$$

assuming $E/G = 2.5$ for steel.

Therefore, applying the end conditions and substituting in eq. (12)

$$\theta_1 = -\frac{T}{E\delta b^2} \left[\frac{12\,\epsilon^3 t}{(1 + t)} \times \frac{(2 + 6.5r + 0.019s + 0.013sr)}{(3 + 0.97s + 0.003s^2)} - 2.5\frac{\epsilon}{t} \right]$$

and

$$\theta_2 = +\frac{T}{E\delta b^2} \left[\frac{12\,\epsilon^3 t}{(1 + t)} \times \frac{(2 + 6.5r + 0.019s + 0.013sr)}{(3 + 0.97s + 0.003s^2)} + 2.5\frac{\epsilon}{t} \right] \qquad (21)$$

CALCULATION OF STIFFNESS

Stiffness of a beam in torsion is defined as

$$K = \text{torque/angle of rotation}$$

However, in the case of thin-walled structures the whole section does not rotate through the same angle. The horizontal and vertical walls rotate through different angles due to distortion, as shown in Fig. 5b. From this figure it is clear that the effect of distortion on one wall

is additive to the angle of rotation due to pure torsion and on the other it is opposite. Therefore, it is possible to write two types of angle of rotation

$$\theta = \frac{\theta_1 + \theta_2}{2} \quad \text{and} \quad \psi = \frac{\theta_2 - \theta_1}{2}$$

corresponding to the angle of rotation due to pure torsion and contour distortion respectively. The values of θ calculated from eq. (21) gives

$$\theta = \frac{\theta_1 + \theta_2}{2} = \frac{2 \cdot 5 \, T\epsilon}{2E\delta b^2} \left(\frac{1}{t} + 1 \right)$$

$$= \frac{Tl \, (b + h)}{2G\delta b^2 h^2} \tag{22}$$

The value of θ is the same as calculated by the theory of pure torsion.[12]

Therefore, the torsional stiffness of a box beam could be characterized by two factors,
namely:

(i) Torsional stiffness against shear deformation $= T/\theta$ and

(ii) Torsional stiffness against distortion of contour $= T/\psi$

EXPERIMENTAL INVESTIGATIONS

In order to verify the above analysis tests were carried out on box-type beams of varying
lengths and constant cross-section $4 \times 4 \times \frac{3}{16}$ in wall thickness.

Tests were carried out in a similar manner as described earlier for case D when studying
the effect of end fixings.

The horizontal (V_1) and vertical (V_2) displacements of the edge of the beam were measured
at both ends of the beam by means of clock gauges graduated in 1/10,000 in divisions.
Although the headstock was prevented from rotating, the difference of deflections at the
two ends of the beam were recorded for accuracy.

From the displacement V_1 and V_2, θ_1 and θ_2 were calculated from the formulae

$$\theta_1 = 2V_1/h; \qquad \theta_2 = 2v_1/b.$$

DISCUSSION OF RESULTS

The experimental and theoretical values of

$$\theta = \frac{\theta_1 + \theta_2}{2} \quad \text{and} \quad \psi = \frac{\theta_2 - \theta_1}{2}$$

for different values of $\epsilon (= l/h)$ for the beam under investigation are plotted in Fig. 6.
This shows an agreement of the values to an extent of $\pm 8\%$.

From these results, it is clear that for beams in torsion having a flexible contour, the angle
of rotation is much greater than that calculated on the basis of the theory of pure torsion.
When considering a beam under torsional forces the distortion of contour should therefore be considered.

When beam ends are rigid and the contour of the section is inflexible, as in short beams
with thick transverse walls closing the ends, the contour deformation is small compared

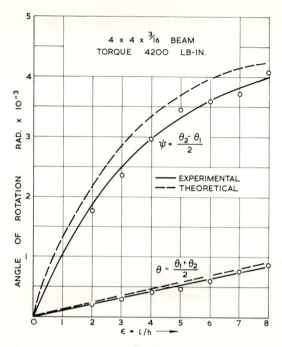

Fig. 6. Torsion test.

to shear deformation and the criterion for stiffness against torsion can be taken due to shear deformation only. However, in cases where the section is allowed to deform, as in long flexible beams with open ends at which the torque is applied, both shear and contour deformations must be considered.

The use of two factors of torsional stiffness, viz. pure torsional stiffness and contour deformation stiffness, defines completely the torsional stiffness of a structure. The former is used when there is no deformation, while the latter is used when ends are completely free to distort.

CONCLUSIONS

When a basic beam section is buttressed with another rectangular section the static stiffness-to-weight ratio and also the first natural frequency under bending loads are increased with respect to the corresponding basic beam. This is due to the fact that the overall height of the section is increased.

Under static torsional loads the effect of the stiffness of components which are not located in the main stream of shear stresses depends on the torsional stiffness of these components in free torsion. This, however, when the members are of similar proportions to the main members, is usually small enough to be neglected, and the torsional stiffness of the section depends on the area bounded by this stream of shear stress and its perimeter. Depending upon these two factors, the torsional stiffness-to-weight ratio for some types of buttressings increases, whilst for others it decreases. In the case of buttressing of group II the torsional stiffness-to-weight ratio is increased with respect to corresponding beams of group I. On the other hand, the stiffness-to-weight ratio is decreased for beams of groups III and IV.

For buttressed beams the natural frequency and damping ratio under dynamic torsion decrease with respect to the basic structure. This is due to the fact that as h/b increases, the torsional natural frequency and damping ratio decreases.

When comparing the buttressed beam with the simple rectangular beam of equal overall dimensions, the static stiffness-to-weight ratios and natural frequencies under bending and torsion decreases. However, the damping ratio under dynamic torsional loads is increased in the case of buttressed beams.

The positioning of buttressing from centre to one side has little effect upon stiffnesses, natural frequencies and damping ratios for both bending and torsional loads.

As the depth of the beam increases a greater number of plate vibrations fall before the first bending natural frequency of the beam as a whole. This means that when beams of large depth are used stiffening ribs must be provided on the panels. In addition to reducing panel vibration these stiffening ribs, if positioned transversely across the section, will prevent contour deformation of the cross-section and hence increase the overall stiffness of the cross section.

ACKNOWLEDGEMENTS

This work was carried out as part of the research programme of the Machine Tool Engineering Division of the Department of Mechanical Engineering, University of Manchester Institute of Science and Technology.

The authors are indebted to Professor F. Koenigsberger for his helpful criticisms and comments during this investigation. Acknowledgements are also due to the Science Research Council for financial assistance which has made the work possible.

REFERENCES

1. K. LOWERFIELD. The Stiffness of Box Sections. Translated from 3 *Fokoma* **II**, D. 129–140.
2. M. W. BADAWI and R. H. THORNLEY. Comparison of Static and Dynamic Characteristics of Closed, Box Section Beams With and Without Core Holes and For Some Other Beam Configurations. *Int. Jnl. Mach. Tool Des. Res.* **6**, 1966, 199–226.
3. H. OPITZ. Performance Testing of Machine Tools. Conference on Technology of Engineering Manufacture, I.Mech.E., 1958.
4. R. H. THORNLEY. Machine Tool Structures Using Model Techniques. The Manchester Ass. of Engrs. Session 1964–65, No. 3.
5. F. M. STANSFIELD. Some Notes on the Use of Perspex Models for the Investigation of Machine Tool Structures. 6th Int. Mach. Tool Des. Res. Conf., Manchester, 1965.
6. R. H. THORNLEY, R. CONNOLLY, M. M. BARASH and F. KOENIGSBERGER. The Effect of Surface Topography upon the Static Stiffness of Machine Tool Joints. *Int. Jnl. Mach. Tool Des. Res.* **5**, 1965, 57–76.
7. R. CONNOLLY and R. H. THORNLEY. The Significance of Joints on the Overall Deflection of Machine Tool Structures. 6th Int. Mach. Tool Des. Res. Conf., Manchester, 1965.
8. R. H. THORNLEY and P. KUMAR. Statische und dynamische Eigenschaften einiger kastenformiger Elemente von Werkzeugmaschinen. 6. Int. Werkzeugmaschinentagung, Dresden, 1968.
9. V. Z. VLASOV. *Thin Walled Elastic Beams.* Moscow, 1959. Translated from Russian by Israel Program for Scientific Translators, Jerusalem, 1961.
10. V. V. KAMINSKAYA and E. A. KUNIN. Investigation and Calculating the Rigidity of Universal Boring Machine Columns. *Machines and Tooling,* **XXXI**, 1960.
11. V. V. KAMINSKAYA, D. N. RESHETOV, Z. M. LEVINA and I. STANING. *Korpusuye detali Metallorexhushchikh Stankov (Raschet i Konstruicovanie).* Mashgiz, 1960.
12. J. PRESCOTT. *Applied Elasticity.* Dover Publications Inc., 1961.

ANALYTIC AND EXPERIMENTAL STUDY OF CERTAIN TYPES OF MACHINE TOOL SUBASSEMBLIES

RAFFAELLO LEVI, SERGIO ROSSETTO and ANNA VERRINI

Istituto di Tecnologia Meccanica, Politecnico di Torino, Italy

SUMMARY

A number of comparable machine tool subassemblies are evaluated in terms of their dynamic properties.

The results of an analysis performed on a digital computer are compared with the results of vibration tests performed on the actual prototypes. Information is also given on the computing program and the instrumentation system used.

1. INTRODUCTION

Digital computing techniques are now currently used for the prediction of machine tool dynamic performance. Their main advantage is that of providing designers with information similar to that obtained from actual tests, saving much of the time and expenditure involved in building and testing models. It is furthermore easy to assess the effect of a modification, as special techniques were developed[1] for this purpose.

However, at the present time computer aided design cannot supersede completely test methods, as the answers to some questions may come only from the test floor. Computer analysis is suited ideally for tackling solid elements. Methods derived either from the work of Timoshenko[2, 3] and/or lumped constants methods often based on Myklestad's approach[4-10] are currently used. They yield usually accurate results, as far as natural frequencies, stiffness and mode shapes are concerned. However, when actual machines are considered, the effect of joints, bearings and similar elements must be taken into account. The characteristics of these elements are seldom known beforehand with an accuracy approaching that obtained easily in the evaluation of mass and length parameters. Furthermore, these characteristics may have the vicious habit of being inherently nonlinear[11,12,13] so that further approximations are inevitable when they are introduced, out of sheer necessity, in the straight jacket of a linear model. Finally, the sources of damping in machine tool structures are hardly known to the extent of enabling accurate predictions to be made solely on a theoretical basis. Experimental data obtained from similar machines are usually considered but obviously this practice makes the computer dependent on the laboratory for supplying the necessary data.

After looking at this rather dark picture one would be tempted to ask whether the use of computer aided design is worth anything at all. The expense involved is not remarkably small, and considerable preparatory work and computing time are needed for each analysis. Yet the whole process is justifiable also from an economic point of view, provided that the user is fully aware of which question may reliably be answered by the computer, and how the accuracy of the answers is linked to that of the raw data fed into the machine. This means that routine experimental tests must be carried out in order to

779

check, and complete the computed results. Tests must be made until the body of evidence accumulated is large enough to provide a reasonably accurate answer to every likely question concerning the reliability of computed data.

It should be realized that while the authors of the program, who double also as experimenters, are generally aware of the pitfalls of both methods, this is not necessarily the case for the occasional user. His problem is that of getting an answer as fast as possible on some questions concerning either a set of drawings or a prototype, and it hardly matters how the answer is arrived at. The problem is therefore that of showing quickly what may be done, and how accurately, when analyzing and testing machine tool structures.

This is the goal of the case history presented in this paper. It is obviously referred to the computing and testing techniques actually being used by the authors. A short description is given in Appendices 1 and 2 of both these techniques. Their shortcomings are not obviously shared with the more sophisticated methods used in other laboratories. On the other hand, newcomers to this field may spare some time and profit from the lessons learned the hard way in the course of the development of this work.

2. MAIN CHARACTERISTICS OF THE SUBASSEMBLIES STUDIED

Three boring heads were submitted for analysis by the manufacturer, the main purpose of being that determining their dynamic properties.

Figures 1, 2 and 3 show the heads, which will be denoted by the letters A, B and C. The heads share a number of common features. The outer diameter of the central span of the spindle is the same for all three types, while the inner diameter is the same for heads B and C. On the other hand, the main casting is basically the same for heads A and C, head B being a stub version of model C.

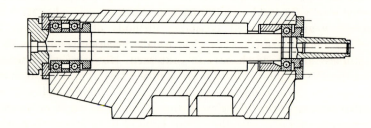

FIG. 1. Schematic section of boring head A.

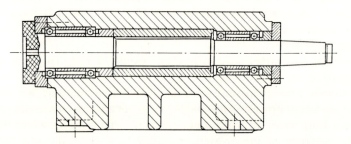

FIG. 2. Schematic section of boring head B.

The layout of the bearings suggests a strong similarity between models B and C; in these models four ball bearings are used, with a rather broad spacing between each pair. Only three bearings are used in model A, where the forward pair is mounted in a rather short housing.

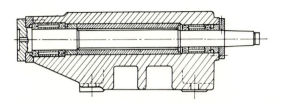

FIG. 3. Schematic section of boring head C.

All three heads are interchangeable. They are mounted on the boring machines according to the job at hand, and for some jobs any one of the three could be used almost indifferently. The interest of finding out their main dynamic properties is thus clear.

As the heads are manufactured in medium size batches, to concentrate on the most advantageous designs might entail larger batch sizes and lower costs.

The problem is also an interesting one for the laboratory, and the computer. While the subassembly is comparatively simpler than a whole machine, yet most of the elements are present—spindle, bearings, main casting—and the effect of each can be suitably assessed, with both of the methods used.

3. DERIVATION OF THE MODEL

As bending vibrations in one plane are analyzed at this stage, the model is a plane one. The heads are broken down in two main systems, namely, rotor and stator, connected together, and to a stationary reference system, by two series of springs. As the influence of external restraints is to be kept down to a minimum, the stiffness of the springs linking the casting to the outer system is taken very low. On the other hand, the stiffness of the springs introduced in lieu of the bearings is nominally the computed stiffness of the bearings, due account being taken of the preload. End fixing, and thus bending, besides axial stiffness of the springs is considered where the bearings oppose not only radial play but also rotation of the corresponding spindle section.

Both main systems, namely, rotor and stator, are approximated by sets of lumped constant spans. Each span is formed by a number of dimensionless stations, in which the inertia properties are concentrated, connected by weightless segments having the required stiffness characteristics. Individual stiffness and inertia values were computed for each system, at each segment and station location. Constant cross sections were considered between adjacent stations, sloping lines being approximated by a "staircase" of short, offset parallel lines. Figures 4 and 5 show the subdivision adopted for stator and rotor of head A. Obviously, the larger the number of segments the better the approximation: however, there is no point in pushing the closeness of the subdivision much beyond the limits shown. No detectable gain in accuracy could be obtained, against a very sizeable increase in analyzing time.

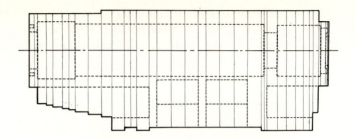

FIG. 4. Subdivision adopted for stator of boring head *A*.

FIG. 5. Subdivision adopted for rotor of boring head *A*.

Figures 6, 7 and 8 show the simplified models for the three heads. Individual masses were omitted for the sake of clarity, the closeness of subdivision being approximately the same for the three heads. It may be remarked that an additional span is shown fixed to the spindle nose in the three aforementioned figures. This span, which simulates the boring bar, is the same for all three heads. Every care was taken in order to fasten rigidly the dummy bar to the spindle during the tests, in order to reduce the disturbances due to the introduction of an extra joint.

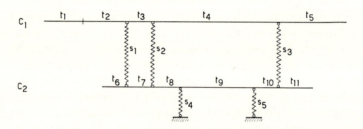

FIG. 6. Simplified model for the boring head *A*. C_1, rotor; C_2, stator; t_i, spans; s_i, springs.

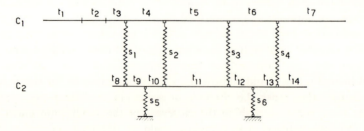

FIG. 7. Simplified model for the boring head *B*. C_1, rotor; C_2, stator; t_i, spans; s_i, springs,

No particular difficulties are usually found in computing the elastic and inertia properties of each segment and station, as long as solid members are analyzed. Bearings sometimes are tougher to deal with, especially when their axial stiffness, and therefore the restraint they oppose to spindle bending, cannot be disregarded. Each spring substituting a bearing must then be considered with both ends built up, and bending stiffness values must be introduced besides the axial stiffness figures.

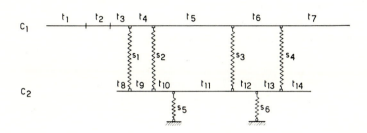

FIG. 8. Simplified model for the boring head C. C_1, rotor; C_2, stator; t_i, spans; s_i, springs.

Attempts were made to evaluate ball bearing stiffness on a theoretical basis, starting from Hertz's theory. Empirical formulae were also used, and both methods yielded reasonable approximations of the actual values[14]. It must, however, be remarked that the stiffness of ball and roller bearings is inherently a nonlinear function of the applied load. The discrepancies between the computed performances of the linear model, and the ones measured on the prototype, can be minimized artificially by operating at low force levels, thus reducing the stiffness range. However, the machine must be expected to behave in a different way when submitted to the service loads. On the whole, it appears that the problem of analyzing quickly and correctly the effects of bearings is still far from being completely solved.

4. COMPUTER ANALYSIS

An outline of the computer program used in these tests is given in Appendix 1. This program was developed for the analysis of static and dynamic bending deformation of plane structures, approximated by lumped constants models. The analysis yields the following information:
—static deformations;
—natural frequencies;
—mode shapes at resonance;
—mode shape under forced vibration;
—kinetic and potential energy at resonance and under forced vibration.
No damping is considered in the program. Mode shapes at resonance are therefore obtained on a non dimensional basis. The same applies to the energy distribution under these conditions. It would be desirable to introduce proper values of the damping coefficient right where the damping action takes place. It is the feeling of the authors that this is seldom possible now, due to the lack of proper information on a sufficiently general scale. Again, nonlinearity in some damping effects limits the extent to which the accuracy of a linear model can be pushed. Nonlinear elastic effects cannot be considered.

Forces due to unbalanced rotors can be accounted for, as well as external loads up to the maximum number of stations of the model. These loads need not to be static only, provided their frequency is the same for all.

For the particular case at hand natural frequencies, mode shapes, response to external excitation and energy distribution were computed for the interval 0·3–3 kHz. As external load a dynamic force of 0·5 kg was considered, applied to station No. 2 of rotor's span No. 1. The natural frequencies computed are given in Table 1. Mode shapes may be traced directly by a data plotter driven by the computer. Figure 9 shows, for example, a calcomp plot of the mode shape for the 3rd natural frequency of head *B*. The mode shapes for the 1st natural frequency of heads *A, B, C* are shown in Fig. 10.

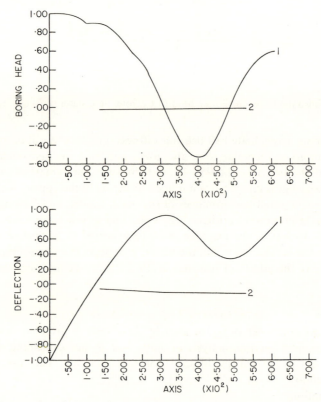

FIG. 9. Calcomp plot of the mode shape (rotation θ and deflection y) for the 3rd natural frequency of boring head *B*.

Figures 11–16 show the computed response to forced vibration, considered at the station where the external load is applied. The computer gives the response in terms of displacement; it is shown here also converted to acceleration, for convenience of comparing it with the test results. As the damping of the system is rather low for most modes (much less than the critical damping value, as found almost invariably on machine tool structures) the conservative model yields quite reasonable values as long as the forcing frequency is not near (say 10%) to a natural frequency.

Besides deflections and angles of rotation, shear forces and bending moments are obtained for each mode shape.

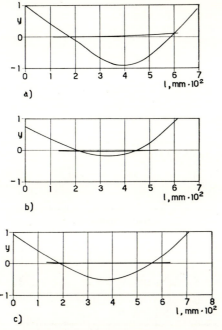

FIG. 10. Computed mode shapes for the first natural frequency: (a) boring head A; (b) boring head B; (c) boring head C.

FIG. 11. Displacement response to forced vibration, computed at station No. 2 of rotor's span No. 1 (head A).

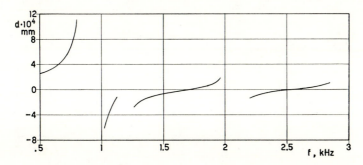

FIG. 12. Displacement to forced vibration, computed at station No. 2 of rotor's span No. 1 (head B).

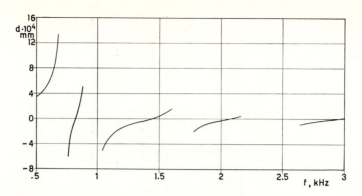

FIG. 13. Displacement response to forced vibration, computed at station No. 2 of rotor's span No. 1 (head *C*).

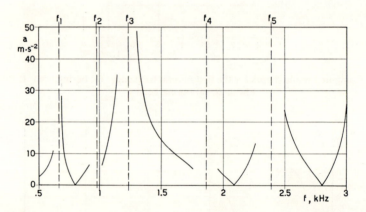

FIG. 14. Acceleration response to forced vibration, computed at station No. 2 of rotor's span No. 1 (head *A*).

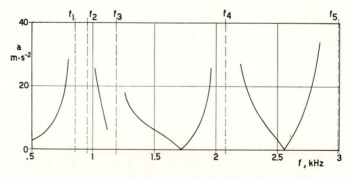

FIG. 15. Acceleration response to forced vibration, computed at station No. 2 of rotor's span No. 1 (head *B*).

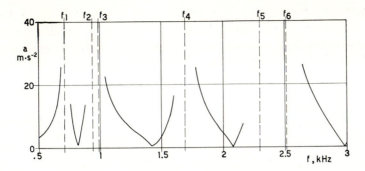

FIG. 16. Acceleration response to forced vibration, computed at station No. 2 of rotor's span No. 1 (head *C*).

5. LABORATORY TESTS

The main purpose of the laboratory tests was that of checking the computer analysis and finding out the main discrepancies between computed and measured results with the aim of tracking their origins and take remedial action. Under these conditions the instrument system was designed with the goal of making a result producing machine comparable in convenience with the computer. Data concerning the instrumentation are given in Appendix 2.

Figure 17 shows the test setup. The boring head is supported by a heavy cast iron plate via flexible rubber pads. A flexible suspension is used also for the exciter, and no coupling via the exciter support and plate could be detected.

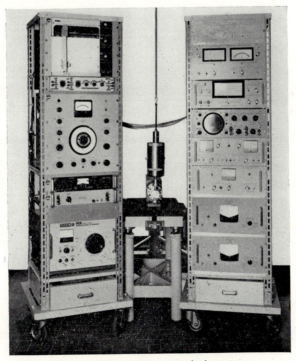

FIG. 17. Instrument system during test.

Figures 18–23 show the frequency response, measured for the three heads. The transducer —a quartz accelerometer—is mounted directly under the point of application of the exciting force. The value of the force is measured by a piezoelectric pickup in series with the exciter spindle, and the oscillator output is servo controlled in order to keep down the force fluctuation throughout the whole frequency range.

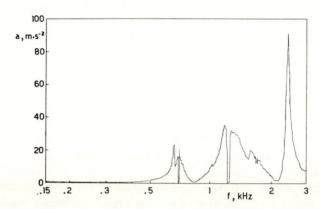

FIG. 18. Acceleration response of the boring head *A*, measured at the point of excitation.

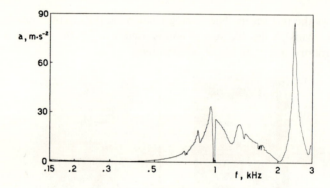

FIG. 19. Acceleration response of the boring head *B*, measured at the point of excitation.

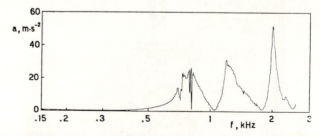

FIG. 20. Acceleration response of the boring head *C*, measured at the point of excitation.

Remarkably high damping is shown to take place at the first measured resonances. This is hardly unexpected as bearing deflection is mainly responsible for these resonances, where its contribution is of the same order of magnitude of structural deformation.

Lower damping is observed for the higher frequencies, where most of the deformation is due to bending of the spindle.

The response curves, which were plotted by an X-Y recorder, show a number of small peaks and kinks which are due to parasitic effects. A relatively high level of environmental random vibration was kept down, but not eliminated altogether by the precautions taken during testing. Further improvements are being made in this direction. Some components were also found to contribute a sizeable amount of noise, and by changing one of them some improvements were later obtained. Nonlinear effects in the prototypes may also be responsible for some effects.

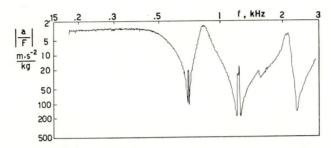

FIG. 21. $|a/F|$ response of the boring head A, measured at the point of excitation.

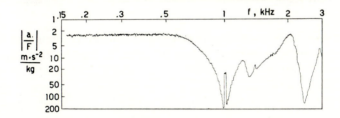

FIG. 22. $|a/F|$ response of the boring head B, measured at the point of excitation.

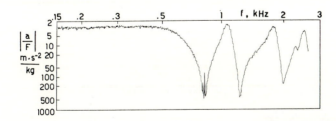

FIG. 23. $|a/F|$ response of the boring head C, measured at the point of excitation.

Bode graphs were also recorded, by separating the response (acceleration) into the in phase and quadrature components. These plots showed in detail the phase relationship between the applied force and measured response, and helped in finding the natural frequencies. The experimentally determined values are always frequencies of amplitude resonance. The larger the damping the larger the discrepancy between measured and computed frequencies. However, the errors due to this source are small as compared with those due to the approximations introduced in the method.

SOME OBSERVATION ON COMPUTED AND MEASURED RESULTS

Table 1 gives the natural frequencies computed in the range between 0·3 and 3 kHz for the three boring heads. This is a rather sweeping statement: the values given are an estimate, accurate within less than one per cent, of the natural frequencies of the lumped constant models. These, in turn, are believed to duplicate with the same order of accuracy the static and dynamic characteristics of the distributed constant model. Most of the errors are introduced in the transition from the physical to the mathematical model. The mechanical properties of the materials used are seldom known with an accuracy better than one per cent, and a number of other factors do not make the situation any better. Simple strength of materials formulae are used, while more involved methods would often be appropriate for representing accurately the elastic properties of the members involved.

TABLE 1. COMPUTED NATURAL FREQUENCIES IN THE FREQUENCY RANGE
BETWEEN 0·3 AND 3 kHz

Boring head	Frequencies, kHz					
A	0·67	0·98	1·23	1·87	2·39	
B	0·86	0·96	1·19	2·08	2·99	
C	0·72	0·94	0·99	1·69	2·29	2·51

In the light of these considerations, the agreement between the computed data of Table 1 and the corresponding experimental values of Table 2 may be considered acceptable. One would indeed be hardly able to justify an agreement consistently better, considering the limitation of the method.

TABLE 2. EXPERIMENTAL RESONANCE FREQUENCIES IN THE FREQUENCY
RANGE BETWEEN 0·15 AND 3 kHz

Boring head	Frequencies, kHz					
A	0·65	0·70	1·15	1·25	1·57	2·45
B	0·94	1·00	1·30	2·10	3·00	
C	0·78	0·81	0·83	1·20	2·00	2·70

The response of the heads to forced vibration shows in details the extent to which computed values can be expected to represent faithfully the effective response of the structure. No quantitative conclusion can be drawn near resonant frequencies, due to the neglect of damping; nevertheless the trend of the frequency response is clearly given, especially as far as the higher frequencies are concerned.

The lower natural frequencies are mainly due to bearing deflection. As the approximation of ball bearings with weightless linear springs is far from being accurate, the discrepancy between computed and measured response is larger in these regions. This may be observed by comparing the plots of Figs. 14 to 16 with corresponding measured graphs of Figs. 18 to 20.

Some inaccuracies of the instrument system are responsible to some extent for these discrepancies. Changes in mechanical impedance of the test structure induced alterations of the exciting force level, and the compressor device fitted on the oscillator could reduce, but not eliminate, these fluctuations in the input signal. Better results were obtained by plotting the ratio of the output signal (acceleration) to the input force, and Figs. 21 to 23 show indeed some difference from the corresponding Figs. 18 to 20. A number of spurious peaks are effectively eliminated by this arrangement.

CONCLUSIONS

Three boring heads were tested in order to determine their dynamic properties and check the computed values obtained through the analysis of a set of mathematical models. The results obtained with both the experimental and the analytic method were compared and some of the discrepancies were explained in terms of lack of agreement between some of the assumptions and the real systems.

It is shown that large errors may be made mainly in the transition from the physical to the mathematical model.

While a number of the predicted values cannot be expected to be verified with an accuracy much better than one part in ten, some useful information is however derived reasonably cheaply and quickly from the computer analysis. Mode shapes point out which element contributes most to the deflection observed at any frequency, and then how to alter as a first step the system in order to get some definite results. For instance, no remarkable improvement can be obtained at the lower frequencies by using a stiffer spindle if the bearings are not stiffened too; the opposite being true in the upper frequency region.

The computed response to forced vibration shows clearly the frequency regions pertaining to each mode shape. While frequency response plots for a given station can be derived rather quickly with an adequate instrument system, measuring and plotting mode shapes is a long and tiresome task. When some elements, such as spindles, are not easily accessible for a good deal of their length, it becomes even more inconvenient to obtain their mode shapes. Here the computer provides a faster and cheaper answer even when the prototype is already available in the laboratory.

It was found that for best results most computed values must be checked against their experimentally determined counterparts, at least until the body of experience accumulated is broad enough to allow immediate detection of results of a doubtful value. Some discrepancies between computed and measured results are not necessarily an evil. Several hypotheses were in fact reviewed in the light of the first failures which spurred quite a few improvements in the critical technique of building the mathematical model from the physical one. Thus while these computing techniques can, and do yield useful results even a short time after completion of the main program, best results are not obtained until the users are made fully aware by repeated experience of the operating characteristics of these methods. Until then there's no way out of checking everything in the laboratory.

ACKNOWLEDGMENTS

Acknowledgments are due to Prof. G. F. Micheletti for his constant encouragement in the course of this work. At the beginning of this work Dr. J. R. Lemon laid out clearly before the authors the principles along which the computer program was to be subsequently developed. Dr. L. Guerri and Ing. A. Benuzzi, C.C.R. Euratom, Ispra, gave constant assistance and solved quite a few awkward problems in the course of the development of the program. Prof. B. F. von Turkovich suggested a number of improvements of the experimental techniques and analytic methods. His guidance is gratefully acknowledged. Financial support from Consiglio Nazionale delle Ricerche enabled us to carry out this work, which is presently being developed in order to extend the capabilities of the methods beyond the limits referred to above.

REFERENCES

1. T. R. Comstock and J. R. Lemon, Effects of Reinforcing Members on the Dynamic Performance of Structures, A.S.M.E. Vibration Conference, Boston, March 1967.
2. S. P. Timoshenko, *Vibration Problems in Engineering*, Van Nostrand, New York, 1955.
3. J. C. Maltbaeck, Classical Beam Method for the Prediction of Vibration Characteristics of Machine Tool Structures, *Advances in Machine Tool Design and Research*, Pergamon, Oxford, 1964.
4. N. O. Myklestad, A New Method of Calculating Natural Modes of Uncoupled Bending Vibration of Airplane Wings and Other Types of Beams, *Journal of the Aeronautical Sciences*, **153**, April 1944.
5. W. T. Thomson, A Note on Tabular Methods for Flexural Vibrations, *Journal of the Aeronautical Sciences*, **62**, January 1953.
6. T. C. Huang and N. C. Wu, Approximate Analysis of Flexural Vibrations of Beams, *Proceedings of the 7th Midwestern Mechanics Conference*, Michigan State University, September 1961.
7. J. K. Sevcik, System Vibration and Static Analysis, A.S.M.E. Paper 63-AHGT-57, March 1963.
8. S. Taylor and S. A. Tobias, Lumped-constants Method for the Prediction of the Vibration Characteristics of Machine Tool Structures, *Advances in Machine Tool Design and Research*, Pergamon, Oxford, 1964.
9. S. Taylor, The Design of Machine Tool Structures Using a Digital Computer, *Advances in Machine Tool Design and Research*, Pergamon, Oxford, 1966.
10. S. Taylor, The Prediction of the Dynamic Characteristics of Machine Tool Structures, Ph.D. Thesis, University of Birmingham, March 1966.
11. R. H. Thornley, R. Connolly, M. M. Barash and F. Koenigsberger, The Effect of Surface Topography upon the Static Stiffness of Machine Tool Joints, *International Journal of Machine Tool Design and Research*, **5**, 1965.
12. R. Connolly and R. H. Thornley, The Significance of Joints on the Overall Deflection of Machine Tool Structures, *Advances in Machine Tool Design and Research*, Pergamon, Oxford, 1966.
13. E. Meldau, Einfluss der Lagerluft auf die Druckverteilung, die statische Tragfahigkeit und die Lebensdauer radialbelasteter Walzlager, *Konstruktion*, **4**, 79, 1952.
14. S. Rossetto, Some Aspects of Static and Dynamic Deformations of Ball Bearings. These Proceedings, p. 605.
15. D. K. Faddeev and V. N. Faddeeva, *Computational Methods of Linear Algebra*, W. H. Freeman, San Francisco, 1963.
16. G. F. Micheletti, S. Rossetto, A. Verrini and R. Levi, Ricerca sulla determinazione delle caratteristiche ottimali di organi di macchine utensili ai fini di consentirne la progettazione automatica. Report to C.N.R., September 1967.
17. G. F. Micheletti and R. Levi, Ricerca sull'automazione delle operazioni di collaudo dinamico delle macchine utensili, Report to C.N.R., September 1967.

APPENDIX I

Main Principles of Computer Analysis

The structure to be analyzed is reduced to a lumped constants model, by dividing it into discrete elements. All inertia properties, such as mass and rotary inertia, are concentrated in dimensionless stations, connected by weightless segments having the required elastic properties. A set of stations and elements is thus used to represent the physical model.

At both ends of each segment a state vector $\mathbf{z}$ is defined by the following components: deflection y, rotation θ, shear v, bending moment m. By considering the relationship between the components y, θ, v, m at the ends of a segment, the following equation is derived in matrix form:

$$\mathbf{z}^L{}_i = \mathbf{F}_i\, \mathbf{z}^R{}_{i-1} \tag{1}$$

where the state vector at the first member is at the left (L) end of the ith station, and the second one at the right (R) end of the $(i-1)$th station.

In a similar way another equation is derived:

$$\mathbf{z}^R_i = \mathbf{P}_i \, \mathbf{z}^L_i + \Delta\mathbf{z}_i \tag{2}$$

between the state vector at the end of a segment and that at the beginning of the following one. The matrix $\mathbf{F}_i$ is called the field transfer matrix, and $\mathbf{P}_i$ the point transfer matrix; both are four by four. Externally induced displacements (or forces or moments) are introduced with the vector $\Delta\mathbf{z}_i$ in eq. (2).

We obtain directly by substitution from eqs. (1) and (2):

$$\mathbf{z}^R_i = \mathbf{B}_i \, \mathbf{z}_{i-1} + \Delta\mathbf{z}_i \tag{3}$$

where

$$\mathbf{B}_i = \mathbf{P}_i \times \mathbf{F}_i$$

A succession of segments each connected to the following one by a station is called a span. The equation linking the state vector at the end of a span to that at the beginning is derived easily in terms of eq. (3):

$$\mathbf{z}_n = \mathbf{U}\mathbf{z}_o + \mathbf{d} \tag{4}$$

where $\mathbf{U}$ is the product of the relevant $\mathbf{B}_i$ matrices, and $\mathbf{d}$ is given by the vectors $\Delta\mathbf{z}_i$. Thus if a structure is formed by s spans, a set of s equations of the form of eq. (4) may be written. By introducing the boundary conditions at the joints between spans (which link the state vector at the end of a span with the one at the beginning of the following span) a system of $4s$ equations with $8s$ unknowns is derived. The number of unknowns is reduced from $8s$ to $4s$ by writing the state vectors at the end of each span in terms of eq. (4). Eventually a system of the following form is obtained

$$\mathbf{R}\mathbf{x} = \mathbf{c} \tag{5}$$

where the vector of the unknowns $\mathbf{x}$ is formed by the $4s$ components of the state vectors as defined at the beginning of each span. Vector $\mathbf{c}$ is formed by the quantity derived from the vectors $\mathbf{d}$ of each span. Finally the elements of matrix $\mathbf{R}$ are functions of the square of the frequency Ω.

The computer program forms the matrix $\mathbf{R}$ in terms of the mechanical and geometrical characteristics of the model, with the information concerning boundary condition kept down to the minimum. Natural frequencies of the structure correspond to eigenvalues of the matrix $\mathbf{R}$. They are found by a point by point search of the null values of determinant $|\mathbf{R}|$ conducted with Gauss's triangularization method.[15]

Some modifications were made to the method in order to reduce the computing time. This is needed as the matrix $\mathbf{R}$ can reach the order of 100, as the maximum number of spans which may presently be dealt with is 25. The system is then solved by backward substitution in the triangularized system. When mode shapes at resonance are to be computed the homogeneous system:

$$\mathbf{R}\mathbf{x} = 0 \tag{6}$$

must be solved, that is the eigenvectors corresponding to the relevant natural frequencies must be computed. One of the variables is an arbitrarily given unit value, as no damping is considered. It is clear that the computed mode shapes cannot then give quantitative information on the effective amplitudes of the real ones. Relative amplitudes only can be considered at resonance.

The response of the model to forced excitation at one or more stations can be analyzed with the program. To this purpose the forcing frequency is entered as Ω in the matrix $\mathbf{R}$, while the vector **c** contains the terms due to the moduli of the exciting forces in the vectors Δz_i corresponding to the stations excited. More detailed information concerning the program is to be found in ref. 16.

It should be remarked that the program in its present version does not have any optimization feature of the input data. The problem of designing and operating such programs is known to be remarkably complex. Our program will give information on the static and dynamic performances of the structure analyzed only as far as its characteristics are represented by the lumped constants model.

APPENDIX 2

Some Information on the Instrument System

The instrument system presently used enables vibration tests to be carried out in a semi-automatic way. It consists of a set of instruments, connected together as shown in the simplified block diagram of Fig. 24.

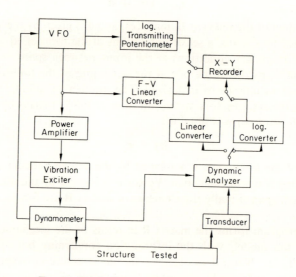

Fig. 24. Block diagram of instrument system.

A variable frequency oscillator can sweep back and forth through any selected frequency range between 5 Hz and 10 kHz. The scanning rate is adjustable. Two types of power amplifiers are provided, for use with either a 1 kg exciter or a 25 kg exciter. A piezoelectric dynamometer in series with the exciter spindle provides a signal for monitoring the force signal and driving the automatic compressor device. This should keep constant the force level, or more realistically cut down the force fluctuations brought about by changes in mechanical impedance of the structure under test. The signal from the dynamometer is also used as a reference for the phase measuring system.

Several types of vibration transducers may be used according to the specific tests to be made. While integrating and derivating networks are available to transform the output signal as required, they are presently seldom used due to the sizeable errors they are likely to introduce, as well as the signal attenuation. This explains the use of acceleration plots in this paper, an accelerometer being used on the heads tested.

The transducer's signal, amplified as required, is then filtered by a tracking filter driven by the exciting frequency. Most spurious signals are removed at this stage. The 60 db dynamic range of the tracking filter makes it unnecessary to switch range during the sweep, which can thus proceed almost unattended once the system is set.

A number of operations may be performed on the filtered signal. Its phase angle may be measured with reference to that of the exciting force by a phasemeter, or it may be resolved into phase and quadrature components for plotting either polar plots or Bode's diagrams.

Eventually, linear or logarithmic converters are selected in order to derive a slowly varying d.c. signal for recording on a two-pen X-Y recorder. The logarithmic converter allows the ratio of two signals to be plotted. This is a convenient feature, for it allows for those force fluctuations which cannot be eliminated by the compressor device presently being used. Either a linear or logarithmic frequency scale may be selected. The instrument system is shown in Fig. 17. Some of the plots which can be obtained with it are shown in Figs. 18–23. Figure 25 is a plot versus frequency of the force signal when the compressor is out of action, its effect being shown in Fig. 26, obtained under otherwise comparable conditions.

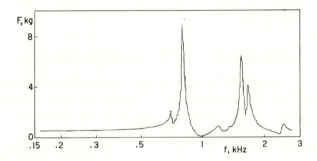

FIG. 25. Plot of the force signal versus frequency when the compressor is out of action.

Plots of phase and quadrature components proved quite helpful in locating the less marked resonant frequencies, occurring with a heavy damping. This may be observed, for example, in Fig. 27. More detailed information on the system is to be found elsewhere.[17]

The system shown was assembled using instruments made by several firms. Several arrangements were necessary to make each one compatible with the others, and reducing noise. While quite a considerable time was spent in assembling and tuning, it was felt that there were worse ways of spending a good deal of that time and labour. The confidence gained in building the system enabled us to cut down considerably the trouble shooting time when malfunctions occurred.

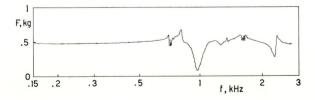

FIG. 26. Plot of the force signal versus frequency with the compressor operating.

This is hardly the case when more involved factory-assembled systems fail to perform properly once in the user's laboratory, a down time of several months being not unheard of in such circumstances.

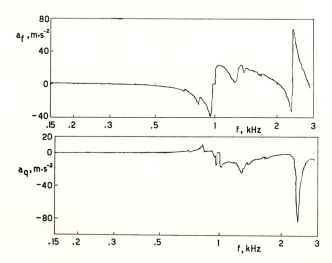

FIG. 27. Plot of phase and quadrature components of the acceleration versus frequency.

Also, an intimate knowledge of the effective operating characteristics of each subassembly enabled us to form realistic estimates of the overall accuracy. While it is no doubt possible to obtain much better results by point by point techniques, automatic sweep frequency tests seldom yield an accuracy much better than 10% or so. And this is quite adequate for many kind of tests.

COMPUTER-AIDED DESIGN OF HYDROSTATIC BEARINGS FOR MACHINE TOOL APPLICATIONS

PART I. ANALYTICAL FOUNDATION

OTTO DECKER and WILBUR SHAPIRO

Friction and Lubrication Laboratory, Franklin Institute Research Laboratories,
Philadelphia, Penn., U.S.A.

SUMMARY

The role of the high-speed digital computer in the design and analysis of incompressible, fluid-film bearings and bearing systems is presented with special emphasis on machine-tool applications. The more important features and capabilities of the computer programs are described including the class of application problems that can be solved. The significance of the support structure characteristics on bearing performance is reviewed. A combined structures and bearing computer program is briefly described. The analytical foundations are presented in Part I. Several applications are covered in detail, in Part II, including description of the general problem, the analytical approach and sample results.

INTRODUCTION

Paralleling the rapid pace of technological advancement has been an appreciation of the role that support bearings play in affecting overall equipment performance. The significant influence of the bearings has been manifest for a wide spectrum of machinery, from high-speed turbines with shafts rotating at hundreds of thousands of revolutions per minute to massive satellite-tracking radio telescopes with slew rates measured in revolutions per day. It has become increasingly difficult to select the bearings from a catalogue or to extrapolate the "old standards" that have worked well in the past to a new machine that extends the boundaries of contemporary requirements. The necessity to integrate the bearing design at an early stage of the overall system development has often resulted in improved performance and machine reliability. It is now becoming recognized that solutions to the Reynolds equations based on invariable lubricant film geometry are no longer adequate for the design of efficient bearing systems. Computer programs have been developed which take into account deviations in film geometry due to moment loading of the bearing housing, distortions of the housing or shell due to high film pressures or elastic deflections of the supports, and due to assumed partial failure of the lubricant supply system. Solutions are available for steady-state operating conditions and for dynamic conditions when, for instance, the orbit of a shaft or the dynamic response of a tool-point may be required, and when the bearing housing, support structure, foundation characteristics, and other external influences must be included.

It is a recognized fact that bearings and their lubrication have played a key role in the realization of our present-day highly mechanized civilization. Nowhere is this more evident, for example, than in the thousands of different varieties of machine tools required to fabricate the millions of components for equipment and devices upon which we are all so dependent these days. As a direct result of continual urgency to upgrade equipment performance, designers have been virtually forced into either "beefing-up" the old models or, much more rewarding, developing new techniques and methods which offer the promise of making a "giant step" forward.

The advent of the use of numerical solution methods, coupled with the use of high-speed digital computers to obtain practical and exact solutions represents such a "giant step" forward.

The use of digital computer solutions in aiding in the design of hydrostatic bearings in machine tool applications is the subject of this paper.

ANALYSIS

Steady State Analysis

A number of design manuals for fluid-film bearings have been published.[1–3] These manuals provide a wealth of bearing design information and are usually adequate for conventional applications. However, the need for complex configurations (e.g. hybrid bearings which include rotational effects) and, in many instances, the necessity to consider the effects of bearing misalignment and structural deflections is growing proportionally with increasing machinery performance requirements. Complex problems of this nature can now be analyzed by numerical techniques, programmed for use on a high-speed digital computer. The computer programs are very general and can be applied to a wide variety of applications.

1. Steady State Exact

(a) *General Description.* A most versatile and useful program in the field of laminar, incompressible lubrication, is the so-called Steady State Exact Program (SSE). Its primary function is to determine steady-state performance characteristics of individual bearings of practically any configuration. The program handles all the common mechanisms of pressure generation namely, hydrodynamic or self-acting, hydrostatic or externally pressurized, and a combination of the two or hybrid.† Essentially, the bearing area is subdivided into a grid pattern and the pressure determined at each grid point by simultaneous satisfaction of the Reynolds lubrication equation and continuity of flow through the supply circuit and bearing film. Typical utilization of this program is for analyzing hydrostatic spindles as shown in Fig. 1.

(b) *Film Thickness Distribution.* No complications are added by including eccentricity and misalignment into the film thickness distribution. Also the film thickness can be specified externally at each grid point if, for example, structural deformations are to be included (see later section).

(c) *Coordinate Reference Frames.* The program has wide generality with respect to coordinate reference frames, and operable programs are available for the following configurations:
(a) Flat slab bearings.
(b) Cylindrical or journal bearings.
(c) Circular or sector thrust bearings.
(d) Conical
(e) Spherical $\Big\}$ combination radial and thrust bearings.

(d) *Hydrodynamic Profiles.* Since the film thickness distribution is computed by separate subroutine great flexibility in the selection of hydrodynamic profiles exist. To change profiles it is a simple matter of instructing the program to select a different film thickness sub-

---

†The important aspects of this program are the inclusion of velocity effects and non-uniform film thickness as in the case of a high-speed highly loaded spindle. The program numerically solves the Reynolds lubrication equation for given geometric bearing parameters.

routine. In addition to the usual distributions such as a journal with eccentricity and mis-alignment, or a thrust sector at a specified tilt, some of the more uncommon but useful profiles include:

(a) Rayleigh steps parallel and normal to the direction of motion.

(b) Convex crowned profiles for sector thrust bearings.

(c) Tapered land sectors.

(d) Pocket type sectors.

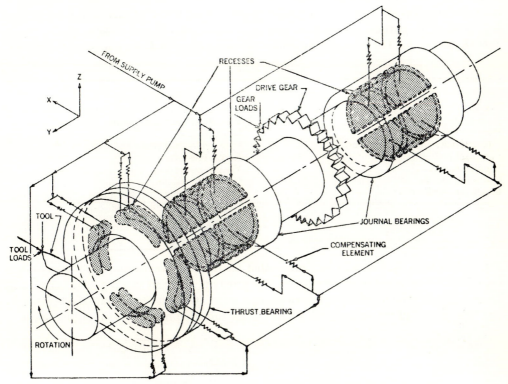

FIG. 1. Diagrammatic arrangement of a hydrostatically lubricated spindle supported in two journal bearings and one thrust bearing.

(e) *Boundary Conditions and Symmetry Options.* The cyclic boundary condition represent-ative of complete circular journal or thrust bearings, or sectors whereby the boundaries are formed by interrupted regions of ambient pressure are both available. In addition symmetry options exist in either or both of the coordinate directions. For example, a plain journal bearing subjected to eccentricity but not misalignment has symmetry about a plane that passes through the center of the bearing and the displaced center of the journal. By ana-lyzing only half the bearing a finer grid could be used for increased accuracy, or if the same grid interval was maintained, as for a complete bearing, only half the number of grid points are needed and the speed of computation is increased.

(f) *Hydraulic Supply Circuitry.* In its most powerful form the program obtains the hybrid bearing performance. In this instance the entire feed circuitry is included and the complete problem solved. Capillary, orifice and flow control valve compensation are optional for each individual recess. As many as ten recesses per bearing can be accommodated.

(g) *Typical Example.* A simple use of the SSE program is to solve for load and flow coefficients for hydrostatic bearings. To compute coefficients for an average pad requires about 30 sec on a high-speed computer such as an IBM 7094 or a Univac 1107. Pure hydrodynamic performance is readily obtained for preselected eccentricities and misalignments.

A good example of the use of this program for a hybrid case was the analysis conducted for the bearing shown on Fig. 2. The bearing consists of three sectors, each sector separated by an axial groove. At the leading edge of each sector is a Rayleigh step to augment the hydrostatic load capacity with hydrodynamic action. The step is bounded on three sides to inhibit side leakage and increase load capacity. Downstream of each step is an orifice-fed recess which introduces externally pressurized fluid into the film region.

(h) *Input–Output.* The input and output for the above described case is indicated in Table 1.

TABLE 1.

Input	Output
Bearing geometry	Pressure at every grid point
Film thickness distribution	Load capacity
(eccentricity and misalignment)	Flow (individual recess and total)
Lubricant properties	Recess pressures
Supply pressure	Viscous power loss
Operating speed	Center of pressure
Restrictor coefficient	Bearing righting moment about orthogonal axes
	through center of bearing
	Minimum film thickness
	Attitude angle

It should be noted that all input and output for this program employs non-dimensional quantities for increased generality. Using this program, complete bearing performance was obtained, for this relatively complicated configuration, over a range of journal eccentricities and orientations of the sectors with respect to the direction of the journal displacement.

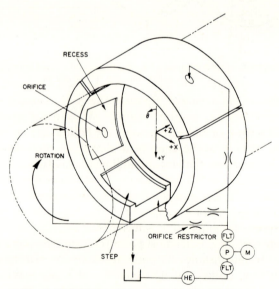

FIG. 2. Hybrid journal bearing with Rayleigh step.

(i) *Dynamic Spring and Damping Coefficients.* There is another important use of this program, of which the title of the program gives little hint. Its name implies the absence of dynamic considerations, yet the program handily determines dynamic spring and damping coefficients due to normal relative velocity of the opposed bearing surface. These coefficients are extremely valuable for determining the dynamic response of structures supported by fluid-film bearings (e.g. space tracking antennas, centrifuge space simulators, machine tools), and in the general area of rotor dynamics. This type of information would be very helpful for determining the threshold of chatter for machine tools.

(j) *General Theory.* A summary of the general theory of SSE program is described in Appendix A and Ref. 4. The Reynolds equation is solved by the formation of coefficient matrices that act upon column vectors of the pressures progressively from one boundary to the other without requiring an iterative scheme. In their treatment multiple recesses are handled by component solutions.

The steady state exact program treats the bearing elements as rigid bodies. If the elastic deflections of the housing and structure are appreciable then they must be considered since the fluid-film thickness may only be one or two thousandths of an inch. In such a case the steady state exact program is used to arrive at an initial bearing geometry. This geometry is then modified as required based on the coupled lubrication-elasticity program which is discussed later.

2. *Table Program*

(a) *General Description.* Table was developed primarily to analyze slideway bearing systems. Consider the system used to support the lathe carriage shown in Fig. 3. Here a complex of 12 bearings restrains the carriage in all degrees of freedom except in the direction of translation in the axial direction (*y*-axis). Note the rather odd supply system circuit. Instead of feeding all pads from a common supply manifold through individual flow restrictors, flow control valves are introduced to feed the bearing pairs that support loads in the *x* and *z* directions. It is well known that greater lubricant film stiffness can be achieved by individual pump or flow control valve compensation than by orifice or capillary compensation. Thus, the system achieves greater stiffness in these directions without undue complication. An important feature of this program is the built-in capability of being able to vary the supply circuitry and the individual types of compensation, to augment stiffness in preferred directions, by merely changing input quantities. The complete bearing complex with the associated hydraulic circuit is analyzed as a system. A point on the body is selected as the origin of the system. For example, in Fig. 3 the tool point is selected as the origin since this is the point where loads and motions are of interest. It is optional to specify as input either the loads or displacements, in all six degrees of freedom, at the origin. If the loads are input then the displacements are produced as output and vice versa. For most machine tool applications, tool loads are the given quantities and tool movements the desired output.

In the example presented in Fig. 3, a complex of 12 bearings is used to restrain the carriage in all degrees of freedom except in the direction of translation in the axial direction (*y*-axis). The lubricant is fed from a single pump to each of the bearings through individual restrictors or compensating elements. The function of the restrictors is to prevent excessive flow unbalances to the bearings; they are also the means of introducing oil-film stiffness into the system.

In the case of the lathe carriage shown in Fig. 3 relative motion was along the *y*-axis with possible cutting loads in the x and z directions applied at the tool point. There were 12 reaction points to any cutting load. The approach was to initially assign finite displacements and rotations about the three axes within the clearance volume and to evaluate the resulting forces at the tool point. Based on the results obtained, the computer assigned new displacements and rotations. This was repeated a number of times until the computer load agreed with the given load. Many calculations were required since motions in one direction produced reactions and moments in other directions and, which in turn depended on the supply circuitry and preload pressures, all of which could be variable.

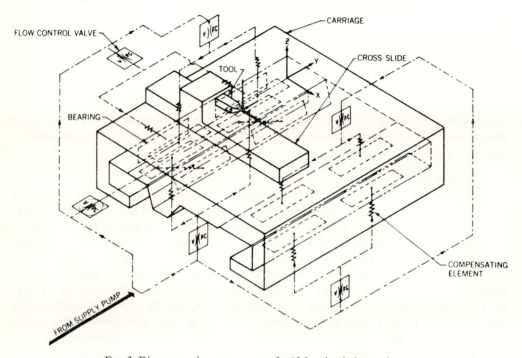

Fig. 3. Diagrammatic arrangement of a 12-bearing lathe carriage.

The program is not restricted to flat bearing configurations but has direct applicability to turntables and spindles such as shown in Fig. 1, provided the spindle operates in the near-concentric position. Any type of compensating element (orifice, capillary, or flow control valve) can be specified for each of the individual bearings.

An important feature of this program is the built-in capability of being able to vary the supply circuitry and the individual types of compensation as considered necessary to satisfy stiffness requirements. Using the computer program it is a simple matter to vary the supply and compensation system and obtain results very quickly.

(b) *Input–Output.* Table 2 below shows the pertinent input and output of this program.

Although the tool motions are most important there are still many other performance quantities associated with the hydrostatic bearing system as indicated by the additional output in Table 2.

(c) *Sample Problem.* The complete performance for the example of the system shown on Fig. 3 is presented in Table 3. The data shown are all by-products of the analysis and represent a wealth of information that would require a very complex extensive instrumentation setup to determine by test.

Also included in Table 3 is a complete stiffness matrix. The columns of the matrix represent loads on the carriage in the 6 degrees of freedom, and the rows represent corresponding displacements. The numbers represent a stiffness whose units are determined by the intersection of the row and column. For example, the intersection of the θ_y row with the W_y

TABLE 2.

Input	Output
Initial bearing film thickness	Body displacements or alternatively body loads in
Lubricant viscosity	6 degrees of freedom
General information regarding the supply circuitry	Bearing loads
(Type of restrictors and how they are coupled)	Bearing clearances
Bearing locations	Recess pressures
Bearing orientations	Supply pressures
Bearing load factors	Bearing flows
Bearing flow factors	Oil-film stiffness matrix
Restrictor coefficients	
Body loads or alternatively body motions	

column represents the ratio of the change in load in the y direction due to a rotation about the y–y axis. The units would then be lb/rad. The diagonal (principal) elements of the matrix represent the more common values of lb/in and in-lb/rad.

(d) *Assumptions and Restrictions.* The primary assumption made in the analysis is that load and flow coefficients for any particular bearing remain constant. This assumption is quite valid for situations in which nonuniform clearance effects are insignificant. Generally, for sizeable multipad systems, the span to clearance ratio of the bearings is large enough to preclude any marked variations in clearance uniformity. The program is also valid for analyzing spindle systems that operate near the concentric position. Other restrictions in the analysis are

(1) The fluid is incompressible.

(2) Hydrodynamic effects are neglected.

(3) Absolute viscosity of lubricant is constant.

A summary of the general theory of the program is described in Appendix B and Ref. 5.

3. *Table Series Program*

(a) *General Description.* A ramification of the program Table is the program Table Series. It permits analysis of a system of hydrostatic bearings that include internal systems fed by a primary system.

(b) *Typical Example.* Consider the machine tool spindle shown on Fig. 4. A rotating sleeve is supported by a system of hydrostatic thrust and journal bearings. Concentrically mounted within the sleeve is a shaft that is caused to rotate by a key connecting the sleeve and shaft. The inner shaft, however, can translate within the sleeve so that the tool point can be moved forward or rearward. To insure accurate and smooth translation of the inner shaft, it was decided to support this shaft by hydrostatic pads machined into the sleeve. Since the sleeve is rotating, feeding the bearing contained in it is not conveniently accomplished from the stationary housing. However, feeding the interior bearing from a recess of the exterior

TABLE 3. ORIFICE RESTRICTORS FED FROM SIX FLOW CONTROL VALVES. PAIRED FLOWS HEAVY CUT LOADS

Bearing no.	Bearing clearance (mils)	Bearing normal loads (lb)	Recess pressure (psi)	Supply pressure (psi)	Individual bearing flow (in³/sec)	Displacement components	Displacement of origin	Load components	Load
1	1.2030	443.1438	115.9389	319.6987	0.134122	U_x	-0.1832×10^{-3} in	W_x	-1500 lb
2	1.1703	472.4496	123.6071	319.8861	0.131638	U_y	0.0 in	W_y	0.0 lb
3	0.8103	1172.9663	306.8835	440.1282	0.108460	U_z	-0.294×10^{-3} in	W_z	-2000 lb
4	1.1400	610.8472	159.8161	440.0842	0.157300	θ_x	0.3879×10^{-4} rad	M_{xx}	10000 in-lb
5	1.1897	525.8949	137.5900	406.0447	0.153950	θ_y	0.7347×10^{-6} rad	M_{yy}	4000 in-lb
6	0.8600	1011.1675	264.5520	406.1571	0.111810	θ_z	-0.3850×10^{-5} rad	M_{zz}	-2000 in-lb
7	0.7283	1649.3704	431.5252	510.4266	0.110738				
8	0.9215	1139.8389	298.2164	570.4246	0.155022				
9	0.7991	1213.6660	317.5318	448.7500	0.107632				
10	1.1288	632.5296	165.4889	448.7151	0.158128				
11	1.2009	509.0358	133.1792	399.2478	1.153264				
12	0.8712	978.6216	256.0370	399.3842	0.112496				

Stiffness matrix

Load displacement	W_x (lb)	W_y (lb)	W_z (lb)	M_{xx} (in-lb)	M_{yy} (in-lb)	M_{zz} (in-lb)
U_x (in)	-9.716×10^6	0.0	-4.254×10^6	2.853×10^6	31.481×10^6	-4.595×10^6
U_y (in)	0.0	0.0	0.0	0.0	0.0	0.0
U_z (in)	-4.285×10^6	0.0	-22.37×10^6	4.573×10^6	39.515×10^6	-3.102×10^6
θ_x (rad)	2.404×10^6	0.0	3.533×10^6	-268.2×10^6	-12.56×10^6	-41.74×10^6
θ_y (rad)	31.63×10^6	0.0	39.20×10^6	-1.943×10^6	-1306×10^6	14.17×10^6
θ_z (rad)	-3.838×10^6	0.0	-2.364×10^6	41.72×10^6	10.03×10^6	-104.6×10^6

bearings provides a simple solution to the problem. As shown in Fig. 4 the front interior bearings are supplied from the recesses of the fixed thrust bearing.

Analysis of this system is difficult because of the interrelationships between the inner and outer bearing systems. However, with the program Table Series parametric studies and system performance is easily accomplished. The input and output for this program closely parallels that of Table.

4. Structural Deformations

Often the support structure will deflect in the same order of magnitude as the oil film thickness. In such cases it is necessary to combine a structural deflection analysis, in the form of a digital computer program with the lubrication program. The two programs are sometimes matched by interrupted manual interpretation, and in more complex cases completely automatically.

A structural foundation analysis computer program providing deflections of rings, plates and shells has been linked with a lubrication program and automatically iterated until convergence of the pressure and film thickness distributions was attained. The sequence of computation starts from an assumed pressure distribution over the bearing areas found from the steady state exact part of the coupled programs. With the structural program, deformations of the pad and bearing are calculated and superimposed on the film thickness, which is then used to compute a new pressure distribution from the lubrication program. The process is continued until convergence of film thickness and pressure occurs.

Details of the coupled lubrication-elasticity program will be presented in a future paper. The program is sufficiently comprehensive and general to apply not only to very large structures, but to cover every conceivable geometry of bearing and support down to gas-lubricated turbogenerator bearings of $\frac{1}{4}$ in diameter rotating at 500,000 rpm in an atmosphere of gaseous hydrogen at cryogenic temperatures.

APPENDIX A

Matrix-Column Method of Solving Reynolds Equation and Solution of Feeding Problem by Component Solutions

Nomenclature

C	reference clearance, in
f_j	restrictor coefficient of jth recess $= Q_j/(P_{s_j} - P_{R_j})^{s_j}$
f_L	coefficients of general second order equation
F_r	dimensionless resultant flow for rth flow condition
h	local clearance, in
H	$= h/C$
L	length of pad, in
N	number of recesses
N_f	number of flow conditions
N_i	number of rows in grid
N_j	number of columns on grid
p	local pressure, lb/in²

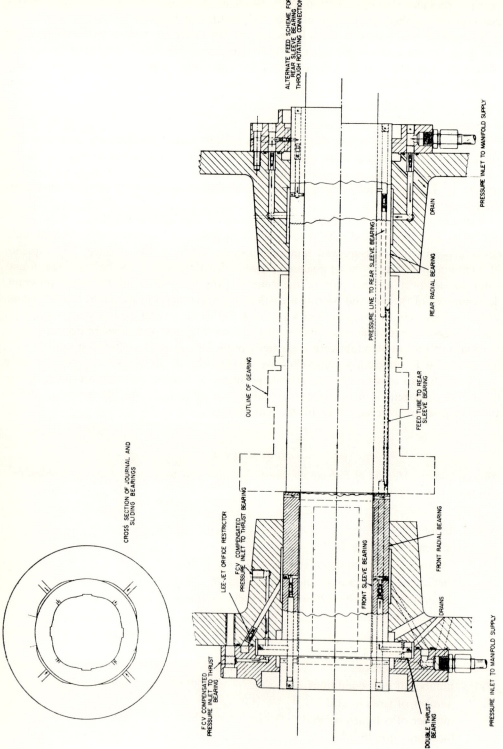

FIG. 4. Preliminary spindle layout.

p_a ambient pressure, lb/in²

p_{rj} recess pressure of jth recess, lb/in²

p_R reference pressure, lb/in²

p_{sj} supply pressure to jth recess, lb/in²

P dimensionless pressure $= p - p_a /p_R - p_a$

P_i dimensionless pressure due to unit pressure in ith recess

P_{rj} dimensionless recess pressure of jth recess $= p_{rj} - p_a/p_R - p_a$

P_{sj} dimensionless supply pressure to jth recess $= p_{sj} - p_a/(p_R - p_a)$

P_Λ dimensionless pressure generated by Λ component solution

q_j flow out of jth recess, in³/sec

Q_{ij} dimensionless flow out of a path surrounding the jth recess due to unit pressure in ith recess

$Q_{\Lambda j}$ dimensionless flow out of a path surrounding the jth recess due to Λ component solution

R_i dimensionless recess pressure in ith recess used for computing component solutions

s_j characteristic flow exponent of restrictor for restrictor feeding jth recess

S_r dimensionless resultant pressure for rth pressure condition

U relative velocity, in/sec

$\hat{\mathbf{V}}$ unit velocity vector

a_i ith influence coefficient (equals ith recess pressure)

$[\gamma_{rj}]$ coefficient matrix combining supply circuit feed and pressure conditions

Λ speed parameter $= 6\mu UL/(p_R - p_a)C^2$

μ absolute viscosity of lubricant, lb-sec/in²

ϕ independent variable of general second order partial differential equation $= P$ for lubrication problem

$\mathscr{Q}_j$ dimensionless total flow out of jth recess $= q_j \cdot 12\mu/C^3(p_R - p_a)$

$\mathscr{Q}_T$ dimensionless flow out of all recesses $= \sum \dfrac{q_j\,12\mu}{C^3(p_R - p_a)}\Big|_{j=1\to N}$

Note: Vector quantities are designated by heavy type.

Introduction

The numerical treatment of practical lubrication problems requires the solution of large sets of difference equations with simple boundary conditions applied to boundaries of complex shapes. Since the use of direct matrix inversion routines is not practical, the solution of such problems has been primarily accomplished by means of various relaxation schemes. A method is presented which, inspired by the treatment of tri-diagonal matrices, takes advantage of the form of the boundary conditions to solve the problem exactly and accurately without requiring iteration. The resulting gain in computation speed and reliability of the answers together with great simplicity of programming and generalization make this method very attractive.

A line of treatment particularly suitable for hydrostatic and hybrid bearings with multiple recesses where the external feeding network determines the final answer to the problem is also described. The generalized solution of any feeding problem without need for the repeated solution of the bearing film equations makes for its distinct usefulness as a design tool.

The presentation of this Appendix is organized so that it may be used as a guide to the assembly of an in-house program for the solution of a vast class of hydrostatic, hydrodynamic, or hybrid fluid film bearing problems.

The basic method does not lend itself too readily to the treatment of floating $\partial P/\partial n = 0$ conditions as they are met in cavitated regions. Therefore, such effects are either not considered or simply accounted for by the $p = $ constant condition.

I. *Lubrication Problem: Handling by Components.* The lubrication equation governing the pressure of the lubricant in the clearance space is, in dimensionless form

$$\nabla . (H^3 \nabla P) = \Lambda \hat{\mathbf{V}} . \nabla H \tag{1}$$

The boundary conditions applicable to the problem are:
(a) $P = 0$ at the boundary of the bearing region exposed to ambient pressure
(b) $P = $ const. $= R_i$ in recesses
(c) $dP/dn = 0$ on lines of symmetry (where dP/dn is the derivative in the direction locally normal to the line of symmetry)
(d) $P(X, Y) = P(X + \delta, Y)$

$$\frac{dP}{dX}(X, Y) = \frac{dP}{dX}(X + \delta, Y)$$

for cyclic conditions in direction X with period δ. Analogous cyclic conditions also apply in the Y direction.

Note that the constant values in condition (b) may be different from recess to recess and are determined by matching the bearing flow problem with the feeding problem. Therefore they are unknown at this stage.

The general solution of eq. (1) if the recess pressures are not known can be stated as

$$P(X, Y) = \sum_{i=1}^{N} a_i P_i(X, Y) + \Lambda P_A(X, Y) \tag{2}$$

where
$$\nabla . (H^3 \nabla P_i) = 0 \qquad\qquad i = 1, N \tag{3}$$

with all conditions of type (a), (c), (d) satisfied and $R_i = 1$, all other R's $= 0$ and

$$\nabla . (H^3 \nabla P_A) = \hat{\mathbf{V}} . \nabla H \tag{4}$$

with all conditions of type (a), (c), (d) satisfied and

$$R_j = 0; j = 1 \rightarrow N$$

Once the solutions P_i, $i = 1 \rightarrow N$ and P_A are obtained by a method such as the one expounded in section III, the flow out of each recess corresponding to each solution should be computed by integration around a closed path encompassing the wanted recess only:

$$Q_{ij} = \oint_{L_j} H^3 \nabla P_i . \hat{\mathbf{n}} dl_j \qquad i = 1, N$$

$$\left.\begin{array}{c}\\ \end{array}\right\} j = 1, N \tag{5}$$

$$Q_{Aj} = \oint_{L_j} [H^3 \nabla P_A - \hat{\mathbf{V}} H] . \hat{\mathbf{n}} \, dl_j$$

where L_j is a closed path around the jth recess.

This information is then used in combining component solutions to satisfy the feeding problem. Indeed, the total flow out of each recess is

$$\mathcal{Q}_j = \sum_{i=1}^{N} a_i \, Q_{ij} + Q_{Aj} \tag{6}$$

The solution of the feeding problem provides the influence coefficients a_i, $i = 1, N$.

II. *General Feeding Problem: Solution Using Components.* If the flow $\mathcal{Q}_j$ is coming from the feeding lines to the jth recess through a compensating element, the dimensionless flow-pressure relation is:

$$\mathcal{Q}_j = f_j \, (P_{sj} - P_{rj})^{s_j} \tag{7}$$

where the dimensionless recess pressure, P_{rj}, is actually the value of the influence coefficient (only P_{rj} contributes toward a non-zero pressure in the jth recess).

P_{sj} is the dimensionless supply pressure upstream of the compensation element.

In general, the feeding hook-up to a bearing containing N recesses can be described in N conditions involving the flows $\mathcal{Q}_j$ and the supply pressures P_{sj}. Such conditions can be described by

$$\left\{ \begin{array}{ll} \sum\limits_{j=1}^{N} \gamma_{rj} \, \mathcal{Q}_j = F_r & r = 1, N_f \\[2em] \sum\limits_{j=1}^{N} \gamma_{rj} \, P_{sj} = S_r & r = N_f + 1, N \end{array} \right. \tag{8}$$

The $N \times N$ matrix γ_{rj} and the right-hand side vector $\left\{ \begin{array}{c} F_r \\ S_r \end{array} \right\}$

specify the problem completely. For example if the bearing has six recesses and is fed by two positive displacement pumps so that the first supplies recesses 1, 3, 5 through compensation and the second supplies recesses 2, 4, 6 through compensation; the six corresponding conditions are:

1. $\mathcal{Q}_1 + \mathcal{Q}_3 + \mathcal{Q}_5 = F$ (pump output)
2. $\mathcal{Q}_2 + \mathcal{Q}_4 + \mathcal{Q}_6 = F$ (pump output)
3. $P_{s1} = P_{s3}$
4. $P_{s1} = P_{s5}$
5. $P_{s2} = P_{s4}$
6. $P_{s2} = P_{s6}$

Therefore $[\gamma] = \begin{bmatrix} 1 & 0 & 1 & 0 & 1 & 0 \\ 0 & 1 & 0 & 1 & 0 & 1 \\ 1 & 0 & -1 & 0 & 0 & 0 \\ 1 & 0 & 0 & 0 & -1 & 0 \\ 0 & 1 & 0 & -1 & 0 & 0 \\ 0 & 1 & 0 & 0 & 0 & -1 \end{bmatrix}$

$$\left\{ \begin{array}{c} F_r \\ S_r \end{array} \right\} = \left\{ \begin{array}{c} F \\ F \\ 0 \\ 0 \end{array} \right\}$$

Using the expressions for $\mathcal{Q}_j$ from (6) and for P_{sj} from eq. (7), eq. (8) becomes

$$\begin{cases} \sum_{j=1}^{N} \gamma_{rj} \left(\sum_{i=1}^{N} a_i \, Q_{ij} + \Lambda \, Q_{\Lambda j} \right) = F_r, & r = 1, N_f \\[2em] \sum_{j=1}^{N} \gamma_{rj} \left(\left[\dfrac{\sum_{i=1}^{N} a_i \, Q_{ij} + \Lambda \, Q_{\Lambda j}}{f_j} \right]^{1/s_j} + a_j \right) = S_r, & r = N_f + 1, N \end{cases} \tag{9}$$

Letting

$$C_j = f_j^{-1/s_j} \left(\sum_{j=1}^{N} a_i \, Q_{ij} + \Lambda \, Q_{\Lambda j} \right)^{((1/s_j) - 1)} \tag{10}$$

the system (9) becomes, in matrix form,

$$[\Gamma] \, \{a\} = \{\Theta\} \tag{11}$$

where

$$\Gamma_{ri} = \begin{cases} \sum_{j=1}^{N} \gamma_{rj} \, Q_{ij}, & r = 1, N_f \\[2em] \sum_{j=1}^{N} \gamma_{rj} \, [Q_{ij} \, C_j + \delta_{ij}], & r = N_f + 1, N \end{cases} \tag{12}$$

and

$$\Theta_r = \begin{cases} F_r - \sum_{j=1}^{N} \gamma_{rj} \, Q_{\Lambda j}, & r = 1, N_f \\[2em] S_r - \sum_{j=1}^{N} \gamma_{rj} \, Q_{\Lambda j} \, C_j, & r = N_f + 1, N \end{cases} \tag{13}$$

To solve,

(a) C_j is initially assumed to be given by

$$C_j = f_j^{-1}, \qquad j = 1, N;$$

(b) system (11) is formed and solved by matrix inversion

$$\{a\} = [\Gamma]^{-1} \, \{\Theta\}; \tag{14}$$

(c) eq. (10) is used to evaluate modified values of $C_j, j = 1, N$.
Comparison with the previous values of C_j and satisfaction of a truncation criterion makes the process terminate; otherwise go back to step (b).

Note that C_j does not vary for any recess whose compensating element has $s_j = 1$ (viscous restrictor, i.e. capillary tube).

The above described process converges very rapidly to high accuracy requirements. Moreover it enables the designer to try several feeding conditions on the same bearing geometry without solving Reynolds equation repeatedly.

When a satisfactory answer has been reached the obvious integrations for load components can be carried out on the final value of P as indicated by eq. (2).

Solution of Linear System by Column Method. Equations (1), (3) and (4) expressed in cartesian, polar, or spherical coordinates, can be put in the general form

$$f_1 \frac{\partial^2 \varphi}{\partial X^2} + f_2 \frac{\partial^2 \varphi}{\partial Y^2} + f_3 \frac{\partial^2 \varphi}{\partial X \partial Y} + f_4 \frac{\partial \varphi}{\partial X} + f_5 \frac{\partial \varphi}{\partial Y} + f_6 \varphi = f_7 \tag{15}$$

where

$$f_i = f_i(X, Y) \qquad i = 1, 7$$

If the X, Y region of interest is included in the smallest rectangular region containing it and ϕ is represented by its values at the nodes of an $N_i \times N_j$ rectangular grid, eq. (15) can be approximated numerically by three point central difference formulae and be written as the system

$$[C_j]\{\phi_j\} + [E_j]\{\phi_{j-1}\} + [D_j]\{\phi_{j+1}\} = \{R_j\}, \quad j = 1, N_j \qquad (16)$$

where

$$\{\phi_j\} = \left\{ \begin{array}{c} \varphi_{1j} \\ \varphi_{2j} \\ \vdots \\ \varphi_{N_ij} \end{array} \right\}$$

The treatment of the boundary conditions is most important. The grid should include all points where φ is known (recesses and outer boundaries) and all lines of symmetry. Moreover, cyclic joints should be represented so that each unknown point appears only once (if φ is cyclic in the i-direction the conditions are $\varphi_{1j} = \phi_{N_{i+1}j}$, (not on grid) and φ_{0j} (not on grid) $= \varphi_{N_i, j}$. Similarly for the j direction.

The matrices $[C_j]$, $[E_j]$, $[D_j]$ and $\{R_j\}$, $j = 1, N_j$ should be formed so that all conditions expressing the fact that the function φ is known at some points or symmetric about some lines or cyclic in the i-direction are automatically satisfied and no points exterior to the grid are over multiplied by non-zero terms. For example, if (i, j) is a point where the value of φ is fixed equal to 1

$$\left. \begin{array}{l} [C_j]_{ik} = \delta_{ik} \text{ (Kronecker delta)} \\ [E_j]_{ik} = 0 \\ [D_j]_{ik} = 0 \\ \{R_j\}_i = 1 \end{array} \right\} \quad K = 1, N_i$$

Obviously eq. (16) cannot be used to satisfy identically cyclic conditions in the j direction because these involve more than three neighbouring columns of φ.

The solution procedure is then the following:

Let

$$\{\phi_{j-1}\} = [A_j]\{\phi_j\} + \{B_j\} + [F_j]\{\phi_{N_j}\} \qquad (17)$$

By substitution into eq. (16) the following recurrence relations are obtained

$$\begin{array}{l} [A_{j+1}] = -[T_j][D_j] \\ \{B_{j+1}\} = [T_j]\langle\{R_j\} - [E_j]\{B_j\}\rangle \\ [F_{j+1}] = -[T_j][E_j][F_j] \end{array} \qquad (18)$$

where

$$[T_j] = [[C_j] + [E_j][A_j]]^{-1}$$

To start the process the following definitions are valid:

$$\begin{array}{l} [A_1] = [0] \\ \{B_1\} = \{0\} \end{array}$$

(a) If $j = 1$ is not at a joint (i.e. if φ is not cyclic in the j-direction)

$$[F_1] = [0]$$

Note that in this case $[F_j] = 0$ for all j and the $\{\phi_N\}$ term need not be carried in eq. (17).

(b) If $j = 1$ is a joint

$$[F_1]_{ik} = \delta_{ik}$$

Equations (18) are used for $j = 1$, N_j. Then the solution is gotten from the following procedures:

(a) $j = N_j$ is not at a joint

Use eq. (17) for $j = N_{j+1,2}$. Note $\{\phi_{N+1}\}$ is arbitrary since it is multiplied by $[A_{N_j+1}]$ which is a zero matrix.

(b) If $j = N_j$ is at a joint

Define

$$\left. \begin{aligned} [R_{N_j}] &= ([1] - [F_{N_j+1}])^{-1} [A_{N_j+1}] \\ \{S_{N_j}^{i}\} &= ([1] - [F_{N_j+1}])^{-1} [B_{N_j+1}] \end{aligned} \right\}$$

then use

$$\left. \begin{aligned} [R_{j-1}] &= [A_j] [R_j] + [F_j] [R_N] \\ \{S_{j-1}\} &= [A_j] \{S_j\} + \{B_j\} + [F_j] \{S_N\} \end{aligned} \right\}$$

for $j = N_j$, 2.

$$\{\phi_1\} = [[1] - [R_1]]^{-1} \{S_1\}$$

while the rest of the solution is then given by

$$\{\phi_j\} = [R_j] \{\phi_1\} + \{S_j\}$$

for $j = 2$, N_j

This method presents great advantages over relaxation and boundary influence methods since it solves the problem exactly in a finite number of steps and it is not equivalent to the determination of homogeneous solutions. Indeed the influences of neighbouring columns are not multiplied out until all boundary conditions are satisfied.

Since the greatest consumption of computation time occurs in the inversions necessary to determine the matrices $[T]$ (formula (18)), care should be exercised to realize the most economical set-up. For this purpose, the computation time can be taken to be proportional to

$$N_j N_i^3$$

which shows the importance of selecting the column direction so that

$$N_j > N_i$$

APPENDIX B

Methods of Determining Performance of Systems Supported by Multiple Hydrostatic Bearings

Nomenclature

a_{fj}	load factor of jth bearing
A_{pj}	projected area of jth bearing, in^2
C_d	orifice coefficient of discharge
C_j	clearance of jth bearing, in
C_{0j}	initial clearance of jth bearing, in
d_{cj}	capillary diameter — jth bearing, in
d_{oj}	orifice diameter — jth bearing, in
f_j	restrictor coefficient of jth bearing, in^5/(lb-sec) or in^4/(lb$^{1/2}$-sec)

F_j flow to jth bearing, in³/sec

k_j $q_{f(j)}/(A_{p(j)}\mu)$, 1/(lb-sec)

K_j $a_{f(j)}\,A_{p(j)}$, in²

K_{li} stiffness in lth degree of freedom from motion in ith degree of freedom

l_{cj} capillary length — jth bearing, in

$[M]$ load coefficient matrix $= [a]^T$

$\mathbf{n}_j$ unit normal vector of jth bearing

N number of bearing

NF number of flow conditions

NFR number of degrees of freedom of motion

P percentage change in clearance for determining stiffness

P_{rj} recess pressure of jth bearing, psi

P_{sj} supply pressure of jth bearing, lb/in²

q_{fj} flow factor of jth bearing

Q_f flow resultant for fth flow condition, in³/sec

$\mathbf{r}_j$ position vector of jth bearing, in

s_j characteristic exponential of flow

S_l supply pressure resultant for lth pressure condition, psi

$\mathbf{U}$ translation vector of origin, in

W_i total load in ith degree of freedom, lb

$[a]$ clearance transformation matrix

β_j load on jth bearing, lb

$[\gamma]$ flow condition coefficient matrix

$\{\delta\}$ translation-rotation vector

$$\delta_{1,\,2,\,3} = U_{x,\,y,\,z}, \text{ in}$$
$$\delta_{4,\,5,\,6} = \theta_{x,\,y,\,z}, \text{ rad}$$

ΔC_j change in bearing clearance, in

$\boldsymbol{\theta}$ rotation vector of origin, rad

$[\theta]$ bearing characteristic coefficient matrix

$\{\Lambda\}$ flow, pressure condition, vector

μ absolute lubricant viscosity, lb-sec/in²

$[\nu]$ supply circuit coefficient matrix

ρ lubricant density, lb-sec²/in⁴

$[\omega]$ pressure condition coefficient matrix

Note: Repeated subscripts, unless parenthesized, are accumulated according to cartesian tensor analysis.

I. *Introduction*

The determination of performance parameters for a body supported by multiple hydrostatic bearings requires solution of a set of simultaneous equations equal to the number of separate bearing pads; further complication is introduced by the fact that the equations are not always linear. If a number of bearing configurations, supply circuits, type of compensation, and loading conditions are to be investigated, hand computations become impractical and digital computer techniques are necessary.

The prominent variables influencing system performance include bearing configuration, location and clearances, hydraulic flows and pressures, and the type of compensation and supply circuit network employed. Selection of most of these variables is dictated by geo-

metry, available commercial hydraulic equipment, and structural and manufacturing prac-
ticalities. To offset the restriction on performance of these constraints the compensation
and supply circuitry can often be manipulated. It is well known that greater lubricant
film stiffness can be achieved by individual pump or flow control valve compensation than
by orifice or capillary compensation. By proper combination of the several means of compensation, increased stiffness can be achieved in preferred directions without introducing
undue complexity. The analysis following develops the governing equations for determining
performance of hydrostatic systems and indicates how solutions can be obtained with the
aid of a high-speed digital computer.

II. *Analysis*

1. *Assumptions*

   (a) The primary assumption made in the analysis is that load and flow coefficients[2]
   for any particular bearing remain constant. This assumption is quite valid for situations
   in which non-uniform clearance effects are insignificant. Generally, for sizeable multipad systems, the span to clearance ratio of the bearings is large enough to preclude
   any marked variations in clearance uniformity.

   (b) The fluid is incompressible.

   (c) Hydrodynamic effects are neglected.

   (d) Absolute viscosity of lubricant is constant.

2. *Development of Governing Equations.* A point on the body is selected as the origin of the
system. The clearance of the jth bearing is

$$C_j = C_{0j} + \Delta C_j \tag{1}$$

ΔC_j, the change in clearance, is caused by a combination of translations along the coordinate axes and rotations about the coordinate axes. In vector notation,

$$\Delta C_j = (\mathbf{U} + \theta \times \mathbf{r}_j) . \mathbf{n}_j \tag{2}$$
$$j = 1 \to N$$

Substituting (2) into (1), expanding, and combining into a generalized matrix form, for
convenience of programming, leads to the relationship

$$C_j = C_{0j} + a_{jl}\, \delta l \tag{3}$$
$$j = 1 \to N$$
$$l = 1 \to NFR$$

where $\delta_{1,\ 2,\ 3} = U_{x,\ y,\ z}$ are the translations along the coordinate axes, and $\delta_{4,\ 5,\ 6} = \theta_{x,\ y,\ z}$
are the respective rotations about the coordinate axes. $[a_{jl}]$ is the coefficient matrix which is
used to convert translations and rotations at the origin to changes in clearance at the bearings. A typical row of the $[a]$ matrix would be

$$[a_j] = [n_{jx},\ n_{jy},\ n_{jz},\ r_{(j)y}\, n_{(j)z} - r_{(j)z}\, n_{(j)y},$$
$$r_{(j)z}\, n_{(j)x} - r_{(j)x}\, n_{(j)z},\ r_{(j)x}\, n_{(j)y} - r_{(j)y}\, n_{(j)x}] \tag{3a}$$

The subscript j refers to a particular bearing. The subscript x, y or z refers to the respective x, y, or z component of the parameter. Parenthetic repeated subscripts are not dummy
subscripts and thus the terms are not to be accumulated in the usual sense of tensor notation.

The flow from each bearing can be expressed as

$$F_j = k_{(j)}\, \beta_{(j)}\, C_{(j)}^3 \tag{4}$$

where

$$k_j = q_{f(j)}/(A_{p(j)}\, \mu) \tag{4a}$$

and

$$\beta_j = K_{(j)}\, Pr_{(j)} \tag{4b}$$

with

$$K_j = a_{f(j)}\, A_{p(j)} \tag{4c}$$

By virtue of continuity, the flow through the compensating elements or restrictors feeding the bearing must equal the flow through the bearing. For capillary and orifice compensation, this flow is given by:

$$F_j = f_{(j)}\left(P_{s(j)} - \frac{\beta_{(j)}}{K_{(j)}} \right)^{s_{(j)}} \quad j = 1 \to N \tag{5a}$$

For flow control value compensation

$$F_j = \text{constant} \tag{5b}$$

where the exponent

$$\begin{aligned} s_j &= 1 \text{ for capillary compensation} \\ s_j &= 1/2 \text{ for orifice compensation} \end{aligned}$$

and

$$f_j = \frac{\pi d_{c(j)}^4}{128\, \mu l_{c(j)}} \qquad\qquad \text{for capillaries}$$

$$f_j = \frac{C_d \pi d_{oj}^2}{4}\sqrt{\left[\frac{2}{\rho}\right]} \qquad\qquad \text{for orifices}$$

Solving for P_{sj} and substituting the right-hand side of (4) for F_j we obtain

$$P_{sj} = \left(\frac{k_{(j)}\, \beta_{(j)}\, C_{(j)}^3}{f_{(j)}} \right)^{1/s_{(j)}} + \frac{\beta_{(j)}}{K_{(j)}} \quad j = 1 \to N \tag{6}$$

Since β_j is the desired quantity it is extracted from the exponential term and (6) is put into the form

$$P_{sj} = \left[\left(\frac{k_{(j)}\, C_{(j)}^3}{f_{(j)}} \right)^{1/s_{(j)}} \cdot \beta_{(j)}^{(1/s_{(j)} - 1)} + \frac{1}{K_{(j)}} \right] \beta_{(j)} \quad j = 1 \to N \tag{6a}$$

Equations (4) and (6a) constitute a set of equations which can be solved for β_j provided consistency between the number of unknowns and equations can be established. In general the known number of individual supply pressures P_{sj} plus the known number of individual pad flows F_j will be less than the number of bearings. Indeed, if F_j and P_{sj} were known for all bearings the problem would be essentially solved. Further information can be obtained by examination of the pressure and flow conditions of the entire bearing system.

The system flow and pressure conditions are related to those of the individual pad as follows:

$$Q_f = \gamma_{fm}\, F_m \qquad \begin{aligned} m &= 1 \to N \\ f &= 1 \to NF \end{aligned} \tag{7a}$$

$$s_l = \omega_{lm}\, P_{sm} \qquad \begin{aligned} l &= NF + 1 \to N \\ m &= 1 \to N \end{aligned} \tag{7b}$$

where

$$f + l = N \tag{7c}$$

To illustrate these equations consider a system of four bearings fed from one common supply. If the total displacements of the pump is used to supply the bearings, then the sole specified flow condition is that the total pump flows equals the sum of the flows to the individual bearings. The remaining three pressure conditions are obtained by equating the bearing supply pressures of three different bearing pairs. Equations (7a) and (7b) then take on the form

$$
\begin{Bmatrix} Q_f \\ \hline 0 \\ 0 \\ 0 \end{Bmatrix} = \left[\begin{array}{cccc} 1 & 1 & 1 & 1 \\ \hline 1 & -1 & 0 & 0 \\ 1 & 0 & -1 & 0 \\ 1 & 0 & 0 & -1 \end{array} \right] \left[\begin{array}{c|ccc} F_1 & P_{s1} & P_{s1} & P_{s1} \\ F_2 & P_{s2} & 0 & 0 \\ F_3 & 0 & P_{s3} & 0 \\ F_4 & 0 & 0 & P_{s4} \end{array} \right]
\tag{8}
$$

The rows on top or columns left of the broken line are flow conditions, while those below or right of the line are pressure conditions. It is convenient to define the vector on the left hand side of the equation as

$$
\{A\} = \left(\frac{Q_f}{S_l} \right)
\tag{9}
$$

Also the left-hand matrix on the right-hand side of (8) is defined as

$$
[\nu] = \begin{bmatrix} \gamma_{fm} \\ \omega_{lm} \end{bmatrix}
\tag{10}
$$

where again the dotted line signifies separation of the specified flow (top) and supply pressure (bottom) conditions.

If the expressions for F_m and P_{sm} indicated in eqs. (4) and (6a) are substituted into (7a) and (7b) the following matrix equations can be developed:

$$
\{A\} = [\theta] \{\beta\}
\tag{11}
$$

where

$$
[\theta_{ij}] = \left[\frac{\gamma_{i(j)} \, k_{(j)} \, C^3_{(j)}}{\omega_{i(j)} \, [(k_{(j)} \, C^3_{(j)}/f_{(j)})^{1/s_{(j)}} \cdot \beta_{(j)}^{(1/s_{(j)}-1)} + 1/K_{(j)}]} \right] \begin{array}{l} i = 1 \to NF \\ i = NF+1 \to N \\ j = 1 \to N \end{array}
\tag{11a}
$$

It should be noted here that the resulting set of simultaneous equations is linear when al[l] the exponents are, s_j equal 1. Physically this means that capillary compensation is employed throughout. The equations will also be linear if flow control valve compensation is used throughout, since then the flow condition equations make up the entire set and exponentials of β are not involved.

The reaction forces of the bearing, accumulated at the origin, are obtained by

$$
W_i = M_{ij} \beta_j
\tag{12}
$$
$$
i = 1 \to NFR
$$
$$
j = 1 \to N
$$

where $W_{1,\,2,\,3} = x, y, z$ force components of $\beta_j n_j$ (12a)
and $W_{4,\,5,\,6} = x, y, z$ moment components of $\mathbf{r} \times \boldsymbol{\beta} . \mathbf{n}$ (12b)

A typical column of the $[M]$ matrix would be

$$
[M_j] = \begin{bmatrix} n_{1j} \\ n_{2j} \\ n_{3j} \\ r_{2(j)} n_{3(j)} - r_{3(j)} n_{2(j)} \\ r_{3(j)} n_{1(j)} - r_{1(j)} n_{2(j)} \\ r_{1(j)} n_{2(j)} - r_{2(j)} n_{1(j)} \end{bmatrix}
\tag{13}
$$

It is noted that the $[M]$ matrix is the transpose of the $[\alpha]$ matrix used in determining the bearing clearances (eq. (3)).

3. *Method of Solution.* The solution is obtained by inversion of eq. (11)

$$\{\beta\} = [\theta]^{-1} \{\Lambda\} \tag{14}$$

Since the theta matrix contains an exponential of β, eq. (14) is solved by an iterative scheme, in which the exponential term of β is treated as a coefficient rather than a variable. The general procedure is as follows:

(a) Form bearing clearance from (3).
(b) Guess $\beta_{(j)}^{(1/s_{(j)}-1)}$.
(c) Form $[\theta]$ by (11a).
(d) Obtain β_j by (14).
(e) Calculate $\beta_{(j)}^{(1/s_{(j)}-1)}$.
(f) Repeat (c)–(e) until desired truncation on $\beta_{(j)}^{(1/s_{(j)}-1)}$ is achieved.
(g) Calculate forces and moments by (12).

The input necessary for the above procedure includes:

(a) Initial clearances for all bearings C_{oj}.
(b) Each bearing position with respect to the origin, $\mathbf{r}_j$.
(c) The direction of each bearing normal, $\mathbf{n}_j$.
(d) Rotations and translations of the body about the origin, $\boldsymbol{\delta}_i$, in each degree of freedom.
(e) Bearing load and flow coefficients and projected areas, a_{fj}, q_{fj}, A_{pj}.
(f) Restrictor coefficients, f_j
(g) Compensating exponent, s_j

It is very often required to solve the inverse problem to that described above; namely to find body movements due to given applied loads rather than loads due to applied movements. Solution to this problem necessitates calculating the stiffness of the system in all degrees of freedom so that from some starting position an accurate estimate of the body displacement can be made to find the body position at which the bearing reactions will accumulate to balance the applied loads. The estimated displacements in all degrees of freedom is determined by

$$\{\Delta\delta_i\} = [K_{li}]^{-1} . \{W_l \text{ (applied)} - W_{lj} \text{ (calculated)}\} \tag{15}$$
$$i = \text{degree of freedom } 1 \rightarrow NFR$$
$$l = \text{degree of freedom } 1 \rightarrow NFR$$

The stiffness is double subscripted since for a displacement in one direction there is a corresponding change in load or moment in that direction plus in every other direction. Thus, K_{li} implies the change in load or moment in the lth degree of freedom resulting from displacement in the ith direction.

At the newly displaced body position the resultant bearing reactions are compared with the applied loads, and the process is continued until the desired truncation is satisfied. When formulating this procedure, the question arises as to how to decide the proper movement of the body in each degree of freedom in order to evaluate the stiffnesses. For any body position the effect of movement in a particular degree of freedom, on each bearing is, in general, different. Because of non-symmetric bearing locations and variations of the initial clearances some bearings will react much more sensitively than others. To ensure convergence of the entire iterative scheme, large fluctuations should be avoided. If the movements are too small, however, large errors in stiffness can result because the numerical value of the load variation may be smaller than the digital accuracy of the process. The philosophy used in the analysis was to examine the percentage change in clearance of each bearing, due to a movement of the body in a particular degree of freedom, and select the displacement of the body in that direction to limit the largest percentage change in clearance to a specified value. The change in clearance is given by

$$\Delta C_j = a_{ji} \Delta \delta_j \tag{16}$$
$$i = 1 \rightarrow NFR$$
$$j = 1 \rightarrow N$$

(*Note*: only one $\Delta \delta_i \neq 0$ at a time.) The percentage change in clearance per unit of body motion is from (16)

$$\frac{\Delta C_j}{C_j \Delta \delta_i} = \frac{a_{ji}}{C_j} \tag{17}$$
$$j = 1 \rightarrow N$$
$$i = 1 \rightarrow NFR$$

Each bearing is investigated in turn to find the most sensitive bearing and the absolute maximum percentage change in clearance is set equal to a specified value.

$$P = \left(\frac{\Delta C_i}{C_i} \right)_{max} \tag{18}$$

Substitution into [17] yields the body displacement as

$$\Delta \delta_i = \frac{P.(C_j)_{max}}{(a_{ji})_{max}} \tag{19}$$

Investigating all degrees of freedom in this manner permits acceptable body displacements for evaluating stiffness in all degrees of freedom.

The general solution procedure is similar to that described for the known body displacement case with the exception that the body displacement is determined from an initial value which is incremented in accordance with (15). Also bearing resultant loads are compared with desired loads and the process continued until specified truncation on loads is accomplished.

REFERENCES

1. H. C. RIPPEL, *Cast Bronze Hydrostatic Bearing Design Manual*, The Franklin Institute Research Laboratories, published by Cast Bronze Bearing Institute, Inc., 1964.
2. H. C. RIPPEL, *Cast Bronze Bearing Design Manual*, The Franklin Institute Research Laboratories, published by Cast Bronze Bearing Institute, Inc. 1959.
3. H. C. RIPPEL, *Cast Bronze Bearing Design Manual*, The Franklin Institute Research Laboratories, published by Cast Bronze Bearing Institute, Inc., 1967.
4. V. CASTELLI and W. SHAPIRO, Improved Method for Numerical Solutions of the General Incompressible Fluid Film Lubrication Problem, *ASME Trans., Jnl. of Lubrication Technology*, **89**, Series F, No. 2, April 1967, pp. 211–218.
5. W. SHAPIRO, V. CASTELLI and S. HELLER, Determination of Performance Characteristics of Hydrostatic Bearing Systems that Support a Rigid Body, *ASLE Trans.*, **9**, 1966, 272–282.

COMPUTER-AIDED DESIGN OF HYDROSTATIC BEARINGS FOR MACHINE TOOL APPLICATIONS

PART II. APPLICATIONS

Otto Decker and Wilbur Shapiro

Friction and Lubrication Laboratory, Franklin Institute Research Laboratories,
Philadelphia, Penn., U.S.A.

SUMMARY

The role of the high-speed digital computer in the design and analysis of incompressible, fluid-film bearings and bearing systems is presented with special emphasis on machine-tool applications. The more important features and capabilities of the computer programs are described including the class of application problems that can be solved. The significance of the support structure characteristics on bearing performance is reviewed. A combined structures and bearing computer program is briefly described. The analytical foundations are presented in Part I. Several applications are covered in detail, in Part II, including description of the general problem, the analytical approach and sample results.

NOMENCLATURE

Ma_f	load coefficient
C	concentric film thickness, in
D	bearing diameter
d_o	orifice diameter, in
F_G	gear load, lb
H	pump power, h.p.
HP_B	pumping horsepower
HP_s	viscous shear and pressure breakdown horsepower loss
h	film thickness, in
h_{max}	maximum film thickness, in
h_{min}	minimum film thickness, in
l	recess length, in
L	bearing length, in
M	moment, in-lb
$M_{xx, yy, zz}$	moment applied at origin about xx, yy and zz axis respectively, in-lb
n	number of recesses
p_a	ambient pressure, psia
p_r	recess pressure, psia
p_{rmax}	maximum recess pressure, psig
p_{rmin}	minimum recess pressure, psig
p_s	supply pressure, psig
Q	flow, gpm
Q	flow, in³/sec
Q_g	flow, gpm
q_f	flow coefficient
R	radial load, lb, bearing radius, in
$S_{x, y, z}$	principal stiffness in x, y and z directions respectively, in-lb/rad

$S_{xx, yy, zz}$ principal torsional stiffness about xx, yy and zz axes respectively, in-lb/rad
T thrust load
T_a average fluid temperature, °F
$U_{x, y, z}$ displacements at origin in x, y and z directions, respectively, in
W load, lb
$W_{x, y, z}$ load applied at origin in x, y and z directions respectively, lb
a misalignment, rad
β ratio of recess to supply pressure
δ eccentricity ratio, attitude angle
$\delta_{x, y, z}$ displacements at origin in x, y and z directions respectively, in
ΔT total temperature rise, °F
ΔT_B temperature rise due to pressure breakdown, °F
ΔT_S temperature rise due to viscous shear, °F
l recess length, in
Λ compressibility parameter = $6\mu\omega R^2/P_a C^2$
μ local fluid viscosity, lb-sec/in²
μ_a average fluid viscosity, lb-sec/in²
ω shaft speed, rad/sec
θ_p pad angle, degrees
θ_r recess angle, degrees
θ_s sill angle, degrees
$\theta_{xx, yy, zz}$ rotations at origin about x, y, and z axes respectively, rad

GENERAL APPLICATION AREAS

The role, advantages and potential of both incompressible and compressible flow hydro-static bearings in machine tool applications has been reviewed in Refs. 1, 2 and 3. Typical of where hydrostatic bearings are presently being used and where computer-aided design information can be of the utmost importance in keeping design, analysis and development cost to a minimum are the general application areas summarized in those references.

The fluids used in machine tool application are by no means restricted to lubricating oils or hydraulic fluids. The Heald Machine Company has developed a hydrostatic, water-lubricated, precision work head spindle for use on a Heald internal grinder.[2]

Compressible fluids, such as compressed air, have a very definite role to play in the development of adequate bearings for today's and tomorrow's machine tools.[3] Indeed, particularly in England, compressible fluid, externally pressurized bearings are already having some impact on the machine tool trade.[4] In addition to its chief advantage of low bearing temperature rise, a compressible fluid offers the advantage over oil of not having to be recirculated. The problems to be faced, however, are bearing stability and proper selection of bearing materials. Based on experience to date, some people feel that air will replace oil in machine tool hydrostatic bearings even in moderately heavy load applications. This is exciting to contemplate because it means that roughing and finishing to ultra-precision tolerances could be accomplished on the *same* machine—a very worthwhile goal.

APPLICATIONS

Several of the significant examples of computer-aided design applications of hydrostatic bearings and bearing systems are presented below. The application examples consider the structure as rigid. A future paper will present additional examples in which it was necessary to include the influence of the structure in the analysis and design of the bearing systems.

The discussion of the computerized design includes a description of the general problem, the analytical approach, sample results and actual performance where, available. The purpose of this section is to provide the reader with an appreciation of the advantage of the computerized analyses. In each application the fundamental theoretical problems are of general contemporary interest, and the solutions could not have been practically accomplished without the computational assistance of computer programs.

1. *Vertical Boring Bar Machine*

(a) *General Description.* A large manufacturer of special machines was selected to build a huge vertical boring bar machine for boring holes in large vessels. The various slideways and turntables were so large that hydrostatic support was necessary in order to accurately position the components for machining.

The general configuration of the machine is shown on Fig. 1. There were essentially four separate hydrostatic support systems as follows:

(a) Workpiece thrust bearings (adapter ring).
(b) Rotary table thrust and journal bearings.
(c) Horizontal saddle slideways bearings.
(d) Cross-rail saddle slideways bearings.

The primary function of the bearings is to support gravity loads imposed during posi-

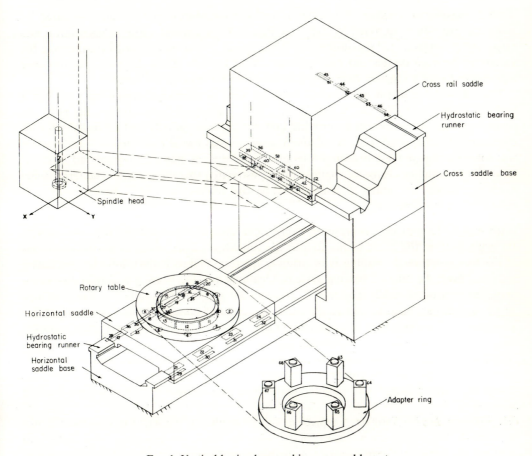

FIG. 1. Vertical boring bar machine—general layout.

tioning of the workpiece for the next operation. During the normal machine cycle the bearings are deactivated and the rotary table and slideway components are rigidly clamped in position. Excess capacity was built into the horizontal and cross-rail slideway systems to permit additional machining loads to be supported by the hydrostatic bearings and still contain the resulting tool point movement within the limitations of accuracy specifications. Thus, if chatter, and unsuitable finish or precision results, the option of energizing the slideway bearings during machining, to improve overall damping and stiffness characteristics, is possible.

(b) *Requirements*

(a) Load on horizontal slideway bearings—74,064 lb
(b) Load on cross-rail slideway bearings—122,260 lb
(c) Load on rotary table—45,094 lb
(d) Workpiece weight—28,000 lb
(e) Positioning accuracy—± 0.002 in
(f) Repeatability—±0.001 in
(g) Ambient temperature—70 to 100°F
(h) Lubricant viscosity—320 SUS @ 100°F

(c) *Techniques of Analysis.* The slideway pad configurations were initially estimated by using the procedures outlined in Ref. 1 of Part I. Flow control valve compensation was employed throughout for maximum film stiffness and overall flexibility to compensate for variations between design and final product. After initially sizing the pads, performance was checked over a range of operating conditions with the program Table.

The rotary table thrust and journal bearings were proportioned using Ref.1 of Part I and the programs Table and Steady State Exact. Eight pads were selected for both the journal and thrust assemblies to provide uniform stiffness radially and torsionally for all directions of the applied load. A relatively large journal radial clearance of 8 to 10 miles was specified to ease the fabrication tolerances for the 52 in diameter journal and to maintain small percentage clearance changes due to estimated thermal growth. The large clearances necessitated the design of small journal bearing recesses in order to keep flow requirements at a reasonable level.

(d) *Results.* In applying the program Table, a coordinate reference frame was selected with an origin fixed in the plane of the tool at the centerline of the boring head, since motions of the tool point under loading is of primary interest. Figure 1 illustrates the coordinate reference axes, the relative bearing systems locations and the bearing pad reference numbers. The horizontal saddle (pad nos. 17 through 38) and the cross slide carriage (pad nos. 39 through 62) are each supported by preloaded pairs of plane bearings. The rotary table is supported by a preloaded journal bearing (pad nos. 9 through 16) and a uni-directional load thrust bearing (pad nos. 1 through 8).

In determining the performance of the bearing systems, the supported bodies were subjected to loadings caused by component weights in the z-direction and overturning moments about the x and y axes. Other possible loads were also considered in the journal bearing analysis.

d.1. *Slideway Bearing Systems.* A typical sample of results for the slideway systems is shown on Table 1. This information was extracted from the output of runs made with Table and does not include all of the information available from the program. Only the maximum and

TABLE 1. CROSS RAIL BEARING SYSTEM PERFORMANCE SUMMARY

Run no.	201*	202	203	204	205	206	207**	208**
T_a °F	100	85	70	70	85	100	100	100
$\mu_a \times 10^6$, lb-sec/in²	9.18	15.31	24.1	24.1	15.31	9.18	9.18	9.18
$W_x \times 10^{-6}$, lb	0	0	0	0	0	0	0	0
$W_y \times 10^{-6}$, lb	0	0	0	0	0	0	0	0
$W_z \times 10^{-6}$, lb	−0.1222	−0.1222	−0.1222	−0.1062	−0.1062	−0.1062	−0.0497	−0.0497
$M_{xx} \times 10^{-6}$, in-lb	0	0	0	0	0	0	0	0
$M_{yy} \times 10^{-6}$, in-lb	−8.577	−8.577	−8.577	−6.465	−6.465	−6.465	−4.473	−4.473
$M_{zz} \times 10^{-6}$, in-lb	0	0	0	0	0	0	0	0
$\delta_x \times 10^{+6}$, in	0	43	69	52	20	−37	106	142
$\delta_y \times 10^{+6}$, in	0	0	0	0	0	0	0	0
$\delta_z \times 10^{+6}$, in	0	451	720	645	351	−163	31	2538
$\theta_{xx} \times 10^{+6}$, rad	0	n	n	n	n	n	n	n
$\theta_{yy} \times 10^{+6}$, rad	0	−3	−4.6	−3.5	−1.3	+2.5	−7.0	−9.4
$\theta_{zz} \times 10^{+6}$, rad	0	n	n	n	n	n	n	0
$S_x \times 10^{-9}$, lb/in	0.114	0.191	0.300	0.300	0.191	0.114	0.114	0.114
$S_y \times 10^{-9}$, lb/in	0	0	0	0	0	0	0	0
$S_z \times 10^{-9}$, lb/in	0.254	0.405	0.616	0.616	0.404	0.276	0.238	0.029
$S_{xx} \times 10^{-9}$, in-lb/rad	159	236	363	365	237	159	141	19.4
$S_{yy} \times 10^{-9}$, in-lb/rad	2458	3806	5865	5989	3882	2510	2277	317.8
$S_{zz} \times 10^{-9}$, in-lb/rad	69.1	115	181	181	115	69.1	69.1	69.1
$h_{min} \times 10^3$, in	3.00	2.73	2.58	2.58	2.73	2.79	2.52	3.00
$h_{max} \times 10^3$, in	3.00	3.25	3.40	3.40	3.26	3.21	3.44	4.88
p_{rmin}, psig	250	327	451	449	325	205	166	58
p_{rmax}, psig	250	551	1037	1035	551	312	420	250
ΣQ, in³/sec	55.76	→	→	→	→	→	55.76	28.37
ΣQ_g, gpm	14.50	→	→	→	→	→	14.50	7.38

* Denotes on design data.

** Denotes installation data.

n = Negligible or insignificant value.

minimum values of recess pressures are indicated in Table 1, and the stiffness matrix was abridged to the more important diagonal terms of the matrix. The starred run is for design operation. The other runs represent off-design operation at different viscosities and loadings. Information is also supplied for installation purposes when only partial gravity loads are imposed.

d. 2. *Rotary Table Bearings.* The rotary table bearings required some additional studies to arrive at a final configuration. The problem was to liberalize the machined clearance to accommodate thermal growth, while simultaneously maintaining a reasonable stiffness value and limiting the flow required to acceptable levels. The flow through a journal pad in the concentric position is determined from the following expression:

$$Q = q_f a_f p_r \, (h^3/\mu)$$

where Q = flow, in³/sec, p_r = recess pressure, psi,
              q_f = flow coefficient, h = concentric clearance, in,
              a_f = load coefficient, μ = absolute viscosity, lb-sec/in².

The variables p_r, h and μ were specified as

$$p_r = 250 \text{ psi}$$
$$h = 0.010 \text{ in}$$
$$\mu = 9.18 \times 10^{-6} \text{ lb-sec/in}^2$$

The desired limit of flow per pad was 2 gpm. Both a_f and q_f are geometry dependent and in order to determine the proper multiple of these factors it was necessary to run the program Steady State Exact with different recess sizes. Also configurations with and without axial drain grooves between recess domains were investigated. Table 2 shows the results from a series of runs with various recess to pad geometric relationships.

The tabulation indicated that equal sill bearings required too much flow, and were thus abandoned. It was also determined, that for the small recesses required to limit flow there was no appreciable flow reduction or load increase accomplished by the elimination of axial bleed slots. The slots provided convenient drainage passages and thus were maintained. Figure 2 is a plot of load (a_f) and flow coefficients (q_f) and flow factors ($a_f q_f$) vs. the angular extent of the recess (θ_r). From this curve a value of θ_r was selected to limit flow to the desired value. The load coefficient was also determined and was found to be satisfactory with regards to load capacity and stiffness.

This design study resulted in some critical decisions, because remachining and refitting of the large journal bearing would be a costly process. With the aid of computer programs the necessary information to determine the journal bearing pad configuration was directly obtained. Figure 3 shows the final geometry of a journal pad.

(e) *Hardware Performance.* The hardware is operational and cutting metal within specification limits. Thermal problems were initially encountered because of the high oil temperature. The saddle distorted excessively and caused binding of the bearings. The problem was eliminated by changing to a lighter oil and operating at lower oil temperatures. Considering the structure as a rigid member in this case caused some developmental problems. Excessive structural deformations were encountered in the cross rail saddle, but since the deformations were repeatable they could be compensated without loss of accuracy. The rotary table bearings performed without difficulty.

(f) *Computer Charges.* The computer charges for this project amounted to 4.2% of the total cost.

Computer-aided Design of Hydrostatic Bearings for Machine Tool Applications. II 825

TABLE 2. ROTARY TABLE JOURNAL BEARING

Run no.	Equal sill bearing	Axial drain grooves	l/L	θ_r	a_f	q_f	$a_f \times q_f$
3001	Yes	Yes	0.1	—	0.4538	2.028	0.9203
3002	↑	↑	0.2	—	0.5260	2.174	1.1435
3003			0.3	—	0.5783	2.364	1.3671
3004	↓		0.4	—	0.6342	2.533	1.6064
3005	Yes		0.5	—	0.7090	2.988	2.1185
3006	No		3/28	17.5	0.329	1.862	0.6126
3007	↑	↑		15.0	0.301	1.84	0.5538
3008				12.5	0.272	1.825	0.4964
3009				10.0	0.243	1.795	0.4362
3010				7.5	0.213	1.775	0.3781
3011				5.0	0.182	1.75	0.3185
3012		Yes		2.5	0.148	1.735	0.2568
3013		No		17.5	0.347	1.758	0.6100
3014		↑		15.0	0.316	1.749	0.5527
3015				12.5	0.284	1.738	0.4936
3016				10.0	0.253	1.726	0.4367
3017				7.5	0.221	1.711	0.3781
3018	↓	↓		5.0	0.188	1.693	0.3183
3019	No	No	3/28	2.5	0.154	1.671	0.2573

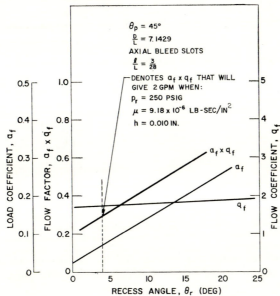

FIG. 2. Rotary table journal bearings—determination of recess angle.

2. Hydrostatic Boring and Milling Spindle

(a) *General Description.* A manufacturer of a precision horizontal boring and milling machine decided to develop a prototype hydrostatic spindle for one of their line of machines. The configuration of the spindle is shown on Fig. 4 of Part I and described in the section dealing with the Table Series Program of Part I. Figure 4 of Part I was an early configuration that was substantially modified as the design progressed. The interior bearings were eliminated as was the flow control valve compensation for the thrust bearings. Figure 4

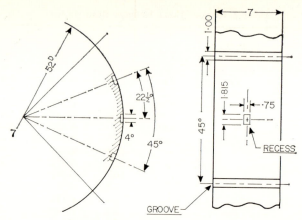

FIG. 3. Rotary table journal bearing—final configuration.

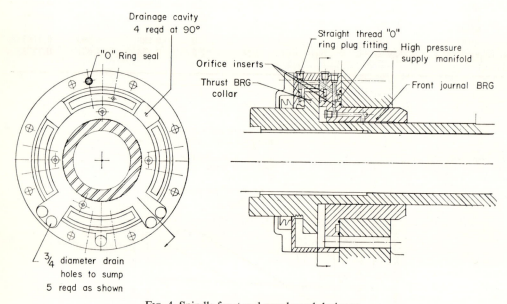

FIG. 4. Spindle front end supply and drainage.

shows the final design of the front end journal and thrust bearing. Orifice compensation was used throughout and the drain volume was considerably liberalized.

(b) *Requirements.* The general requirements were as follows:

(a) Spindle speed—2000 rpm max.—double the conventional rolling element spindle design.

(b) Thrust load—0–10,000 lb.

(c) Radial load—0–2000 lb acting 0 to 20 in from the front surface of the spindle housing.

(d) Pumping power—less than $7\frac{1}{2}$ h.p.

(e) Tool point motions—less than 0.0002 in with 250 lb tool force and 20 in overhang based on rigid shaft.

(f) Maximum supply pressure—1000 psig.

(g) Lubricant viscosity—1.23 $\times$ 10^{-6} lb-sec/in^2 at 100°F.

(c) *General Approach.* At the outset, extensive studies were made to determine bearing configurations that would:

(a) Provide good load capacity and stiffness.

(b) Maintain flow levels to limit pumping horsepower.

(c) Not overheat at high speed (one of the desirable requirements was to double existing shaft speed).

Bearings with and without axial drain grooves were investigated at various recess sizes, supply pressures and machined clearances. The thrust bearing was subjected to the same treatment although drainage considerations required the use of radial slots.

System performance was obtained by treating the shaft as a rigid member subjected to radial eccentricities and angular tilts. Each position of the shaft resulted in a combined eccentricity and tilt of the journals and thrust collar for each of the journal bearings and the thrust bearing. By appropriately superposing bearing loads and moments tool point loads were determined.

TABLE 3. CONCENTRIC POSITION PERFORMANCE AS FUNCTION OF RECESS TO BEARING LENGTH RATIO

Equal Sill Bearing $(D(\theta_p - \theta_r) = 2L(1 - l/L)$

$\mu = 1.23 \times 10^{-6}$ lb-sec/in^2; $h = 0.0015$ in; $p_s = 1500$ psig; $\beta_0 = 0.5$

Axial grooves	l/L	0.4		0.6		0.8	
	Speed	1200 rpm	2400 rpm	1200 rpm	2400 rpm	1200 rpm	2400 rpm
Yes	HP_s	1.587	6.349	1.195	4.780	0.6445	2.578
No	HP_s	1.711	6.843	1.180	4.719	0.6154	2.462
Yes	Q_g, gpm	1.99	1.99	4.09	4.09	10.18	10.18
No	Q_g, gpm	1.43	1.43	2.46	2.46	5.27	5.27
Yes	HP_B @ 85% eff	2.05	2.05	4.21	4.21	10.47	10.47
No	HP_B @ 85% eff	1.47	1.47	2.53	2.53	5.42	5.42
Yes	ΔT_s, Deg. F	9.42	37.70	3.46	13.832	0.750	2.998
	ΔT_s, Deg. F	14.15	56.61	5.68	22.73	1.383	5.53
Yes	ΔT_B, Deg. F	10.35	10.35	10.35	10.35	10.35	10.35
No	ΔT_B, Deg. F	10.35	10.35	10.35	10.35	10.35	10.35
Yes	ΔT, Deg. F	19.77	48.05	13.81	24.18	11.10	13.34
No	ΔT, Deg. F	24.50	66.96	16.03	33.08	11.73	15.88

(d) *Journal Bearing Studies.* The program Steady State Exact was utilized extensively in order to obtain an optimum configuration. Comparative studies were made between bearings with and without axial slots. Table 3 indicates the significant results of the comparative studies. From this information some conclusions of general interest could be made for bearings of similar geometries to the one studied.

(a) Flows and pumping horsepower are less without axial grooves and the differences increase as the recess size becomes larger with respect to overall bearing size.

(b) The combined viscous drag and pressure flow loss are greater for the unslotted bearing. for small recesses. As the recess size increases the losses in the slotted bearings becomes slightly larger due to greater friction loss from pressure drop.

(c) Temperature rise is greater for the unslotted bearing because of smaller flows.

No meaningful comparative studies were made with regard to ultimate load capacity and stiffness. The pre-load for the unslotted bearing is obviously greater. In some unrelated supplementary studies, it was determined that with some geometric proportions, stiffness of the unslotted bearing was very sluggish near the concentric position. Whether the same behavior would occur in slotted bearings was not determined. A significant result from the computerized studies was that for certain geometries the non-linear effects become significant at small eccentricities, and analyzing the journal bearing using constant load and flow coefficients could be considerably in error even though operation was near the concentric position. For this reason it was necessary to use the program Steady State Exact, which solves the governing lubrication equations without approximation, to accurately determine performance.

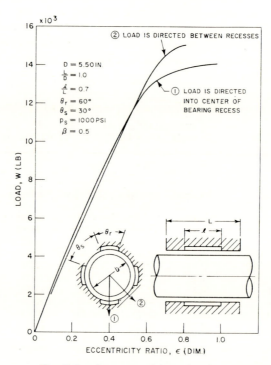

FIG. 5. Load vs. eccentricity ratio—front end journal bearing.

A series of computer runs were made in which the shaft was moved radially in the bearing. Since loads on the bearing are rotating, eccentricities were taken in directions into and between recesses. Bearing performance of load, flow, and recess pressures as functions of bearing eccentricity ratios were determined. Figure 5, Load vs. Eccentricity ratio, shows curves for cases with the load directed into the recess and with load directed between recesses. At the lower eccentricity ratios, the load capacity is slightly higher for loading into a recess. At higher eccentricity ratios, $\epsilon > 0.5$, the load capacity is higher for loading between recesses. For either case, film stiffnesses are almost the same. Figure 6, Flow and pump power vs. Eccentricity ratio, shows that the total bearing flow changes about $- 11\%$ over the full eccentricity ratio range for loading into the recess, and about $- 7\%$ for loading between recesses. Figure 7, Bearing recess pressure vs. Eccentricity ratio, shows how the bearing recess pressures vary for the load conditions. No angular tilts are applied. The unloaded side recess pressure falls off at a less rapid and non-linear rate when compared with the increase in recess pressure of the loaded side. This is the basic reason why linear coefficient approximations are in error as the linear approach would predict symmetric recess pressure curves about the concentric position pressure. Figure 8, Bearing recess pressure vs. Eccentricity ratio shows the effects of angular tilts of the shaft in the bearing with the radial load applied into a recess. Shaft misalignment in the bearings results in decreased load capacity.

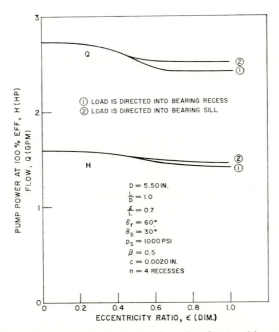

FIG. 6. Pump power and flow vs. eccentricity ratio—front end journal bearing.

(e) *Thrust Bearing.* Sustaining thrust loads was not a particularly difficult problem for the space and pressure available to the thrust bearing. Inhibiting flow with a reasonable installation clearance was the primary problem. A trade off was necessitated whereby stiffness was off optimum in order to minimize flow consumption. This was accomplished by utilizing a smaller restrictor and thus reducing concentric recess pressures. Normally a design supply to recess pressure ratio of 0.5 is employed to optimize stiffness. In this instance the ratio was reduced to 0.2. The orifice diameter was still well above the normal limit generally used to exclude clogging by contamination. Figures 9 and 10 are typical performance curves obtained from computer studies. Figure 11 shows the performance of the opposed pair at various eccentricity ratio and thrust collar inclinations.

(f) *System Performance.* The individual bearing performance information was superposed to provide overall system performance. Figure 12 shows Load vs. Moment for various eccentricity ratios and angular tilts. Moments are about an origin as shown on Fig. 13. Figure 14 depicts loads and moments about the origin transposed to the tool point for various tool point overhangs. These curves also include gear loads between the bearings. The dashed lines represent tool point displacements when the gear load and tool loads

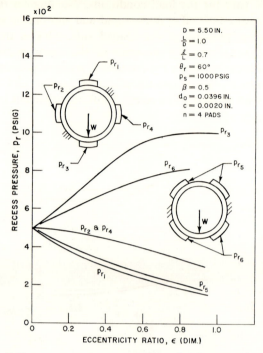

Fig. 7. Recess pressure vs. eccentricity ratio—front end journal bearing.

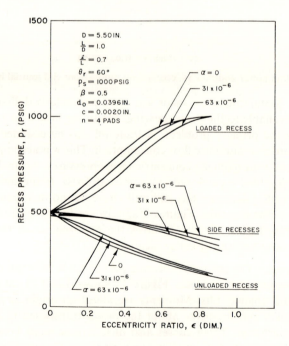

Fig. 8. Recess pressure vs. eccentricity ratio and misalignment—front end journal bearing.

are of opposite phase. The solid lines are for gear and tool loads in phase. It is seen that even at maximum overhang and heavy loads rigid body displacements are very reasonable.

Table 4 presents a summary of the performance characteristics and shows how the requirements were achieved with the bearings designed by computer.

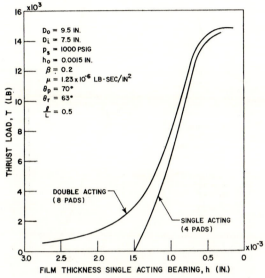

FIG. 9. Thrust load vs. film thickness.

(g) *Present Status.* Design drawings have been completed and are presently being reviewed prior to production of a prototype spindle. The results of the analytical investigation are most encouraging and it is anticipated that all design goals will be achieved.

(h) *Computer Charges.* The computer costs for this project amounted to about 7% of the total.

TABLE 4. SYSTEM PERFORMANCE

Lubricant (Mobil Velocite No. 6)	
Flow rate	$= 9.63$ gpm
Recess viscosity	$= 1.23 \times 10^6$ reyn at $100°$F
Supply pressure	$= 1000$ psig
Pump power at 100% efficiency	$= 5.62$ h.p.
Viscous power loss at 2000 rpm	$= 6.10$ h.p.
Temperature rise at 2000 rpm	$= 14°$F of which $7°$F is due to $\Delta p = 1000$ psig
Spindle	
Radial stiffness	$= 21 \times 10^6$ lb/in
Angular stiffness	$= 3.67 \times 10^9$ in-lb/rad
Thrust stiffness	$= 10 \times 10^6$ lb/in
Tool point motions neglecting elastic deformations (rigid body displacements)	$= 0.0009$ in with 2000 lb tool force and 20 in overhang
	$= < 0.0001$ in with 250 lb tool force and 20 in overhang

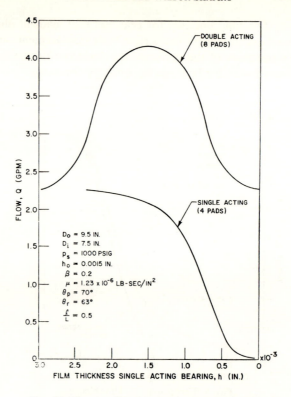

FIG. 10. Flow vs. film thickness—thrust bearings.

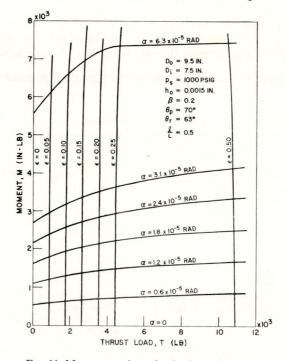

FIG. 11. Moment vs. thrust load—thrust bearings.

CONCLUSIONS

Experience has proven that the use of computer programs not only provides solutions, but also allows accurate and expeditious parametric and design studies to be performed which greatly facilitates the process of arriving at sound design decisions. Once embarked on computer methods the development of new programs and upgrading of existing ones becomes perpetual. As time goes on each program increases in generality and becomes a

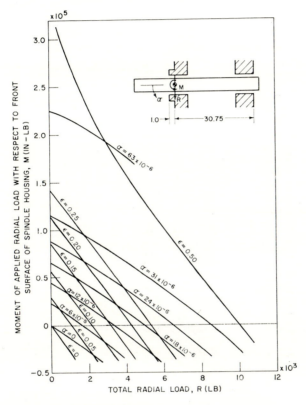

FIG. 12. System performance—load vs. moment.

more powerful aid. Opportunities for researching many virgin problem areas suddenly appear, and theoretical solutions to problems once considered impossible to solve before the use of computers now became almost commonplace.

Many thousands of calculations are necessary to design the bearing systems for precision machinery when structural deformations, dynamic characteristics and variable supply circuitry are taken into account in addition to the normal hydrodynamic and hydrostatic effects. With the aid of existing computer programs, it is possible to make these calculations in a matter of seconds.

With the technology available today and the severe requirements that must be met by the lubrication engineer, it is dangerous and costly to "scale-up" old designs or to extrapolate performance from curves that served extremely well a few years ago. The expense of modification, testing and experimentation of such bearing systems is often unjustifiable in the light of the advanced analytical tools available. Without these tools lubrication engineers were forced to take the "hit-and-miss" course, resulting in long, difficult and expensive development periods. However, the increasing accessibility of high-speed digital computers and the increasing sophistication of numerical analysis evident over the last few years, now allows the previously long period of development to be drastically reduced to mere minutes of computer time, so that the predominantly experimental approach is obsolete. As the pace of technological development continues to accelerate the utilization of these computer-aided bearing design techniques is the only way the lubrication engineer can ensure that it will not be the bearing retarding the development of new machinery and equipment.

In closing it should be emphasized that the bearings are still not designed by a computer but by engineers *using a computer as an aid*. The engineer must still call on his own practical experience and judgment in deciding what are acceptable values for the variables.

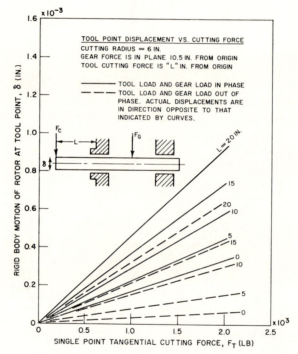

Fig. 13. System performance—tool point displacement vs. load.

REFERENCES

1. H. C. Rippel, Role of Hydrostatic Bearings in Manufacturing, Franklin Institute Research Laboratories Report No. 32TR67-8, presented at the Horizons in Manufacturing Technology Seminar, Institute of Science and Technology, University of Michigan, 9 May 1967.
2. H. C. Rippel and D. D. Fuller, Advantages offered by Hydrostatic Bearings in Machine Tool Applications, Franklin Institute Research Laboratories Report No. 32TR68-5, presented at ASME Design Engineering Conference, Chicago, April 1968.
3. O. Decker, Potential of Gas-Lubricated Bearings in Ultra-Precision Machine Tool Applications, Franklin Institute Research Laboratories Report No. 32TR68-7, presented at ASME Design Engineering Conference, Chicago, April 1968.
4. H. L. Wunsch, Air Bearing Applications to Machine Tools and Measuring Instruments, for presentation at the 2nd International Symposium on Gas-Lubrication, ASME Spring Symposium, June 1968, Las Vegas.

A COMPARISON BETWEEN AIR BEARING SPINDLES AND BEARINGS OF OTHER TYPES

H. RENKER

Truelweg 4, 3600 Thun/Schweiz

1. INTRODUCTION

Aerodynamic and aerostatic bearings are finding increasing application in advanced design techniques. Whilst the main field of application of the aerodynamic bearing is in the manufacture of gyroscopes and turbines, the aerostatic bearing is becoming more and more popular in machine tool design. One of the main applications which has recently been given wide publication is for the spindles of precision grinding machines. This entirely new bearing system has been widely and enthusiastically recommended for all spindle applications on grinding machines.

Such recommendations are, however, often made without giving due consideration to technical and economic implications. The object of this paper is to present a critical comparison of the different bearing applications on cylindrical grinding machines. The comparison is based on tests carried out on both aerostatic bearings from various manufacturers and hydrodynamic bearings with automatic clearance control.

Roundness measurements were made on a Talyrond instrument and the examples of work were selected from a wide range of precision manufacturers.

2. DEFINITION OF WORKPIECE QUALITY

The definition of the quality of a precision ground component can vary and will depend on the actual function of the component. For clarity the possible deviations from an ideal workpiece have to be explained.

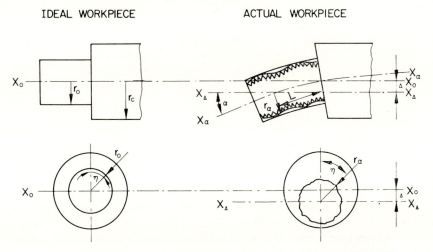

FIG. 1. Presentation and definition of possible deviations on a component.

Nature of Deviation:
   2.1 Deviation from an ideal dimension.
       (Difference between r_0 and r_a)
   2.2 Deviation from an ideal axis.
       (Difference Δ and α between axes X_0, $x\Delta$ and xa)
   2.3 Deviation from the macrogeometrical shape.
       (Variations in radius r_a as a function of length L)
   2.4 Deviations from the microgeometrical shape.
       (Variations of surface roughness as a function of angle η)
       (Variations of surface roughness as a function of length L)

3. THE INFLUENCE OF THE APPLIED BEARING SYSTEM ON THE WORKPIECE QUALITY

3.1 *Centres of Component and Machine*

When a workpiece is ground between dead centres, the quality of the centres in the component and the mating centres on the machine have a direct influence on the quality obtained.

The resultant errors will appear mainly as deviations from the macrogeometrical shape as described in para. 2.3. Chatter marks in the component centres will produce generally an error in the form of a certain polygon. Generally speaking, errors of shape in the component centres will have the greater influence since the machine centres are usually produced to a much higher standard of accuracy and in any case will be lapped more round in use. The most common error found on the machine centres is ovality whereas errors in the component centres may take any shape.

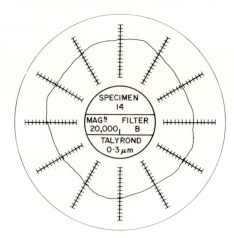

FIG. 2. Out-of-roundness of 12 μin produced by slight chatter in the component centres.

3.2. *Workhead Spindle and Bearings* (*Spindle Diameter 3 in*)

3.2.1. For live spindle grinding, either in the chuck or collets, it is essential that the spindle runs true and has a high rate of stiffness. The workhead spindle on a precision grinding machine should run true within 10 μ in. This accuracy can be achieved with either hydro-dynamic or aerostatic bearings.

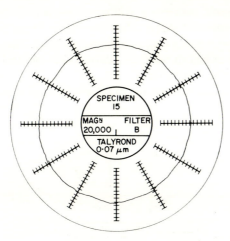

FIG. 3. Out-of-roundness 2·8 μin produced on the same component after removing chatter marks from the centres.

Hydrodynamic bearings however do not require any ancillary oil or pressure supply system. It is possible to design a hydrodynamic bearing in such a way that the supporting fluid film operates in a closed circuit which is put into motion by the spindle itself. Since the speed range of a workhead spindle is relatively small it is not necessary to worry about heat development. To maintain the high rotational accuracy mentioned it is necessary to incorporate an automatic device for controlling the clearance in the bearing.

Aerostatic bearings must be manufactured to perfect dimensional sizes to obtain the above rotational accuracy. Since this type of bearing operates with an open pressure circuit, extremely efficient filtering is required at the air inlet.

3.2.2. The second main consideration in selecting a workhead spindle bearing is the actual

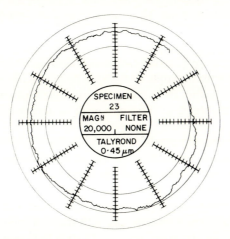

FIG. 4. Out-of-roundness 18 μin caused by lack of stiffness of spindle bearing (400,000 lb/in) and component with interrupted diameter.

stiffness of the bearing itself. Insufficient stiffness in the bearing (measured in lb/in radially and axially) will cause errors of shape. These errors are found particularly when grinding heavy components or components with interrupted surfaces. Heavy components are subject to errors of shape as described in para. 2.1 (deviation from ideal dimension) and para. 2.2 (deviation from ideal axis). Components with interrupted diameters will cause errors as described in para. 2.3 (deviation from the macrogeometrical shape).

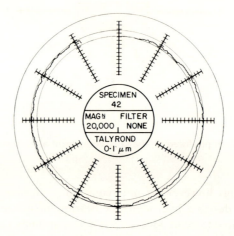

FIG. 5. Out-of-roundness 4 μin caused by component with interrupted diameter as shown in Fig. 4. Stiffness of spindle bearing 1,500,000 lb/in.

Hydrodynamic as well as hydrostatic bearings fulfil the requirements with regard to stiffness. On the hydrodynamic workhead spindle used for the tests, the stiffness can be increased up to 1,700,000 lb/in. It is possible of course to obtain even higher values with hydrostatic bearings dependent on the oil pressure applied.

The aerostatic bearings used in the tests have a maximum stiffness of 400,000 lb/in using a supply pressure of 88 lb/in². This value is not sufficient for high accuracy requirements.

3.3. *Spindle and Bearings of an External Grinding Wheelhead* (*Spindle Diameter* 2½ *in*)

3.3.1. The same considerations apply to this spindle as described in para. 3.2. Rotational accuracy should again be within close limits (10 μin) otherwise the wheel will not cut regularly and deviations in the shape of the component as described in paras. 2.3 and 2.4 will occur.

3.3.2. The stiffness of the bearing becomes increasingly important as the weight of the wheel increases. A grinding wheel must be considered as a porous body which is both ejecting coolant by centrifugal force and at the same time absorbing it by capillary action. This phenomenon occurring during the grinding process causes a continuous variation in the state of balance. The forces thus produced must be flattened out by the bearing stiffness.

The stiffness ratio of 587,500 lb/in obtained with the hydrodynamic bearing is adequate whereas the stiffness of the aerostatic bearing 250,000 lb/in was found to be satisfactory only when using wheels with diameters less than 12 in. The aerostatic bearing is therefore technically sufficient within this limit but it is less economical in operation.

3.4. *Spindle and Bearings of an Internal Grinding Wheelhead (Spindle Diameter 2 in)*

This spindle operates at relatively high speeds (25,000–50,000 rpm) and is equipped with wheels of low volume. The out of balance forces occurring during grinding therefore are negligible. A comparison was made between a spindle with aerostatic bearings and a ball bearing spindle.

The aerostatic spindle gave better results with regard to roundness, surface finish and quietness of operation. The superior surface finish can be partially accounted for by the fact that an aerodynamic spindle produces under some circumstances an axial vibration. This axial vibration is dependent on the dimensions of the axial air bearing and the air consumption (supply pressure). A disadvantage of the aerostatic bearing is the relatively low load limit it can sustain as compared with the ball bearing spindle. A momentary overload can cause seizure of the aerostatic bearing whereas the ball bearing spindle can easily absorb short overloading.

4. CONCLUSIONS

4.1. *Workhead*

The aerostatic bearing, due to its relatively low stiffness ratio, is suitable for only limited applications. For general use and for difficult operations the hydrostatic and hydrodynamic bearings are more suitable. Hydrostatic bearings, however, have the disadvantage of relatively high capital and operating costs. They also sometimes are less precise in true running.

4.2. *External Grinding Wheelhead*

For this application the aerostatic bearing is again only suitable for small and light grinding wheels. Large wheels and heavy wheels demand a high stiffness ratio which can better be obtained economically with hydrodynamic bearings.

4.3. *Internal Grinding Wheelhead*

Internal grinding spindles with speeds in excess of 5000 rpm provide the suitable field of application for the aerostatic bearing.

Rotational accuracy and stiffness are adequate for this purpose and the quality of surface finish obtainable is also adequate as compared with other types of bearing.

HYDROSTATIC GAS BEARINGS: DESIGN FOR STABILITY

M. G. JONA

Istituto per le Ricerche di Tecnologia Meccanica, Vico Canavese, Torino, Italy

SUMMARY

Three theories for predicting a hydrostatic gas bearing's stability are discussed, and compared with experimental data. The influence of a tuned resonator on the bearing's stability is described. Experimental data are presented, in order to show how critical the tuning of the resonator is.

1. INTRODUCTION

Various systems have been proposed for obtaining low-friction slideways. Among these, the use of roller bearings first, and of incompressible fluid hydrostatic bearings later, has become quite common among machine tool builders. Compressible fluid hydrostatic bearings, have been, up to now, treated with a certain amount of distrust; consequently such bearings have only had limited applications, either for cases in which no accuracy is required[1] or for cases in which the load on the bearing varies but slowly with time.[2]

Hydrostatic bearings using a compressible fluid, hydrostatic air bearings especially, have a few important advantages over hydrostatic oil bearings. First, the air which has passed through the bearing can be discharged directly in the atmosphere which acts as a practically infinite reservoir (with, consequentially, infinite thermal capacity). Second, air at an adequate pressure is generally available in the workshop.

On the other hand, hydrostatic gas bearings have a few undeniable drawbacks. In order to attain a high stiffness, very low gap heights must be used. Very strict tolerances are thus required on the construction of the slideways. Unfortunately the stiffest bearing designs are prone to dynamic instability, a fact which has often discouraged machine-tool builders from using hydrostatic gas bearings.

It is the purpose of this paper to show that much can be done to optimize the dynamic behaviour of hydrostatic gas bearings. It is felt that, even when optimized from a dynamic point of view, hydrostatic gas bearings will not be the best solution to all problems in which low friction is required, but certainly the field of their economic applicability will be much extended.

2. THE DYNAMIC STABILITY OF A RECESSED HYDROSTATIC GAS BEARING

Various theories dealing with the dynamic stability of hydrostatic gas bearings have been published. Their purpose is to predict the stable operating regions for hydrostatic gas bearings. Three theories dealing with the stability of a recessed bearing will be examined here.

A schematic of the recessed bearing is shown in Fig. 1. It is generally known that the

841

smaller the recess volume, $(\pi/4)D_s^2 h$, the greater the chances of having a stable bearing.[3, 4] It is also known that a decrease of the recess diameter, D_s or, to a certain extent, of the recess depth, h, reduces the bearing stiffness. Hence the interest in knowing exactly what maximum recess volume can be tolerated, without falling in unstable operating conditions.

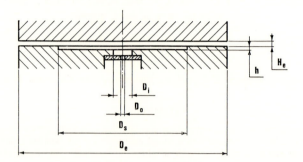

FIG. 1. Schematic drawing of a recessed hydrostatic gas bearing.

Three stability theories, by Licht,[5] by Licht, Fuller and Sternlicht[6] and by Turnblade,[7] are discussed here. More complete theories have been proposed,[8] but, from the design engineer's point of view, the complex calculations involved do not justify the difference in the results.

It is interesting to notice that, although the first two theories are similar, the third is qualitatively quite different.

The first two theories, as a matter of fact, give the maximum gap height which will allow stable operation, for each given bearing geometry and for each supply pressure P_0. Turnblade's theory instead, for the same conditions, will define some unstable bands of gap height. This is in good agreement with experimental evidence, which shows that the same bearing might be stable for very high loads (and, consequentially, small gap heights), becomes unstable for smaller loads, and might again become stable for very low loads. However, this second stable zone is quite far from the usual operating conditions, referring to very low loads and stiffnesses. The second stable zone is practically of no interest to the design engineer.

Another fundamental difference can be observed when examining the formulae which give the stability conditions, according to the different theories.

The bearing diameters only appear as ratios in the first two theories, while the third includes a representative bearing diameter, in the formula for the bearing transit time. Essentially, according to the first two theories, the use of one bearing with a high load-carrying surface or of a few smaller bearings of similar geometry (ratio between recess diameter and bearing diameter, recess depth), adding up to the same total area, would be quite indifferent. According to Turnblade's theory, many small bearings would be the best choice from a dynamic point of view.

From the design engineer's point of view, there is a serious drawback to Turnblade's theory. This requires, in fact, the calculation of the bearing's natural frequency. Though the gap height, H_e, under a given load F, can be predicted within a reasonable accuracy, the uncertainties on calculated stiffness tend to be rather larger. A further difficulty towards the calculation of the bearing's natural frequency arises from the fact that the relationship between H_e and F is not linear, stiffness not being constant.

The bearing frequencies measured by the author were generally lower than the calculated values, even to a ratio 1:2. The stiffness introduced in the calculations has been derived from the slope of the tangent to the F, H_e curve. This approach would be acceptable for the calculation of the frequency of very low amplitude vibrations (μm order of magnitude). The observed vibration amplitudes were greater by an order of magnitude.

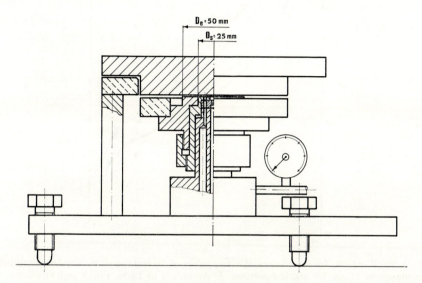

FIG. 2. Experimental bearing.

In order to test the dynamic stability of a gas bearing, a very simple experimental bearing can be seen in Fig. 2. Its main features are: constant diameters and variable recess depth; single inlet orifice, which can be easily changed in order to test different orifice diameters, D_0; a cavity connected to the recess through a narrow hole. This hole can be closed in order to obtain a simple bearing, according to the schematic in Fig. 1. The stability limits for the bearing in these conditions have been calculated for three different values of the recess depth, h, according to the theories given in refs. 5 and 6. The actual limits can be found in Fig. 3. Figures 4, 5 and 6 show the experimental evidence. It can be seen that the theories used for these calculations will give rather conservative estimates of the stable range of the bearing. It can also be seen that such an estimate is much more conservative for high pressures ($P_0 = 6$ kg/cm²) than for low pressures ($P_0 = 3$ kg/cm²). In particular it can be noticed that, according to the experimental results of Figs. 4 and 5, the calculated stability limits are far too conservative, presenting, as unpractical, some fairly good bearing designs.

The vibration frequencies for unstable bearings increase with increasing supply pressures, at constant loads, and decrease with increasing loads at constant supply pressures. Figure 7 shows some oscillographs in which the bearing frequency varies between 80 c/s and 135 c/s.

3. EFFECT OF A SIMPLE RESONATOR ON THE BEARING'S STABILITY

The use of a tuned resonator to increase bearing stability has been first proposed by Sixsmith[9] and it is surprising to see how little attention the literature has since given to the subject. The resonator proposed by Sixsmith is a tuned cavity, connected through a hole

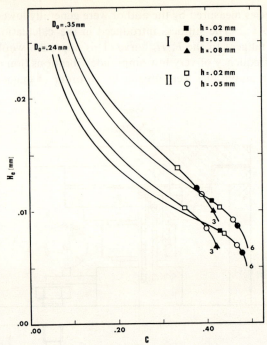

FIG. 3. Stability limits for three values of h, two values of D_0 and two of P_0. The limit indicates the maximum height for stable operation. I, According to Licht, Fuller and Sternlicht.[6] II, According to Licht.[5] The numbers near the curves indicate the supply pressures P_0 (kg/cm²).

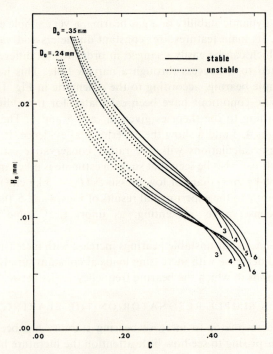

FIG. 4. Experimental stability results, no resonator. The numbers near the curves indicate the supply pressures P_0 (kg/cm²). Recess depth $h = 0.02$ mm.

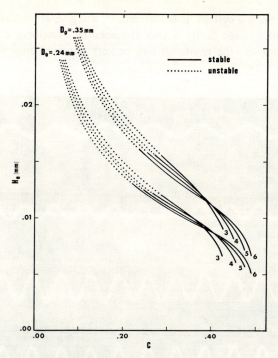

FIG. 5. Experimental stability results, no resonator. The numbers near the curves indicate
the supply pressures P_0 (kg/cm²). Recess depth $h = 0.05$ mm.

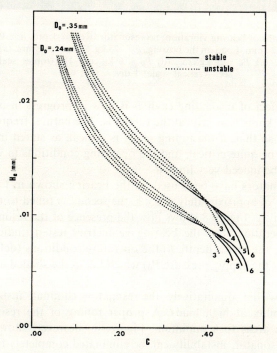

FIG. 6. Same as Figs. 4 and 5. Recess depth $h = 0.08$ mm.

to the bearing recess. Such a system, it is claimed,[10] will give almost any ratio of damping to stiffness. The use of a tuned cavity is also discussed by Sheinberg and Tabachnikov.[11] According to this reference, the resonator can be very effective only within a limited frequency range.

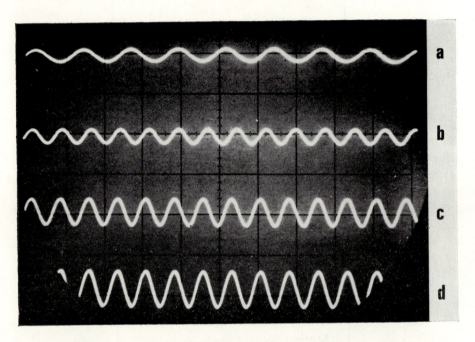

FIG. 7. Oscillographs of bearing vibration, no resonator. Recess depth, $h = 0.08$ mm. Orifice diameter, $D_0 = 0.24$ mm. Load on the bearing, $F = 25$ kg. Supply pressure, (a) $P_0 = 3$ kg/cm², (b) $P_0 = 4$ kg/cm², (c) $P_0 = 5$ kg/cm², (d) $P_0 = 6$ kg/cm². Horizontal scale: 1 div = 0.01 sec. Vertical scale: 1 div = 0.01 mm.

Research on the effect of resonating cavities is now in progress in our laboratory. It is important, in fact, to know how critical the tuning is. The natural frequency of the bearing varies with the load, so that, if the tuning really is critical, as stated in ref. 11, the design of the resonator will be quite critical, and the operating conditions in which the resonator will be effective will be indeed very limited.

Two different resonators have been used on the bearing shown in Fig. 2, the first has a resonating frequency of approximately 400 c/s, the second is tuned to the frequency range of the bearing instability. Figure 8 shows how the presence of the resonators influences the regions of stable operation. For the bearing geometries tested (indicated by the orifice diameter, D_0, and by the recess depth, h) the operating conditions (defined by the load on the bearing F, and by the supply pressure P_0) which fall in the shaded area on the diagram, are stable.

The diagrams show that, qualitatively, the resonators eliminate instability at low loads, but introduce instability at high loads. A proper tuning of the resonator (see Fig. 8c) reduces the unstable areas to a minimum. Figure 8b, however, shows that, even with a very poorly tuned resonator, instability can be eliminated completely for low values of the recess depth.

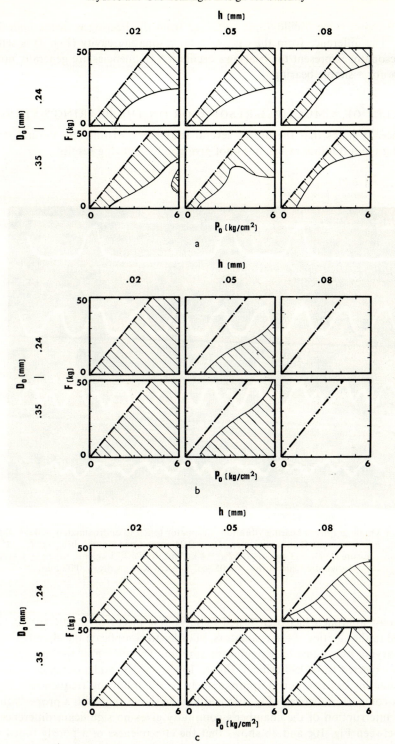

FIG. 8. Regions of stable and unstable operation for the experimental bearing. The shaded area is the stable region. (a) Without resonator. (b) Resonator tuned to approximately 400 c/s. (c) Resonator tuned to approximately 100 c/s.

Figure 9 shows some oscillographs obtained from the bearing with the high frequency resonator. The difference from the case with no resonating cavity (Fig. 7) is self-evident. When a resonator is present the bearing's oscillation frequencies are generally much lower than those for a simple bearing.

4. EFFECT OF A MULTIPLE RESONATOR ON THE BEARING'S STABILITY

Reference 11 states that the use of several resonators, of different natural frequencies overlapping the instability range, does not give good stabilizing results.

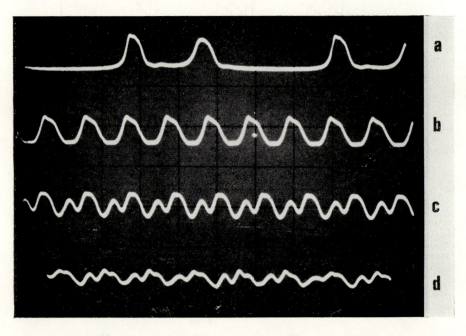

FIG. 9. Oscillographs of bearing vibration. Resonator tuned to approximately 400 c/s. Recess depth, $h = 0.08$ mm. Orifice diameter, $D_0 = 0.24$ mm. Load on the bearing, $F = 25$ kg. Supply pressure (a) $P_0 = 3$ kg/cm², (b) $P_0 = 4$ kg/cm², (c) $P_0 = 5$ kg/cm², (d) $P_0 = 6$ kg/cm². Horizontal scale: 1 div = 0.05 sec. Vertical scale: 1 div = 0.02 mm.

A different approach has been tested here. A few pieces of plastic tubing have been introduced in the cavities of the resonators, in order to complicate the chamber's geometry, without varying its volume to a significant amount. The tuning of the resonator is consequentially expected to be less critical. Figure 10 shows the results obtained with three different resonating cavities, tuned approximately to the following frequencies: 50, 100 and 400 c/s. A comparison between Fig. 10b and Fig. 8c shows that with a properly tuned resonator the interruption of the chamber's continuity gives no significant difference. A comparison between Fig. 10c and 8b shows that the effectiveness of a poorly tuned resonator is much increased by the interruption of the chamber's continuity. Further research on this subject is in progress.

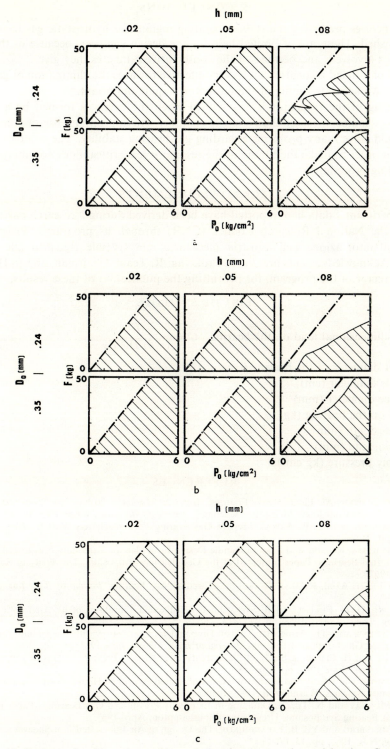

FIG. 10. Regions of stable and unstable operation for the experimental bearing. Complex resonators: (a) Tuned to approximately 50 c/s. (b) Tuned to approximately 100 c/s. (c) Tuned to approximately 400 c/s.

5. CONCLUSIONS

A few theories predicting the stable operating regions for hydrostatic gas bearings have been published. These are of limited help for the design engineer, because of the difficult calculations involved, and because of the conservative estimates they give.

The use of a tuned resonator can be very effective towards the elimination of gas-bearing instability. An accurate tuning of the resonator is quite important.

The tuning of the resonator is less critical when the cavity is formed by a few communicating chambers, rather than a single chamber. An appropriate use of complex resonators could solve many problems regarding gas-bearing stability. The complex geometry required in order to make the tuning of the resonator less critical does not imply any complex building procedure.

ACKNOWLEDGEMENTS

The experimental data here reported have been derived during a research partially sponsored by the National Research Council (CNR) through its program: "Programma di ricerca sull'automazione nell'industria meccanica con speciale riguardo alle macchine utensili." Acknowledgements are due to Prof. Ing. R. Teani, Chairman, and to Dr. Ing. A. Palagi, Director of the program, for permitting the publication of these results.

LIST OF SYMBOLS

C non-dimensional load on the bearing $C = \dfrac{F}{(\pi/4)\, D_e^2\, P_0}$

D_e outer diameter of the bearing (cm)
D_s recess diameter (cm)
D_0 orifice diameter (mm)
F load on the bearing (kg)
h recess depth (mm)
H_e bearing gap (mm)
P_0 supply pressure (kg/cm^2)

REFERENCES

1. SYDNEY GLOSSOP, Air Floats Heavy Fixtures. *American Machinist*, 20 Nov. 1967, pp. 110–112.
2. HARRY MOTE, Coordinate Measuring Machines. *American Machinist*, 1 Jan. 1968, pp. 65–76.
3. B. STERNLICHT and E. B. ARWAS, Modern Gas-Bearing Turbomachinery, Part I. *Mech. Engng*. Jan. 1966, pp. 24–29.
4. P. L. HOLSTER, Reliable and Easy to Handle Design Formulae for Externally Pressurized Gas Thrust and Journal Bearings. Paper presented at the Gas-Bearing Symposium, University of Southampton, April 1967.
5. LAZAR LICHT, Axial, Relative Motion of a Circular Step Bearing. *Journal of Basic Engng.*, 1959, pp. 109–117.
6. L. LICHT, D. D. FULLER and B. STERNLICHT, Self-Excited Vibrations of an Air-Lubricated Thrust Bearing. *Trans. ASME*, 1958, pp. 411–414.
7. R. C. TURNBLADE, The Molecular Transit Time and Its Correlation with the Stability of Externally Pressurized Gas-Lubricated Bearings. *Journal of Basic Engng.*, June 1963, pp. 297–303.
8. L. LICHT and H. ELROD, A Study of the Stability of Externally Pressurized Gas Bearings. *Journal of Applied Mech.*, June 1960, pp. 250–258.
9. H. SIXSMITH, The Theory and Design of a Gas Lubricated Bearing of High Stability. 1st Int. Symp. Gas Lubricated Bearings. Washington, D.C., 1959.
10. A. J. MUNDAY and N. TULLY, Damping in Externally Pressurized Gas Bearings. Paper presented at the Gas Bearing Symposium, University of Southampton, April 1967.
11. S. A. SHEINBERG and YU. B. TABACHNIKOV, The Design of Air-Lubricated Flat Slideways. *Machines & Tooling*, Nov. 1967, pp. 13–17.

THE BEHAVIOUR OF A MULTIRECESS HYDROSTATIC JOURNAL BEARING IN THE PRESENCE OF SEVERE SHAFT BENDING

P. B. Davies and T. A. Andvig

SUMMARY

The elastic bending deflections of machine tool main spindles supported by hydrostatic journal bearings may, in some designs, become comparable with the radial displacements of the spindles within the bearings themselves. The behaviour of an assembly consisting of a four recess orifice compensated hydrostatic journal bearing and a shaft which deformed severely under the imposed radial loads was investigated experimentally.

The bearing's performance was predicted theoretically by assuming that the shaft moved parallel to the axis of the bearing. The predictions were then applied in the presence of shaft bending by taking the displacement at the mid-plane of the bearing as the effective displacement. Satisfactory prediction of the bearing's behaviour was demonstrated on this basis even when the displacement of the shaft at the ends of the bearing differed by as much as $\pm 30\%$ of the displacement at its mid-plane.

1. INTRODUCTION

Hydrostatic journal bearings have many features which make them differ from their hydrodynamic counterparts but particularly noticeable is the complexity of the internal geometry of the various designs of journals that have been proposed. Of the many designs that have been used the multirecess type first studied in depth by Raimondi and Boyd[1] has proved to be the most successful and is characterized by two prominent constructional features, i.e.

(a) no drainage grooves exist between adjacent recesses, and

(b) the total area of the relatively deep recesses constitute a large proportion of the developed area of the whole journal.

A qualitative appraisal of these features indicates that the first tends to reduce the total flow rate of fluid required to operate the bearing while the second provides large areas of high-pressure fluid which yield good load capacity.

Since the early investigation reported in ref. 1 many other workers[2-4] have indicated how the original analysis may be refined and extended so that the knowledge now available is sufficient for dependable design procedures to be established.[5-7]

However, comparatively little experimental verification of these largely theoretical studies have been published with the exception of the results presented in ref. 1 for the tests conducted without shaft rotation and those of ref. 4 for a limited range of conditions with shaft rotation. The tests reported in ref. 1 were completed with both orifice and capillary compensation but the latter was used exclusively in ref. 4.

The application of hydrostatic journal bearings in machine tools is well documented and justified[8-10] and some comparative studies have been made with other types of bearings.[11] Additionally, Opitz[12] has emphasized the importance of shaft flexibility in causing premature metal-to-metal contact at the front end of bearings in machine tool mainspindle assemblies but no consideration appears to have been made of its influence on the performance of the hydrostatic journal bearings themselves.

This paper describes the results obtained from tests on an orifice compensated hydrostatic journal bearing with four recesses under the action of steady radial loads in the presence of severe shaft bending. The bearing in question supported a flexible shaft whose proportions may be considered unrepresentative of modern machine tool practice and a somewhat extreme situation was created as the eccentricity of the shaft at the ends of the bearing differed considerably under some states of loading. The results presented therefore refer to a situation which is unlikely to occur with such severity in a machine tool installation. In fact, predictions based on a theory which assumes uniform eccentricity along the length of the journal were found to be satisfactory when the effective eccentricity of the deformed shaft was taken to be that at the mid-plane of the bearing.

The increasing use of hydrostatic journal bearings in machine tools and other types of machinery is likely to result in situations where significant shaft bending occurs and the results presented here show that its effects are not as disastrous as would perhaps be expected.

2. NOMENCLATURE

The following symbols are broadly the same as those in ref. 1.

D shaft diameter

l_p length of recess

l length of axial lands

l_c length of circumferential lands

c radial clearance between shaft and journal when eccentricity is zero

d_o orifice diameter

C_d discharge coefficient of orifice

p_s supply pressure

ρ density of fluid

μ dynamic viscosity of fluid

N speed of shaft, revolutions per unit time

W radial load carried by bearing

e eccentricity ratio of shaft

α attitude angle of shaft, measured from line of loading

r zero-eccentricity pressure ratio

$$m = \frac{l_p l}{D l_c}, \text{ aspect ratio of bearing}$$

$$\lambda = \left(\frac{C_d\, d_o\, l\mu}{c^3\, D}\right)^2 \frac{72\pi^2}{\rho\, p_s}, \text{ dimensionless design variable}$$

$$\omega = \frac{24\pi\, N\, l\mu}{c^2\, p_s}, \text{ dimensionless speed variable}$$

$$\bar{W} = \frac{W}{D\,(l_p + l)\, p_s}, \text{ dimensionless load variable}$$

$\bar{W}_i,$ an increment in $\bar{W}$

The following symbols are used in the Appendix:

a, f suffixes referring to the aft and fore ends of the journal bearing respectively (see section 3)

d, s shaft displacements measured by transducers in the vertical and horizontal directions respectively

v, h shaft displacements within the journal bearing in the vertical and horizontal directions respectively

w eccentricity of shaft

W_e load applied to free end of shaft

Symbols without suffixes refer to mid-plane of bearing.

3. DESCRIPTION OF TEST ASSEMBLY

(a) *The Hydrostatic Journal Bearing*

The salient dimensions of the journal used in the tests are shown in Fig. 1 and described in the nomenclature of ref. 1. The radial clearance between the shaft and journal, c, was determined by conducting flow tests at zero eccentricity and found to be $3 \cdot 3 \times 10^{-3}$ in. Similarly, the four orifices used were simultaneously calibrated and their mean coefficient of discharge found to be $0 \cdot 69$ for a bore diameter of $0 \cdot 040$ in. The aspect ratio of the bearing, m, as defined in ref. 1 was $1 \cdot 39$.

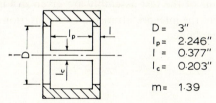

$$D = 3''$$
$$l_p = 2 \cdot 246''$$
$$l = 0 \cdot 377''$$
$$l_c = 0 \cdot 203''$$

$$m = 1 \cdot 39$$

FIG. 1. Details of test bearing.

(b) *Shaft and Journal Assembly*

The tests described in this paper were conducted on the assembly shown schematically in Fig. 2. This consisted of a solid steel shaft A of approximately 3 in diameter supported at one end by a self-aligning double row ball bearing set in a housing B which was mounted on a baseplate C. The hydrostatic journal bearing D was machined from a block of mild steel roughly cubical in shape and positioned on the baseplate near the free end of the shaft. A floating housing E was provided with a pair of preloaded ball bearings and installed on

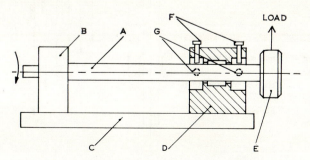

FIG. 2. General arrangement of test assembly.

the shaft so that vertical radial loads could be applied to the free end of the shaft by a screwjack through a calibrated spring. Provision was made in the housing B for adjustment of the position of the centreline of the shaft relative to that of the journal D to be made.

The journal block D included moutings in which inductive displacement transducers F and G were held to measure movements of the shaft relative to the journal in the vertical and horizontal directions respectively. Brass rings were shrunk onto the shaft in the planes of the transducers to improve their performance. The end of the journal nearest the loading device was referred to as the fore end of the bearing and the opposite end as the aft end. The journal bearing itself was orientated so that the direction of loading corresponded to that defined as type A in ref. 1, i.e. into a recess.

A variable speed hydrostatic transmission driven by an electric motor was coupled to the end of the shaft supported in housing B. This enabled the shaft to be revolved at any speed between zero and 3000 rpm and the coupling did not exert any bending restraint on the shaft. The oil used to operate the journal bearing was an industrial hydraulic oil with a kinematic viscosity of 23 and 15·5 cS at 20° and 30°C respectively. The oil was supplied by a constant displacement vane pump at pressures up to 200 psi through a combined pressure regulating and bypass valve to a manifold which distributed the fluid to the four orifices within the bearing. A flow meter enabled the total flow rate of oil entering the bearing to be measured and various pressure gauges and thermocouples were provided to ensure proper operation of the assembly.

The important dimensions of the test assembly are given in Fig. 3 and are referred to in detail in the Appendix where the interpretation and compensation of the transducer readings for shaft bending is fully explained.

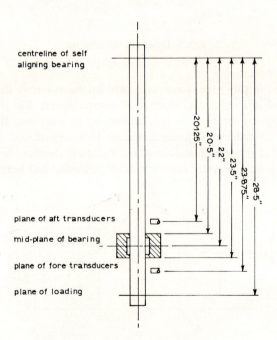

FIG. 3. Dimensions of test assembly.

4. DESCRIPTION OF TESTS

As described in ref. 1, the operating characteristics of a given bearing of the type under discussion are determined by two dimensionless parameters, a design variable λ and a speed variable ω. The first of these has been termed a design variable as it is a function of the dimensions of the bearing, the properties of the fluid being used, and details of the compensating elements employed. However, this parameter is directly related to a variable which has a clear physical meaning, namely the ratio between the pressure in the recesses of the bearing and the supply pressure when the shaft is concentric within the journal. This ratio will be referred to as the zero eccentricity pressure ratio or simply the pressure ratio, r, of the bearing and recourse to the theory of ref. 1 reveals that it is related to the design variable λ for a given bearing with four recesses by the following expression:

$$\frac{r}{\sqrt{(1-r)}} = \frac{\sqrt{\lambda}}{\pi}$$

Because of its readily measurable form and easily appreciated physical meaning the variable r was chosen to classify the tests conducted and results are presented for discrete values of $r = 0{\cdot}4$, $0{\cdot}5$, $0{\cdot}6$, and $0{\cdot}7$. No such clear physical meaning could be attached to the speed variable ω so shaft speeds of 0, 500, 1500, and 3000 rpm were selected after some experience as a revealing range of conditions could then be investigated within the capabilities of the apparatus.

The radial load, W, applied to the bearing was expressed non-dimensionally as

$$W = \frac{W}{D\,(l_p + l)\,p_s}$$

where p_s was the current supply pressure and the other variables are defined in Fig. 1. Now, in order to obtain constant increments in this dimensionless variable for constant increments in applied load it was obviously necessary to conduct all tests at a unique supply pressure for each pressure ratio studied. Because orifices were used as compensating elements the design variable λ, and hence the pressure ratio r, were dependent on the viscosity of the oil which was, of course, determined by the temperature of the oil. Consequently, for each pressure ratio there was a unique oil temperature that ensured these desirable conditions and the variations in oil temperature that were tolerated in practice resulted in the pressure ratio actually used deviating by less than $\pm 10\%$ of their nominal values. Thus, the results presented here were obtained at supply pressures which were constant for each of the four pressure ratios studied and are noted alongside the graphical presentations described in section 5.

Considerable amounts of experimental data had to be recorded and processed to determine the deflected form of the shaft within the bearing; this work was facilitated by use of a data logging system and a digital computer.

5. PRESENTATION OF EXPERIMENTAL RESULTS

The results of the tests described in the previous section are presented graphically in two sets of curves (Figs. 4 and 5). Those in Fig. 4 show the experimentally measured movement of the shaft at the mid-plane of the journal in terms of eccentricity ratio, e, and attitude angle, α, measured from the line of loading for the four pressure ratios and shaft speeds selected for investigation. For example, Fig. 4a refers to tests completed at a nominal

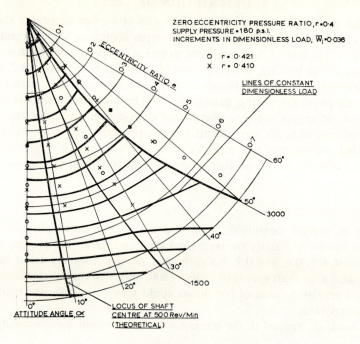

Fig. 4a

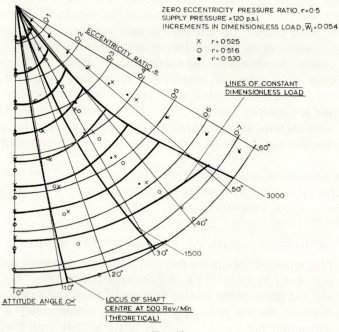

Fig. 4b

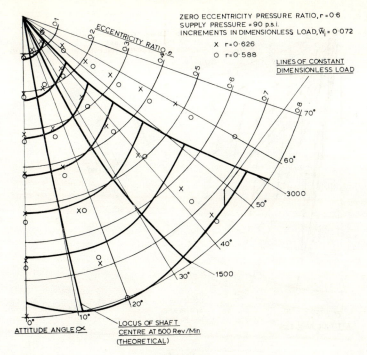

FIG. 4c

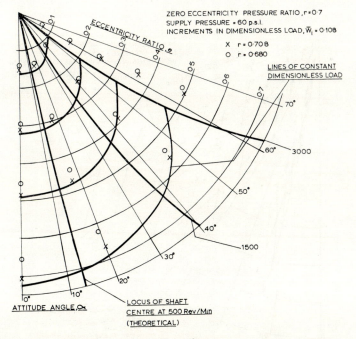

FIG. 4d

FIG. 4. Experimentally measured shaft displacement at mid-plane of bearing compared with that predicted theroretically.

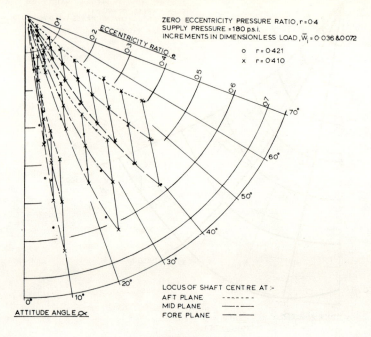

FIG. 5a

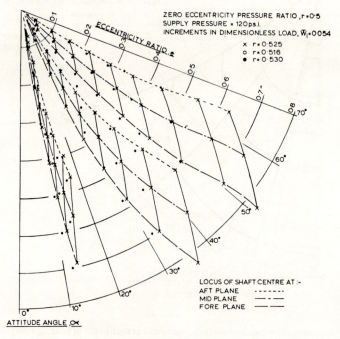

FIG. 5b

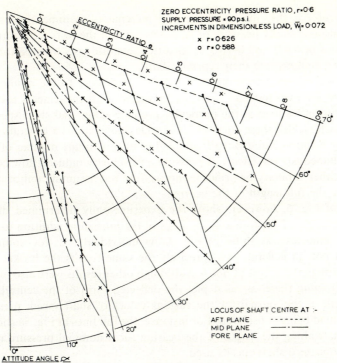

FIG. 5c

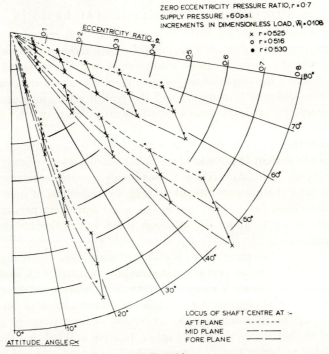

FIG. 5d

FIG. 5. Experimentally measured shaft displacement along length of bearing.

pressure ratio of $r = 0.4$, the corresponding increments in dimensionless load W_i was 0·036 for a constant supply pressure of $p_s = 180$ psi and the tests were actually performed at pressure ratios of 0·421 and 0·410. At least two separate tests were performed at each set of conditions to demonstrate their reproducibility.

The theoretical curves drawn on these figures correspond to the nominal experimental conditions used for each test and show the predicted shaft movements for constant increments in dimensionless load at the particular pressure ratios and shaft speeds in question. The theory which yielded these curves is due to Davies[13] and is a generalized development of that used by Raimondi and Boyd[1] adapted to deal with any number of recesses and to consider simultaneously the effects of pressure and velocity induced flow between adjacent recesses with either capillary or orifice compensation but assuming uniform eccentricity of the shaft along the full length of the bearing.

The curves of Fig. 5, however, show the experimentally determined displacements of the shaft at the edges of the fore and aft lands of the journal in addition to that at its mid-plane for the same tests as those of Fig. 4. As individual tests for nominally the same conditions did not, in general, yield precisely the same results the broken lines of Fig. 5 were drawn through the points of one test although other results were also plotted as shown. The full lines joining these curves depict the deflected form of the centreline of the shaft along the length of the journal for the various values of shaft speed, pressure ratio, and dimensionless load being considered. For instance, the full lines in Fig. 5a show the deflected form of the shaft along the length of the bearing for a nominal pressure ratio of $r = 0.4$ and the three non-zero shaft speeds used.

6. DISCUSSION OF EXPERIMENTAL RESULTS

Study of Figs. 4 and 5 indicate that the tests were acceptably reproducible in spite of the fact that the zero eccentricity pressure ratios used for each series of tests did not, in general, coincide exactly with the nominal values. Furthermore, the determination of the deflected form of the shaft along the length of the bearing involved calculations based on readings obtained from four different transducers and the typical errors in eccentricity ratio for similar tests were less than about 0·01 for eccentricity ratios less than 0·5, i.e. less than about 3×10^{-5} in.

Consideration of the curves drawn in Fig. 4 reveals that for any given load the measured eccentricity ratio of the shaft at the mid-plane of the bearing was predicted with reasonable accuracy by the theory outlined in section 5. This is only true for eccentricity ratios of up to about 0·5 for the tests at a pressure ratio of 0·4 (Fig. 4a) regardless of shaft speed. For tests at higher pressure ratios, say $r = 0.7$ in Fig. 4d, good agreement persisted for values of eccentricity ratios of up to about 0·7 although there was noticeable divergence at the higher shaft speeds.

A prominent feature of all these results is that the theory consistently underestimated the attitude angle of the shaft by an amount that generally increased with shaft speed.

The deflected shapes of the shaft along the length of the bearing for these tests are shown in Fig. 5 and as intimated considerable variation of eccentricity ratio occurred along the length of the bearing. In the worst case this amounted to approximately $\pm 30\%$ of that at the mid-plane of the bearing as shown in Fig. 4a. Bearing in mind that the lower pressure ratios corresponded to higher supply pressures the actual load required to attain a given eccentricity ratio at the mid-plane of the bearing increased as the pressure ratio decreased.

Consequently, greater bending deflections of the shaft occurred at the lower pressure ratios as can be seen by comparing Fig. 5a with Fig. 5d. This is consistent with there being better agreement between the measured and predicted performance of the bearing for higher values of eccentricity ratio at the higher pressure ratios than at the lower pressure ratios as proportionally less shaft bending occurred during the former.

Another feature of interest was that the bearing became increasingly less stiff with increasing shaft speed at low pressure ratios (Fig. 4a) while at high pressure ratios an increase in stiffness occurred as shaft speed increased (Fig. 4d). This tendency is predicted theoretically and a more detailed discussion of this phenomenon is given in ref. 13.

To emphasize the fact that good prediction of eccentricity ratio was provided by the theory the curves of Figs. 6 and 7 were drawn to show the variation of eccentricity ratio with dimensionless load at the three planes of interest regardless of attitude angle for the extreme pressure ratios of 0·4 and 0·7. On this basis it is seen that the predicted eccentricity ratio at the mid-plane of the bearing corresponded favourably with that measured experimentally. For a pressure ratio of 0·4 (Fig. 6) shaft speed appeared to have little effect on this correlation in spite of the large bending deflections of the shaft. The effect of shaft speed was more marked for the pressure ratio of 0·7 (Fig. 7) as the errors increased with increasing speed although smaller bending deflections existed in this case. In fact, Figs. 7c

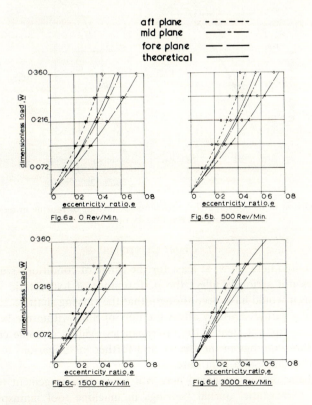

FIG. 6. Comparison of theoretical and experimental behaviour of bearing in terms of dimensionless load and eccentricity ratio for $r = 0·4$.

and 7d show considerable departures at eccentricity ratios greater than about 0·6 while the results of Fig. 7a show good correlation up to nearly 0·8.

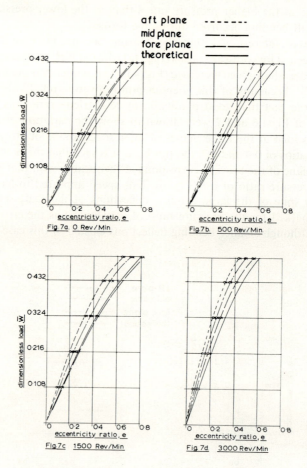

FIG. 7. Comparison of theoretical and experimental behaviour of bearing in terms of dimensionless load and eccentricity ratio for $r = 0·7$.

7. CONCLUSIONS

The results presented here show, at least for the range of conditions studied, that severe shaft bending does not seriously affect the performance of hydrostatic journal bearings. This is surprising as it would have been thought that the varying shaft eccentricity along the length of the bearing would seriously affect the flow of fluid within the bearing and hence its performance. It is not yet clear if the discrepancies evident in some cases were due to shaft bending or shortcomings in the theory and further investigations are in progress. It is hoped to report on these at a later date.

It is hoped that the results presented here will be an encouragement to those who are considering using hydrostatic journal bearings in machine tool mainspindle assemblies where such severe shaft bending is unlikely to occur and eccentricity ratios are usually limited.

8. ACKNOWLEDGEMENTS

The authors wish to express their gratitude to Professor A. W. J. Chisholm of the Department of Mechanical Engineering, University of Salford, for introducing them to the study of hydrostatic bearings in machine tools and for his continuing help and encouragement during the course of their work. Thanks are also extended to Mr. B. J. Crowe, who was largely responsible for the manufacture and instrumentrtion of the experimental apparatus.

REFERENCES

1. A. A. RAIMONDI and J. BOYD, An Analysis of Orifice and Capillary-compensated Hydrostatic Journal Bearings. ASME–ALSE Joint Conference on Lubrication, Baltimore, October 1954.
2. H. MORI and H. YABE, A Theroetical Investigation on Hydrostatic Bearings, *Bull. JSME*, **6**, No. 22, 1963, 354–363.
3. J. N. SHINKLE and K. G. HORNUNG, Frictional Characteristics of Liquid Hydrostatic Journal Bearings. *ASME Trans.*, Series D, *J. Basic Eng.* **87**, No. 1, 1965, 163–169.
4. A. K. KHER and A. COWLEY, The Design and Performance Characteristics of a Capillary Compensated Hydrostatic Journal Bearing, *Proc. 8th M.T.D.R. Conf.*, *University of Manchester*, 12–15 September 1967.
5. H. C. RIPPEL, Design of Hydrostatic Journal Bearings—Part 10—Multirecess Journal Bearings, *Machine Design*, **35**, 5 Dec. 1963, pp. 158–162.
6. F. M. STANDSFIELD, The Design of Hydrostatic Journal Bearings, *Proc. 8th M.T.D.R. Conf.*, *University of Manchester*, 12–15 September 1967.
7. H. C. RIPPEL, Review of Hydrostatic Lubrication. Franklin Institute Sponsored Seminar on Computer Aided Design of Hydrostatic Bearing Systems, London 1967.
8. R. S. HAHN, Some Advantages of Hydrostatics in Machine Tools. ASLE Annual Meeting, Chicago, 26–28 May 1964.
9. G. N. LEVESQUE, Error Correcting Action of Hydrostatic Bearings. ASME Lubrication Symposium, New York, 6–9 June 1965.
10. H. C. RIPPEL, Role of Hydrostatic Bearings in Manufacture. Horizons in Manufacturing Seminar, Institute of Science and Technology, University of Michigan, Ann Arbor, 9 May 1967.
11. W. B. ROWE, Experience with Four Types of Grinding Machine Spindles, *Proc. 8th M.T.D.R. Conf.*, *University of Manchester*, 12–15 September 1967.
12. H. OPITZ, Pressure Pad Bearings. Conf. on Lubrication and Wear, Inst. Mech. Engs. Session 2, Fluid Film Lubrication, paper 8, London, September 1967.
13. P. B. DAVIES, A General Analysis of Multirecess Hydrostatic Journal Bearings. University of Salford, Department of Mechanical Engineering, to be published.

APPENDIX

Determination of Shaft Eccentricity in the Presence of Shaft Bending

The important dimensions which describe the test assembly are given in Fig. 3, in this all distances are measured from the centre of the self-aligning ball bearing referred to in section 3b. Because this bearing had a radial stiffness considerably higher than that of the hydrostatic journal bearing it was assumed that it effectively pinned the end of the shaft relative to the baseplate as far as transverse displacements were concerned. The locations of the displacement transducers, bearing centre, bearing lands, and line of action of the applied

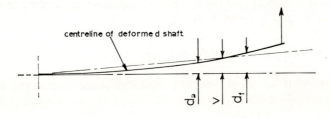

FIG. A1. Deflected form of shaft.

radial load were required to enable corrections to be made to the transducer readings in order to determine the eccentricity of the shaft at the planes of interest along the length of the bearing.

Clearly, the radial load applied to the bearing under test was simply 28·5/22 times the load at the free end of the shaft, i.e. $W = 1·2954\ W_e$.

The deflections of the shaft measured by the displacement transducers in the fore and aft positions included the effects of bending and rigid body rotation of the shaft as indicated in Fig. A1.

The apparent deflection of the shaft relative to the journal measured by the aft vertical transducer, d_a, was clearly decreased while that measured by the fore vertical transducer, d_f, was increased. By use of geometry and elementary bending theory, assuming Young's Modulus for the steel shaft to be 30×10^6 psi, it is easily shown that the actual deflection at the centre of the bearing, v, can be found as:

$$v = \frac{d_a}{0·9149} + \frac{W_e}{900} \text{ , from the aft transducer}$$

and

$$v = \frac{d_f}{1·0853} - \frac{W_e}{849} \text{ , from the fore transducer}$$

where the deflections are measured in thousandths of an inch and the load, W_e, in pounds weight.

Similar considerations show that the vertical deflection at the outside edge of the aft and fore lands, v_a and v_f respectively, are given by:

$$v_a = \frac{v}{1·075} - \frac{W_e}{1196·4}$$

and

$$v_f = \frac{v}{0·9418} + \frac{W_e}{993·8}$$

Because no horizontal force was applied to the shaft it was unnecessary to compensate for bending in this direction.

Thus, the horizontal displacement of the shaft at the mid-plane of the bearing, h, was related to the displacements recorded by the aft and fore horizontal transducers, s_a and s_f respectively, by the following expressions.

$$h = \frac{s_a}{0·9149} \text{ , from the aft transducer}$$

and

$$h = \frac{s_f}{1·0853} \text{ , from the fore transducer}$$

From the foregoing it is clear that the horizontal displacements at the outside edges of the aft and fore lands, h_a and h_f respectively, are:

$$h_a = \frac{h}{1·075} \quad \text{and } h_f = \frac{h}{0·9418}$$

Now, from Fig. A2 it is evident that the eccentricity of the shaft at the mid-plane of the bearing, w, is:

$$w = \sqrt{(v^2 + h^2)}$$

and the attitude angle, a, in this location is

$$a = \tan^{-1}\left(\frac{h}{v}\right)$$

Similar expressions can be derived for the other eccentricities and attitude angles (Fig. A2).

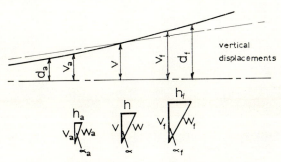

FIG. A2. Deformed shape of shaft along bearing.

In practice, the displacements found separately from the fore and aft transducers were averaged to give a mean displacement at the mid-plane of the bearing in both the vertical and horizontal directions. These mean values were then used to calculate the displacements at the lands. The eccentricities were converted into eccentricity ratios by simply dividing them by c, the radial clearance.

EFFECT OF PAD GEOMETRY, SURFACE FINISH AND HYDRODYNAMIC ACTION ON THE PERFORMANCE OF SLIDING EXTERNALLY PRESSURIZED RECTANGULAR BEARING

M. A. EL. KADEEM, E. A. SALEM and H. R. EL SAYED

Production and Mechanical Engineering Departments, Alexandria University

SUMMARY

This paper supplements the authors' paper "Effect of pad geometry and surface finish on the performance of a stationary externally pressurized rectangular bearing", in which they presented an analysis of a stationary rectangular, externally pressurized bearing which was lubricated by air incompressible fluid, and compensated by a capillary restrictor. Both papers are based on the same assumptions.

The present study explains the effect of pad geometry, initial pad tilt, and bearing surface finish on the performance of a sliding rectangular bearing.

INTRODUCTION

Numerous applications of externally pressurized slider bearings have been suggested. The main advantages of such bearings are the low frictional characteristics and the elimination of any stick between the bearing parts, and hence reduction of wear.

Too little attention is paid to finding out an analytical solution of such a bearing and to reveal its features. However, Scales[1] had carried out some comparative cutting tests on a horizontal milling machine to study the effect of externally pressurizing the pads of the milling machine table. He did not give any design data. Loxham,[2] though he considered the application of hydrostatic bearings to machine tools, did not mention the effect of the bearing velocity. Harkanyi[4] did not consider the effect of the bearing velocity in his analysis of the hydrostatic slideways. On the other hand, many analytical solutions to the problems of the application of the hydrodynamic theory of lubrication to plane or tapered sliders are published.[3, 5, 6]

It is the purpose of this paper to discuss the effect of the hydrodynamic action on the performance of externally pressurized sliding pad bearings, and to present an investigation into the effect of the bearing surface finish and pad geometry on the performance of the bearing.

1.1. *The Equation Governing the Flow*

If the inertia and body forces of the incompressible lubricant are neglected, and steady state conditions prevail, Navier-Stoke's equation in the x direction[7] reduces to

$$\frac{d^2u}{dz^2} = \frac{1}{\mu}\frac{dP}{dx} \tag{1.1}$$

1.2. The Bearing Characteristic Number

Integrating eq. (1.1) twice, and substituting the following boundary conditions,

$$u = 0 \quad \text{at } z = 0 \quad \text{and } u = U \quad \text{at } z = h$$

therefore

$$u = \frac{1}{2\mu} \frac{dP}{dx} [z^2 - zh] + \frac{U}{h} z \tag{1.2}$$

and the volume rate of flow is given by

$$Q = \frac{lhU}{2} - \frac{lh^3}{12\mu} \frac{dP}{dx} \tag{1.3}$$

where l is the bearing length,
       h the film thickness,
       U the sliding velocity, and
       μ the viscosity of the lubricant.

From eq. (1.3), it can be seen that the volume rate of flow in the bearing consists of two parts.

1. The first is due to the sliding velocity U.
2. The second is due to the recess inlet pressure P_i.

Consequently, a slider bearing should be treated as consisting of two main regions. The first is that in which $B_1 \leqslant x \leqslant B_2$, and the second is that in which $- B_1 \geqslant - x \geqslant - B_2$. In each region, the volume rate of flow also consists of the two previously mentioned parts. In the first region the two parts are added, while in the second region they are subtracted.

Since the flow in region 2 is the difference between the two parts of the volume rate of flow, the flow in that region will vanish under certain circumstances.

Under such prescribed circumstances, the bearing will fail, because when no flow exists in region 2, the pressure under this region will be atmospheric, and free surface may be formed, consequently this region will have no load-carrying capacity.

Therefore the limiting working conditions of a sliding bearing are those which equalize the two parts of the volume rate of flow in the region $- B_1 \geqslant - x \geqslant - B_2$.

These conditions can be formulated in a dimensionless form to give a performance indicating number, referred to as the bearing characteristic number and denoted by H_n where

$$H_n = \frac{Pi\, h^2}{6\mu U B_2} \tag{1.4}$$

and this will be discussed later on. Equation (1.4) is shown in Fig. 6.

DEVIATION OF THE BASIC THEORETICAL EQUATIONS

1.3. Case I. Rectangular Pad of Constant Film Thickness

It is well known that the hydrodynamic effects appear only when a slight inclination exists between the related moving bearing parts. In this section a bearing with a constant film thickness is considered and constant pressure is assumed in the bearing recess. Consequently no hydrodynamic action will occur and the pressure distribution along the lubricant film will be given by

$$Px_1 = \frac{P_1}{(B_2 - B_1)} \{B_2 - x\}^{(8)} \qquad \text{for the region} \quad B_1 \leqslant x \leqslant B_2 \qquad (1.5)$$

$$Px_2 = \frac{P_1}{(B_2 - B_1)} \{B_2 + x\}^{(8)} \qquad \text{for the region} - B_1 \geqslant - x \geqslant - B_2 \quad (1.6)$$

Equations (1.5) and (1.6) are valid only so far as the bearing characteristic number H_n is considered.

Figure 1 shows the bearing arrangement.

The pad is sliding with a constant velocity U, while the bed is stationary. The lubricant is supplied through a supply hole located at the centre of the pad. The pressure in the recess is assumed to be constant and equal to the inlet pressure P_i.

1.3.1. *Total Volume Rate of Flow*. The volume rate of flow at any section is defined by eq. (1.3). The total volume rate of flow is therefore

$$Q_t = Q_1 + (- Q_2)$$

where $\quad Q_1 =$ flow in region 1 $\qquad B_1 \leqslant x \leqslant B_2$

$\qquad\quad Q_2 =$ flow in region 2 $\qquad\quad - B_1 \geqslant - x \geqslant - B_2$

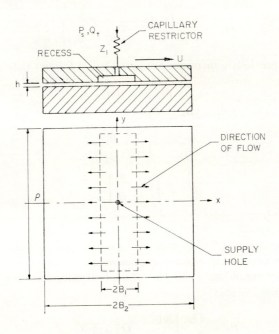

FIG. 1. Plain rectangular bearing.

Integrating eq. (1.3) with respect to x and introducing the following boundary conditions in the above equation:

$$\text{at } X = B_1 \quad P_x = P_i \quad \text{and} \quad \text{at } X = B_2 \quad P_x = 0$$

$$\text{at } X = -B_1 \quad P_x = P_i \quad \text{and} \quad \text{at } X = -B_2 \quad P_x = 0$$

Therefore

$$Q_1 = \frac{lh^3}{12\mu} \left\{ \frac{6\mu U}{h^2} + \frac{P_i}{(B_2 - B_1)} \right\} \qquad B_1 \leqslant x \leqslant B_2 \tag{1.7}$$

$$Q_2 = \frac{lh^3}{12\mu} \left\{ \frac{6\mu U}{h^2} - \frac{P_i}{(B_2 - B_1)} \right\} \qquad -B_1 \geqslant -x \geqslant -B_2 \tag{1.8}$$

and hence

$$Q_t = \frac{lh^3 P_i}{6\mu (B_2 - B_1)} \tag{1.9}$$

1.3.2. *The Dimensionless Group.*[8] From eqs. (1.5), (1.6) and (1.9), it can be seen that P_{x1}, P_{x2} and Q_t are independent of the velocity U, and are the same as those of stationary bearing.[8]

Hence, it can be stated that so far as H_n is considered, the performance of the sliding pad bearing of constant film thickness is the same as that of the stationary bearing.[8] Therefore the dimensionless groups for the plain rectangular sliding pad bearing are also the same as those of the stationary bearing.[8]

1.3.3. *The Bearing Characteristic Number.* From eq. (1.8), the volume rate of flow in region 2 is given by

$$Q_2 = \frac{lh^3}{12\mu} \left[\frac{6\mu U}{h^2} - \frac{P_i}{(B_2 - B_1)} \right]$$

According to the previously mentioned suggestions, it follows that

$$\frac{P_i}{(B_2 - B_1)} \geqslant \frac{6\mu U}{h^2}$$

Therefore

$$\frac{P_i h^2}{6\mu U B_2} \geqslant \left(1 - \frac{B_1}{B_2} \right)$$

$$H_n \geqslant \left[1 - \frac{B_1}{B_2} \right] \tag{1.10}$$

this is shown in Fig. 7.

1.4. *Case Rectangular Stepped Pad Bearing with Constant Film Thickness*

Figure 2 shows the considered bearing arrangement.

1.4.1. *Pressure Distribution and Total Flow Rate.* Following the same procedure explained in section (1.3.1), it follows that

$$P_x = \frac{12\mu}{lh^3} \left[\frac{lhU}{2} - Q \right] x + C$$

Where C is a constant of integration the following can be determined from the boundary conditions:

(1) For the region $\qquad B_1 \leqslant x \leqslant B_3$

$$\text{at } X = B_1 \qquad P_x = P_i \text{ and } h_x = h_2$$

$$\text{at } X = B_3 \qquad P_x = P_3 \text{ and } h_x = h_2$$

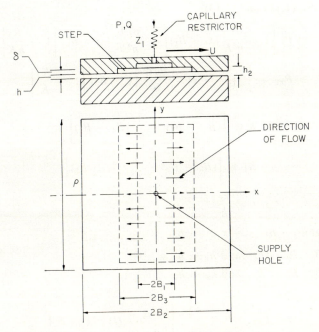

FIG. 2. Stepped rectangular bearing.

Therefore

$$P_{x1} = P_i + \frac{12\mu}{lh_2{}^3} \left[\frac{lh_2 U}{2} - Q_1 \right] [x - B_1] \qquad (1.11)$$

$$P_3 = P_i + \frac{12\mu}{lh_2{}^3} \left[\frac{lh_2 U}{2} - Q_1 \right] [B_3 - B_1] \qquad (1.12)$$

(2) For the region $\qquad B_3 \leqslant x \leqslant B_2$

$$\text{at } X = B_3 \qquad P_x = P_3 \text{ and } h_x = h$$

$$\text{at } X = B_2 \qquad P_x = 0 \text{ and } h_x = h$$

Therefore

$$P_{x2} = \frac{12\mu}{lh^3} \left[\frac{lhU}{2} - Q_1 \right] [x - B_2] \qquad (1.13)$$

$$P_3 = \frac{12\mu}{lh^3} \left[\frac{lhU}{2} - Q_1 \right] [B_3 - B_2] \qquad (1.14)$$

Due to the continuity of the flow and from eqs. (1.12) and (1.14), we get

$$P_i = \frac{12\mu}{lh^3} K Q_1 + \frac{6\mu U}{h^2} K_1 \tag{1.15}$$

$$Q_1 = \frac{lh^3}{12\mu K} \left[P_i - \frac{6\mu K_1}{h^2} U \right] \tag{1.16}$$

From eqs. (1.16), (1.11) and (1.13),

$$P_{x1} = P_i \left[1 - \frac{1}{n^3 K} (x - B_1) \right] + \frac{6\mu U}{n^2 h^2} [x - B_1] \left[1 + \frac{K_1}{nK} \right] \tag{1.17}$$

$$P_{x2} - \frac{P_i}{K} [B_2 - x] + \frac{6\mu U}{h^2} [x - B_2] \left[1 + \frac{K_1}{K} \right] \tag{1.18}$$

Where

$$K = \left[(B_2 - B_3) + \frac{1}{n^3} (B_3 - B_1) \right]$$

$$K_1 = \left[(B_3 - B_2) - \frac{1}{n^2} (B_3 - B_1) \right]$$

$$n = \frac{h_2}{h}$$

(3) For the region $- B_1 \geqslant - x \geqslant - B_3$

at $X = - B_1$ $P_x = P_i$ and $h_x = h_2$

at $X = - B_3$ $P_x = P_3$ and $h_x = h_2$

Therefore

$$P_{x3} = P_i + \frac{12\mu}{lh_2{}^3} \left[\frac{lU h_2}{2} - Q_2 \right] [x + B_1] \tag{1.19}$$

$$P_3 = P_i + \frac{12\mu}{lh_2{}^3} \left[\frac{lU h_2}{2} - Q_2 \right] [B_1 - B_3] \tag{1.20}$$

(4) For the region $- B_3 \geqslant - x \geqslant - B_2$

at $X = - B_3$ $P_x = P_3$ and $h_x = h$

at $X \equiv - B_2$ $P_x = 0$ and $h_x = h$

Therefore

$$P_{x4} = \frac{12\mu}{lh^3} \left[\frac{lhU}{2} - Q_2 \right] [x + B_2] \tag{1.21}$$

$$P_3 = \frac{12\mu}{lh^3} \left[\frac{lhU}{2} - Q_2 \right] [B_2 - B_3] \tag{1.22}$$

Due to the continuity of the flow, and from eqs. (1.20) and (1.22) we get

$$P_i = - \frac{12\mu K}{lh^3} Q_2 - \frac{6\mu U}{h^2} K_1 \tag{1.23}$$

$$Q_2 = - \frac{lh^3}{12\mu K} \left[P_i + \frac{6\mu K_1}{h^2} U \right] \tag{1.24}$$

from eqs. (1.24), (1.19) and (1.21),

$$P_{x3} = P_i \left[1 + \frac{1}{n^3 K} (B_1 + x) \right] + \frac{6\mu U}{n^2 h^2} (B_1 + x) \left[1 + \frac{K_1}{hK} \right] \qquad (1.25)$$

$$P_{x4} = \frac{P_i}{K} [B_2 + x] + \frac{6\mu U}{h^2} \left(1 + \frac{K_1}{K} \right) [B_2 + x] \qquad (1.26)$$

From eqs. (1.16) and (1.24),

$$Q_t = \frac{l h^3}{6\mu K} P_i \qquad (1.27)$$

Figure 8 shows the pressure distribution along the fluid film for this bearing. From this figure it can be seen that the bearing has an asymmetrical pressure distribution about its centre. The pad will therefore be acted upon by a turning moment, whose magnitude depends on B_3, U and n.

To reduce the effect of the turning moment, a further condition is introduced into eq. (1.17). This is:

$$\text{at } x = \frac{B_3}{B_2} \qquad P_x = P_i$$

Therefore

$$n = 1 + \frac{Hn}{(1 - B_3/B_2)} \qquad (1.28)$$

Equation (1.28) gives the value of $n = h_2/h$, corresponding to a certain value of H_n, which improves the pressure distribution along the fluid film in the bearing (Fig. 9).

1.4.2. *The Total Load-carrying Capacity.* The total load-carrying capacity of the bearing at a given inlet pressure P_i can be determined by integrating the pressure of each region over the area of that region. Therefore

$$L = 2P_i B_1 l + \int_{B_1}^{B_3} P_{x1} l dx + \int_{B_3}^{B_2} P_{x2} l dx + \int_{-B_1}^{-B_3} P_{x3} (- l dx) + \int_{-B_3}^{-B_2} P_{x4} (- l dx) = P_i AG \quad (1.29)$$

where G is a shape factor

$$\frac{1}{B_2} \left[B_3 - \frac{1}{2n^3 K} (B_3 - B_1)^2 + \frac{1}{2K} (B_2 - B_3)^2 \right]$$

and A is the projected area of the pad.

1.4.3. *The Dimensionless Groups.*[8] From eqs. (1.27) and (1.29) it can be seen that Q_t and L are independent of the velocity U, and are the same as those for stationary bearing.[8]

Hence it can be stated, so far as H_0 is considered, that the performance of the stepped sliding pad bearing of constant film thickness is the same as that of the stationary bearing.[8]

Therefore the dimensionless groups for the stepped rectangular sliding pad bearing are also the same.[8]

1.4.4. *The Bearing Characteristic Number*. From eq. (1.24), the volume rate of flow in the region $-B_1 \geqslant -x \geqslant -B_2$ is

$$Q_2 = -\frac{lh^3}{12\mu K}\left[P_i + \frac{6\mu K_1}{h^2}U\right]$$

Therefore

$$P_i \geqslant -\frac{6\mu K_1}{h^2}U$$

and hence

$$H_n \geqslant \left[\left(1 - \frac{B_3}{B_2}\right) + \frac{1}{n^3}\left(\frac{B_3}{B_2} - \frac{B_1}{B_2}\right)\right] \qquad (1.30)$$

This is shown in Fig. 7.

1.5. *Case* 3. *Initially Tilted Rectangular Pad Bearing*

Figure 3 shows the considered bearing arrangement.

1.5.1. *Pressure Distribution and Total Flow Rate*. Following the same procedure explained in section 1.31,

$$P_x = \frac{-6\mu U}{k(h+kx)} + \frac{6\mu Q}{lk(h+kx)^2} + C$$

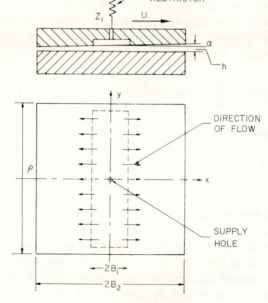

FIG. 3. Tilted rectangular pad bearing.

where C is a constant of integration, the following can be determined from the boundary conditions.

(1) For the region $B_1 \leqslant x \leqslant B_2$

at $x = B_1$ $P_x = P_i$
at $x = B_2$ $P_x = 0$

Therefore

$$P_{x1} = \frac{6\mu U}{k} \left[\frac{1}{(h + kB_2)} - \frac{1}{(h + kx)} \right] + \frac{6\mu Q_1}{lk} \left[\frac{1}{(h + kx)^2} - \frac{1}{(h + kB_2)^2} \right] \tag{1.31}$$

$$P_i = \frac{6\mu U}{k} D_3 + \frac{6\mu Q_1}{lk} D_1 \tag{1.32}$$

Therefore

$$Q_1 = \frac{lk}{6\mu D_1} \left[P_i - \frac{6\mu U}{k} D_3 \right] \tag{1.33}$$

from eqs. (1.31) and (1.33)

$$P_{x1} = \frac{P_i}{D_1} \left[\frac{1}{(h + kx)^2} - \frac{1}{(h + kB_2)^2} \right] + \frac{6\mu U}{k} \left[\left(\frac{1}{(h + kB_2)} - \frac{1}{(h + kx)} \right) \right.$$
$$\left. + \frac{1}{D_4} \left(\frac{1}{(h + kx)^2} - \frac{1}{(h + kB_2)^2} \right) \right] \tag{1.34}$$

where

$$D_1 = \left[\frac{1}{(h + kB_1)^2} - \frac{1}{(h + kB_2)^2} \right]$$

$$D_3 = \left[\frac{1}{(h + kB_2)} - \frac{1}{(h + kB_1)} \right]$$

$$D_4 = \left[\frac{1}{(h + kB_2)} + \frac{1}{(h + kB_1)} \right]$$

(2) For the region $- B_1 \geqslant - x \geqslant - B_2$

at $x = - B_1$ $P_x = P_i$
at $x = - B_2$ $P_x = 0$

Therefore

$$P_{x2} = \frac{6\mu U}{k} \left[\frac{1}{(h - kB_2)} - \frac{1}{(h + kx)} \right] + \frac{6\mu}{lk} Q_2 \left[\frac{1}{(h + kx)^2} - \frac{1}{(h - kB_2)^2} \right] \tag{1.35}$$

$$P_i = \frac{6\mu U}{k} D_5 + \frac{6\mu}{lk} Q_2 D_2 \tag{1.36}$$

$$Q_2 = \frac{lk}{6\mu D_2} \left[P_i - \frac{6\mu U}{k} D_5 \right] \tag{1.37}$$

From eqs. (1.35) and (1.37)

$$P_{x2} = \frac{P_i}{D_2} \left[\frac{1}{(h + kx)^2} - \frac{1}{(h - kB_2)^2} \right] + \frac{6\mu U}{k} \left[\left(\frac{1}{(h - kB_2)} - \frac{1}{(h + kx)} \right) \right.$$
$$\left. + \frac{1}{D_6} \left(\frac{1}{(h + kx)^2} - \frac{1}{(h - kB_2)^2} \right) \right] \qquad (1.38)$$

where

$$D_2 = \left[\frac{1}{(h - kB_1)^2} - \frac{1}{(h - kB_2)^2} \right]$$

$$D_5 = \left[\frac{1}{(h - kB_2)} - \frac{1}{(h - kB_1)} \right]$$

$$D_6 = \left[\frac{1}{(h - kB_2)} + \frac{1}{(h - kB_1)} \right]$$

From eqs. (1.33) and (1.37), we get

$$Q_t = \frac{P_i lk}{6\mu} D_7 + lUD_8 \qquad (1.39)$$

$$P_i = \frac{6\mu}{lkD_7} (Q_t - lUD_8) \qquad (1.40)$$

where

$$D_7 = \left[\frac{1}{D_1} - \frac{1}{D_2} \right]$$

$$D_8 = \left[\frac{1}{D_4} - \frac{1}{D_6} \right]$$

1.5.2. *The Total Load-carrying Capacity.* Integrating the pressure of each region over the area of that region, hence

$$L = 2P_i B_1 l + \int_{B_1}^{B_2} P_{x1} l dx + \int_{-B_1}^{-B_2} P_{x2} (- l dx)$$
$$= P_i AG + R$$

where G is a shape factor

$$G = \frac{1}{B_2} \left[B_1 + \frac{1}{2D_1} \left(\frac{(B_1 - B_2)}{(h + kB_2)^2} - \frac{D_3}{k} \right) - \frac{1}{2D_2} \left(\frac{(B_2 - B_1)}{(h - kB_2)^2} - \frac{D_5}{k} \right) \right] \qquad (1.41)$$

$$R = \frac{6\mu U l}{k} \left[(B_2 - B_1) D_9 - \frac{1}{k} \log \left(\frac{D_{10}}{D_{11}} \right) - \frac{D_3}{kD_4} + \frac{D_5}{kD_6} \right.$$
$$\left. - (B_2 - B_1) \left(\frac{1}{(h + kB_2)^2 D_4} + \frac{1}{(h - kB_2)^2 D_6} \right) \right] \qquad (1.42)$$

1.5.3. *The Stiffness.* When the bearing is compensated by a capillary restrictor of resistance Z_1, where

$$Z_1 = \frac{128\mu l}{\pi d^4}$$

and since

$$P_s - P_i = Z_1 Q_t$$

where P_s is the supply pressure, then, by substituting the values of Z_1 and Q_t, the pressure ratio (P_i/P_s) is given by

$$\frac{P_i}{P_s} = \frac{6\mu}{P_s} \left[\frac{(P_s - Z_1 l U D_8)}{(6\mu + Z_1 l k D_7)} \right] \tag{1.43}$$

Therefore, the stiffness of the bearing

$$\lambda = -6\mu A \frac{d}{dh} \left[G \cdot \frac{(P_s - Z_1 l U D_8)}{(6\mu + Z_1 l k D_7)} \right] - \frac{d}{dh} R \tag{1.44}$$

1.5.4. *The Power Consumption.* The power dissipated in the bearing E_b and the total power required to operate the bearing E_t are

$$E_b = \frac{lk P_i^2}{6\mu} D_7 + P_i l U D_8 \tag{1.45}$$

$$E_t = \frac{lk P_i P_s}{6\mu} D_7 + P_s l U D_8 \tag{1.46}$$

1.5.5. *The Dimensionless Groups.*[8] The dimensionless groups for this bearing are given by:
(1) Load per unit area

$$I = G + \frac{R}{P_i A} \tag{1.47}$$

(2) Load per unit flow

$$J = [P_i A G + R] \left[\frac{6}{(P_i k D_7 + 6\mu U D_8)} \right] \tag{1.48}$$

(3) Stiffness per unit supply pressure

$$F = -\frac{h}{P_s A} \left[6\mu A \frac{d}{dh} \left(G \frac{(P_s - Z_1 l U D_8)}{(6\mu + Z_1 l k D_7)} \right) + \frac{d}{dh} R \right] \tag{1.49}$$

(4) Stiffness per unit power

$$H = \frac{-6\mu}{Z_1 l} \left[6\mu A \frac{d}{dh} \left(G \cdot \frac{(P_s - Z_1 l U D_8)}{(6\mu + Z_1 l k D_7)} \right) + \frac{d}{dh} R \right] \left[\frac{1}{(P_i l k D_7 + 6\mu U l D_8)} \right] \tag{1.50}$$

1.5.6. *The Bearing Characteristic Number.* From eq. (1.37), the volume rate of flow in the region $-B_1 \geqslant -x \geqslant -B_2$ is

$$Q_2 = \frac{lk}{6\mu D_2} \left[P_i - \frac{6\mu U}{k} D_s \right]$$

Therefore

$$P_i \geqslant \frac{6\mu U}{k} D_5$$

and hence

$$H_n \geqslant \frac{h^2}{kB_2} \left[\frac{1}{(h - kB_2)} - \frac{1}{(h - kB_1)} \right]$$ (1.51)

this is shown in Fig. 7.

1.6. *Effect of Bearing Surface Finish*

In this section, the effect of bearing surface finish is considered. The bed is assumed to have surface irregularities of considerable magnitude, while the sliding pad is assumed to have irregularities of small magnitude to the extent that they can be neglected on account of the bed roughness. These conditions are quite similar to the reverse case in which the surface roughness of a stationary pad is considered and the roughness of a sliding bed is neglected. The last case will be chosen to ease the mathematical treatment.

The effect of surface irregularities on the pressure distribution along the fluid film relates to the possible formation of a series of hydrodynamic wedges due to both the shape of the

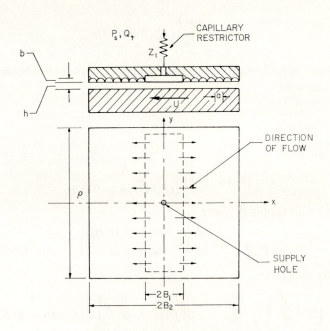

FIG. 4. Surface roughness of bearing pad.

irregularities and the sliding motion of the pad. Consequently, a regular wavy surface will not yield any improvement on the load carrying capacity of the bearing, due to the possible formation of the same number of positive and negative wedges.

However, the experimental work of Salama[9] showed clearly the positive effect of the bearing surface finish. As a preliminary detection of the effect of surface finish on the performance of a sliding pad bearing was made, only the positive effect of each wave will be considered.

1.6.1. *Effect of Surface Roughness.* Figure 4 shows the considered bearing arrangement. The film thickness at any point is defined by the following equations:

(1) For the region $B_1 \leqslant x \leqslant B_2$

(a) for lines having a positive slope

$$h_x = h + \frac{2b}{a}(x - B_1) - 2b(N - 1)$$

(b) for lines having a negative slope

$$h_x = h + \frac{2b}{a}\{Na - (x - B_1)\}$$

(2) For the region $-B_1 \geqslant -x \geqslant -B_2$

(a) for lines having a positive slope

$$h_x = h + \frac{2b}{a}\{Na + (x + B_1)\}$$

(b) for lines having a negative slope

$$h_x = h - \frac{2b}{a}(x + B_1) - 2b(N - 1)$$

1.6.1.1. *Pressure distribution and total flow rate.* The velocity distribution of the lubricant through the bearing is expressed by eq. (1.1).

The total volume rate of flow is given by

$$Q_t = Q_1 + (-Q_2)$$

where Q_1 = flow rate in the region $B_1 \leqslant x \leqslant B_2$
            Q_2 = flow rate in the region $-B_1 \geqslant -x \geqslant -B_2$

and

$$Q = -\frac{lhU}{2} - \frac{lh^3}{12\mu}\frac{dP}{dx}$$

and the pressure distribution under a y line of positive slope in the region $B_1 \leqslant x \leqslant B_2$

$$P_x = \frac{3\mu\,aU}{b[h + (2b/a)(x - B_1) - 2b(N - 1)]} + \frac{3\mu a\,Q_1}{bl[h + (2b/a)(x - B_1) - 2b(N - 1)]^2} + C$$

Considering the first wave, then

$$P_x = \frac{3\mu aU}{b[h + (2b/a)(x - B_1)]} + \frac{3\mu aQ}{bl[h + (2b/a)(x - B_1)]^2} + C$$

To find the hydrodynamic pressure generated under this line, the following boundary conditions are used

$$\text{at } x = B_1 \qquad\qquad P_x = 0$$
$$\text{at } x = B_1 + a/2 \quad P_x = 0$$

Therefore

$$P_x = \frac{3\mu a Q}{bl}\left[\frac{1}{[h + (2b/a)\,(x - B_1)]^2} - \frac{1}{h^2}\right] - \frac{3\mu a U}{b}\left[\frac{1}{h} - \frac{1}{[h + (2b/a)\,(x - B_1)]}\right] \quad (1.52)$$

$$Q = -\,lU\left[\frac{h(h + b)}{(2h + b)}\right] \quad (1.53)$$

Therefore

$$P_x = \frac{-3\mu a U}{b}\left[\left(\frac{h(h + b)}{(2h + b)}\right)\left(\frac{1}{(h + (2b/a)(x - B_1)^2} - \frac{1}{h^2}\right)\right.$$

$$\left. + \left(\frac{1}{h} - \frac{1}{[h + (2b/a)\,(x - B_1)]}\right)\right] \quad (1.54)$$

Therefore, the pressure distribution along the fluid film in the bearing can be expressed by considering the stationary case[8] and (eq. 1.54), so that

(1) For the region $B_1 \leqslant x \leqslant B_2$

  (a) *Pressure distribution under lines having a positive slope:*

$$P_{x1N} = P_i\left[1 + \frac{1}{K_1}\left(\frac{2(N - 1)}{(h + b)^2} - \frac{(2N - 1)}{h^2} + \frac{1}{\{h + (2b/a)\,(x - B_1) - 2b(N - 1)\}^2}\right)\right]$$

$$- \frac{3\mu a U}{b}\left[\left(\frac{h(h + b)}{(2h + b)}\right)\left(\frac{1}{[h + (2b/a)\,(x - B_1) - 2b(N - 1)]^2} - \frac{1}{h^2}\right)\right.$$

$$\left. + \left(\frac{1}{h} - \frac{1}{[h + (2b/a)\,(x - B_1) - 2b(N - 1)]}\right)\right] \quad (1.55)$$

  (b) *Pressure distribution under lines having a negative slope:*

$$P_{x2N} = P_i\left[1 + \frac{1}{K_1}\left(\frac{2N}{(h + b)^2} - \frac{(2N - 1)}{h^2} - \frac{1}{[h + (2b/a)\,\{Na - (x - B_1)\}]^2}\right)\right] \quad (1.56)$$

(2) For the region $-\,B_1 \geqslant -\,x \geqslant -\,B_2$

  (a) *Pressure distribution under lines having a positive slope:*

$$P_{x4N} = P_i\left[1 + \frac{1}{K_1}\left(\frac{2N}{(h + b)^2} - \frac{(2N - 1)}{h^2} - \frac{1}{[h + (2b/a)\,\{Na + (x + B_1)\}]^2}\right)\right]$$

$$- \frac{3\mu a U}{b}\left[\left(\frac{h(h + b)}{(2h + b)}\right)\left(\frac{1}{[h + (2b/a)\,\{Na + (x + B_1)\}]^2} - \frac{1}{h^2}\right)\right.$$

$$\left. + \left(\frac{1}{h} - \frac{1}{[h + (2b/a)\,\{Na + (x + B_1)\}]}\right)\right] \quad (1.57)$$

  (b) *Pressure distribution under lines having a negative slope:*

$$P_{x3N} = P_i\left[1 + \frac{1}{K_1}\left(\frac{1}{[h - (2b/a)\,(x + B_1) - 2b(N - 1)]^2}\right.\right.$$

$$\left.\left. - \frac{(2N - 1)}{h^2} + \frac{2(N - 1)}{(h + b)^2}\right)\right] \quad (1.58)$$

and also

$$Q_1 = \frac{blP_i}{3\mu a K_1} - lU \left[\frac{h(h + b)}{(2h + b)}\right] \tag{1.59}$$

where

$$Q_t = \frac{2blP_i}{3a\mu K_1} \tag{1.60}$$

$$K_1 = \frac{2(B_2 - B_1)}{a} \left[\frac{1}{h^2} - \frac{1}{(h + b)^2}\right]$$

1.6.1.2. *The total load-carrying capacity.* Integrating the pressure of each line over the area of that line, and by summation,

$$L = 2P_i \, B_1 \, l + l \sum_{N=1}^{N=m} \int_{x_{11}}^{x_{12}} P_{x1_N} \, dx + l \sum_{N=1}^{N=m} \int_{x_{21}}^{x_{22}} Px_{2_N} \, dx - l \sum_{N=1}^{N=m} \int_{x_{31}}^{x_{32}} P_{x3_N} \, dx - l \sum_{N=1}^{N=m} P_{x4_N} \, dx$$

where $x_{11} = - x_{31} = B_1 + a(N - 1)$

$x_{12} = x_{21} = - x_{32} = - x_{41} = B_1 + a(N - \tfrac{1}{2})$

$X_{22} = - x_{42} = B_1 + a(N)$

$m = (B_2 - B_1)/a$

The load-carrying capacity of this bearing can also be given by

$$L = L_1 + 2 \, [(B_2 - B_1)/a] \int_{B_1}^{B_1 + a/2} P_x \, ldx$$

where P_x is defined by eq. (1.54) and L_1 is the total load-carrying capacity of the stationary bearing.[8]

Therefore

$$L = P_iAG - 2 \frac{(B_2 - B_1)}{a} \cdot \frac{3\mu alU}{b} \left[\frac{a}{2h} - \frac{ab}{2h(2h + b)} - \frac{a}{2b} \log \left(\frac{h + b}{h}\right)\right] \tag{1.61}$$

$$= P_iAG + R_2$$

where

$$G = \frac{1}{B_2} \left[B_1 + \sum_{N=1}^{N=m} \left(a - \frac{a^2(2N - 1)}{2(B_2 - B_1)}\right)\right] \tag{1.62}$$

$$R_2 = 2 \left(\frac{B_2 - B_1}{a}\right) \cdot \frac{3\mu al}{b} U \left[\frac{- a}{2h} + \frac{ab}{2h(2h + b)} + \frac{a}{2b} \log \left(\frac{h + b}{h}\right)\right] \tag{1.63}$$

1.6.1.3. *The stiffness.* When the bearing is compensated by a capillary restrictor of resistance Z_1, the pressure ratio will be

$$\frac{P_i}{P_s} = \left[\frac{3\mu a K_1}{3\mu a K_1 + 2lbZ_1}\right]$$

and the stiffness of the bearing is

$$\lambda = \lambda_1 + 2 \left(\frac{B_2 - B_1}{a}\right) \cdot \frac{3\mu al}{b} U \left[\frac{- a}{2h^2} + \frac{ab \, (4h + b)}{2h^2 \, (2h + b)^2} + \frac{a}{2h \, (h + b)}\right] \tag{1.64}$$

where

$$\lambda_1 = 12\mu\, P_s AG\,(B_2 - B_1)\left[\frac{1}{h^3} - \frac{1}{(h+b)^3}\right]\left[\frac{2lbZ_1}{(3\mu aK_1 + 2blZ_1)^2}\right]$$

1.6.1.4. *The power consumption.* The power dissipated in the bearing E_b and the total power required to operate the bearing E_t are

$$E_b = \frac{2lbP_i^2}{3\mu aK_1} \tag{1.65}$$

and

$$E_t = \frac{2lbP_iP_s}{3\mu aK_1} \tag{1.66}$$

1.6.1.5. *The dimensionless groups.*[8] The dimensionless groups for this bearing are

(1) Load per unit area

$$I = \frac{1}{B_2}\left[B_1 + \sum_{N=1}^{N=m}\left(a - \frac{a^2(2N-1)}{2(B_2 - B_1)}\right)\right] + \frac{R_2}{P_iA} \tag{1.67}$$

(2) Load per unit flow

$$J = \frac{3a\,K_1\,AG}{2b} + \frac{3a\,K_1\,R_2}{2bP_i} \tag{1.68}$$

(3) Stiffness per unit supply pressure

$$F = F_1 + \frac{2h}{P_sA}\left(\frac{B_2 - B_1}{a}\right)\cdot\frac{3\mu al}{b}\,U\left[\frac{-a}{2h^2} + \frac{ab(4h+b)}{2h^2(2h+b)^2} + \frac{a}{2h(h+b)}\right] \tag{1.69}$$

where

$$F_1 = 12\mu hG\,(B_2 - B_1)\left[\frac{1}{h^3} - \frac{1}{(h+b)^3}\right]\left[\frac{2lbZ_1}{(3\mu aK_1 + 2blZ_1)^2}\right]$$

(4) Stiffness per unit power $= H$

$$H = H_1 + \frac{9\mu^2a^2K_1}{b^2Z_1P_il}\,U\left(\frac{B_2 - B_1}{a}\right)\left[\frac{-a}{2h^2} + \frac{ab(4h+b)}{2h^2(2h+b)^2} + \frac{a}{2h(h+b)}\right] \tag{1.70}$$

where

$$H_1 = \frac{24\mu B_2G(B_2 - B_1)}{(3\mu aK_1 + 2blZ_1)}\left[\frac{1}{h^3} - \frac{1}{(h+b)^3}\right]$$

1.6.1.6. *The bearing characteristic number.* From eq. (1.59), the volume rate of flow in the region $B_1 \leqslant x \leqslant B_2$ is given by

$$Q_1 = \frac{blP_i}{3\mu aK_1} - IU\left[\frac{h(h+b)}{(2h+b)}\right]$$

Therefore

$$\frac{blP_i}{3\mu aK_1} \geqslant IU\left[\frac{h(h+b)}{(2h+b)}\right]$$

Therefore

$$H_n \geqslant \left[\frac{aK_1h^3(h+b)}{2bB_2(2h+b)}\right] \tag{1.71}$$

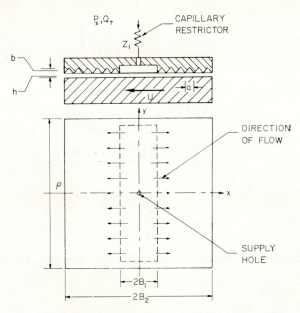

FIG. 5. Surface waviness of bearing pad.

1.6.2. *Effect of Surface Waviness.* Figure 5 shows the considered bearing arrangement.
1.6.2.1. *Pressure distribution and total flow rate.* Following the same procedure explained in section 1.3.1

$$P_x = - \int \frac{6\mu U d_x}{h^2} - \int \frac{12\mu Q}{lh^3} \, d_x$$

For the regions starting at x_{11} and ending at x_{12} where

$$x_{11} = B_1 + a(N-1) \qquad x_{12} = B_1 + a(N - \tfrac{1}{2})$$

where N is the wave order

$$h_x = b \left[\omega + \sin \pi \left(\frac{x - a(N-1) - B_1}{a} \right) \right]$$

therefore

$$P_x = \frac{-6\mu U}{b^2} \int \frac{d_x}{[\omega + \sin \pi \{(x - a(N-1) - B_1)/a \}]^2}$$

$$- \frac{12\mu Q}{lb^3} \int \frac{d_x}{[\omega + \sin \pi \{(x - a(N-1) - B_1)/a \}]^3}$$

Using the previously defined substitutions,[8] therefore

$$P_x = \frac{-6\mu U}{b^2} \cdot \frac{a}{\pi} \cdot \frac{1}{(\omega^2 - 1)^{\frac{3}{2}}} \, [\omega\gamma + \cos \gamma]$$

$$- \frac{12\mu Q}{lb^3} \cdot \frac{a}{\pi} \cdot \frac{1}{(\omega^2 - 1)^{\frac{5}{2}}} \, [(\omega^2 + \tfrac{1}{2})\gamma + 2\omega \cos \gamma - \tfrac{1}{4} \sin 2\gamma] + C$$

where

$$\gamma = \sin^{-1}\left[\frac{1 + \omega \sin y_1}{\omega + \sin y_1}\right]$$

$$\cos \gamma = \left[\frac{(\omega^2 - 1)^{\frac{1}{2}} \cos y_1}{(\omega + \sin y_1)}\right]$$

$$\sin 2\gamma = \left[\frac{2(\omega^2 - 1)^{\frac{1}{2}} (1 + \omega \sin y_1) \cos y_1}{(\omega + \sin y_1)^2}\right]$$

$$y_1 = \pi \left[\frac{x - a(N - 1) - B_1}{a}\right]$$

For the first wave

$$y_1 = \pi \left(\frac{x - B_1}{a}\right)$$

To find the hydrodynamic pressure generated under this part, the following boundary conditions are used

at $x = B_1$ $P_x = 0$

$x = B_1 + a/2$ $P_x = 0$

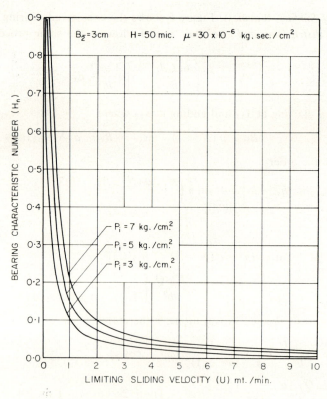

FIG. 6. The bearing characteristic number.

Therefore

$$Q = \frac{-bl\,(\omega^2 - 1)}{2}\,R_5 U \tag{1.72}$$

$$P_x = \frac{-6\mu U}{b^2} \cdot \frac{a}{\pi} \cdot \frac{1}{(\omega^2 - 1)^{\frac{3}{2}}}\,[R_5\,\{R_3 - E_1\gamma - 2\omega\cos\gamma + \tfrac{1}{4}\sin 2\gamma\} + \{\omega\gamma + \cos\gamma - R_4\}] \tag{1.73}$$

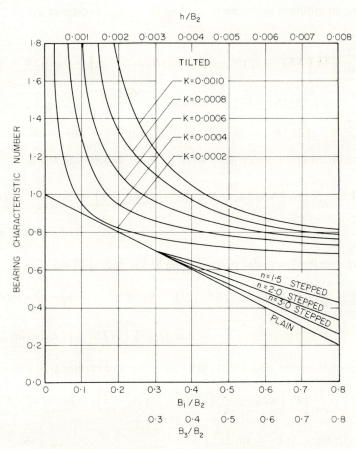

FIG. 7. The bearing characteristic number for the different cases.

where

$$R_5 = \frac{R_4 - \omega_2^\pi}{R_3 - \frac{\pi}{2}E_1} \quad R_4 = \omega E_2 + E_4 \quad R_3 = E_1 E_2 + 2\omega E_4 - \tfrac{1}{2}E_5$$

$$E_1 = (\omega^2 + \tfrac{1}{2})$$

$$E_2 = \sin^{-1}\frac{1}{\omega}$$

$$E_4 = \frac{(\omega^2 - 1)^{\frac{1}{2}}}{\omega}$$

$$E_5 = \frac{(\omega^2 - 1)^{\frac{1}{2}}}{\omega^2}$$

Therefore, the pressure distribution along the fluid film in the bearing can be expressed by considering the stationary case[8] and eq. (1.73). Therefore,

(1) Pressure distribution under parts starting at x_{11} and ending at x_{12}

$$P_{x1_N} =$$

$$P_i \left\{ 1 - \frac{1}{R_1} \left[E_1 \left(NE_2 - \frac{\pi}{2} (N-1) - E_3 (N-1) - \gamma \right) + 2\omega \left\{ (2N-1) E_4 - \cos \gamma \right\} \right. \right.$$

$$\left. \left. - \tfrac{1}{4} \left\{ 2 (2N-1) E_5 - \sin 2\gamma \right\} \right] \right\} \frac{-6\mu U}{b^2} \frac{a}{\pi} \frac{1}{(\omega^2 - 1)^{\frac{1}{2}}}$$

$$[R_5 \left\{ R_3 - E\gamma - 2\omega \cos \gamma + \tfrac{1}{4} \sin 2\gamma \right\} + \left\{ \omega\gamma + \cos \gamma - R_4 \right\}] \quad (1.74)$$

(2) Pressure distribution under parts starting at x_{21} and ending at x_{22}

$$P_{x2_N} = P_i \left\{ 1 - \frac{1}{R_1} \left[E_1 \left(NE_2 - \frac{\pi}{2} N - (N-1) E_3 - \phi \right) + 2\omega \left\{ (2N-1) E_4 + \sin \phi \right\} \right. \right.$$

$$\left. \left. - \tfrac{1}{4} \left\{ 2 (2N-1) E_5 + \sin 2\phi \right\} \right] \right\} \quad (1.75)$$

(3) Pressure distribution under parts starting at x_{31} ending at x_{32}

$$P_{x3_N} = P_i \left\{ 1 - \frac{1}{R_1} \left[E_1 \left(NE_2 - \frac{\pi}{2} (N-1) - E_3 (N-1) - \psi \right) + 2\omega \left\{ (2N-1) E_4 \right. \right. \right.$$

$$\left. \left. \left. - \cos \psi \right\} - \tfrac{1}{4} \left\{ 2 (2N-1) E_5 - \sin 2\psi \right\} \right] \right\} \quad (1.76)$$

(4) Pressure distribution under parts starting at x_{41} and ending at x_{42}

$$P_{x4_N} = P_i \left\{ 1 - \frac{1}{R_1} \left[E_1 \left(NE_2 - \frac{\pi}{2} N - (N-1) E_3 - \epsilon \right) + 2\omega \left\{ (2N-1) E_4 \right. \right. \right.$$

$$\left. + \sin \epsilon \right\} - \tfrac{1}{4} \left\{ 2(2N-1) E_5 + \sin 2\epsilon \right\} \right] \right\} - \frac{-6\mu U}{b^2} \cdot \frac{a}{\pi} \cdot \frac{1}{(\omega^2 - 1)^{\frac{1}{2}}} \left[\left\{ R_6 - \omega\epsilon + \sin\epsilon \right\} \right.$$

$$\left. + \frac{R_6}{R_7} \left\{ E_1 \epsilon - 2\omega \sin \epsilon + \tfrac{1}{4} \sin 2\epsilon - R_7 \right\} \right] \quad (1.77)$$

Also

$$Q_t = - \frac{lb^3 \pi (\omega^2 - 1)^{\frac{5}{2}} P_i}{6\mu a R_1} \tag{1.78}$$

$$Q_1 = \frac{\pi lb^3 (\omega^2 - 1)^{\frac{5}{2}}}{12\mu a R_1} P_i - \frac{lb (\omega^2 - 1) R_5}{2} U \tag{1.79}$$

where

$$R_6 = \omega E_3 - E_4$$

$$R_7 = E_1 E_3 - 2\omega E_4 + \tfrac{1}{2} E_5$$

$$\psi = \sin^{-1} \left[\frac{1 + \omega \sin y_3}{\omega + \sin y_3} \right]$$

$$\phi = \sin^{-1} \left[\frac{(\omega^2 - 1)^{\frac{1}{2}} \sin y_2}{(\omega + \cos y_2)} \right]$$

$$\epsilon = \sin^{-1} \left[\frac{(\omega^2 - 1)^{\frac{1}{2}} \sin y_4}{\omega + \cos y_4} \right]$$

$$y_2 = \pi \left\{ \frac{x - a(N-1) - (B_1 + 9/2)}{a} \right\}$$

$$y_3 = -\pi \left\{ \frac{x + a(N-1) + B_1}{a} \right\}$$

$$y_4 = -\pi \left\{ \frac{x + a(N-1) + (B_1 + 9/2)}{a} \right\}$$

$$R_1 = \left(\frac{B_2 - B_1}{a} \right) \left[E_1 E_2 - \frac{\pi}{2} E_1 - E_1 E_3 + 4\omega E_4 - E_5 \right]$$

and x_{11}, x_{12}, x_{21}, x_{22}, x_{31}, x_{32}, x_{41} and x_{42} are defined in section 1.6.1.2.

1.6.2.2. *The total load-carrying capacity.* Integrating the pressure under each part and over the area of the part and by summation

$$L = 2 P_i B_1 l + l \sum_{N=1}^{N=m} \int_{x_{11}}^{x_{12}} P_{x1_N} dx + l \sum_{N=1}^{N=m} \int_{x_{21}}^{x_{22}} P_{x2_N} dx - l \sum_{N=1}^{N=m} \int_{x_{31}}^{x_{32}} P_{x3_N} dx - l \sum_{N=1}^{N=m} \int_{x_{41}}^{x_{42}} P_{x4_N} dx$$

(1.80)

where P_{x1_N}, P_{x2_N}, P_{x3_N} and P_{x4_N} are defined by eqs. (1.74), (1.75), (1.76) and (1.77).

1.6.2.3. *The power consumption.* The power dissipated in the bearing E_b and the total power required to operate the bearing E_t are

$$E_b = - \frac{lb^3 \pi (\omega^2 - 1)^{\frac{5}{2}} P_i^2}{6\mu a R_1}$$

(1.81)

and

$$E_t = - \frac{lb^3 \pi (\omega^2 - 1)^{\frac{5}{2}} P_i P_s}{6\mu a R_1}$$

(1.82)

1.6.2.4. *The bearing characteristic number.* From eq. (1.79) the volume rate of flow in the region $B_1 \leqslant x \leqslant B_2$ is

$$Q_1 = \frac{lb}{2} (\omega^2 - 1) \left[\frac{b^2 \pi (\omega^2 - 1)^{\frac{3}{2}} P_i}{6\mu a R_1} - R_5 U \right]$$

Therefore

$$\frac{b^2\pi\,(\omega^2 - 1)^{\frac{3}{2}}}{6\mu a R_1}\,P_i \geqslant R_5 U$$

Therefore

$$H_n \geqslant \frac{a R_1 h^2}{B_2 \pi b^2\,(\omega^2 - 1)^{\frac{3}{2}}}\,R_5 \qquad\qquad (1.83)$$

DISCUSSION AND CONCLUSIONS

Discussion

Figures 8, 9 and 10 show the pressure distribution along the fluid film in the stepped and initially tilted bearings. From these figures can be seen the following:

1. The stepped bearing has an asymmetrical pressure distribution about its centre (Fig. 8). Therefore the pad will be acted upon by a turning moment whose magnitude depends upon U and the bearing dimensions.

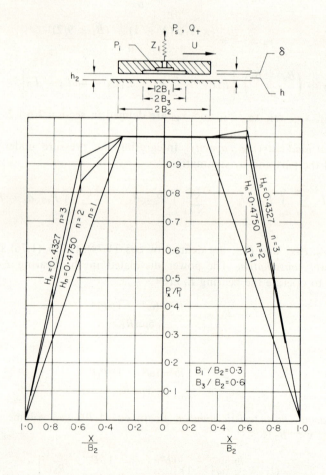

Fig. 8. Pressure distribution along the fluid film for stepped bearing.

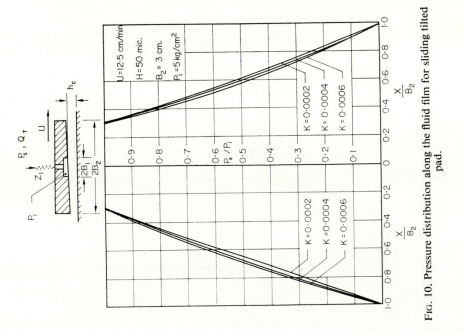

FIG. 10. Pressure distribution along the fluid film for sliding tilted pad.

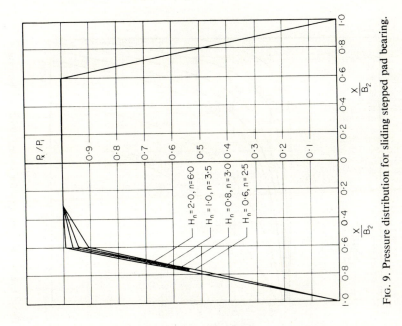

FIG. 9. Pressure distribution for sliding stepped pad bearing.

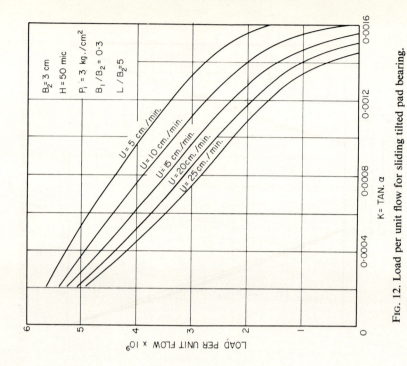

Fig. 12. Load per unit flow for sliding tilted pad bearing.

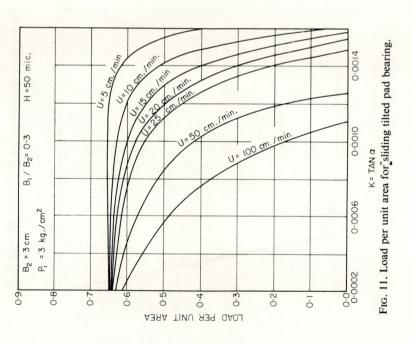

Fig. 11. Load per unit area for sliding tilted pad bearing.

2. The pressure distribution along the fluid film in the initially sliding tilted pad bearing is also asymmetrical about the pad centre.

The dimensionless groups for the plain and stepped rectangular bearings are previously drawn and discussed.[8] Figures 11, 12, 13, 14 and 15 and Tables 2, 3 and 4 show the dimensionless group of the other cases considered. From the figures and tables it can be seen that:

1. The stiffness per unit power and the stiffness per unit supply pressure has a minimum value depending on k, U and the bearing dimensions (Figs. 13 and 14).

TABLE 1. VALUES OF H_n FOR INITIALLY TILTED PAD

$B_2 = 3$ cm; $B_1/B_2 = 0.3$; $H = 50\,\mu$; $k_{max} = h/B_2 = 0.001666$.

K	a min sec		H_n	U_{lim} (cm/sec)		
				$P_i = 3$ kg/m^2	$P_i = 5$ kg/m^2	$P_i = 7$ kg/m^2
0.0002	0	41.4	0.79	0.167	0.280	0.392
0.0004	1	22.8	0.96	0.145	0.242	0.340
0.0006	2	4.2	1.18	0.117	0.196	0.274
0.0008	2	45.6	1.52	0.092	0.153	0.214
0.0010	3	27.0	2.14	0.065	0.108	0.151

TABLE 2. EFFECT OF SURFACE ROUGHNESS ON LOAD PER UNIT AREA

$B_2 = 3$ cm; $l/B_2 = 5$; $\mu = 30 \times 10^{-6}$ kg sec/cm^2; $B_1/B_2 = 0.3$; $H = 50\,\mu$; $P_i = 5$ kg/cm^2.

b (mm)	a (mm)	U (cm/min)					
		5	10	15	20	25	400
0.02	0.2	0.650001	0.650003	0.650004	0.650005	0.650007	0.650106
	0.5	0.650003	0.650006	0.650010	0.650013	0.650016	0.650263
	1.0	0.650006	0.650013	0.650019	0.650026	0.650033	0.650527
0.06	0.2	0.650004	0.650007	0.650011	0.650014	0.650018	0.650283
	0.5	0.650009	0.650017	0.650026	0.650035	0.650044	0.650706
	1.0	0.650170	0.650035	0.650053	0.650070	0.650088	0.651413
0.10	0.2	0.650005	0.650011	0.650016	0.650021	0.650026	0.650423
	0.5	0.650013	0.650026	0.650039	0.650053	0.650066	0.651057
	1.0	0.650026	0.650052	0.650079	0.650105	0.650132	0.652112
0.15	0.2	0.650007	0.650014	0.650021	0.650028	0.650035	0.650558
	0.5	0.650017	0.650035	0.650052	0.650069	0.650087	0.651395
	1.0	0.650034	0.650069	0.650104	0.650139	0.650174	0.652790
0.20	0.2	0.650008	0.650016	0.650025	0.650033	0.650041	0.650659
	0.5	0.650020	0.650041	0.650061	0.650082	0.650103	0.651648
	1.0	0.650041	0.650082	0.650123	0.650184	0.650206	0.653295
0.30	0.2	0.650009	0.650020	0.650030	0.650040	0.650049	0.650790
	0.5	0.650024	0.650049	0.650074	0.650098	0.650123	0.651975
	1.0	0.650049	0.650098	0.650184	0.650197	0.650246	0.653950

N.B. Under the same conditions, the load per unit area for stationary plain bearing = 0.65000.

Table 3. Effect of Surface Roughness on Load per Unit Flow $\times 10^9$

$B_2 = 3$ cm; $l/B_2 = 5$; $\mu = 30 \times 10^{-6}$ kg sec/cm²;
$B_1/B_2 = 0.3$; $H = 50 \mu$; $P_i = 5$ kg/cm².

b (mm)	a (mm)	U (cm/min)					
		5	10	15	20	25	400
0·02	0·2	5·56097	5·56098	5·56100	5·56101	5·56102	5·56186
	0·5	5·56099	5·56102	5·56104	5·56107	5·56110	5·56322
	1·0	5·56101	5·56107	5·56113	5·56118	5·56124	5·56547
0·06	0·2	4·98297	4·98300	4·98303	4·98305	4·98308	4·98511
	0·5	4·98301	4·98308	4·98315	4·98321	4·98328	4·98836
	1·0	4·98308	4·98321	4·98335	4·98348	4·98362	4·99377
0·10	0·2	4·50454	4·50457	4·50461	4·50465	4·50468	4·50743
	0·5	4·50459	4·50468	4·50477	4·50486	4·50495	4·51182
	1·0	4·50468	4·50468	4·50504	4·50523	4·50501	4·51915
0·15	0·2	4·01266	4·01270	4·01274	4·01279	4·01283	4·01606
	0·5	4·01272	4·01283	4·01293	4·01304	4·01315	4·02122
	1·0	4·01283	4·01304	4·01326	4·01347	4·01369	4·02983
0·20	0·2	3·61033	3·61037	3·61042	3·61047	3·61051	3·61394
	0·5	3·61040	3·61051	3·61062	3·61074	3·61085	3·61943
	1·0	3·61051	3·61074	3·61097	3·61120	3·61142	3·62859
0·30	0·2	2·99445	2·99456	2·99460	2·99465	2·99469	2·99800
	0·5	2·99458	2·99469	2·99485	2·99495	2·99503	3·00356
	1·0	2·99469	2·99495	2·99515	2·99537	2·99560	3·00266

N.B. Under the same conditions, the load per unit flow for stationary plain bearing = 5.89680×10^9.

Table 4. Effect of Surface Roughness on Stiffness per Unit Supply Pressure

$B_2 = 3$ cm; $l/B_2 = 5$; $\mu = 30 \times 10^{-6}$ kg sec/cm²; $B_1/B_2 = 0.3$; $H = 50 \mu$; $P_s = 7$ kg/cm².

b (mm)	a (mm)	U (cm/min)					
		5	10	15	20	25	400
0·02	0·2	0·274690	0·274693	0·274695	0·274698	0·274701	0·274909
	0·5	0·274694	0·274701	0·274708	0·274715	0·274722	0·275241
	1·0	0·274701	0·274715	0·274728	0·274742	0·274756	0·275796
0·06	0·2	0·284513	0·284520	0·784527	0·284535	0·284542	0·285078
	0·5	0·284524	0·284542	0·284559	0·284577	0·284595	0·285936
	1·0	0·284542	0·284577	0·284613	0·284649	0·284685	0·287366
0·10	0·2	0·293382	0·293392	0·293402	0·293412	0·293423	0·294198
	0·5	0·293397	0·293423	0·293449	0·293474	0·293500	0·295437
	1·0	0·293423	0·293474	0·293526	0·293578	0·293629	0·297503
0·15	0·2	0·303235	0·303248	0·303261	0·303274	0·303287	0·304268
	0·5	0·303254	0·303287	0·303320	0·303352	0·303385	0·305839
	1·0	0·303287	0·303352	0·303418	0·303483	0·303548	0·308456
0·20	0·2	0·311839	0·311854	0·311869	0·311884	0·311899	0·313015
	0·5	0·311861	0·311898	0·311936	0·311973	0·312010	0·314800
	1·0	0·311898	0·311973	0·312047	0·312122	0·312196	0·317776
0·30	0·2	0·325768	0·325784	0·325805	0·325858	0·325834	0·327080
	0·5	0·325795	0·325834	0·325876	0·325957	0·325959	0·329079
	1·0	0·325834	0·325957	0·326000	0·326084	0·326567	0·332407

N.B. Under the same conditions, the stiffness per unit supply pressure for stationary plain bearing = 0·269396.

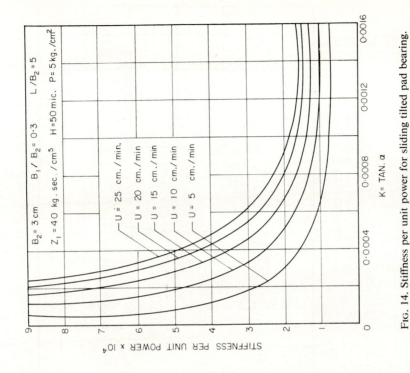

Fig. 14. Stiffness per unit power for sliding tilted pad bearing.

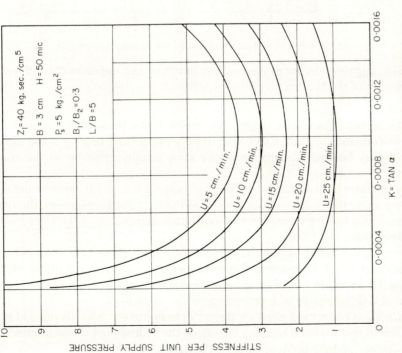

Fig. 13. Stiffness per unit supply pressure for sliding tilted pad bearing.

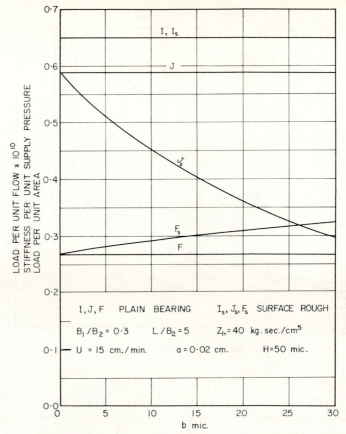

Fig. 15. Effect of surface roughness on the bearing performance.

2. Surface roughness decreases the load per unit flow and improves the stiffness per unit supply pressure of the bearing.

Conclusions

From the previous discussion, figures, and tables, it can be concluded that:

1. The performance of a plain sliding externally pressurized bearing is the same as that of the stationary bearing working under the same conditions so far as the H_n value is concerned.

2. The performance of a stepped sliding pad bearing is also the same as the stationary one working under the same conditions, so far as the H_n value is concerned, though the pad is acted upon by a turning moment whose magnitude depends upon the velocity U and the bearing dimensions.

3. In general, the sliding velocity U reduces the load per unit area, load per unit flow, stiffness per unit supply pressure and stiffness per unit power of the initial tilted sliding pad bearing. However, the minimum value of the stiffness per unit supply pressure and power is attained at a certain value of k depending on the value of U.

4. Surface finish has a small effect on the performance of the plain sliding pad bearing.

This is due to the two main resisting factors on the hydrodynamic effect, namely, surface finish dimensions and the limiting value of the sliding velocity U.

REFERENCES

1. C. A. SCALES, The Performance of Hydrostatic Slideway Bearings in a Milling Machine. NEL Report No. 26, March 1962.
2. J. LOXHAM, The Application of Hydrostatic Bearings to High Precision Machine Tools.
3. F. R. ARCHIBALD, A Simple Hydrodynamic Thrust Bearing. *Trans. A.S.M.E.*, **72**, No. 4, May 1950.
4. I. HARKANYI, Determination of Optimum Conditions of Capillary Compensated Hydrostatic Slideways. *Acta Technical (Budapest)*, **49**, 1964.
5. A. CAMERON and W. L. WOOD, Parallel Surface Thrust Bearing. *Proc. 6th International Congress for Applied Mechanics*, 1946.
6. A. CHARNES and E. SAIBEL, On the Solution of the Reynolds Equation for Slider Bearing Lubrication, I. *Trans. A.S.M.E.*, **74**, 1952.
7. O. PINKUS and B. STERNLICHT, *Theory of Hydrodynamics Lubrication*. McGraw-Hill, 1961.
8. M. A. EL KADEEM, E. A. SALEM and H. R. EL. SAYED, Effect of Pad Geometry and Surface Finish on the Performance of Stationary Externally Pressurized Rectangular Bearing. To be published.
9. M. E. SALAMA, The Effect of Surface Finish on the Performance of Parallel Surface Thrust Bearings. Ph.D. Thesis, University of Manchester, 1949.

A NEW METHOD FOR BONDING STRIP SLIDEWAYS

A. DAPIRAN

Istituto di Ricerche in Tecnologia Meccanica Vico Canavese, Torino

SUMMARY

A new method for bonding strip slideways is presented. The procedure has been studied in order not to require particular skill from the operator. The accuracy which can be obtained without further machining of the slideways is investigated experimentally.

1. INTRODUCTION

The surface hardness of machine tool slideways is quite critical in determining their wear, and various methods have been used by machine tool builders in order to increase the wear resistance of the ways.

In particular, various methods for applying strips of materials having particular properties to the ways have been used. The properties usually sought for are high surface hardness or low friction. Depending on the thickness of the strips applied to the slideways, these can be fastened by screws or glued. One of the purposes of these procedures is that of an easy servicing of the machine tool: the machine must be stripped of its worn slideways and equipped with new ones as fast as possible and with the least possible effort.

Unfortunately the methods proposed require the ways to be machined after their application.[1,2] If this machining operation could be avoided, the slideway cost could be greatly reduced. Many procedures based on adhesive bonding also require a particular training of the operator. The pot life of many adhesives, before they start to harden, is quite limited, and only a skilled operator can lay the strips on the ways with the required cleanliness and accuracy in the limited time available.

The use of strips of thin sheet material has a few undeniable advantages. One of these is the ready availability of the material in the required hardness ranges and sizes (in particular long lengths are no problem) and with close tolerances on thickness. This last property is required if the machining of the finished slideway must be avoided.

In the following a method for bonding steel strips to machine tool slideways is presented. Two purposes have been kept in mind: that of avoiding the necessity of machining the finished slideway, and that of defining a simple procedure which does not require a skilled operator.

2. APPLICATION METHOD PROPOSED

The usual methods for strip application require; first, the coating of the base surface with resin; second, the application of the strip. This is then loaded in order to obtain the required contact pressure, and cured under load. The layer of adhesive gives a poorly defined reference surface, and consequentially the planarity of the bonded slideway is rather poor.

897

Some builders avoid this trouble by increasing the slideway area and using part of the strip, directly in contact with the base surface, as slideway, and an area adjacent to that as bonding surface. A cross section of a slideway built according to this concept is shown in Fig. 1. The glue is introduced in the groove A and, when the slip is pressed to the base

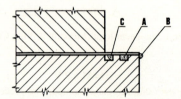

FIG. 1. Scheme of the method for bonding strip slideways used by Scharmann & Co.

surface, the glue is squeezed out and covers part of the mating surfaces. The excess glue is either squeezed towards the outer border B, or towards the center of the way. In order to avoid introducing some glue in the part that will be used as sliding surface, a second groove C is machined, in order to receive the excess glue. This method is patented by Scharmann & Co. Rheydt (Rhld) (P.A. 579792 and 1880547). The disadvantages of this system are essentially the need of great surfaces, and the fact that only a fraction of the surface is used for the bonding, thus reducing the strength of the bonds.

The principal features of the method proposed here are the following:

(a) The strip is put in tension on the base slideway previous to the introduction of the adhesive.
(b) The strip is compressed against the base surface previous to the glueing.
(c) The adhesive is injected under pressure between the two surfaces.

The thermal conductivity of the resins which can be used for bonding slideways is rather low. Consequentially, in operation, the temperature of the strip material can be significantly higher than the temperature of the basic structure. The tension given to the strip should be great enough to ensure that the bonded strip will always be under tension, even in operating conditions, when the surface material will tend to expand with respect to the machine structure. This precaution will ensure that the thermally induced tangential stresses on the bonds will be lower in operating conditions than in rest. Two examples of fixtures to apply tension to the strip are shown in Fig. 2. These should be attached at both ends of the slideway.

After being put in tension on the base slideway, the strip is compressed against it. This can be done in two essentially different ways. A plain metal rule of high stiffness can be pressed on the strip, compressing it against the base structure; in this case, when the adhesive is injected under pressure the strip will tend to "copy" the rule's surface. A thick sheet of rubber can be interposed between the rule and the strip, creating something comparable to an hydrostatic pressure; in this case the strip will tend to copy the base surface.

The adhesive must then be injected between the two surfaces. Supply holes must be left in the machine tool structure, and grooves must be machined in the base slideway surface, in order to allow the adhesive to cover the whole contact surface. A typical base surface

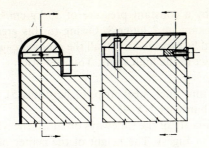

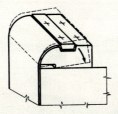

FIG. 2. Examples of fixtures used to apply tension to the strips.

can be seen in Fig. 3. The distance between the grooves, the number of supply holes and the supply pressure must be adequate to the viscosity of the adhesive.

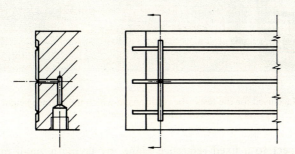

FIG. 3. Typical base surface.

3. EXPERIMENTAL TESTS

Experimental tests have been run, in order to obtain information on the influence of a few variables on the surface quality. The variables examined are:

(A) The method of loading the strip:

    A1 directly by the stiff metal rule;

    A2 with a rubber sheet interposed between the strip and the rule.

(B) The surface finish of the base slideway:

    B1 ground surface;

    B2 scraped surface.

(C) The cross-section of the grooves:

    C1 cross-section 1 mm;^{2}

    C2 cross-section 2 mm.^{2}

The strip has been loaded by a constant pressure of 35 kg/cm.², and the adhesive supply pressure has been kept constant at 7 kg/cm². Owing to this great difference between the pressures the adhesive should be expected to fill all the grooves caused by surface roughness, leaving metal to metal contact in the high spots. This condition is expected to insure a high bearing stiffness.

The strip used was a 0·3 mm thick steel strip, of hardness 51 HRc. The tests were run with Araldite AY 103 with hardener HY 951.

In order to control the planarity of the slideways obtained the whole slideway was divided in 36 parts, as can be seen in Fig. 4. The height of the center of each of these areas was

FIG. 4. Base surface used for the tests: the positions in which the surface height readings have been made are indicated.

measured, with respect to a fixed reference plane, by laying a small metal block on the surface and measuring its height. The measurement is thus not influenced by surface roughness. For each test the measurement has been made before and after bonding the strip.

The test results are shown in Figs. 5 through 12. The broken lines indicate the base surface, and the full lines the height of the final surface, minus the nominal strip thickness. Thus if the strip thickness were exactly constant and equal to the nominal, the distance between the two lines would be the gap between the two metal surfaces. This gap should be completely filled with adhesive. The strip thickness is not exactly constant; the tolerances on the thickness are of a few micrometers over rather long lengths of material, and should be quite higher over the small lengths used in these tests. So, even if the gap height indicated in the figures might be off by a few micrometers, the accuracy of the variation in gap height shown should be considerably better.

In order to have an idea of the planarity of all the surfaces obtained, and to check the parallelism between this plane and the original "base" surface, the best plane approximating the measured surface has been calculated by a least squares procedure. The position of this "regression plane" has been indicated with a dotted line in Figs. 5 through 12.

A first qualitative result is that though the surfaces obtained in conditions A2 are generally rather good at the center, they were found to be relatively poor at the border. The same

FIGS. 5–12 Surface profiles obtained in the tests.

— — — — base surface.

————— final surface.

. . . . "regression plane".

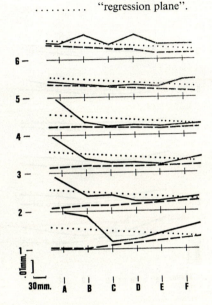

FIG. 5. Test conditions: A2, B1, C1 (see list of variables tested).

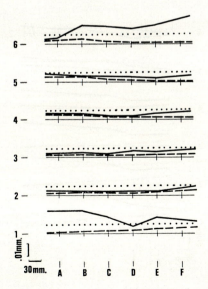

FIG. 6. Test conditions: A2, B1, C2 (see list of variables tested).

thing cannot be said for the surfaces obtained in conditions A1. A further observation of the bonds between base surface and strip have explained this fact. The adhesive did not reach the border of the surface. This is probably due to the fact that the compression applied, which can be approximately considered an hydrostatic pressure, was much higher than the adhesive supply pressure. Further tests are being run to determine the optimum pressure ratio, as a function of the distribution and cross-section of the grooves machined in the base surface.

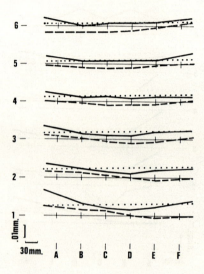

FIG. 7. Test conditions: A1, B1, C1 (see list of variables tested).

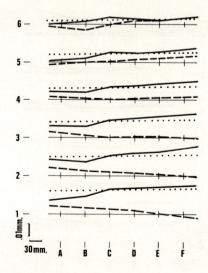

FIG. 8. Test conditions: A1, B1, C2 (see list of variables tested).

Apart from this drawback, which can be eliminated by an accurate calibration of the pressures involved, the surfaces obtained are generally of the same quality of the base surfaces. Consequentially they could be used for slideways without requiring further machining. In order to have a quantitative measurement of the planarity of the surfaces obtained by this method, the sum of squares of the distances of each measured point from the "regression plane" were calculated. This could be used as an estimate of the planarity of the surface obtained. However, in order to present an estimate more familiar to the

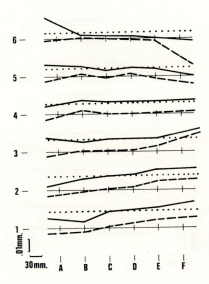

FIG. 9. Test conditions: A2, B2, C1 (see list of variables tested).

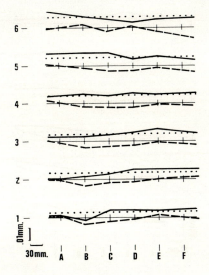

FIG. 10. Test conditions: A2, B2, C2 (see list of variables tested).

engineer, the root mean square deviation from the "regression plane" was used. The values are shown in Table 1.

Using the second order interaction as an estimate of our error, none of the six components of the sum of square of deviations (three main factors and three first order interactions) is significant to a 90% confidence level, and only two interactions are significant to a confidence level over 80%. The amount of data available is insufficient to draw conclusions.

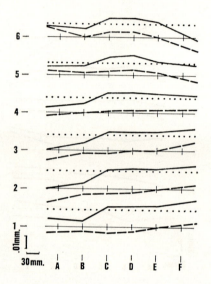

FIG. 11. Test conditions: A1, B2, C1 (see list of variables tested).

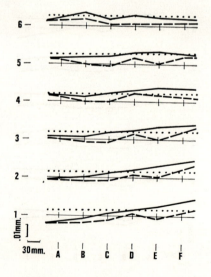

FIG. 12. Test conditions: A1, B2, C2 (see list of variables tested).

However, there is an indication that the interactions between the method of loading the strip and the roughness, and between the method of loading the strip and the cross-section of the grooves might, by replicating the tests, be significant. When the strip is loaded by the stiff metal rule, better results are obtained with a ground base surface and small grooves. When it is loaded by "hydrostatic" pressure better results have been found with a scraped surface and larger grooves.

TABLE 1

		B			
		1		2	
		C		C	
		1	2	1	2
A	1	3·990	5·180	4·878	5·303
	2	6·433	4·967	4·604	3·293

Research now in progress at R.T.M. will define the optimum conditions for bonding the strips when loaded with an "hydrostatic" pressure. This bonding condition is very interesting in the case of long slideways, in which the use of stiff metal rules might become costly and difficult.

The wear of strip slideways will also be investigated. Friction measurements are also in program.

CONCLUSIONS

The proposed procedure has given slideways surfaces which are comparable, from the point of view of surface planarity, to those obtainable by scraping or grinding. It is felt, as a consequence, that the surfaces thus obtained can be used as slideways without further machining.

The procedure has proved easy enough for unskilled labour to learn efficiently and quickly. The surfaces presented in the test were bonded by an unskilled labourer who had seen the whole procedure only once before. No particular difference has been found between the first surfaces he bonded and the last.

REFERENCES

1. K. P. DOLGOV and E. G. NIZHNIK, Design and Repair of Slip Slideways. *Machines and Tooling*, 1967, No. 3, 22–25.
2. A. L. F. CELINDER, Controlling Friction on Tool Ways. *American Machinist*, 4 Dec., 1967, p. 146.

COMPUTER DESIGN OF ELASTIC MULTI-COMPONENT CYLINDRICAL PRESSURE VESSELS SUBJECTED TO CYCLIC INTERNAL PRESSURE

R. J. Pick, D. J. Burns

University of Waterloo, Ontario

and

B. Lengyel

Imperial College, University of London

SUMMARY

A computerized, heuristic search method is described for minimizing the wall thickness of elastic, multi-ring shrink-fit vessels subjected to cyclic internal pressure. The design constraints considered are: fatigue, brittle fracture, assembly temperatures, material ruling sections, machining tolerances and no yielding during assembly or operation. The method is particularly useful when three of more rings are required and some of the rings have different elastic constants.

INTRODUCTION

Taper or shrink-fit multi-ring compound cylinders have been used extensively for extrusion dies, intensifier barrels and compressor cylinders which are subjected to a large number of pressure cycles. The development of commercial hydrostatic extrusion processes, that require very high cyclic pressures, has led to a renewal of interest in the design of compound cylinders, satisfying fatigue conditions imposed by the batch loading. [1,2]

Material and cost limitations frequently lead the designer to attempt to minimize the wall thickness of such vessels. Direct methods for minimizing the wall thickness of compound cylinders consider only one of several design constraints, for example simultaneous yielding of the components under pressure [3,4] or required fatigue life of the components. [2] Designs formulated in either of these ways must be checked and perhaps modified to ensure that the other criterion is satisfied; that yielding will not occur during the assembly operation; that interference fits give acceptable assembly temperatures; that machining tolerances will not lead to unacceptable stresses; that material ruling sections are not exceeded and that brittle materials carry little or no tension. Since all requirements must be satisfied simultaneously, modifications made by trial and error [5] can be time-consuming and give a solution which is acceptable but not necessarily optimum. This is particularly true when more than three components are required and the components have different elastic constants.

This paper describes a computerized, heuristic search method which considers all of the aforementioned constraints and determines the minimum wall thickness of the shrink-fit multi-ring cylinder required for a given bore diameter and cyclic internal pressure. There is no fundamental reason why this heuristic search method cannot be used to optimize taper-fit vessels; vessels containing autofrettaged components [6] or vessels of the ring-fluid-segment type. [2]

M.T.C—2G

DESIGN CRITERIA

The program is written to analyse long, elastic cylinders made by shrinking together concentric cylinders, which are stress-free prior to assembly. It is assumed that the vessel operates at room temperature and that fluctuations in room temperature do not lead to significant thermal stresses. However there is no reason why thermal stress calculations cannot be inserted into the programme when the transient and steady state temperature distributions can be specified.

When calculating the stresses produced by (a) the interference fits and (b) the operating pressure, a state of plane stress (longitudinal stress is zero) or generalized plane strain (longitudinal strain is constant) is assumed for the centre section of each component. To use generalized plane strain for calculating residual stresses it is necessary to estimate the additional longitudinal strains which are likely to be induced during the assembly as a result of the components gripping each other before they have reached equilibrium temperature.[7]

The program can be used with either the Tresca maximum shear stress or Maxwell–Mises octahedral shear stress criteria for yielding.

As a criterion for avoiding brittle fracture, limits are set to the maximum tensile stress which each component can carry. For materials such as tungsten carbide no tensile stresses are allowed. For materials which have no definite transition temperature, linear-elastic fracture mechanics and estimates of initial flaw size and cyclic crack growth rates can be used to determine the limit for the maximum tensile stress.[8] For materials which have a sharply defined transition temperature, the Pellini fracture analysis diagram can be used to set a stress limit for a given operating temperature.[9]

There is still considerable uncertainty as to the stress parameters which control the fatigue behaviour of thick-walled cylinders, particularly when the cylinders are made from very high strength materials and/or are subjected to high mean stress.[10,11] Three fatigue criteria are incorporated in this program. It has been assumed that the fatigue strength of relatively low strength, ductile materials in the elastic range is dependent primarily on the maximum range of shear stress. The mean shear stress and maximum normal stress on the planes of maximum range of shear stress are included as secondary factors.[12] Lack of experimental data made it necessary to assume that very high strength materials would not fail in fatigue, if they were loaded in compression only and subjected to below yield stresses.[1]

Factors of safety are applied directly to the fatigue, yield and brittle fracture limits; these factors can be varied from component to component.

HEURISTIC SEARCH METHOD

A preliminary selection of materials and a quick estimate of the minimum number of component cylinders required for a given bore pressure and radius can be made by using the method developed by Manning[3,4] to find designs which give simultaneous yielding of all components at the operating pressure. This minimum number of components and the yield strength, section thickness, fatigue, brittle fracture and assembly temperature characteristics of the chosen materials are the starting point for the heuristic search method.

The first step in the search method (Fig. 1) is to choose the interface radii and outside radius, r_0, by setting the wall thicknesses of all components at their minimum acceptable values. The second step is to calculate the stress ranges corresponding to this set of radii and the given internal pressure. These stress ranges are then used with the fatigue, brittle fracture and yield characteristics of the chosen materials to set limits for the "operating"

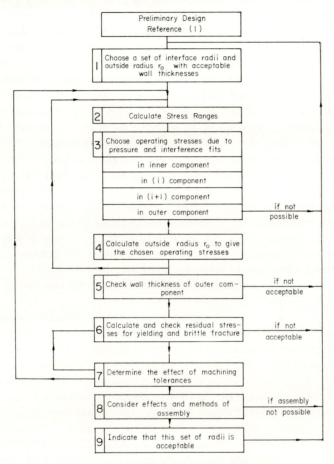

FIG. 1. Heuristic search method.

stresses produced by the combination of internal pressure and interference fit. The program considers each component in turn, beginning with the inner component and working outwards. The operating stresses in a component are placed at the maximum acceptable level within the yield, brittle fracture and fatigue limitations, by the choice of the interference fit on the outside of that component; this obviously specifies the pressure on the bore of the next component. The pressure on the outside of the outer component cannot be adjusted and therefore acceptable operating stresses can only be attained by adjusting the outside radius. Since this outside radius will disagree with r_0, steps 2 to 4 are repeated with the adjusted outside radius until the disagreement is insignificant. Using the new outside radius the wall thickness of the outer component is checked to see if it is acceptable (step 5).

If it is not possible to satisfy all stress limitations or to determine an outside radius within acceptable wall thicknesses the program produces a new set of radii and repeats steps 2–5. New sets of radii are produced and checked until a set satisfies steps 2 to 5. The procedure by which sets of radii are produced will be described later.

In step 6 the residual stress distribution is checked to see that it does not violate the yield and brittle fracture criteria. If there is a violation the program produces a new set of radii and repeats the search method until steps 2 to 6 are satisfied.

Step 7 examines the maximum effect that machining tolerances might have on the residual and operating stress distributions. For any component the combination of interference fits giving the highest residual stress is when all interference fits inside the component are at their minimum value and all interference fits outside the component are at their maximum value. These critical combinations of interference fits are fed back into step 6 to check that the tolerances do not give residual stresses which violate the yield or brittle fracture criteria. If there is no violation the second part of step 7 examines the effect that machining tolerances might have on the operating stress distributions.

For any particular component the combination of interference fits which gives the highest operating stresses is when all interference fits interior to the component are at their maximum values and all interference fits exterior to the component are at their minimum values. These critical combinations of interference fits are used to calculate the maximum increases in operating stress that could occur in each component. These increases in operating stress violate one or more of the design criteria considered previously in step 3. Therefore a new set of lower operating stresses must be chosen to offset the increases that may occur because of machining tolerances. In other words the iteration starts again at step 3 with an additional restriction on operating stresses and repeats until the machining tolerances do not lead to violations of the design criteria.

The next step is to check that the vessel can be assembled without violating yield, brittle fracture and assembly temperature criteria. The latter are the maximum temperatures to which each component material can be heated or cooled. Although the computer can be programmed to examine all permutations for the order of assembly of the components, there are preferred orders which it checks first. For example, if the two inner components are made from brittle materials it first considers assembling inwards from the third component and then outwards from the sub-assembly. The temperature difference calculated for each assembly operation includes allowances for machining tolerances, the current diameter of any subassemblies and the minimum clearance required to facilitate assembly. If it is not possible to assemble the cylinder the program produces a new set of radii and continues the search.

If it is possible to assemble the cylinder the computer indicates that this set of radii and interference fits are acceptable. It then proceeds to examine the remaining sets of radii.

As each set of radii is examined, the outside radius calculated in step 4 is compared with the value obtained for the first acceptable design. If the calculated value is larger, then the set of interface radii is rejected as the resulting design could not be closer to the optimum, i.e. the steps 4 to 8 are not completed. If the proposed set of interface radii completes step 8 and has a smaller outside radius, then it is stored in place of the first design and used for further comparisons. Thus when the search is completed the set of interface radii with the minimum outside radius will have been retained. If on completing the search an acceptable design has not been found, the designer must start again with a preliminary design study for an assembly containing one additional component and then repeat the search program.

The actual method of producing all possible sets of radii within a region is best illustrated by an example. Each design produced must differ from all others by an increment δr on at least one of the dimensions. Obviously if δr is large the number of possible sets of interface radii within the acceptable wall thicknesses will be small. Thus the value of δr will determine the coarseness of the search and is decreased after the initial search.

To illustrate the procedure for selecting sets of radii, assume that the preliminary design calculations have shown that a minimum of three components is required. The first set of radii, r_1, r_2, r_3 and r_0 is derived from the given bore radius r_1 by setting the wall thicknesses of all components at their minimum values. Sets of radii can now be generated in two ways. First by increasing the interface radius between the outer and middle components in steps until the middle component is at its maximum wall thickness. This gives sets of radii r_1, r_2, $r_3 + \chi \delta r_3$, where χ is the number of steps taken, δr_3 is the size of the step and $\chi < n$ where $n\delta r_3$ is the difference between the maximum and minimum wall thicknesses acceptable for the middle component.

When $\chi = n$, further sets of radii are generated by increasing the interface radius between the inner and middle components in steps until the inner component is at its maximum wall thickness. After each step, δr_2, the procedure described in the previous paragraph is repeated. This gives sets of radii r_1, $r_2 + y\delta r_2$, $r_3 + y\delta r_2 + \chi\delta r_3$, $r_0 + y\delta r_2 + \chi\delta r_3$ where y is the number of steps taken and $y < p$, where $p\delta r_2$ is the difference between the maximum and mimimum wall thicknesses for the inner component.

When acceptable sets of radii have been found the size of the steps δr_2 and δr_3 can be reduced and a search is made in the area of the acceptable set of radii which gave the minimum outside radius. The program is efficient because the optimum is approached smoothly even for reasonably large values of δr_2 and δr_3.

The portion of the program producing the sets of interface radii is independent of other portions of the program. This allows modification of the design criteria without disturbing the search method.

EXAMPLE

Consider the design of an open-ended cylinder with a 4 in bore diameter, to withstand a cyclic internal pressure of 250,000 lb/in². Assume that preliminary calculations have shown that a minimum of five components will be required. Assume that the designer has decided to use tungsten carbide for the two inner components because of its high compressive strength and evidence that it will have an acceptable fatigue life if it is not subjected to tensile stresses.[1] Also consider making the three other components of the same alloy steel. The fatigue strength of these components has been estimated by reference to data showing the effect of the maximum tensile stress on the fatigue behaviour of a $2\frac{1}{2}\%$ Ni-Cr-Mo steel.[1] It is assumed that Tresca's yield criteria is applicable.

In view of the uncertainty about the longitudinal strain distributions produced during the assembly operation[7,10] the program was first run using plane stress to calculate residual stress and generalized plane strain to calculate stress amplitudes. During this first run the requirement that tungsten carbide components must not be subjected to tensile stresses had to be relaxed to allow longitudinal tensile stresses. If machining tolerances are not considered the cylinder designed to the above specifications using a search increment, δr, of 0·5 in has the dimensions shown in Table 2. If a machining tolerance of $\pm (0\cdot001 + r(0\cdot0001))$ inches is allowed on an interference fit at radius r, there is a significant change in the minimum outside radius and the interface radii associated with the minimum (Table 3).

TABLE 1. DESIGN DATA. ALL UNITS IN LBS AND INCHES

Component No.	1 (Inner)	2	3	4	5
Shear yield stress	220000	179200 (1/2 compressive yield)	100000	100000	100000
Maximum allowable tensile stress	0	0	160000	160000	160000
Maximum acceptable wall thickness	1·5	2·5	4·0	5·0	6·0
Minimum acceptable wall thickness	·5	·5	1·0	2·0	3·0
Poisson's ratio	·31	·31	·28	·28	·28
Modulus of elasticity	35×10^6	35×10^6	30×10^6	30×10^6	30×10^6
Longitudinal residual strains	− ·0044	− ·00425	− ·0025	− ·00030	+ ·00120

TABLE 2. DESIGN WITH MINIMUM OUTSIDE RADIUS: NOT CONSIDERING MACHINING TOLERANCES, USING GENERALIZED PLANE STRAIN FOR STRESS AMPLITUDES AND PLANE STRESS FOR RESIDUAL STRESS. ALL UNITS IN LBS AND INCHES

Component	1 (Inner)	2	3	4	5
Outside radius	3·0	4·5	6·0	8·5	12·18
Radial interference fit at bore	—	·006	·0144	·0159	·0246
Operating tangential stress at bore	0	0	50182	90340	131687
Residual tangential stress at bore	− 273077	− 127777	− 2626	57861	112322
Operating radial stress at bore	250000	180556	130401	90899	45432
Residual radial stress at bore	0	75855	90278	71104	38751

Assembly Operation:
1. Assemble component 3 over 2, temperature difference required = 279 centigrade degrees
2. Assemble components 2, 3 over 1, temperature difference required = 475 centigrade degrees
3. Assemble Component 4 over 1,2,3, temperature difference required = 431 centigrade degrees
4. Assemble component 5 over 1,2,3,4, temperature difference required = 465 centigrade degrees.

TABLE 3. DESIGN WITH MINIMUM OUTSIDE RADIUS: CONSIDERING MACHINING TOLERANCES, USING GENERALIZED PLANE STRAIN FOR STRESS AMPLITUDES AND PLANE STRESS FOR RESIDUAL STRESS. ALL UNITS IN LBS AND INCHES

Component	1 (Inner)	2	3	4	5
Outside radius	3·0	4·5	6·5	9·5	13·943
Radial interference fit at bore	—	·0064	·0146	·0208	·0283
Operating tangential stress at bore	− 33372	− 18404	37892	87136	127894
Residual tangential stress at bore	− 303132	− 143787	13202	60138	112691
Operating radial stress at bore	250000	189825	142208	95318	46799
Residual radial stress at bore	0	84203	100754	77960	41236

Assembly Operation:
1. Assemble component 3 over 2, temperature difference required = 284 centigrade degrees
2. Assemble components 2, 3 over 1, temperature difference required = 524 centigrade degrees
3. Assemble component 4 over 1,2,3, temperature difference required = 450 centigrade degrees
4. Assemble component 5 over 1,2,3,4, temperature difference required = 468 centigrade degrees.

TABLE 4. DESIGN WITH MINIMUM OUTSIDE RADIUS: CONSIDERING MACHINING TOLERANCES, USING GENERALIZED PLANE STRAIN THROUGHOUT. ALL UNITS IN LBS AND INCHES

Component	1 (Inner)	2	3	4	5
Outside radius	3·0	4·5	6·5	9·5	14·12
Radial interference fit at bore	—	·0057	·0105	·0152	·0221
Operating tangential stress at bore	— 37842	— 21065	37258	86497	127298
Residual tangential stress at bore	— 307335	— 146254	— 13697	59629	112221
Operating radial stress at bore	250000	191067	143844	96693	47978
Residual radial stress at bore	0	85371	102283	79219	42296
Operating axial stress	— 246584	— 217864	— 107718	— 14729	55335
Residual axial stress	— 249274	— 220554	— 107474	— 14485	55579

Assembly Operation
1. Assemble component 3 over 2, temperature difference required = 208 centigrade degrees
2. Assemble components 2,3 over 1, temperature difference required = 440 centigrade degrees
3. Assemble component 4 over 1,2,3, temperature difference required = 342 centigrade degrees
4. Assemble component 5 over 1,2,3,4, temperature difference required = 365 centigrade degrees

In both cases the tungsten carbide components are apparently required to carry a longitudinal tensile stress of about 2650 lb/in², which is unacceptable. Fortunately this tensile stress will almost certainly be eliminated by compressive longitudinal stresses introduced in the inner components by the order and method of assembly (Tables 2 and 3). A rough estimate of these compressive longitudinal stresses/strains can be made using the procedure suggested by Jorgensen.[7]

The program can then be run using generalized plane strain and the estimated longitudinal strains. Other estimates of longitudinal strains may also be used to demonstrate their effect. In this example their general effect is to reduce the required interference fits and assembly temperature differences. This suggests that the estimate of longitudinal strains based on data from Table 3 and Jorgensen's procedure are too high. Therefore one half of these values were used (Table 1) and the program run again (Table 4). Consideration of residual longitudinal strains still causes a significant change in the interference fits necessary to minimum outside radius. As mentioned previously they ensure that brittle inner components give the do not carry longitudinal tensile stress.

The investigation of this example required a total time of approximately 15 min computing time on an IBM 360/75 computer.

REFERENCES

1. B. LENGYEL, D. J. BURNS and L. V. PRASAD, Design of Containers for a Semi-continuous Hydrostatic Extrusion Production Machine, *Proc. 7th Int. Machine Tool Design and Research Conference*, Pergamon Press, 1967, p. 319.
2. T. C. GERDEEN and R. J. FIORENTINO, Analysis of Several High Pressure Design Concepts. Conf. on High Pressure Engineering, *Proc. Instn. Mech. Engrs., London*, **182**, part 3C, 1967–68, pp. 11–21.
3. W. R. D. MANNING, The Design of Compound Cylinders for High Pressure Service, *Engineering*, **163**, 1947, 349–352.
4. K. E. BETT and D. M. NEWITT, The Design of High Pressure Vessels, *Chemical Engineering Practice*, Butterworth Scientific Publications, 1958, vol. 5, p. 196.
5. K. E. BETT and D. J. BURNS, Design of Elastic Multi-component Compound Cylinders. Ibid., ref. 2, pp. 22–29.
6. T. L. SKINNER, R. D. DANIELS and C. M. SLIEPCEVICH, Design of Vessels for Commercial Service at Extreme Pressures, *British Chemical Engineering*, **8**, 1963, 245.
7. B. CROSSLAND and D. J. BURNS, Behaviour of Compound Steel Cylinders subjected to Internal Pressure, *Proc. Instn. Mech. Engrs. London*, **175**, 1961, 1083.

8. E. T. Wessel, W. G. Clark and W. K. Wilson, Engineering Methods for the Design and Selection of Materials against Fracture. Clearinghouse for Federal, Scientific and Technical Information, AD 801005, 1966.

9. A. G. Pickett and S. C. Grigory, Prediction of the Terminal-failure Behaviour of Pressure Vessels. A.S.M.E. preprint 67-Met-2 presented at Metals Engineering Conference, Houston, Texas, April 1967.

10. D. J. Burns and J. S. C. Parry, Effect of Mean Shear Stress on the Fatigue Behaviour of Thick-walled Cylinders. Ibid., ref. 2, pp. 72–80.

11. B. A. Austin, A. N. Reiner and T. E. Davidson, Low Cycle Fatigue Strength of Thick-Walled Pressure Vessels. Ibid., ref. 2, pp. 91-105.

12. W. J. Frost and D. J. Burns. Effect of Oil and Mercury at High Pressure on the Fatigue Behaviour of Thick-walled Cylinders of EN25 Steel. Ibid., ref. 2, pp. 65–71.

COMPUTER-OPTIMIZED GENERATING CYCLE FOR CUTTING HYPOID PINIONS

D. N. Curtis, G. M. Spear and M. L. Baxter, Jr.

Gleason Works, Rochester, New York, U.S.A.

SUMMARY

This paper describes a computer-aided design technique for obtaining an optimum cycle based on minimum cutter wear in the generation of spiral bevel and hypoid pinions.

The chips produced in this type of gear cutting are polyhedrons with characteristics varying over a wide range, depending on the instantaneous relative position and velocity of cutter and work. The geometry of these individual chips and a cutter wear equation of exponential form are the two basic elements of this procedure. A quantity defined as the cutter wear number is calculable from this information; this quantity is to be minimized. This is done by changing the functional relationship of roll-feed with respect to time in an iterative process until a minimum value of the cutter wear number is obtained.

The resulting ideal roll-feed function must be compared with the dynamic requirements of the cam operated machine and modified, if necessary.

Comparison of a cutter wear number with that of a known cycle gives a measure of predicted cutter wear. The use of a properly designed cycle can reduce cutter wear by one-half or more from that of a uniform motion cycle.

INTRODUCTION

The history of hypoid pinion manufacture has been one of steady improvement in both quality and production rate. This paper describes a computer-aided design procedure to achieve further improvement in production by determining an optimum cutting cycle based on minimum cutter wear.

In the mass production of hypoid pinions, the first operation is a roughing cut which generates the approximate tooth shape. A small amount of uniform stock remains to be removed in subsequent finishing. The major portion of the metal removal work is done during the roughing operation and cutter wear is an important factor limiting the production rate. The importance of this factor has been magnified by current progress toward extremely fast production rates in the newest hypoid pinion generator designs.

Hypoid pinions have a unique tooth shape, as illustrated in Fig. 1. The tooth is curved in both the profile and lengthwise directions with tapering depth and thickness. This shape is the result of exact mathematical relationships essential to the production and operation of these pinions with their mating gears. Accordingly, the shape is ideally suited to analysis by digital computer. Other examples of computer-aided design relating to hypoid pinions are the determination of fatigue life and wear,[1] and the analysis of the running qualities of a pair of gears.[1,3]

The generating process used to rough cut hypoid pinions produces chips which vary substantially in size and shape throughout the cycle. This special circumstance provides an opportunity to analyze the chips by computer and modify certain conditions to obtain a more uniform distribution. It will be shown that this distribution gives minimum total cutter wear.

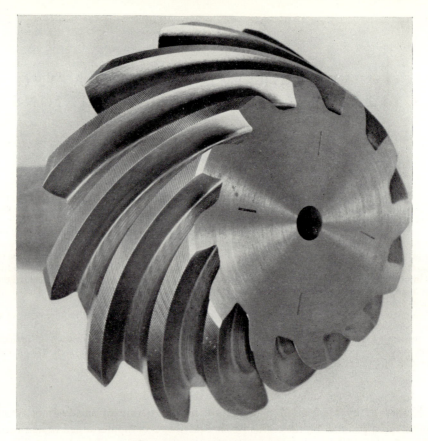

Fig. 1. A typical hypoid pinion.

METAL REMOVAL AND GEAR GENERATION

In the generation of hypoid pinions, the cutter represents a tooth of an imaginary generating gear which is made to roll in the proper relationship with the work piece. A typical machine set-up is shown in Fig. 2. The cutter spindle assembly is carried on a member which rotates about the imaginary generating gear axis, called the cradle axis. The work piece rotates about its own axis, and the work axis and cradle axis are connected internally in the machine by a train of gears to give the correct relative roll.

The basic elements of the machine are diagrammed in Fig. 3. The cradle and work, which are geared together by the generating train, are powered by a roll cam. The roll cam controls the number of degrees of cradle rotation necessary to complete generation of the work. The depth position cam turns at the same rate as the roll cam and moves the work or cutter in the direction of the cradle axis to provide clearance for indexing. Finally, note that the cutter is driven at a rate which is independent from the rest of the system.

From the foregoing, it is seen that the requirement of maintaining proper generating roll governs the process; but uniform generating roll may not necessarily give the best conditions for metal cutting. The chips vary in thickness, shape, and length; Fig. 4 is a diagram showing typical chip variation occurring with uniform generating roll. The diagram is a representation of two roughing chips attached to the final tooth surface at their original

Fig. 2. Machine arrangement for cutting a hypoid pinion.

location. Note that the chips are L-shaped and that the end chip and side chip change considerably in thickness, as shown in the transverse section. In addition, the length and shape vary substantially as generation moves from one end of the tooth to the other.

The special nature of this cutting is also apparent from the transverse section of the diagram. The cutter blades are imbedded in a relatively long, narrow space between the teeth. A face-mill cutter is used, having alternate inside and outside cutting edges, and the metal removal process can be classified as alternate-blade slot-milling. The blade design and arrangement of this cutter must provide for adequate chip clearance and chip flow.

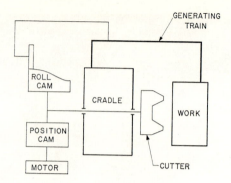

Fig. 3. Basic elements of a hypoid pinion generating machine.

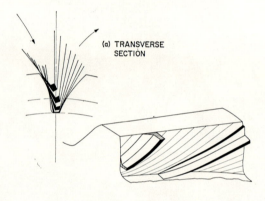

(b) CHIP DISTRIBUTION

Fig. 4. Schematic diagram showing typical chip variation.

NEED FOR AN OPTIMIZING PROCEDURE

Cutter wear can be reduced by making each cutter blade do more nearly the same amount of work. This can best be accomplished by making the chip distribution as uniform as possible. Since the relative generating roll between cradle and work cannot be changed, this leaves only the rate of roll supplied by the roll cam available for revision. A varying rate of roll together with a constant rate of cutter rotation will give a different pattern of chip distribution. The roll cam, then, provides a roll-feed relationship in this process. The purpose of the procedure described in this paper is to find the roll-feed path as the function of time which will give the least total wear on the cutter.

The optimum roll-cam path is assumed to be independent of both the overall cycle time and the speed of the cutter. For any given pinion design, there exists an optimum roll cam for which the cycle time and cutter speed can be varied depending on the economics of the cutting process. This will then result in the minimum cutter wear for that particular speed and cycle time.

Minimum cutter wear, of course, makes possible an increased production rate. The exact balance between cutter wear and production rate is dependent on the costs in the particular shop involved, but an improvement in either item represents a direct saving.

Cams with a varying rate of roll are used on existing generating machines to achieve fast cutting cycles on this basis. The cam path currently used for automotive pinion production was determined by experimental cutting on a typical part and the results made to apply on the average to the whole range of automotive pinions. The method of computer analysis described here is intended to predict the optimum path for any given pinion design. Together with new machines capable of extremely fast production times and having a simplified roll cam system, this computer technique gives the advantage of determining an optimum cam for each individual pinion design.

The multi-bladed face-mill cutters used in pinion roughing cut without a timed relationship between the cutter and work. Under these conditions, a particular blade cuts at many different positions along the pinion tooth throughout its life. Wear, therefore, tends to be distributed uniformly over the full complement of blades. The wear equation, to be discussed later, relates wear to depth of cut raised to a power substantially greater than one. For this reason, a scattering of heavy cuts influences tool life more than a series of lighter cuts. It is because of this effect that an optimized cam design is necessary.

OVERALL COMPUTER PROCEDURE

The computer-aided design procedure for obtaining the optimum roll-feed cam path presently consists of four separate steps, each comprising a separate computer program. By the use of a larger computer, these could be combined into one continuous program. Figure 5 shows the current arrangement.

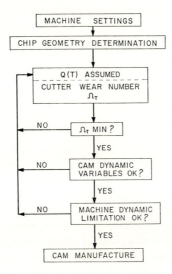

FIG. 5. Flow diagram of overall computer procedure.

Once the pinion design is complete, the first step is to obtain machine settings for producing the part, which includes determining the required amount of generating roll. The second step is to find the size and shape of the chips produced by these machine settings, based on a uniform rate of roll. Next, a third computer program is used to calculate a cutter wear number (CWN) which indicates the relative amount of cutter wear. Changes are made in the roll-feed function, $Q(T)$, until a minimum value is found for the CWN. The roll-feed function obtained in this manner is then analyzed in a fourth step by a separate computer program to determine the actual cam design. This is further checked against machine dynamic requirements and, if satisfactory, the new cam is ready for manufacture.

The second and third steps are the concern of this paper and will be described in more detail.

DETERMINATION OF CHIP GEOMETRY

The size and shape of the chips for a uniform rate of roll are determined using a computer program based on a mathematical model of the generating machine. The exact machine settings and workpiece dimensions are used as input to this program and, therefore, approximations are eliminated. Figure 6 shows the initial definition of the vectors used in this mathematical model.

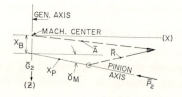

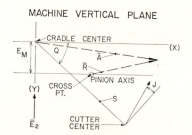

Fig. 6. Vector representation for chip geometry program.

The lower view is in the machine vertical plane, looking into the face of the cradle, with the cradle center at the origin of the X and Y axes. The upper view is in the machine horizontal plane and shows the locations of the generating gear axis and the work axis. Directions of unit vectors are defined as shown. Of principal importance are the vectors $\bar{A}$, from the cradle axis to a point on the cutter, and $\bar{R}$, from the work axis to a corresponding point on the work. The basic operation of the program consists of locating a point on the boundary of the tooth at which the chip dimension is desired and defining this point in terms of $\bar{R}$. The program is then made to iterate until $\bar{A}$ is found for the corresponding point on the cutter.

The program steps the cradle angle by a predetermined increment through successive positions from the start of roll to the end of roll. At each cradle position, the program examines the topland, inner (toe), and outer (heel) boundaries of the tooth and, by varying the cutter phase angle, determines which two of these boundaries apply at that particular cradle position. It is known that the maximum thickness of the chip will occur at the tooth boundaries. At each boundary, the program calculates the end chip thickness, the side chip thickness at the tip of the cutter and periphery of the gear blank, and the width of the side chip. The manner in which these dimensions are found is illustrated in Fig. 7. From the vector system in the program, the relative linear velocity, $\bar{V}$, at any point on the cutter blade can be calculated. The relative motion vector at the tip of the cutter can be projected

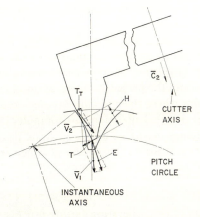

FIG. 7. Relative velocities in chip formation.

in the direction of the cutter axis to give the relative velocity producing the end chip, and in the direction normal to the blade to give the component producing the side chip, as follows:

$$E = \bar{V}_1 \cdot \bar{c}_2$$
$$T = \bar{V}_1 \cdot (-\bar{n}_2)$$

where $\bar{c}_2 =$ unit vector along cutter axis,

$\bar{n}_2 =$ unit vector normal to cutting edge.

In a similar manner, the side chip at the periphery of the gear blank is found by projecting the relative motion vector at that point in the direction normal to the blade:

$$T_T = \bar{V}_2 \cdot (-\bar{n}_2)$$

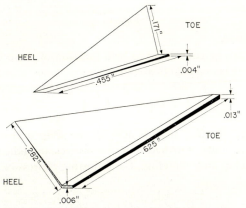

FIG. 8. Chip dimensions calculated by computer.

Because the teeth are being cut from a blank, a positive value for any of these projected quantities indicates that metal is being removed. Since the number of degrees of cradle rotation is known for the time between each blade, these quantities can be directly converted to chip thickness. A negative value indicates that cutting has stopped at that point. This information is used to determine the width of the side chip, H, in Fig. 7. Since T_T is negative and T is positive, a proportion is established and the point where the side chip is zero can then be found.

The program also determines the length of the end chip by finding the difference between the cutter phase angle positions at the two boundaries of the tooth. This angular value is converted directly to a linear dimension. In many cases, the value of the end chip is negative at one of the two boundaries and the program, by iteration, finds the intermediate value of cutter phase angle at the point where the end chip becomes zero.

Dimensions for two typical chips calculated by this computer program are shown in Fig. 8. These are two outside blade chips occurring at two separate positions in the generating roll for a spiral bevel pinion. They illustrate the results obtained from this program and give another indication of the variance in chip geometry.

METAL CUTTING EXPERIMENTS

In view of the special nature of this generating cutting process, it was desirable to conduct an experimental investigation to verify that an optimum relationship could be found. This was done as one phase of a long-range study of the metal cutting process applied to gear teeth. The test arrangement is shown in Fig. 9. Several forms of blade wear and blade failure were encountered in these tests and each form required a different set of criteria for optimization. The tool materials which proved most satisfactory exhibited the conventional wear-land type of failure when dulling.

The Taylor equation relates tool life and tangential velocity of the tool as follows:

$$VT^n = C$$

where V = tangential velocity of tool,
$\quad\quad\ T$ = tool life,
$\quad\quad\ C$ = constant,
$\quad\quad\ n$ = constant.

A variation of this equation used by Shaw[4] extends the concept to include chip thickness in the following manner:

$$ET^{n'} = C'$$

where E = chip thickness.

Tool life, therefore, can be considered to be a function of the product of these two variables in the range that these constants are effective,

$$T = \left(\frac{C}{V}\right)^{1/n} \cdot \left(\frac{C'}{E}\right)^{1/n'}$$

Since a relative measure of cutter wear is desired, the reciprocal of T will be used and C and C' will be chosen as unity. Furthermore, the test results indicate that wear is proportional to the length of chip and, therefore, the following expression is obtained for relative cutter wear:

$$\Omega_i = \frac{1}{T} = L(V)^{\frac{1}{n}} (E)^{\frac{1}{n'}}$$

FIG. 9. Instrumentation used for metal cutting studies.

where Ω_i = wear on an individual blade per cut,
          L = length of chip.

The cutter wear number (CWN) is the summation of the wear produced on each blade by cutting each chip and is represented by the following equation:

$$\text{CWN} = \Omega_T = \sum_{i=1}^{n} \Omega_i$$

A study of wear patterns obtained from tests has shown that significant wear occurs first behind the cutting edge in the region just above the tip of the tool. Figure 10 is a micro-photograph of a typical test blade showing this condition. It has further been concluded that the parameter which most directly influences this wear area is the thickness of the end chip. Accordingly, the calculated length of end chip and the calculated thickness of end chip are the values substituted for L and E in the wear equation.

COMPUTER PROGRAM LOGIC FOR RELATIVE WEAR

The basic procedure for computing the CWN is outlined in Fig. 11. Input to the computer consists of the following three classes of information:

1. *Production data.* General information about pinion design, cutter design, cycle time, and cutting speed.

FIG. 10. Wear land adjacent to cutting edge on a face-mill cutter blade. Magnification approximately 10×.

2. *Chip dimension data.* A series of end chip length and thickness values for uniform increments of cradle angle. This is the output deck of cards from the chip geometry program described earlier in this paper.

3. *Cam path data.* The initial roll-feed function with respect to time, $Q(T)$. The function input consists of velocity and acceleration information at important values of cam turning angle. Stated in other terms, this input is the velocity vs. cam turning angle diagram for the roll-feed cam.

The program uses this input to first calculate the CWN for a uniform roll cam using the end chip thickness and length data. A table look-up and straight-line interpolation are used to find thickness and length values at cradle angles between those given as input.

The program next calculates the CWN for the roll-feed function, $Q(T)$ given as input. Here, the cam follower displacement and velocity are determined at each blade interval and these values are converted to angular velocity and acceleration of the cradle.

The table look-up and interpolation routine is then repeated to find the correct chip length and the proper chip thickness, after modifying the chip thickness by a cradle velocity factor Output is a CWN for the outside blades and a CWN for the inside blades. If desired, the individual values for each intermediate blade can be displayed.

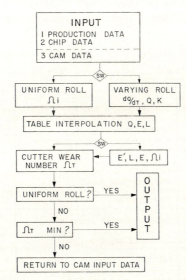

FIG. 11. Flow diagram for computing CWN.

The calculation procedure used in the program can best be explained by describing individual steps in the form of a hand calculation, as shown in Tables 1 and 2 and Figs. 12 and 13. Table 1 shows the first four blades in the CWN calculation for the uniform roll

TABLE I
"CWN" CALCULATION—UNIFORM ROLL

1	2	3	4	5
BLADE NO.	CRADLE ANGLE	END CHIP THICKNESS	CHIP LENGTH	CWN
N	Q	E	L	Ωi
1	0	0	0	0
2	1.1°	0.002"	0.138"	0.000"
3	2.2°	0.005"	0.275"	0.002"
4	3.3°	0.008"	0.413"	0.020"

$$\Omega i = L\,(V)^{\frac{1}{n}} \cdot (E)^{\frac{1}{n'}}$$

$$n = 0.20 \qquad n' = 0.31 \qquad n'' = 1.00$$

FIG. 12. CWN calculation for uniform roll.

portion of the program. Columns 2, 3 and 4 are typical values from the output of the chip geometry program. As stated previously, these values can be found for each cradle angle at which a blade cuts by interpolating between values spaced at larger uniform intervals in the actual chip geometry output. It is necessary to calculate the individual blade wear, Ω_i, for each blade since the CWN will be the sum taken over all the blades used.

Table 2 shows the procedure used when calculating the CWN for a given roll-feed function. Each blade, line 1, corresponds to a cam rotation angle, line 2, which is a measure of time. From the velocity diagram, or other data defining $Q(T)$, the instantaneous velocity of the cradle is found, line 3. Also, the cradle angle displacement is found, line 4. By referring to the uniform roll table, the value of end chip thickness, line 5, and end chip length, line

	TABLE II	
	"CWN" CALCULATION — VARYING ROLL	
1	BLADE NO.	N
2	CAM ROTATION ANGLE	T
3	CRADLE VELOCITY	$\dfrac{dQ}{dT}$
4	CRADLE ANGLE $\quad Q = \int \dfrac{dQ}{dT}$	Q
5	END CHIP AT CRADLE ANGLE (FROM TABLE I)	E'
6	CHIP LENGTH	L
7	PROPORTIONALITY CONSTANT $K = \dfrac{\text{CRADLE VELOCITY (ACTUAL)}}{\text{CRADLE VELOCITY (TABLE I)}}$	K
8	END CHIP THICKNESS $\quad E = K \times E'$	E
9	CWN $\quad \Omega_i = L\,(V)^{\frac{1}{n}} \cdot (E)^{\frac{1}{n'}}$	Ω_i

Fig. 13. CWN calculation for varying roll.

6, can be found for this cradle angle. This is the final value for chip length but the chip thickness must be modified by a proportionality constant, line 7, which relates the instantaneous cradle velocity for $Q(T)$ to the cradle velocity used with uniform roll. End chip thickness is then found, line 8, and the CWN from line 9.

OPTIMIZING TECHNIQUE

Originally, the roll-feed function giving the minimum CWN was found by making repeated runs of the program. The operator revised the input function $Q(T)$ each time, based on his best judgment of the previous CWN results. This procedure may appear cumbersome at first glance, but it represented a major advance when compared with the shop trial-and-error method which it replaced. Figure 14 shows a series of steps used to reduce cutter wear for a hypoid pinion. The upper diagram gives the roll-feed functions used and the lower diagram shows the corresponding plots of end chip thickness. Curve A represents uniform roll of the cradle. Curves B and C are typical successive applications by this method which resulted in substantially uniform end chips and, therefore, an increase in calculated cutter life. Representative computer runs have shown that cutter life is increased from 25 to 55% when an optimum cam is used instead of a uniform roll cam.

As stated earlier, uniform distribution of end chip thickness plays the greatest part in reducing wear because chip thickness is raised to a power substantially greater than one. An optimizing technique within the computer program has been devised to determine the optimum $Q(T)$ without requiring operator assistance. This technique is principally based on the fact that a uniform end chip thickness is very closely related to the optimum roll-feed function.

Essentially, this new optimizing technique consists of reversing the order of the steps described previously for Table 2 and arriving at the $Q(T)$ which most nearly satisfies the requirement for uniform end chip thickness. This $Q(T)$ is then modified in subsequent steps

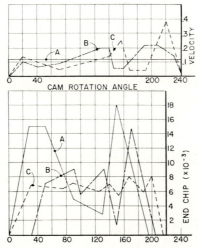

FIG. 14. Graphs representing changes in cutter wear.

to provide beginning and ending portions suitable for a real cam, consistent with a minimum CWN.

The nature of the steps taken in this latest technique have been determined from the following considerations. The end chip thickness can be expressed as:

$$E = \frac{dQ}{dT} \cdot f(Q)$$

where $f(Q)$ is represented by the table of values found by the chip geometry program and dQ/dT is the velocity at any instant. In the case of uniform roll, dQ/dT is a constant.

When the end chip becomes uniform, these functions equal a constant:

$$C = \frac{dQ}{dT} \cdot f(Q)$$

This differential equation can be solved for T, which is the cam rotation angle:

$$T = \int \frac{f(Q)}{C} \, dQ$$

The resulting values of T and Q represent the new roll-feed function, $Q_1 (T)$. If the total displacement Q does not match the cradle roll required for the machine, the area under the velocity curve, dQ/dT vs. time, is adjusted while keeping the shape of the curve. Finally, the computer program adds the beginning and ending portions of the velocity curve for the real cam and adjusts the overall curve until a minimum CWN is found.

EXPERIMENTAL RESULTS

The velocity term and its exponent $1/n$ in the wear equation can be considered constant for the purposes of this paper, since the optimum cycle depends only on relative chip size. In cases where comparisons of relative wear are to be made between different cutter velocities, a value of n equal to 0·20 is used.

In order to find a realistic value for n', a series of cutting trials were run at the Gleason Works. A conventional 12-tooth hypoid pinion of forged SAE 4027 steel (143-163 Brinell) was the workpiece and cutters were considered dull when the wear land was approximately 0·030 in. These trials showed that, on the average, 201 parts could be cut with a varying rate of roll cam before dulling the tool, as compared with 156 parts cut with a uniform roll cam. This represents an increase of 29%.

The coefficient n' was then adjusted in the computer program to agree with this result, giving a value of n' equal to 0·31. This value lies within the range of published results but it can, of course, be modified after further testing.

CONCLUSIONS

This computer-aided design procedure is a means of finding the optimum roll-feed relationship for producing a given hypoid pinion when the cutter design is known. The optimum condition is defined as minimum cutter wear, and this can result in either reduced cutter costs or higher production rate.

Additional advantages of this procedure are:

1. Eliminates shop trials in determining the optimum roll-feed function.
2. Permits comparison between various production estimates at the planning stage.

As more cutting data is obtained from metal cutting studies and from production in the field, refinements and improvements will be made to the computer program. These may include modifications of the values of the exponents and modifications of the technique for adapting the optimum roll-feed function to the real cam path. In addition, this basic technique may be further developed to apply to other bevel gear cutting processes.

It is hoped that this paper will serve not only to provide an insight into an interesting computer application for hypoid pinion production, but may also be of assistance to those considering applications in other areas of design.

REFERENCES

1. W. Coleman, A Scoring Formula for Bevel and Hypoid Gear Teeth. ASME Paper No. 65-Lub-8 October 1965.
2. M. L. Baxter, Jr., Exact Determination of Tooth Surfaces for Spiral Bevel and Hypoid Gears. AGMA Paper No. 139.02, October 1966.
3. G. M. Spear and M. L. Baxter, Jr., Adjustment Characteristics of Spiral Bevel and Hypoid Gears. ASME Paper No. 66-MECH-17, October 1966.
4. M. C. Shaw, Historical Aspects Concerning Removal Operations on Metals. Second Buhl International Conference on Materials, Carnegie Institute of Technology, Pittsburgh, Pennsylvania, March 1966.

THE STEELMAKERS' APPROACH TO MACHINABILITY

C. MOORE

Head, Machinability Studies, The Park Gate Iron & Steel Co. Ltd.

SUMMARY

The output and future trends in free machining steels are given to emphasize the care required in the chemical control during manufacture in the usual production unit of 100 tons of steel.

The setting up of a facility, based on a single spindle automatic lathe, to give commercially usable machining data for steels within En 1A, En 1B range and the reliability of results, is discussed. The relationship between machining and sulphide morphology in the bar is considered and the association of a 10% to 15% decrease in machinability with the incidence of a high proportion of small sulphide inclusions is noted.

The behaviour of sulphides during manufacture is discussed with a view to producing the desired structures more consistently in the final bar. Theories are advanced for the phenomena observed.

INTRODUCTION

Before dealing with the research aspects of free cutting steels it is important to put into perspective the tonnages used, the future trends of usage and the established methods of control during manufacture.

Currently, the annual production of non-alloy free cutting steel billets is in the order of $\frac{1}{2}$ million tons with an average increase in demand of about 20,000 tons each year (see Fig. 1).

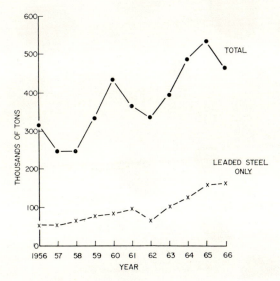

FIG. 1. Production of free cutting steel.

Whether this trend will continue is a matter of conjecture as other forming processes may become more attractive as the years pass. These could be casting, forging, cold forming, weld assembly and powder metallurgy.

Some idea of future trends can be obtained from statistics on machine tool production[1] (see Fig. 2). Metal cutting is the most popular method of shaping at the moment though the trend over the last decade had been downwards with forming increasing in usage. Metal cutting is, however, still the predominant method of shaping and will be for some time to come.

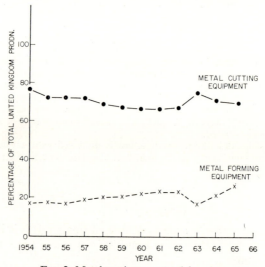

FIG. 2. Metal cutting vs. metal forming.

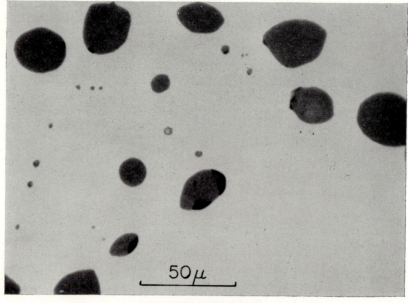

FIG. 3. Type 1 sulphides.

CONTROL IN STEELMAKING

The controls necessary are particularly relevant to low carbon free machining steels of the En 1A, En 1B type. It should be remembered that it is composition that can be controlled, and only indirectly the machinability of the final product.

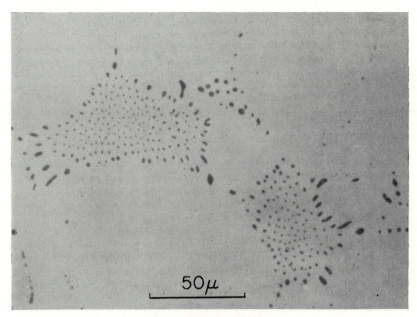

FIG. 4. Type 2 sulphides.

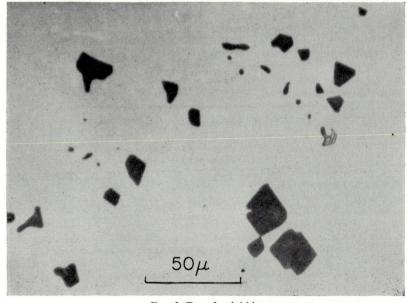

FIG. 5. Type 3 sulphides.

Oxygen and Carbon

Although oxygen is not specified, it has the most significant effect on the morphology of the sulphide inclusions. The sulphides can occur in three forms:[2]

(1) *Type* 1 (Fig. 3).

These are the desired type. Globular in the as cast steel and large. They occur at high levels of oxygen.

(2) *Type* 2 (Fig. 4).

These are the eutectic sulphides—arrays of small non-spherical inclusions which are detrimental to machining. They occur at intermediate levels of oxygen.

(3) *Type* 3 (Fig. 5).

These are large angular inclusions occurring at low levels of oxygen. They are probably as useful as Type 1 in enhancing machining characteristics, other factors being the same.

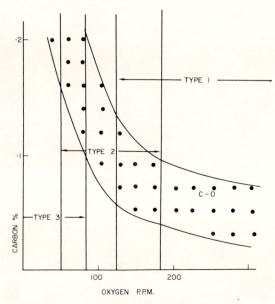

FIG. 6. Carbon, oxygen and sulphide type diagram.

Oxygen	Determines sulphide shape	High
Carbon	Controls oxygen	Low
Aluminium	Deoxidizer—hard inclusions	Low
Silicon	Deoxidizer—long inclusions	Low
Manganese Sulphur	}MnS inclusions	High
Nitrogen Phosphorus	}Ferrite stiffener	High
Lead	Inclusions	High

FIG. 7. Summary of effect of steelmaking additions.

To obtain the Type 1 sulphides, a high level of oxygen must be achieved in the 100 ton molten bath of steel and the carbon level must be reduced. The relationship between carbon, oxygen and sulphide type is shown in Fig. 6.

At melt-out, the carbon level is usually high and the oxygen low. The oxygen in the metal is raised either by blowing oxygen into the molten bath as in the Kaldo or LD or by increasing the oxygen of the slag in open hearth furnaces by addition of iron oxides as scale. The higher the slag oxygen, the higher the metal oxygen consistent with the carbon level. Only when the carbon is below about 0·1 % is the oxygen high enough to ensure the formation of Type 1 sulphides. The bath is now a very low carbon steel to which additions are necessary to make it into a free cutting type. These additions should affect as little as possible the oxygen level in the bath and a "balanced" free cutting steel is produced. The requirement for these and other additions is summarized in Fig. 7.

Aluminium

This is a powerful deoxidant and its addition must be avoided as it reduces the oxygen content and would cause the occurrence of Type 2 sulphides.[3] Secondly, the product of deoxidation—Al_2O_3—occurs as a hard angular inclusion which would reduce tool life in machining.[4]

Silicon

This again is a deoxidant and its primary effect is the same as for aluminium.[3] The deoxidation product is, however, a silicate which is readily deformable giving long thin inclusions in the bar product which detract from machinability.[5] If a killed free cutting steel is required, the use of silicon is preferred to aluminium because it is not as powerful a deoxidant and its product is considered less detrimental than Al_2O_3.

Manganese and Sulphur

These are added to the bath just before tapping or to the ladle during tapping—the manganese as ferro-manganese and the sulphur as rock sulphur or iron sulphide. The amount added depends on the specifications which give sulphur ranges up to 0·6%. The manganese is usually sufficient to give a Mn/S ratio of about 3:1.[6,7,8] The sulphide should be present as a Type 1[2] inclusion.

Nitrogen and Phosphorus

Both are added to the ladle during tapping—the phosphorus as ferro-phosphorus and the nitrogen as high nitrogen ferro-manganese. The usual levels are:

Phosphorus 0·10%
Nitrogen 0·01%

Their effect on machining is beneficial.[4,9] No inclusions are formed and both are in solution in the ferrite whose properties are modified to increase performance.

Lead

From 0·15% to 0·35% is added to the teeming stream from the ladle and it may be added to direct or uphill teemed ingots. The lead is completely insoluble in solid steel and precipitates round the sulphides in leaded re-sulphurized steels and as small globules in leaded only steels.[4,7,9,10,11] Sulphides must be large if the full benefit of the lead (up tp 35% increase in performance) is to be achieved.[7,8]

Tellurium, Selenium, Bismuth

These have been added to free machining steels usually when the steel is already leaded.

Increased machining performance can be obtained but the high cost of the addition and limitations of some machine tools make the recovery of the extra cost of material by reduced machining costs difficult to achieve.[9]

BAR PRODUCTION

Having produced the required composition of the 100 tons of steel, it is then cast into 5 ton ingots. These are rolled to bloom, billets and then to the ordered black bar diameter. One ingot will produce about 50,000 ft of $\frac{1}{4}$ in bar, 3500 ft of 1 in bar, and 1000 ft of 2 in bar.

THE ASSESSMENT OF MACHINING PERFORMANCE

In any method of assessment, the usefulness of the data to the machinist must be the main aim. It is, therefore, necessary to ascertain from the machinist what data he requires. The Park Gate Iron & Steel Co. Ltd., therefore, conducted a survey among its customers before deciding its machinability programme. From the replies, feeds and speeds for given tool lives and surface finishes for selected qualities were required. The tool life must be realistic, i.e. not the time to complete breakdown of the tool but to an amount of wear where finish and dimensional stability of the component are still acceptable and re-grinding does not remove an excessive amount of tool material. The types of wear possible are shown in Fig. 8. Crater-

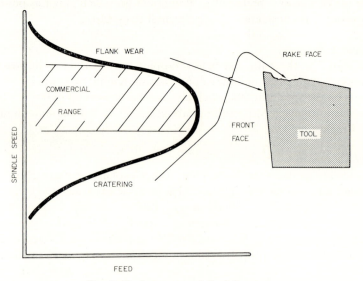

FIG. 8. Tool wear, speed, feed diagram.

ing on the top face requires more grinding for its removal than flank wear. A flank wear of about 0·040 in is a realistic end point.

In deciding on the type of test to be developed at the works, commercial reliability of the machining data to be produced was of prime importance. Extensive research work has been conducted into machinability since the early 1920's but this property is neither simply nor uniquely defined and this presents the first difficulty in establishing a test. The fact that there are no fundamental units of machinability made early researchers choose some simple metal cutting process and define machinability as some particular aspect of their test, e.g.

(1) Maximum spindle speed for a specified tool life at a given feed
(2) Number of holes drilled to a given depth at a fixed speed and feed
(3) Measured tool/workpiece interaction forces.

During such tests, the second difficulty—reproducibility of results—became apparent.

All such tests give machinability as a dimensionless number, usually with 100 as a standard, defined on a material convenient for the researcher. Other materials are then compared with this standard and for different alloys with widely differing machining properties, (e.g. magnesium and "steel") it was found possible to relate machinability with some property of the alloy. It was also possible to predict the index for other metals before testing.

However, when indices were determined for one group of alloys with very different machining properties, the reliability of production decreased and the difference between indices did not necessarily indicate the relative difference in machining. Further, different tests, although each giving consistency within themselves could not be related reliably to one another.

Some tests are, of course, more reliable than others and one of the better short term tests is probably the Taylor V.60 test. Today tests relating machinability to properties during torsion testing, e.m.f. between tool and workpiece, and very rapid tool breakdown (less than 1 min) are being conducted. Though these may be ideal ways of studying the cutting process, machinability can only be expressed comparatively and only a qualitative guide can be given to machinists.

Any type of short term test has, however, another serious problem. By its basic concept, i.e. short term, only a small weight of material is cut. From results from such tests, the machinability of ingots or casts has to be deduced.

It was decided, therefore, that to give commercially reliable data a machining test on a commercial scale must be carried out as has been done in the U.S.A.[12]

Results of feed and speed for given tool lives and surface finishes would then be directly applicable to commercial machines on commercial quantities of steel.

The Test

Machine. The most demanding use of free machining steels is in automatic lathes where not only do the already mentioned criteria apply but also a production run of at least one shift must be possible. This does not mean tool lives of eight hours, as any given tool cuts only for a proportion of the production cycle. This proportion depends on the size and shape of cut required and the surface finish necessary.

It was decided, therefore, to set up a test based on a single spindle automatic lathe producing some component as on a commercial basis, i.e. at the highest possible production rate. In order to define this rate, which forms the basis of comparison of qualities, it is necessary to operate at rates which give tool failure before the 8 hour is completed and rates below the 8 hour end point.

Bar size. The highest tonnages of bars ordered are in the $\frac{5}{8}$ to $1\frac{1}{2}$ in diameter range and it was decided, therefore, that the bar size would be in this range. The larger the bar diameter the greater the capital cost of any machine tool required, the more rigid is the bar and the more difficult is the handling. The final decision was to use $\frac{3}{4}$ in diameter bar.

Component shape. The four basic machining operations are longitudinal turning, forming, centre drilling and parting, and it is possible to shape any component using these operations and appropriate tool geometry. The only exception is the formation of internal or external threads. The shape chosen for the Park Gate test is shown in Fig. 9 with a diagrammatic lay-out for the tools. The four basic operations are used with a simple tool geometry.

Tool material and profile. Our survey showed that the most popular material was the 5% Co type high speed steel. Tools' angles varied from 5° to 15° for top rake with angles from 0° for side rake to 30° for end cutting edge angles. Methods of maintaining the angle also varied. Hand grinding was used in some shops with tool room grinding in others.

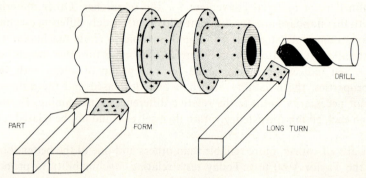

FIG. 9. Component.

In the Park Gate test, a top rake of 10° was chosen for the long turning and forming tools and 0° for the parting tool. Side relief angles of 1°, 5° and 10° were used depending on the tool.

The tool geometry is kept constant by using a Baker-Pera single point tool grinder on a universal grinder. No chip breakers are used so that the inherent breakability of the chips may be studied.

Cutting oil. Commercial usage is almost equally divided between the neat and water soluble type. The neat oil was chosen for the Park Gate test.

Feeds and Speeds

Component shape, tool material, tool geometry and lubricant are always as near the same as is practically possible, so that a standard machining condition exists for the comparison of qualities. Every care has been taken to ensure that the test is as close an approach to this ideal. Even so there is an inherent variability in machining behaviour that requires several tests to be done over a range of feeds and speeds to fully assess a batch of material. It was decided initially to test at three speeds, two feeds, and in duplicate. The duplicated tests were found to be in good agreement so now tests at four speeds and three feeds are conducted to give a broader based picture of performance.

The feeds and speeds used in the investigations begin in the practical range as defined by the users and the practical conditions are taken as the base line and higher feeds and speeds are used to find the upper limit of use of a given batch. The actual testing conditions are not defined before trials on a batch begin. The result at the first feed and speed are assessed before a subsequent condition is fixed.

Measurements on the Components Produced

At the beginning of a test, feed and speed are set on the lathe and remain unchanged throughout the test. The operator has no control over either. The dimensions of the component are set to $\pm$ 0·002 in of the nominal by checking the initial components and resetting the tool traverses as required. Once these limits have been attained, no further adjustment is made during that test.

Every tenth component in order of production is saved with the unused bar stub, for subsequent examination. The following dimensions are logged:

Total length
Length of reduced diameter
Form diameter
Long turn diameter
Bar diameter

The finish produced by the long turn, form and part tools is also measured in μin CLA.

Measurements on the Tools

The tools are examined at the end of a test run—tool wear measurements are not made during a run. The wear scars on the cutting, side and top faces, are measured and the condition of the edge noted. The tool life for a given condition is estimated from the tool wear observations on the two tools.

The relationship between flank wear and tool life is not a simple one.[12] In these experiments, tool life is defined as that time at which the tool profile may be reformed on the tool with a minimum of grinding, generally a wear scar of 0·01–0·04 in. The position is further complicated by the relationship between flank wear and the amount of work done by the tool and the rate at which the work is done[12,13] (see Fig. 10). The relationship as reported

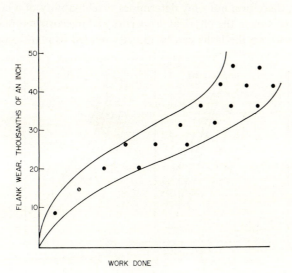

FIG. 10. Tool wear, work done diagram.

by Murphy of the United States Steel Co. generally confirmed by other workers, shows a scatter band width about 50% of the average reading. This anomaly is overcome by the number of tests carried out above and below the chosen end point of 8 hr.

Microscopic Measurements

The bar stub left in the chuck is identified with the quality, speed, production rate, tools and bar order within the test. Such a stub is produced approximately every 50 components, i.e. a 2% sample. The stub is sectioned longitudinally through the bar axis and polished for microscopic examination in the unetched condition at ×300 diameter. A complete traverse is made, the length and breadth of all inclusions intersected is noted.

THE RESULTS

Data Obtained on the Steel Tested

Steelmaking and casting details are obtained on the cast, and where possible, bloom billet and bar samples are obtained as the order is processed. These are examined for inclusion size and shape whose change can be related to a particular part of the process route.

The information is obtained from the product of one ingot of a given cast. The billet are rolled to $\frac{13}{16}$ in diameter round in the 11 in mill and bright drawn to $\frac{3}{4}$ in diameter. About 30 cwt in 5-ft lengths is held in the machinability store.

The whole of this quantity is used in determining the machinability performance of one quality. By the time testing ends, some 60 tools have been used and about 20,000 components have been made—2000 have been saved on which 10,000 dimensions and 6000 surface finishes are measured. 140 micro sections have been prepared and about 30,000 inclusions have been measured.

To supplement the cast data, the physical properties of the bright bar are determined on at least ten bars picked at random. The chemical analysis of these bars is also determined. The Park Gate test, therefore, not only determines machinability of a commercial quantity of steel but also relates this to the physical properties and microstructure of the batch so that significant performance on the lathe can be directly related to plant processing conditions.

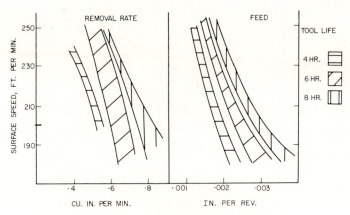

FIG. 11. Performance of an individual tool.

Machining Results

From the tool wear measurements, the useful tool life at any condition is deduced and graphed as in Fig. 11, such a figure is prepared for each tool and the results are summarized in Tables, from which a comparison of qualities for an individual tool can be deduced. An overall comparison of qualities in practical terms is done by comparing the maximum production rates for an 8 hr test at various speeds. The results for En 1A, leaded En 1A and C12L14 are given in Fig. 12.

Computer Analysis of Results

Results for tool life and finishes at various speeds and feeds agreed well with results from customers. A further test of reliability was made by statistically analysing the results. This

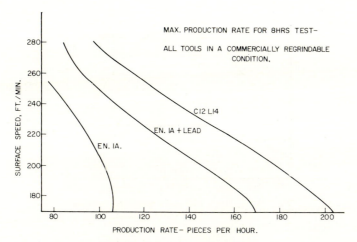

FIG. 12. Comparison of qualities.

has been tried before[7] with limited success, probably because a universal equation between analysis, physical properties and machining was sought.

In the Park Gate studies, chemical analysis, numerical expressions for sulphide inclusion shape and distribution, and machining conditions were related to tool life or finish for each tool, i.e. eight equations. A multiple regression equation from each set of results was compiled and the answer for each individual set computed. The computed answer was then compared with the experimentally determined answer. In more than 50% of the tests, the two answers differed by less than 10% and in more than 80% of tests, the difference was less than 20%. These results confirmed that the Park Gate test had a reasonable level of standardization for comparison of results.

THE RELATIONSHIP BETWEEN MACHINING AND MANUFACTURING

Machining and Inclusion Morphology in Bars

From the traverse of the longitudinal section, the distribution of inclusion aspect ratio, λ, and its length, l, with bar diameter can be deduced. A typical example is given in Fig. 13. From this, it is evident that the shorter inclusions are near the surface and the longer ones in the centre. λ, however, has no such pattern and an average λ is typical of the cross-section. The average λ can be plotted along with the machining variables as a test proceeds and a section of a test log is given in Fig. 14.

From this and all other logs, it is apparent that changes in λ are not reflected by changes in dimensions or surface finishes. No effect could be detected on tool life within the observed range of from about 2·5 to 16 with averages from 4 to 9.

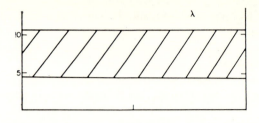

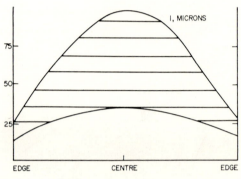

FIG. 13. Variation of λ and l across bar section.

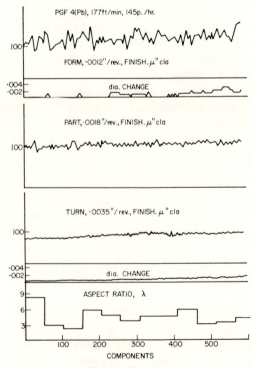

FIG. 14. Test log.

The overall distribution of λ, l and number in the section was determined and two forms so far observed are given in Fig. 15. If the distributions are normal, then differences in machining performance are attributable to the differences in chemical composition even if the averages vary within the observed ranges.

For steels of very similar composition abnormal distributions of inclusion parameters lead to a reduction in machining performance of some 10% to 15%.

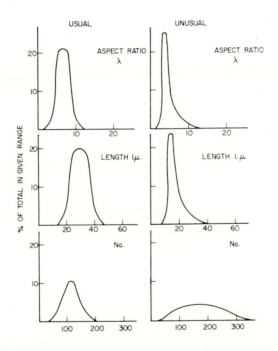

FIG. 15. Distribution of inclusion parameters—bar section.

Since the data showed that λ was not a critical parameter, it was initially decided that the reduction in performance was associated with the high incidence of low length inclusions. This inclusion was examined in the light of how the metal behaves during cutting.

The main action of the sulphide is to provide a film on the rake face of the tool to reduce frictional forces in the contact area. The existence of such a film has been very clearly established by microprobe analysis by Opitz and König[14] and, subsequently, by others.[15] The sulphide is extruded onto the tool face by the compressive stress developed in the chip between the primary shear zone and the limit of the contact area after the inclusion is cut. Flow of the sulphide from the inclusion will depend on the applied stress and on the orifice available for flow.

It can be shown that for a given compressive stress, the mass extruded from an imperfect tube is approximately proportional to the 4th power of the diameter.[16] Hence, for a given stress developed during cutting, small diameter inclusions will contribute far less to film formation than larger diameter inclusions.

The distribution of diameters of all batches was, therefore, carefully established and the patterns determined are shown in Fig. 16. The batch giving a high incidence of inclusions with diameters less than 2 μ in the bar has the reduced machinability and the avoidance of such structures in the bar is clearly necessary if consistently good machining is to be achieved.

To avoid such structures it is necessary to establish how the high incidence of low diameter inclusions occurs during manufacture.

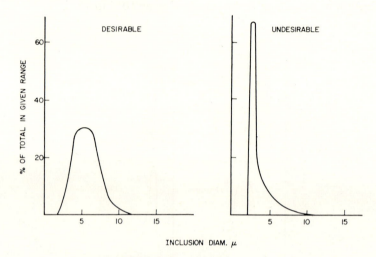

Fig. 16. Distribution of inclusion diameter.

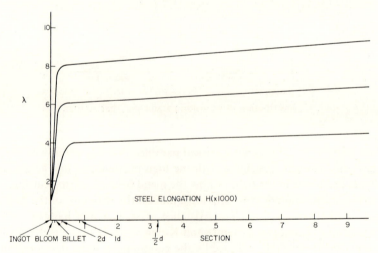

Fig. 17. Change in λ during rolling.

Behaviour of Inclusions During Rolling

On all batches tested and other selected steels, samples were obtained at the bloom, billet, black bar and drawn bar stage for microscopic examination. The inclusion parameters were determined and the change in λ from ingot to bar is shown in Fig. 17. This comparison of λ with H is not theoretically justifiable but it does emphasize the initial rate of change of shape of the inclusion and its marked reluctance to further change after the billet stage, i.e. the diameter of inclusions in the ingot must be controlled if the undesirable high incidence of low diameter inclusions is to be avoided.

Inclusion Morphology in the Ingot

As cast inclusion size and shape is influenced by chemical and physical phenomena. It is accepted that Type 2 sulphides (see Fig. 4) are detrimental to machinability and from the results associating reductions in machinability with small inclusion diameter, it can be deduced that Type 2 sulphides are detrimental because of their small diameters. From long established practice, steelmakers can avoid formation of such sulphides by careful chemical control and promote formation of the required Type 1 sulphides.

This work has, however, shown that small diameter Type 1 sulphides are also detrimental and in order to ensure optimum machinability in the bar product, the steelmaker must carefully control ingot solidification to produce large Type 1 sulphides. It is apparent from the form of the Type 1 sulphide (see Fig. 3) that they form in interdendritic pools after a substantial fraction of the steel has solidified. It is, therefore, the influences on interdendritic pool size that must be established and controlled.

CONCLUSIONS

1. Machining performance is related to the diameter of inclusions, and not to their length or aspect ratio.
2. Inclusions change shape very much less than the bulk change in the steel.
3. Since the diameter of inclusions is the major factor in promoting machinability, and the change in inclusion shape during rolling is relatively small, the production of large diameter inclusions in the ingot is, therefore, of prime importance.

ACKNOWLEDGEMENTS

The author thanks the Directors of the Park Gate Iron and Steel Company Limited for permission to publish this paper and his colleagues who have assisted in obtaining the data.

REFERENCES

1. Home and World Demand for Machine Tools and Other Engineering Equipment to 1970. *Metalworking Production*, 25 July 1966.
2. SIMS and DAHL, *Trans. Amer. Found. Assoc.* **46**, 1938, 132.
3. PALIVODA, The Role of Oxygen in Free Cutting Steels. *Mech. Working of Steel* **2**. A.I.M.E., 1965.
4. TROUPE, A Metallurgical Guide to Machinability. *Metalworking Production*, 11 March 1964.
5. GAYDOS, Morphological Effects of Some Non-metallics in Free Machining Steel. *Mech. Working of Steel* **2**. A.I.M.E., 1965.
6. DU MOND, Free Cutting Steels. *Materials and Methods*, **28**, 1948, 95.
7. SCHRADER and GREIGER, Statistical Study on the Effect of Chemical Composition on Machinability. *Mech. Working of Steel* **2**. A.I.M.E., 1965.
8. RADTKE and SCHREIBER, Relationship between Sulphide Formation and Machinability in Free Cutting Steels. *Steel Times*, 19 August 1966.
9. GARVEY and TATA, Factors affecting the Machinability of Low Carbon Free Machining Steel. *Mech. Working of Steel* **2**. A.I.M.E., 1965.
10. LANE, STAM and WOLFE, General Introductory Review of the Relationship between Metallurgy and Machinability. *I.S.I. Special Report* No. 94, 1967, 1965.
11. PALIVODA, Machinability of Type A Leaded Steels. *Trans. A.S.M.* **50**, 1958.
12. MURPHY, Machinability of Steel. A.S.M.E. Int. Res. Prod. Eng. Conf., 1963, p. 177.
13. SHAW, SMITH, COOK and USUI. *J. of Eng. for Industry*, May 1963.
14. OPITZ and KONIG. *I.S.I. Special Report* No. 94, 1967, p. 35.
15. DR. WILLIAMS, Dept. of Prod. Tech. Bunel Univ. Private communication.
16. DR. MELFORD, Tube Investments Research Laboratories, Hinxton Hall, Saffron Walden. Private communication.

EXPERIMENTS WITH SELF-PROPELLED ROTARY CUTTING TOOLS

N. Ramaswamy Iyer and F. Koenigsberger

Machine Tool Engineering Division,
University of Manchester Institute of Science and Technology

SUMMARY

An interesting development in the field of machining is that of tools having a continuously indexing cutting edge. The cutting tool is free to rotate and has a circular cutting edge which is presented to the workpiece at an inclination with respect to the cutting velocity vector. This produces self-induced rotation and thus continuous indexing of the cutting edge. In this manner a new portion of the cutting edge with an adhering layer of cutting fluid is being continuously brought into action. A rest period is thus provided for the cutting edge and the effective rake angle is increased through the chip being deflected to a direction of greater rake face slope. In this it differs from the action of conventional oblique cutting tools. A rotary tool of about 10 degrees normal rake angle at 40 degrees inclination reduces the cutting power required by about 15% and at the same time increases the life expectancy of the tool by up to 20 times. In the present investigation experiments have been carried out under a wide range of cutting conditions both with stationary and moving cutting edges. A detailed analysis of the chip flow, deformation and cutting conditions is presented. Tool life expectancy relations have been derived from dimensional analysis and compared for stationary and moving cutting edges and the advantages indicated.

SYMBOLS

δ_n normal wedge angle.
γ_n normal rake angle.
λ inclination angle.
ρ' chip flow angle for stationary cutting edge.
δ' effective wedge angle for stationary cutting edge.
γ'_e effective rake angle for stationary cutting edge.
ρ'' chip flow angle for moving cutting edge.
δ'' effective wedge angle for moving cutting edge.
γ''_e effective rake angle for moving cutting edge.
V cutting velocity.
V_t cutting edge velocity.
V'_c chip velocity for a stationary cutting edge.
V'_s shear velocity for a stationary cutting edge.
V''_c chip velocity for a moving cutting edge.
V''_s shear velocity for a moving cutting edge.
V_c chip velocity relative to the moving cutting edge of the tool.

INTRODUCTION

The principal difference between cutting with tools having circular continuously rotating tips and conventional tools is the movement of the cutting edge in addition to the main cutting and feed motions. This additional movement results in a particular portion of the cutting edge being in action for only a brief period, this being followed by a much longer rest period during which thermal energy associated with the cutting process has time to be conducted away into the bulk of the cutter. Moreover, by deflecting the chip towards a direction of greater slope, the effective rake angle is increased thus making the shear process more efficient. A practical way of providing this additional motion is to use a continuous cutting edge in the form of a circular rotary tool. By positioning the cutting edge at an inclination relative to the main cutting direction, the cutting edge motion can be made self-generating (Fig. 1).

Shaw[1] studied the performance of rotary cutting tools which had been equipped with external drive. More recently, Zemlyanskii[2] carried out an investigation with self-rotating tools. However, neither of these authors published a study of the cutting process with stationary *and* moving cutting edge. It would also appear that a large gap in our understanding has to be filled before this method can be put forward as a practical proposition. Further work to evolve a clearer picture of the deformation and tool–chip interface phenomena of rotary tools is in progress and results form the subject of a subsequent paper. The authors are also investigating how this additional motion of the rake face can be used to throw more light on the tool–chip interaction of conventional cutting tools.

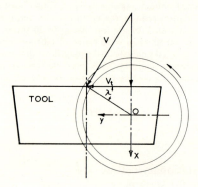

Fig. 1. Relative position of job and tool. X and Y coordinates decide tool inclination.

Effect of a Moving Cutting Edge

When the diameter of the tool is large compared with the width of cut, it can be reasonably assumed that the metal-cutting process represents the case of free oblique cutting. This simplifies the analysis. The validity of this assumption has been verified by studying the chip characteristics and confirmation obtained.

Figure 2 shows the perspective view of an idealized chip-forming process when the cutting edge is oblique to the cutting velocity vector. $PQRS$ represents a rectangular section of the material to be removed as it approaches the cutting edge MO. Under stationary cutting edge conditions the material undergoes deformation resulting in a chip of normal section $P'Q'R'S'$. If an additional motion is given to the cutting edge, the chip, after deformation, is deflected towards the direction of this additional motion. $P''Q''R''S''$ represents the normal section of such a chip. For greater clarity a common shear plan is shown although this is not exactly the case. $E'F'G'H'$ and $E''F''G''H''$ represent the chip sections parallel to the cutting plane $MOQP$ containing the cutting velocity.

Figure 3 shows the velocity configuration under stationary and moving edge conditions.

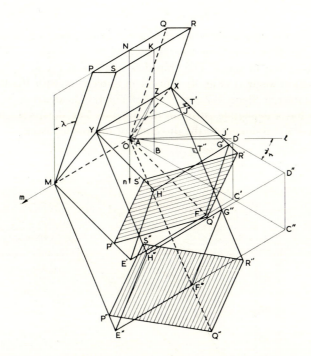

FIG. 2. Idealized chip formation process for tools with stationary and moving cutting edges.

The rake face plane $OMXA$ lies at an angle γ_n (normal rake) to the reference plane lom. The material which approaches the cutting edge with a velocity V, travels after deformation

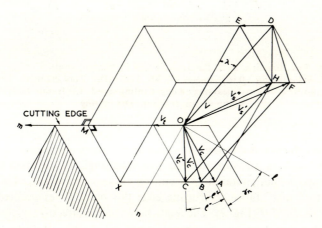

FIG. 3. Velocity configuration.

at velocities V'_c and V''_c respectively, the latter when the cutting edge moves with a velocity V_t tangentially to the rotary motion. V'_s and V''_s represent the corresponding shear velocities. Theoretically the velocity V''_c is given by

$$\overrightarrow{V''_c} = \overrightarrow{V'_c} + \overrightarrow{V_t}$$

but in actual practice the chip is deflected to a lesser extent, because tangential slip takes place. The actual chip flow directions are represented by the angles ρ' and ρ'' in Fig. 3, looking from a fixed point in space. The chip flow angle ρ' is found to be approximately equal to the inclination angle λ (Fig. 4). This is in agreement with the findings of many investigators, e.g. Stabler,[3] Shaw,[4] Zorev,[5] Pal,[6] etc. It indicates that the slight curvature of the cutting edge has a negligible effect on the process. The shear velocity V'_s along the ideal shear plane has a component in the direction of the cutting edge.

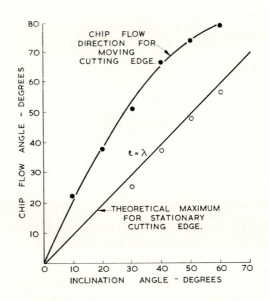

Fig. 4. Chip flow direction as a function of tool inclination angle. Cutting conditions— Job material: Brass tubing 2·0 in diameter (70/30). Tool: HSS. Tool diameter: 2·6 in. Normal rake: 10 deg. positive. Normal clearance: 8 deg. Cutting speed: 123 ft/min. Feed: 0·008 in/rev. Width of cut: 0·125 in. Dry cutting.

Calculation of the Effective Rake Angle

Many research workers measure the effective rake angle in the chip formation plane. This plane is defined as that containing the cutting velocity vector and chip velocity vector. For an oblique tool with stationary cutting edge this plane is represented by $DOBF$ and for a rotary tool by $DOCH$ (Fig. 3). The angle between the direction of the cutting velocity and that of the chip flow gives the true wedge angle δ and $(90 - \delta)$ gives the effective rake angle.

This effective rake angle can be calculated by referring to Fig. 5 in which *lmn* represent the reference axes and *OMDA* the reference plane containing the axes *on* and *om*. The rake face *OBEM* intersects this reference plane along the cutting edge *OM* and the included angle δ_n is the normal wedge angle. Chip flow directions for the stationary and moving

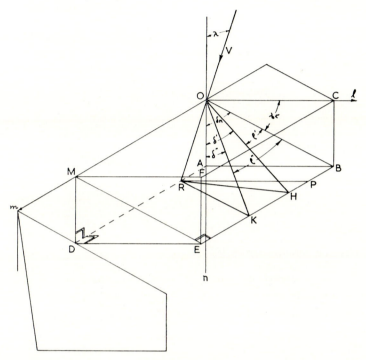

FIG. 5. Tool geometry.

cutting edges are represented by *OH* and *OK* respectively and their chip formation planes intersect the reference and rake face planes along *ROH* and *ROK*. The following relationships can be derived from this figure:

$$\cos \delta' = \cos \lambda \cos \rho' \cos \delta_n + \text{sine } \lambda \sin \rho' \tag{1}$$

$$\cos \delta'' = \cos \lambda \cos \rho'' \cos \delta_n + \sin \lambda \sin \rho'' \tag{2}$$

These equations for different values of λ and ρ for a 10 degree positive normal rake angle are presented graphically in Fig. 6. It is observed that higher chip flow angles go with higher effective rake angles or with lower effective wedge angles.

Chip flow directions were measured by a high-speed photographic technique and checked by chip width measurements. Typical photographs for the two cases under identical conditions are shown in Fig. 7. Measured values for both stationary and moving cutting edge conditions are shown in Fig. 4.

Deformation Studies

Figure 2 showed schematically how rectangular sections of the work material are converted into chips with sections shaped like parallelograms. In practice such regular shapes are seldom observed, as chip sections are always somewhat distorted. Since there is a change in width as well as in thickness, the ratio between the area of the chip and the uncut area is

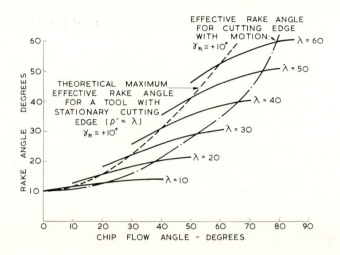

FIG. 6. Relationship between effective rake angle, inclination angle and chip flow angle for a normal rake angle of 10 degrees. Effective rake angles for stationary and moving cutting edge at different inclinations are joined by smooth curves.

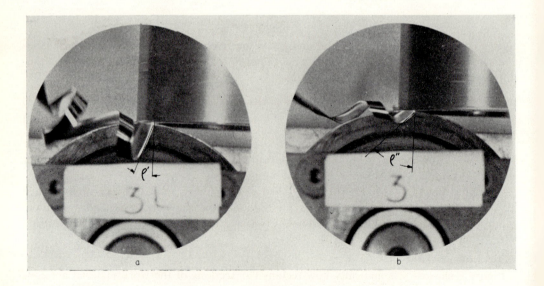

FIG. 7. Photographs taken during the process to measure the chip flow angles. (a) Stationary cutting edge. (b) Moving cutting edge.

defined as the coefficient of deformation (ξ_A). This is identical to the ratio between cutting velocity and chip velocity, a relation which is valid only when there is no change in volume after severe plastic deformation, and this has been reported to be true from research in plasticity.

By measuring the weight and length of a representative chip, ξ_A could be calculated as follows:

$$W = a\,b\,l\,d = a_c b_c l_c d \tag{3}$$

where $W =$ weight of a representative chip of length l_c

$a,b,l,$ = uncut thickness, width and length

$a_c, b_c, l_c =$ corresponding chip dimensions

$d =$ density of the work material.

$$\frac{l}{l_c} = \frac{V}{V_c} = \frac{a_c\,b_c}{ab} = \frac{A_c}{A} = \text{coefficient of deformation by area}$$

or coefficient of deformation by length, otherwise known as the chip length ratio.

Chip measurements were carried out over a wide range of conditions, representative data being plotted in Fig. 8, where the effect of cutting edge inclination is shown. It is observed that an increase of inclination reduces the amount of chip deformation. Also the amount of deformation is less when the cutting edge is moving. Similarly both uncut thickness and cutting velocity have reduced the amount of deformation but to a lesser extent. At high speeds, the actual speed has little effect on the coefficient of deformation.

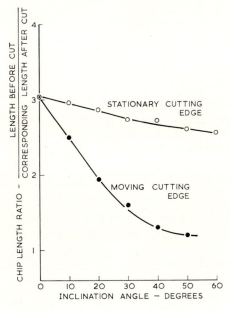

FIG. 8. Coefficient of chip deformation as a function of cutting edge inclination. Job material: Brass tubing 2·0 in diameter (70/30). Tool: HSS. Tool diameter: 2·6 in. Normal rake: 10 deg. Clearance: 8 deg. Cutting speed: 123 ft/min. Feed: 0·008 in/rev. Width of cut: 0·125 in. Dry cutting.

Analysis of Tool Rotation

The cutting edge velocity has been calculated from the rotary tool speed and measured experimentally by means of a calibrated tacho-generator. This has a linear characteristic over a range of speeds from zero to 9000 revolutions with an output of 20·8 V per 1000 rev/min. This output was fed through a 180 ohm resistance to an U.V. recorder with a sensitive moving coil galvanometer.

The ratio between cutting edge velocity and cutting velocity is plotted in Fig. 9 as a function of the tool inclination. If the workpiece material is to drive the tool, this velocity ratio will be sin λ, which is plotted as a continuous curve. Ranges of measured values under different conditions are plotted as a spectrum, the upper and lower limits of these being enclosed within the boundaries of the two dotted lines (Fig. 9). It will be observed that an increase in feed increases the tool speed at all tool inclinations and cutting velocities.

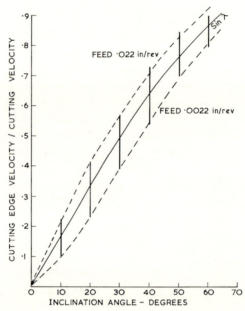

Fig. 9. Ratio of cutting edge velocity to cutting velocity at different tool inclination angle. Cutting speed range: 90 to 180 ft/min. Feed range: 0·0022 to 0·022 in/rev. Other conditions same as for Fig. 8.

Force Relations

The force components acting on the tool in three mutually perpendicular directions were measured using a three-dimensional table dynamometer with resistance strain gauge pick-ups. This dynamometer was designed and developed by Sabberwal.[7] In it 32 strain gauges are mounted on four octagonal rings and the 64 output leads are suitably connected to form 8 full Wheatstone bridges. For the present investigation the electrical scheme has been slightly modified. A twin-output 0–15 V stabilized D.C. power pack has been used to energize the strain gauge bridges and the outputs of the bridges of each channel are added algebraically through operational amplifiers of very low temperature sensitivity. The electrical scheme is shown in Fig. 10.

Typical variations of the three force components under identical conditions and as functions of the inclination angle are shown in Fig. 11. From these the force components in the direction of the cutting velocity which is indicative of the power consumed has been derived and shown in Fig. 12. This force component is from 10 to 15% less when the

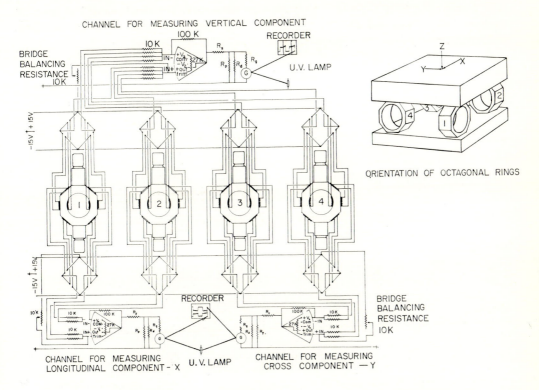

FIG. 10. Electrical circuitry of three-dimensional table dynamometer with strain gauge pick-ups.

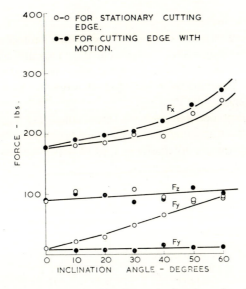

FIG. 11. Forces in three perpendicular directions as a function of tool inclination. (Cutting conditions as in Fig. 8.)

cutting edge is moving than with the cutting edge stationary. The effect of the feed upon the force is almost linear and the effect of the cutting speed very small, especially at higher feed values.

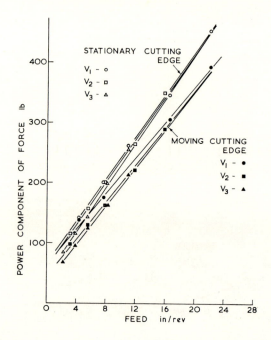

Fig. 12. Major power component of the cutting force as a function of feed for stationary and moving cutting edge. Tool inclination: 30 degrees. V_1, 90 ft/min. V_2, 123 ft/min. V_3, 180 ft/min.

Temperature and Tool-life Studies

It is well known that even a small decrease of the cutting tip temperature can result in a considerable improvement in tool life. Taylor's equation indicates an exponential relationship between cutting speed and tool life. A similar relationship between cutting speed and tool temperature is also reported by other research workers. Kronenberg[8] has derived these relationships by dimensional analysis.

To derive a relationship between tool life expectancy and cutting temperature, the following physical quantities and their dimensions are taken into consideration.

When these physical quantities are analysed a set of dimensionless quantities Q_1 and Q_2

Table 1.

Physical quantity	Symbol	Dimension
Temperature	T_e	θ
Tool life	T_L	T
Chip cross-sectional area	A	L^2
Cutting speed	V	LT^{-1}
Unit cutting force	K_s	$ML^{-1}T^{-2}$
Consolidated heat value	H	$M^{-2}T^{-5\theta-2}$

are obtained

$$Q_1 = \frac{T_e\, H^{\frac{1}{2}}}{T_L^{\frac{1}{2}}\, k_s\, V} \tag{4}$$

$$Q_2 = \frac{A}{V^2 T_L^2} \tag{5}$$

When $\log Q_1$ is plotted as a function of $\log Q_2$, using experimental data, a relationship of the form

$$Q_1 = c_1\, Q_{2m} \tag{6}$$

is obtained. Substituting the values of Q_1 and Q_2 from eqs. (4) and (5) in eq. (6)

$$T_L = \left[\frac{T_e \cdot H^{0.5}}{A^m\, k_s\, c_1\, V^{1-2m}}\right]^{1/(0.5-2m)} \tag{7}$$

When all the parameters except temperature are kept constant this relation is simplified to

$$T_L = c\, T_e^{\,1/(0.5-2m)} \tag{8}$$

The thermal e.m.f. corresponding to the average interface temperature has been measured using a tool–work thermocouple technique and typical values are shown in Fig. 13. The

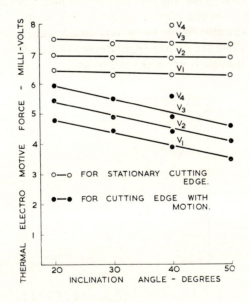

FIG. 13. Average tool–chip interface temperature as a function of tool inclination for different cutting speeds. Job material: En 28 steel. Tool material: K 21 carbide. Tool diameter: 1·25 in. Normal rake: 6·5 deg. negative. Clearance: 6·5 deg. Width of cut: 0·070 in. Feed: 0·006 in/rev. Dry cutting.

inclination angle has very little effect on the interface temperature when the cutting edge is stationary but an appreciable effect when it is moving. Under identical conditions the ratio of tool lives can be calculated from the following equation:

$$\frac{T_{LM}}{T_{LS}} = \left[\frac{T_{em}}{T_{es}}\right]^{1/(0.5-2m)} \tag{9}$$

where T_{LM} = tool life for moving cutting edge,
             T_{LS} = tool life for stationary cutting edge,
             T_{eM} = interface temperature for moving cutting edge,
             T_{eS} = interface temperature for stationary cutting edge.

The exponent m is related to Taylor's tool-life exponent by

$$m = \frac{2(1-n)}{0.5-n} \tag{10}$$

It is known from experience that the tool life exponent n in Taylor's equation varies between 0·25 and 0·08. Corresponding values of m lie between 6·0 and 4·4. For carbide tools a value of m equal to 6·0 can be safely assumed.[9]

Since the thermal e.m.f. is approximately linear with temperature, it can be substituted for the temperature without causing a serious error in the above equation. Ratios of tool lives calculated from eq. (9) are shown in Table 2. It can be seen that a considerable improvement in tool life can be expected especially for inclinations of 30 degrees and above

Disscusion of Test Results

A repeatable process stabilizes to a condition of minimum resistance. In the present case the direction and magnitude of the chip flow velocity assumes a position and value such that the component along the cutting edge is equal to the velocity of the cutting edge. This cutting edge velocity was also found to be very nearly equal to the component of the cutting velocity vector along the cutting edge. Results of chip flow measurements confirm this. A typical velocity triangle is presented in Fig. 14.

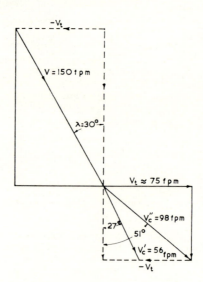

FIG. 14. A typical velocity triangle.

In the section on chip flow, it was shown that the chip flow angle ρ'' and the velocity V''_c are greater than ρ' and V'_c for a stationary cutting edge. These directions were given as observed from the outside. Viewed from a fixed point on the rake face the velocity configuration is different. In order to prove this the velocity of the cutting edge must be subtracted vectorially from the cutting and chip velocity vectors. The uncut material approaches the cutting tool in the direction EO (Fig. 3) and moves out in the direction OA with a velocity V_c. In other words, the process is similar to orthogonal cutting.

In pure orthogonal cutting a force component along the cutting edge should not exist. A study of Fig. 11 reveals that the force component along the cutting edge is extremely small and this is in agreement with the above analysis. The force component measured in

TABLE 2.

λ	Speed ft/min	Stationary cutting edge thermal e.m.f. mv	Moving cutting edge thermal e.m.f. mv	Ratio	Ratio of tool lives
20	186	6·43	4·8	1·34	5·8
	262	6·96	5·44	1·28	4·5
	372	7·49	5·97	1·25	3·8
30	186	6·32	4·45	1·42	8·2
	262	6·90	4·91	1·40	7·5
	372	7·37	5·50	1·34	5·8
40	186	6·32	3·92	1·62	18·0
	262	6·84	4·45	1·54	13·3
	372	7·37	4·91	1·50	11·4
	524	7·96	5·62	1·42	8·2
50	186	6·32	3·51	1·80	34·0
	262	6·90	4·15	1·66	20·9
	372	7·31	4·62	1·58	15·6

Work material: En 28 steel. Tool material: K 21 carbide. Tool diameter: 1·25 in. γ_n: − 6·5 degrees. a_n: 6·5 degrees. Width of cut: 0·070 in. Feed: 0·008 in/rev. Dry cutting.

Note: Ratios of tool lives are calculated for a small length of the cutting edge that is in operation at an instant. To obtain the life of the circular insert this has to be multiplied by the ratio of the total length of the cutting edge to that of the section of the cutting edge operating at every instant.

the X direction is greater for the moving cutting edge than it is for the stationary cutting edge. However, since the force in the Y direction is practically zero for the former, the resultant component in the direction of the cutting velocity vector is lower.

A study of Fig. 6 reveals that the difference in the magnitude of effective rake angles between stationary and moving edge tools increases with increases in the inclination angle. A similar trend is observed for the power component of the cutting force.

In Fig. 9 it is shown that the increase of the feed or the uncut thickness results in an increased tool velocity. This should further result in a larger chip flow angle and thus a larger effective rake angle. This is in agreement with Fig. 12 where the difference in magnitude of the cutting force is shown to increase with the feed.

The heat generated at the cutting edge is less for the moving cutting edge because of less metal deformation. The other reason for the lower temperature and with it improved tool life is that the cutting edge has sufficient time to cool during the idle period. The ratio of the temperatures, i.e. temperature of the stationary cutting edge divided by the temperature of the moving cutting edge decreases with increased cutting speed but increases with the inclination angle.

CONCLUSIONS

By suitably positioning the rotary tool, it is possible to produce a self-generated movement of the cutting edge which is steady and consistent. This increases the chip flow angle and the effective rake angle resulting in the power consumption being lower by up to 15%. The process stabilizes to a condition where there is practically no force component acting tangentially to the cutting edge, any part of which has sufficient idle time to dissipate the heat generated. This improves the tool life appreciably, i.e. an increase which may be up to about 20 times. The chip flow directions can be controlled by varying the tool inclination and by

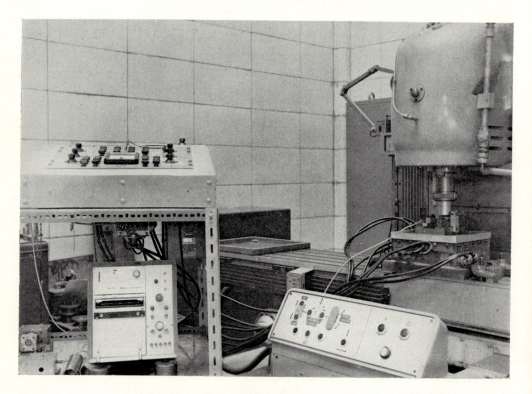

Fig. 15. General set-up.

additional tool rotation which may be necessary at times. The technique makes possible a large variation of rake angles which may be required if optimum cutting conditions are to be created when machining different workpiece materials. Due to the reduced power consumption this technique can be used to increase the productivity of existing machines.

The main problem of the cutting tool designer is the provision of a sufficiently stiff bearing for the rotary cutting tool and development of hydrostatic radial bearings may provide an answer to this problem.

Since the tool is not rigidly clamped to the tool post the quality of the machined surface may be affected. However, this technique is highly efficient for rough machining operations.

For the conditions under which the present investigation was carried out, i.e. for feeds between 0·0022 and 0·022 in/rev, speeds up to 600 ft/min and widths of cut between 0·070 and 0·125 in, no vibrations were observed.

For the series of experiments described, a 40 degree inclination was found to be the optimum. Larger values, though efficient, had the drawback of increasing the length of the cutting edge in operation. At 50 and 60 degrees of tool inclination a small amount of burr was noticed on the tool edge as it left the tube specimen. However, during cylindrical turning this may not be serious. In all tests with the rotary tool no saw-tooth formation on the sides of the chip could be observed and the edges were smooth, whereas during machining with a rigid tool and at higher feeds, the chip edges broke away. This is an indirect indication that the process using the rotary tool had a smoother action.

ACKNOWLEDGEMENTS

The authors wish to thank the staff of the Machine Tool Laboratory for their assistance during the experiments. Thanks are also due to the Indian Institute of Technology, Bombay, for having sponsored the first author for his further studies at the University of Manchester Institute of Science and Technology and to the Commonwealth Scholarship Commission for awarding a scholarship to him.

REFERENCES

1. M. C. SHAW, P. A. SMITH and N. H. COOK, The Rotary Cutting Tool. *Trans. ASME*, August 1952, pp. 1065–1076.
2. V. A. ZEMLYANSKII, Self-induced Rotation of Round Tool Tips. *Russian Engineering Journal, XLVI*, No. 9.
3. G. V. STABLER, The Fundamental Geometry of Cutting Tools. *Proc. Institution of Mechanical Engineers*, **63**, 1951.
4. M. C. SHAW, N. H. COOK and P. A. SMITH, The Mechanics of Three-dimensional Cutting Operations. *Trans. ASME*, **74**, No. 6, 1952.
5. N. N. ZOREV, *Metal-cutting Mechanics*. 1st English edition, Pergamon Press, Oxford, 1966.
6. AJIT PAL and F. KOENIGSBERGER, Some Aspects of the Oblique Cutting Process. *Int. Jnl. of Machine Tool Design and Research*, **7**, 1968, 189–203.
7. F. KOENIGSBERGER, K. D. MARAWAHA and A. J. P. SABBERWAL, Design and Performance of Two Milling Force Dynamometers. *Journal of the Institution of Production Engineers*, Dec. 1958.
8. M. KRONENBERG, *Machining Science and Application*. 1st English edition, Pergamon Press, Oxford, 1966.
9. ROBERT G. BRIERLY and H. J. SIEKMANN, *Machining Principles and Cost Control*. McGraw-Hill, New York, 1964.

HYDRODYNAMIC ACTION AT A CHIP–TOOL INTERFACE

Gabriel J. DeSalvo and Milton C. Shaw

Department of Mechanical Engineering, Carnegie-Mellon University, Pittsburgh, Pa.

SUMMARY

At high cutting speeds a thin structureless film is found to appear between chip and tool. This film is believed to act as a viscous liquid capable of developing positive hydrodynamic pressure. While the presence of such a hydrodynamic film is beneficial from the lubrication point of view it will promote diffusion wear and lead to crater formation due to the slow motion of material adjacent to the tool face. Crater wear is greatly reduced by the deposition of a stationary layer of solid material (sulfide or oxide) on the tool face which acts as a diffusion barrier.

INTRODUCTION

When metal is cut at a relatively low speed, a built-up edge forms on the nose of the tool which changes the tool geometry and usually results in very poor finish. As the cutting speed is increased, the temperature at the chip–tool interface will reach the strain recrystallization temperature and the built-up edge disappears. The side of the chip adjacent to the tool face then resembles a liquid in its behavior; it is extremely ductile and undergoes such enormous strains that it ceases to reveal a crystalline structure when an etched photomicrograph is examined at high magnification. At still higher speeds the layer may transform to austenite which also behaves as a liquid.

Figure 1 shows a quick stop specimen that has been polished and etched. In this case a small piece of carbide was chipped from the tool as it stopped and is shown attached to the underside of the chip. The structureless layer of very high strain which is seen to increase nonlinearly in thickness with position along the tool face, is clearly evident. This layer is frequently referred to as a *flow zone*[2] which consists of either a region of ferritic metal above the strain recrystallization temperature or one of transformed austenite. In either case the zone may be considered to offer little resistance to flow at low strain rates but to be very strain rate dependent. Schaller[1] has suggested that such a layer may be approximated as a Newtonian liquid for which

$$\tau = \mu \frac{\partial \mu}{\partial y} \tag{1}$$

where τ is the shear stress in the layer,
μ is the proportionality constant called viscosity,
$(\partial \mu / \partial y)$ is the rate of shear in the layer.

He further suggested the model shown in Fig. 2 for the flow zone and characterized the action as Couette flow with a constant value of film thickness h. In so doing he inferred that a hydrodynamic pressure would not be developed in the flow zone.

SLIDER BEARING POSSIBILITIES

In Fig. 1 it is evident that h is not constant but increases in the direction of chip flow. Since we have here all of the elements of a slider bearing (two solid surfaces, one of which is moving, with a wedge-shaped liquid film between) it is of interest to inquire whether the bearing is capable of developing a positive hydrodynamic pressure. In order that a slider bearing develop a positive pressure the motion of the moving surface must be directed along the wedge of fluid as shown in Fig. 3. At first glance the motion of the chip shown in Fig. 1 appears to be in the direction of an increasing fluid wedge, which would lead to the development of negative pressure. However, more careful consideration reveals that a

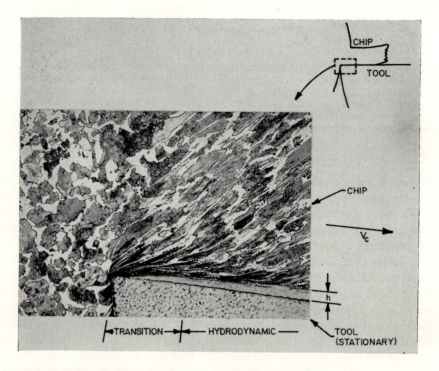

FIG. 1. Partially formed chip obtained by suddenly stopping a cutting operation. Work material, AISI 1053 steel; tool material, P-30 carbide; rake angle, 10°; cutting speed, 200 m/min; undeformed chip thickness, 0.0126 in; width of cut, 0.080 in; cutting fluid, none; magnification, 300 × (after Schaller[1]).

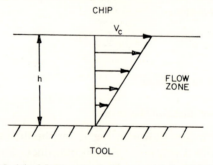

FIG. 2. Model of fluid action in flow zone (after Schaller[1]).

positive pressure will in fact be generated in the flow zone between chip and tool. The explanation of this paradox follows.

Figure 4 shows the elements of a bearing in which the lower surface moves with a velocity (U_0) in the x direction and the upper surface has both x (U_1) and y (V_1) components of velocity. The coordinate system is stationary and the x axis is in the plane of the lower moving surface. In ref. 3 the Reynolds equation for this case is generalized as follows:

$$\frac{1}{6\mu}\left[\frac{\partial}{\partial x}\left(h^3\frac{\partial p}{\partial x}\right) + \frac{\partial}{\partial z}\left(h^3\frac{\partial p}{\partial z}\right)\right] = (U_0 - U_1)\frac{\partial h}{\partial x} + h\frac{\partial(U_0 + U_1)}{\partial x} + 2V_1 \qquad (2)$$

where p is the pressure in the film which is assumed constant across h.

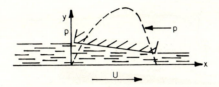

FIG. 3. Typical hydrodynamic slider bearing.

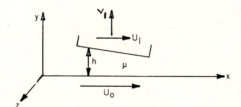

FIG. 4. Generalized hydrodynamic bearing.

When this equation is applied to the conventional case in Fig. 3, the right side of the differential equation (2) becomes

$$RSDE = U\frac{\partial h}{\partial x} \qquad (3)$$

For the bearing inclination shown $(\partial h/\partial x)$ is negative and when integrated a positive pressure distribution as shown in Fig. 3 is obtained.

Figure 5 shows a chip with flow zone. The velocity of the chip (V_c) will be parallel to the tool face. Figure 6a shows the coordinate system in place parallel to the upper boundary of the flow zone. The velocity of the chip is resolved into components parallel and perpendicular to the upper surface of the flow zone (dotted line). If the coordinate system is considered to remain in the plane of this upper surface, the lower (solid) surface will have a y component of velocity as shown in Fig. 6b. Thus, the velocities of Fig. 4 for which eq. (2) pertains will be

$$U_0 = V_c \cos a \cong V_c$$
$$U_1 = 0$$

$$V_1 = -V_c \sin a \cong -V_c \tan a = -V_c\frac{\partial h}{\partial x}$$

since the angle α will always be very small. Thus, from eq. (2)

$$RSDE = V_c \frac{\partial h}{\partial x} + 0 - 2V_c \frac{\partial h}{\partial x} = -V_c \frac{\partial h}{\partial x} \tag{4}$$

Since in this case $\partial h/\partial x$ is positive the right side of the differential equation will be negative as required for the development of positive pressure.

The fact that positive pressure will be developed in the flow zone may be more readily seen by constructing the kinematically equivalent diagram shown in Fig. 7 in which all

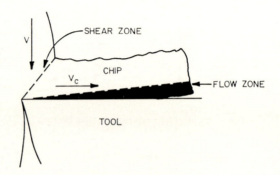

FIG. 5. Moving chip and flow zone.

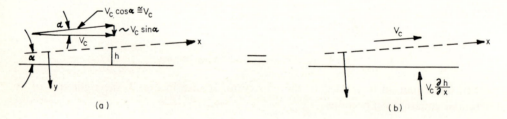

FIG. 6. Velocities pertaining to a flow zone between chip and tool.

elements are translated to the left with a velocity V_c parallel to the tool face. This brings the chip to rest and moves the tool face to the left with a velocity V_c. For the coordinate system shown, Fig. 7 is seen to be equivalent to Fig. 3 and hence positive hydrodynamic pressure is clearly capable of being developed in the flow zone.

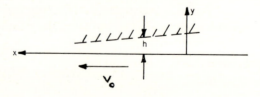

FIG. 7. Kinematically equivalent diagram to Fig. 6a obtained by translating all elements to the left with a velocity V_c.

HYDRODYNAMIC-THERMAL ANALYSIS

The hydrodynamic bearing we are considering here is an unusual one in that one surface is geometrically determined while the geometry of the other surface is determined by a "melting" process. In the following analysis, the usual assumptions of thin film hydrodynamic lubrication are made:

1. Steady state.
2. Newtonian laminar flow with no slip at either solid surface.
3. Inertia terms are negligible.
4. Fluid is incompressible.
5. Pressure variation across the film (in y direction) is negligible.
6. Zero end flow (in z direction).

In addition, it is assumed that a latent heat H (BTU/lb) is absorbed by the material in the chip when it melts. Since the chip is in the extensive member all of the thermal energy is assumed to go into the chip or to remain in the flow zone where it is generated. Negligible thermal energy is assumed to flow into the tool. The temperature of the entire flow zone is assumed to be constant at the "melting" temperature (T_m).

The temperature of the chip leaving the shear zone (Fig. 5) is assumed to have a constant value (T_s) along the shear plane and hence the heat flow from the flow zone into the chip will vary as $(T_m - T_s)$.

Only the main points of the hydrodynamic-thermal analysis of the flow zone will be presented here. Those interested in the details are referred to ref. 4.

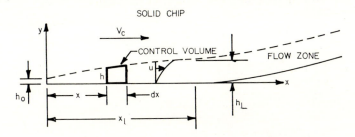

FIG. 8. Control volume used in analysis of flow zone.

The fluid mechanics part of the analysis closely follows conventional bearing analysis. Taking the origin $(x = 0)$ at the point of initial melting on the tool face (Fig. 8) and writing the conservation of mass for the control volume shown in Fig. 8,

$$\int_0^h \rho \, \frac{\partial u}{\partial x} \, dy \, dx = 0 \tag{5}$$

where ρ is the mass density of the fluid, and u is the x component of fluid velocity.

Applying Newton's second law, Newton's law of viscous shear, and the continuity equation, the following "Reynolds" equation is obtained:

$$\frac{\partial}{\partial x} \left(h^3 \, \frac{\partial p}{\partial x} \right) = - 6\mu \, V_c \, \frac{\partial h}{\partial x} \tag{6}$$

As already mentioned, this will lead to positive pressure if $\partial h/\partial x$ is positive as shown in Fig. 8.

After integrating and evaluating the constant of integration

$$\frac{\partial p}{\partial x} = -\frac{6\mu V_c}{h^2} \tag{7}$$

In order to find the pressure at any point x it is necessary to have an expression for h as a function of x. This must be obtained from the thermal part of the analysis.

The thermal energy generated in the flow zone per unit area per unit time will be

$$q_f = \frac{1}{12\mu}\left(\frac{\partial p}{\partial x}\right)^2 h^3 + \frac{V_c^2}{h}\mu \tag{8}$$

The latent heat of melting per unit area per unit time will be

$$q_m = H\rho V_c\frac{\partial h}{\partial x} \tag{9}$$

while the thermal energy per unit area per unit time associated with temperature rise in the chip from the melting point T_m to the average temperature of the flow zone T_w will be

$$q_l = \rho C V_c\left\{h\frac{\partial T_w}{\partial x} + (T_w - T_m)\frac{\partial h}{\partial x}\right\} \tag{10}$$

where ρC is the volumetric specific heat of the chip material. The thermal energy per unit area per unit time flowing into the solid portion of the chip will be

$$q_c = \frac{k(T_m - T_s)}{\sqrt{(\pi a x)}} \tag{11}$$

where

$$a = \frac{k}{\rho C V_c} \tag{12}$$

$k = $ thermal conductivity of the chip material. Making a thermal balance

$$q_f = q_m + q_l + q_c \tag{13}$$

or

$$\frac{h^3}{12\mu J}\left(\frac{\partial p}{\partial x}\right)^2 + \frac{V_c^2\mu}{Jh} = H\rho V_c\frac{\partial h}{\partial x} + \rho C V_c\left[h\frac{\partial T_w}{\partial x} + (T_w - T_s)\frac{\partial h}{\partial x}\right]$$
$$+ \frac{k(T_w - T_s)}{\sqrt{(\pi a x)}} \tag{14}$$

where J is the mechanical equivalent of heat.

Substituting for $\partial p/\partial x$ from eq. (7).

$$\frac{4\mu V_c^2}{Jh} = \rho V_c\left[H + C\left(T_w - T_m\right)\right]\frac{\partial h}{\partial x} + \rho C V_c h\frac{\partial T_w}{\partial x}$$
$$+ \frac{k(T_m - T_s)}{\sqrt{(\pi a x)}} \tag{15}$$

At this point it is expedient to consider special cases. If we first assume that all of the heat is consumed in melting (q_l and q_c are zero) and that the temperature is T_m at all points in the flow zone, then integration of eq. (15) gives:

$$h = \sqrt{(\kappa x + h_0{}^2)} \tag{16}$$

where

$$\kappa = \frac{8\mu V_c}{JH\rho} \tag{17}$$

h_0 = the value of h at $x = 0$.

Equation (16) indicates that h varies as $\sqrt{x}$, which appears to be in excellent agreement with the shape of the flow zone in Fig. 1.

Substituting eq. (16) into eq. (7) and integrating gives

$$p = \frac{6\mu V_c}{\kappa} \ln\left(\frac{\kappa x_1 + h_0{}^2}{\kappa x + h_0{}^2}\right) \tag{18}$$

where x_1 is the value of x where the chip leaves the tool face, at which point p is assumed to be zero.

The shear stress (τ) in the flow zone will be

$$\tau = \mu \frac{\partial u}{\partial y} = \frac{\mu V_c}{h^2}(4h - 6y) \tag{19}$$

and the shear stress on the tool face (where $y = 0$) will be

$$\tau_0 = \frac{4\mu V_c}{h} \tag{20}$$

Other special cases that may be considered include (see ref. 4 for details):

(a) All thermal energy generated in the flow zone goes toward melting and raising the temperature of the liquid ($q_c = 0$). Here the mean temperature of the zone is considered constant at T_w. For this case the solution is the same as the initial case except H is replaced by $H + C(T_w - T_m)$. This has the effect of decreasing κ and hence of decreasing the rate of growth of h with x.

(b) All thermal energy generated in the flow zone goes toward melting and into the chip ($q_l = 0$). In this case the solution is the same as for the initial case except that κ will again be lower and hence the rate of growth of h with x will be reduced.

Thus, thermal energy going into the chip (q_c) and in raising the temperature of the liquid layer above the melting temperature (T_m) both tend to lower κ which in turn lowers the rate of growth of h with x.

EXAMPLE

The following example indicates magnitudes associated with a typical case. The quantity H has been quite arbitrarily assigned based on the transformation of ferrite to austenite while the value of μ is chosen similar to that for a hot glass. The value of h_0 is estimated from Fig. 1 while the other values are based on actual metal cutting results.

Depth of cut: $b = 0.1$ in
Feed: $t = 0.01$ in

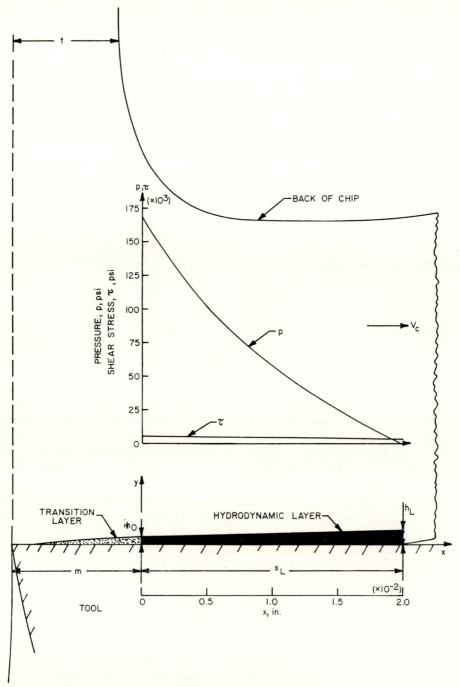

FIG. 9. Variation of hydrodynamic pressure (p) and shear stress (τ) with position along the tool face (x) for special example.

Cutting speed: $V = 600$ fpm
Cutting forces: $F_c = 300$ lb, $F_t = 100$ lb
Chip thickness ratio: $r = 0.46$
Tool–chip contact length: $c = 3 \times 10^{-2}$ in $= m + x_L$
Tool–chip contact width: $b_c = 0.12$ in.

The fully developed hydrodynamic layer is assumed to begin at a distance (m) from the tip of the tool, where the tool–chip interface temperature reaches the transformation value (Fig. 9). Calculation of the shear plane temperature and the interface temperature distribution suggests that the inception of the transformed layer occurs at $m = 1 \times 10^{-2}$ in. The length of the layer, x_L, is given by $c - m$ or 2×10^{-2} in. A value of $h_0 = 5 \times 10^{-4}$ is estimated from Fig. 1, which shows the formation of the hydrodynamic layer at a magnification of $300\times$. The portion of the interface between the tip of the tool and the inception of the hydrodynamic layer includes a transition region. It is within this region that the cutting mechanism changes from dry sliding and the interface temperature increases to the transformation value.

For assumed values of $\mu = 0.0144$ lb-sec/in^2 ($\approx 10^3$ poise) and $H = 60$ BTU/lb we find that the thickness of the layer increases to $h_L = 10.25 \times 10^{-4}$ in at the tool–chip separation point. At the inception of the layer ($x = 0$), the normal pressure and the shear stress on the tool face are calculated to be 170,000 psi and 6350 psi respectively. Figure 9 shows the variation of p and τ with x for this example.

The average forces developed by the hydrodynamic layer are 156 lb normal and 10 lb parallel to the tool face. If within the highly strained transition layer we assume an average normal and shear stress of 100,000 psi the total forces normal and parallel to the tool face become 276 lb and 130 lb respectively. These compare reasonably well with the experimental values of 286 lb and 130 lb derivable from the given data.

If heat is transferred to the bulk of the chip this tends to decrease the rate of growth of the layer and to increase the stresses developed. A non-Newtonian treatment of the layer should increase the shear stress but will also alter heat and flow characteristics so that the overall effect cannot be immediately determined.

DISCUSSION

It has been shown that the structureless flow zone that frequently forms on the face of a cutting tool operating at high speed is capable of generating hydrodynamic pressure if the material behaves as a viscous liquid. The shape of the "lubricant" film depends upon melting and the analysis presented here indicates it should vary parabolically along the tool face. This appears to be in good agreement with experiment (Fig. 1).

The face of the chip will be extremely shiny when cutting with a flow zone present. The flow zone first appears at the speed at which the built-up edge just disappears (V_1 in ref. 5). The flow zone should first appear near the center of chip–tool contact length and move toward the tool point as the cutting speed is increased.

Crater formation is to be expected when a flow zone is present since the real area of chip–tool contact will then approach 100% of the apparent area and time and temperature conditions will be favorable for diffusion wear. The chip material in contact with the tool face will be stationary while that just beyond will move very slowly. This allows relatively long times for cobalt and carbon to diffuse into the chip and for iron to diffuse into the tool. These actions will usually cause an objectionable crater to form. Thus, while the presence of a hydrodynamic flow zone might at first appear to be desirable from the lubrication point of view it will usually give rise to objectionable crater formation due to rapid diffusion induced wear.

Opitz, [6] Trent [2] and others have recently attracted attention to the importance of stationary solid material that collects on the tool face (and to a lesser extent on the tool flank) and prevents crater formation. The most common material of this type is manganese sulfide that is present in free machining steels. Such material is effective at relatively low cutting speeds. Opitz and Koenig [6] have shown that steels deoxidized by use of calcium silicon and ferrosilicon tend to form complex protective oxides on the tool face which neutralize the normal wear-producing tendencies of a flow zone. Such stationary films form progressively and require a full minute or more of cutting to be established. They are frequently nonconductors of electricity and consist of complex oxides that seem to be more readily anchored to tools containing an appreciable percentage of titanium carbide.

Figure 10 shows tool face traces made by Mr. H. Yokouchi at the Carnegie-Mellon University. These tools had been used to cut AISI 1045 steel for different lengths of time. The traces at (a) are for a normally deoxidized steel and show a condition of rapid crater

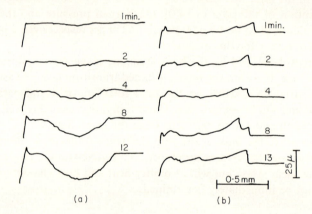

(a)   (b)

FIG. 10. Tool face traces for P-10 carbide tools cutting AISI 1045 steels at 500 fpm, a feed of 0.010 ipr and a depth of cut of 0.100 in. (a) Vacuum deoxidized, (b) special deoxidized.

formation. A liquid-like flow zone was undoubtedly present in this case. The traces at (b) are for steel of the same composition that was specially deoxidized and clearly shows the presence of the protective oxide film described by Opitz and Koenig. [6]

The thickness of the stationary solid protective built-up layer shown in Fig. 10b will usually range from $\frac{1}{2}$ to 5 thousandths of an inch and will be present only over a relatively narrow temperature range. The protective oxide disappears when its melting point is reached.

It thus appears that the hydrodynamic film that is so advantageous in fluid film bearings becomes a source of high crater wear for metal-cutting tools. A stationary protective oxide appears to provide a more refractory diffusion barrier that tends to neutralize the adverse diffusion wear associated with a slowly moving hydrodynamic layer.

ACKNOWLEDGEMENT

A grant from the National Science Foundation was used in support of this work.

REFERENCES

1. E. SCHALLER, Beitrag zur Untersuchung von Spannungen und dynamischen Vorgangen in der Grenz-schicht zwischen Werkzeug und Span bei der Stahlzerspannung mit Hartmetallwerkzeugen. Dissertation T. H. Aachen, 1964.
2. E. M. TRENT, Cutting Steel and Iron with Cemented Carbide Tools, *J. Iron and Steel Institute*, **201**, 1963, 347, 923. 1001.
3. M. C. SHAW and E. F. MACKS, *Analysis and Lubrication of Bearings*, McGraw-Hill, N.Y., 1949.
4. G. J. DeSALVO, Hydrodynamic Behavior at a Tool–Chip Interface. Ph.D Thesis, Carnegie-Mellon University, 1967.
5. K. NAKAYAMA, M. C. SHAW and R. C. BREWER, Relationship Between Cutting Forces, Temperatures, Built Up Edge and Surface Finish, *Annals of CIRP*, **14**, 1966, 211–213.
6. H. OPITZ and W. KOENIG, Basic Research on the Wear of Carbide Cutting Tools, *Machinability*, The Iron and Steel Institute, London, 1967.

EXPERIMENTAL EVALUATION OF TOOL WEAR MECHANISM AND RATE USING ELECTRON MICROPROBE ANALYSIS

Inyong Ham, A. O. Schmidt and R. J. Babcock

The Pennsylvania State University, University Park, Pennsylvania, U.S.A.
General Motors Corporation, St. Louis, Missouri, U.S.A., formerly
Department of Industrial Engineering, The Pennsylvania State University, University Park, Pennsylvania

SUMMARY

The basic wear mechanism of cutting tools was investigated by examining the tool material elements which are transferred to the chips as the tool is machining a specific material. Two primary wear mechanisms, particle transfer by adhesion and diffusion transfer, were explored by scanning the chips with an electron microprobe analyzer. By this method the particle or surface film is detected, the elements present in the wear transfer can be identified, and the size of the particle or the thickness of the film can be measured.

Several tool materials (WC-Co), (WC-TaC-Co), (TaN-ZrB$_2$) and Al$_2$O$_3$ were employed in machining AISI-4340 steel and Rene-41. The analyses of the chips indicate that this method can be used not only for detection, identification and measurement of the tool material elements transferred but also for wear rate measurements in tool life tests.

INTRODUCTION

Tool wear is a major index of performance of a cutting tool since it limits the useful life. Acceptable surface quality and integrity, dimensional accuracy, and the overall economics of a machine are directly influenced by tool wear. Furthermore, continued development of new aerospace alloys, new cutting tool materials, and the more general use of N/C machine tools increase the performance requirements of tools.

The basic mechanism of tool wear has been investigated for many years but is still not fully understood despite the large amount of data and knowledge accumulated. Much remains to be studied to explain fully the causes, process, and nature of wear and to improve tool materials and tool life. Several basic causes of tool wear are currently recognized: (a) abrasion, (b) chipping and fracture, (c) adhesion, (d) diffusion, and (e) oxidation.

Investigators have studied the tool wear mechanism using special experimental methods such as: (a) X-ray spectral analysis of tool–chip contact zone, Loladze,[1] (b) electron microscope and microanalyzer to examine the tool rake surface, Opitz et al.,[2] and (c) electron microprobe analysis of high speed steel tools, Venkatesh.[3] Most of these investigators concentrated on the tool surface or tool–chip interface region rather than the chips alone. However, Schmidt[4] took chips from a carbide steel milling operation and with a spectrograph determined the tool material transferred to the chip.

The study of tools has been related to the determination of the wear rates through the use of tool life tests. These are time consuming and involve extensive machining under specified cutting conditions. The cut is interrupted periodically to measure either flank or crater wear on a carbide tool. This cutting plus measuring process is continued until wear reaches a predetermined limit. From the tool wear data relative to cutting time, specific tool life values are derived.

Consequently, conventional tool life testing is a tedious procedure which consumes large quantities of materials and a great deal of time. Also, such test results do not indicate the wear mechanisms, i.e., how the tool is worn, by what process, etc., but merely the physical dimensions of wear. As new work materials and cutting tool materials are developed and introduced in industry, the evaluation of cutting tool performance in terms of tool wear, rate, process, patterns and mechanisms has become more important.

Several efforts have been made to measure tool wear rates indirectly, rather than by the direct measurement of the wear. One method analyzes the radioactivity of the chips resulting from a cut with a radioactive tool.[5-7] This verified that a major portion of the wear products was transferred and deposited on the chips and was in proportion to the actual tool wear. However, this technique of using a radioactive tool has not become popular as an experimental procedure because of several drawbacks such as safety, difficulty and inaccuracy.

To develop reliable, safe, convenient and economical methods and techniques for the study of tool wear, particularly the study of the wear mechanism and wear rate, the following methods and techniques have proved useful:[8-10]

(a) Electron microprobe analysis and activation analysis in the examination of chips for detection, identification and measurement of tool material transfer.

(b) Study of tool–work interface reactions using an experimental apparatus for simulated frictional interaction under dynamic conditions.

The use of chips for tool wear analysis offers some important advantages over other tool wear test methods. Both electron microprobe and activation analyses provide accurate and sensitive means for detecting and measuring tool wear. The chips can be collected from any desired operation or tool–work pair without interrupting the cutting operation. There is no safety problem in the shop since the chip activation and analysis must take place in a separate laboratory. Lastly, the collected chip specimen can be examined at any convenient time.

ELECTRON MICROPROBE ANALYSIS METHOD

The objectives of electron microprobe analysis of chips are: (a) detection and identification of an element of the tool material transferred to chip, (b) study of wear mechanism and interface reaction (adhesion and diffusion), (c) measurements of particle size or film thickness of the tool material transferred to the chip, and (d) determination of wear rate for tool life data. The schematic diagram of the chip specimen showing how the chip is taken and the area of scanning for electron microprobe analysis is shown in Fig. 1.

(a) *Electron Microprobe—Function and Operation*

The electron microprobe evolved from a combination of three established instruments— the X-ray tube, the electron microscope and the X-ray emission spectrograph. A photograph of the complete electron microprobe instrumentation and a schematic diagram of the basic components and operation of the microprobe are shown in Figs. 2A and 2B, respectively. As indicated in Fig. 2B, when the electron beam strikes the sample several different phenomena can occur. Some of the electrons hitting the sample are back-scattered. The degree of back-scattering is governed by the atomic numbers of the elements present in the

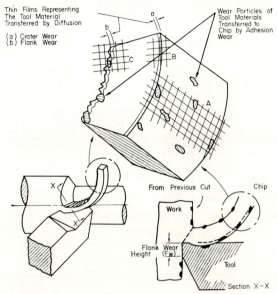

FIG. 1. Schematic diagram of chip specimen and areas of scanning for electron microprobe analysis of that wear.

FIG. 2A. Instrumentation for the electron microprobe analysis.

1. Electron Probe and Spectrometer, Applied Research Lab. Inc.
2. Current Governor, Model CCT–I North Hills Electronic, Inc.
3. Calmag Co., Regulated D.C. Power Supply
        Input 115 v., 60 CPS, 800 Watts,
        Output 3–50 KV., 500μ AMP. Model 6T6
4. G.E. Ratemeter and Scaler
5. Type 561 Oscilloscope, Teckron Co.
6. Specimen

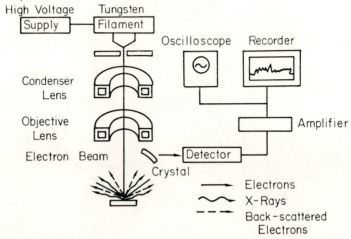

FIG. 2B.   Schematic diagram for production of X-rays in the sample by the electron microprobe.

sample. The higher the mean atomic number the greater the count of back-scattered elec-trons. Electrons which are not back-scattered penetrate into the specimen and lose their energy by collisions with the atoms of the sample. Some of these collisions result in the ejection of the inner shell electrons (K, L and M) of the atoms. When these electrons return to their normal energy levels they emit the characteristic X-rays of the element to which they belong. They are then collected and analyzed by the X-ray spectrometers located in the microprobe.

(b) *Experimental Procedure*

To obtain reliable data from the microprobe analysis the chip samples must be carefully prepared. They must be collected immediately after the cut and not be contaminated by dust or oil. The sample chip is mounted on a plastic disc using an epoxy cement.

The mean atomic numbers of the tool material and the workpiece are calculated by the following relationships:

$$MAN = \sum_{E}[AN(E) . \%(E)]/100\%$$

where MAN = mean atomic number,

$\quad E \quad$ = elements present in the material,

$\quad AN$ = atomic number

$\quad \%$ = weight percent of element in material.

A sample calculation of the mean atomic number and a list of the materials used in this study, along with their mean atomic numbers, are shown in Table 1.

The determination of the mean atomic number enables the detection of any tool particle which has transferred to the chip. Since the mean atomic number is the factor which governs the number of back-scattered electrons, it also influences how the particles of the tool material will appear on the electron back-scatter image. The higher the mean atomic number, the greater the quantity of back-scattered electrons. The region from which the high number of back-scattered electrons originated will appear as a bright area in the electron back-scatter image. The low mean atomic number areas will appear as dark areas since they back-scatter a smaller number of electrons. When selecting tool–workpiece pairs for study, it is desirable to select a combination in which the mean atomic numbers vary by approximately a factor of two. For example, if the tool material has a mean atomic number

TABLE 1. SAMPLE CALCULATIONS OF THE MEAN ATOMIC NUMBERS

(1) AISI-4340 steel:

Element	Atomic number	%	$(AN) \times$ %
C	6	0.40	2.40
MN	25	0.70	17.50
Si	14	0.30	4.20
Ni	28	1·80	50.50
Cr	24	0.25	6.00
Mo	42	0.25	10.50
Fe	26	96.30	2503.80
		Total	2594.80

The mean atomic number of AISI-4340 steel = 2594.80/100 = 25.94.

(2) 60TaN–40ZrB$_2$ tool:

Element	Atomic no.	$(AN)) \times$ %
Ta	73 ⎫	$80 \times 60 = 4800$
N	7 ⎬	
Zr	40 ⎫	$50 \times 40 = \underline{2000}$
B$_2$	5.2 ⎭	6800

The mean atomic number of 60TaN−40ZrB$_2$ = 68.00.

(3) Mean atomic numbers of work and tool materials tested:

Workpieces	Mean atomic number
AISI-4340	25.94
Rene-41	28.00

Tool materials	Mean atomic numbers
40TaN-60ZrB$_2$	62.00
60TaN-40ZrB$_2$	68.00
80TaN-20ZrB$_2$	74.00
C-883 (WC)	71.18
C-370 (WC-TaC)	65.73
CCT-707 Ceramic	10.00
Al$_2$O$_3$	10.00

of fifty-five, then ideally the workpiece should have a mean atomic number of approximately twenty-five or less. By using the factor of two there will be a clear distinction between the areas of tool material which are transferred and the chip itself.

Once the mean atomic numbers for the chip and tool have been determined the analytic lines for the elements under study are selected. In most cases the most prominent and intense line is chosen for each particular element under study, but at the same time it should be free of interference from other lines which may be emitted from the matrix of the chip. A list of the elements investigated and the wave length of their analytic lines are given in Table 2.

TABLE 2. THE ANALYTIC LINES FOR THE ELEMENTS INVESTIGATED

Element	Line	Wavelength (Å)
Ta	$L_{\alpha}1$	1.522
Zr	$L_{\alpha}1$	6.070
W	$L_{\alpha}1$	1.476
W	$M_{\alpha}1$	6.986
Co	$K_{\alpha}1$	1.790
Ni	$K_{\alpha}1$	1.500
Ni	$K_{\alpha}1$	1.659
Al	$K_{\alpha}1$	8.339

Once the chips have been mounted and the analytic lines for the elements to be investigated have been selected the actual analysis can proceed. The curved crystal spectrometers on the microprobe are set to the exact wave length of the analytic lines. The readings taken at these positions will be peak readings. Background readings are obtained by moving the spectrometers off the peak line by $\pm$ 0.03 Å. This small movement will indicate whether the readings obtained at the peak setting can be attributed to the element under study or whether the readings are due to background radiation from the chip matrix. The settings of the spectrometers are then checked against standards of the elements under investigation to insure alignment.

To detect and identify the tool material elements transferred to the chip, the electron beam scanning system is used. The beam scans back and forth over a certain specified area of the stationary chip sample. This area can be varied according to the requirements of the specific analysis. To cover the same area the scanning beam is coupled to an oscilloscope which is triggered by the electron beam in the microprobe.

Thus, the electron back-scatter permits the detection of the area where the tool material was transferred to the chip. If the mean atomic number of the tool material is higher than that of the workpiece, the concentration of tool material on the chip will appear as a white area against a dark background. If the mean atomic number of the tool material is lower than that of the workpiece, the concentrations of the tool material on the chip will appear as dark areas against a lighter background.

Once the beam has been positioned on an area of interest, the individual element spectrometers are connected to the oscilloscope. If the rate meter, an instrument which gives a continuous readout of the X-ray detectors, indicates a large increase in the count-per-second-output over the count when the beam is moved away from the area of interest, the area of interest definitely contains a concentration of the particular elements for which the spectrometers are set. Once the presence of the tool material elements on the chip has been verified the electron back-scatter is again switched to the oscilloscope and a photograph is taken of the display on the scope. Each element spectrometer is then switched to the oscilloscope one at a time and photographs are made of the various element concentrations located in the area of interest. By knowing the magnification of the photograph, the dimensions of the area surveyed during the beam scan and the size of the electron beam, the size of the tool material particles on the chip can be calculated. The particle size can also be calculated from the strip chart by knowing the chart speed and the scanning speed.

To determine if the tool wear is in the form of film transfer to the chip as opposed to particle transfer, another technique is used. The penetration of the electron beam into the specimen is dependent upon the acceleration voltage imposed on the filament of the electron gun. The higher the acceleration voltage, the deeper the penetration of the electron beam into the specimen. If there is a film on the chip it can be detected by using the previously presented relationship between the electron beam penetration and the acceleration voltage on the filament. If a film is present it will be detected by using a relatively high acceleration voltage such as 30 kV. This voltage is high enough to cause the electrons to penetrate beyond the depth of the film; however, as the beam passes through the film, the elements of the film will emit their characteristic X-rays with an intensity proportional to the number of electrons actually being stopped. If the acceleration voltage is reduced to one-half the original voltage, to say 15 kV, the penetration of the beam into the specimen will be greatly reduced and considerably more electron energy will be concentrated on the film. The elements in the film will emit their characteristic X-rays more intensely than they did at the original acceleration voltage.

If there is no film present, the X-ray intensity at 15 kV will be essentially the same as it was at 30 kV. By using this method an upper limit for the thickness of the film can be established. The actual thickness can best be determined by using the electron beam scanning method on a cross-section of the chip. This method was used previously for the detection and identification of tool material particle transfer.

EXPERIMENTAL RESULTS

The first series of tests included the detection and identification of tool material particles. In Fig. 3 the electron beam scan images of Rene-41 chips made by a ($40TaN$-$60ZrB_2$) tool and a (WC-TaC-Co) tool are shown. From the mean atomic number calculations (see Table 1; $40TaN$-$60ZrB_2$ tool-MAN=62, Rene-41-MAN=28, and Fig. 3A), the particles of tool material should appear as light spots against a darker work material background.

In the electron back-scatter photograph (Fig. 3A) several particles of tool material can be seen. The scan field is $60 \mu \times 60 \mu$. The tantalum X-ray image, obtained by connecting the tantalum spectrometer to the oscilloscope, indicates several areas of tantalum concentration, which can be readily identified by clusters of white spots located in the upper portions of the photograph (Fig. 3B). Knowing the size of the scanned area and using the gradations which can be seen on the electron back-scatter image, the particle size can be calculated. Each square of the grid is equal to 7.5μ and each division is equal to 1.5μ. With this information the tantalum and zirconium concentrations are observed to vary from 60 to 250 μm (using the conversion factor $1 \mu = 40 \mu$in). The background X-ray image for the tantalum appears to be quite intense when compared with the background for the zirconium. This is due to the fact that there is a relatively strong nickel line (Ni-$K_{\beta 1}$), with a wave length of 1.005 Å. This is due to the large quantity of nickel, a main alloying element of the work material, which has a wave length close to the analytical tantalum line (Ta-$L_{\alpha 1}$ 1.522 Å).

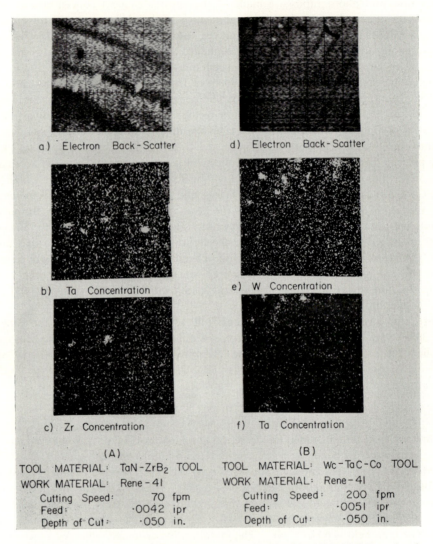

FIG. 3. Electron beam scan images of Rene-41 chips machined by (A) TaN-ZrB₂ tool and (B)WC-TaC-Co tool.

Therefore, as can be seen by a comparison of the tantalum background with the tantalum X-ray image, the only areas which actually contain tantalum are the clusters of white which appear in the upper region of the photograph. From a comparison of the zirconium X-ray image, having an analytic line of 6.070 Å $(L_{\alpha}1)$, with the background photograph, it is readily apparent that the background is much less for the zirconium than for the tantalum. As previously mentioned for the tantalum, only the clusters of white spots are zirconium concentrations. The other spots are due to background X-ray emission from the chip matrix. Comparing the tantalum and zirconium X-ray images it is noted that there is more tantalum transfer than zirconium transfer and that both elements appear to transfer together and not independently of each other. The cutting conditions for this test were selected to give adequate wear, which would aid in the identification of any tool material transferred to the chip.

The second example involving the detection and identification of tool material transferred is with the Rene-41-(WC-TaC-Co) system (Fig. 3B). The electron back-scatter photograph (Fig. 3D) shows several particles of tool material which appear as light areas near the center of the photograph. In the tungsten X-ray image four concentrations of tungsten are readily visible. The tungsten analytic line was the $L_{\alpha}1$, at 1.447 Å. In the tantalum X-ray image two areas of tantalum are visible and their size varies from 90 to 140 μin. From a comparison of the tantalum and tungsten X-ray images it is evident that tungsten and tantalum do not always transfer at the same time. The tungsten is more prone to transfer than is the tantalum.

The results of electron microprobe analysis for the (Al_2O_3)–(AISI-4340) system are shown in Figs. 4 and 5. This combination produced an opposite electron back-scatter relationship. The mean atomic number of the Al_2O_3 tool[10] was less than one half of that of the work material, IASI-4340 (25.94). Therefore, if any tool material was transferred to the chip, it would appear dark against a lighter background.

Figure 4A shows an alumininum concentration from a chip taken at a cutting speed of 1000 fpm. The electron back-scatter (Fig. 4A) has two areas of possible Al concentration. When the Al X-ray image (Fig. 4B) was examined, it was noted that both areas shown in the electron back-scatter are Al concentrations. The size of the large concentration is 320 μin while the size of the small particle is 80 μin.

Figure 4B also deals with an (Al_2O_3)-(AISI-4340) system. The chips for this example were collected at a cutting speed of 700 fpm. The electron back-scatter (Fig. 4C) showed several large areas which could be Al concentrations. When the Al X-ray image (Fig. 4D) was examined heavy concentrations of Al were found in the dark area. Light shadows can be seen on parts of the Al X-ray image photograph. These shadows are the light areas of the electron back-scatter which have been superimposed on the X-ray image to give a clear indication of where the concentrations of tool material were located. The length of the main aluminum concentration was approximately 1000 μin while its width was approximately 300 μin.

When the Fe X-ray images, overlayed on the electron back-scatter, were examined along with the corresponding Al X-ray images, it was found that in the spot of Al concentration there was no Fe concentration. This means that the Al of the ceramic tool can transfer into a hole or cavity in the surface of the chip. This is illustrated in the strip chart from the electron microprobe analysis for a chip of the same system as given in Fig. 5. It can be seen that the concentration of nickel (Ni), which is a main alloying element in AISI-4340 steel, is absent where the Al concentration was recorded. Knowing the chart speed (1 in/mm) and the scanning speed (8 μ/min) the size of the Al particles were found to be 120 μin (Fig. 5a) and 69 μin (Fig. 5b) at different cutting speeds.

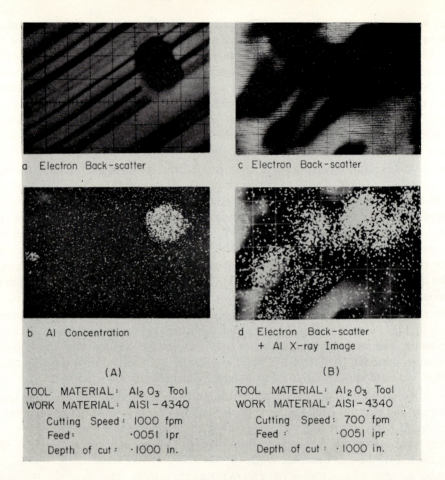

FIG. 4. Electron beam scan images of AISI-4340 chips machined by Al₂O₃ tool at different cutting speeds.

The numerical readout is an integrated output of the X-ray detector. The pulse height analyzer equipment in the microprobe takes the detector output, integrates it over a specified time period, in this case 10 sec, and presents it in numerical form on the scaler. All numerical readout is obtained in this manner. The strip chart is merely a plot of the numerical output of the scaler, but due to such inaccuracies as pen lag and chatter, which can occur in the plotting instrument, the numerical output of the scaler should be used for all numerical calculations. The peak and background readings are obtained by spectrometer adjustment. In order to say that tool material is transferred to the chip, there should be a significant difference between the peak and background readings. This also applies to differences between peak readings at varying acceleration voltages when the detection of a film is desired.

The detection and analysis of the film transfer of the cutting tool material to the chip was investigated in the (WC-Co) + (AISI-4340) system. The objective of this experiment was not only to detect film transfer, but also to indicate from strip chart and numerical

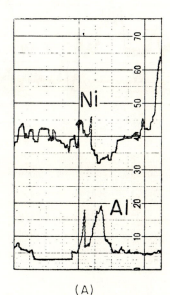

(A)

$[Al_2 O_3 - (AISI-4340)]$ – Peak Readings
500 fpm, ·0051 ipr, ·100 in.

15 kv, ·03 μamps

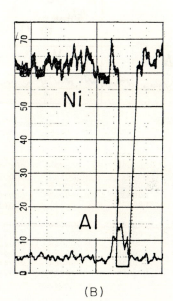

(B)

$[Al_2 O_3 - (AISI-4340)]$ Peak Readings
200 fpm, ·0051 ipr, ·100 in.

15 kv, ·03 μamps

Aluminum	Background	Aluminum	Background
000017	000013	000022	000018
000023	000017	000033	000009
000017	000014	000023	000018
000024	000019	000033	000013
000029	000010	000076	000012
000119	000010	000119	000010
000059	000013	000045	000009
000000	000015	000021	000014
000030	000011	000024	000013
000040	000017	000033	000015

Fig. 5. Strip-charts and numerical read-outs of electron microprobe output with the chips of $[(Al_2O_3) + (AISI-4340)]$ system at (A) 500 fpm and (B) 200 fpm (scan rate: 8μ/in/min).

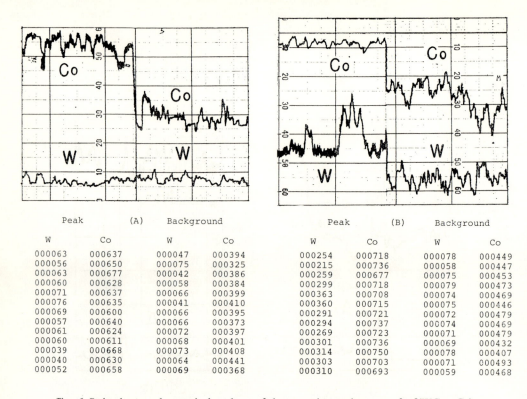

Peak	(A)	Background		Peak	(B)	Background	
W	Co	W	Co	W	Co	W	Co
000063	000637	000047	000394	000254	000718	000078	000449
000056	000650	000075	000325	000215	000736	000058	000447
000063	000677	000042	000386	000259	000677	000075	000453
000060	000628	000058	000384	000299	000718	000079	000473
000071	000637	000066	000399	000363	000708	000074	000469
000076	000635	000041	000410	000360	000715	000075	000446
000069	000600	000066	000395	000291	000721	000072	000479
000057	000640	000066	000373	000294	000737	000074	000469
000061	000624	000072	000397	000269	000723	000071	000479
000060	000611	000068	000401	000301	000736	000069	000432
000039	000668	000073	000408	000314	000750	000078	000407
000040	000630	000064	000441	000303	000703	000071	000493
000052	000658	000069	000368	000310	000693	000059	000468

Fig. 6. Strip-charts and numerical read-out of electron microprobe output for [(WC + Co) (AISI-4340)] system with the chips obtained at (A) 700 fpm and (B) 1000 fpm, both at 0.101 ipr feed, 0.100 in cut, 15kV, 0.03 μA, and scan rate of 8μ/in/min.

read-out that tool wear rate, in terms of the tool material transferred, increases with the cutting speed while the other parameters remain constant. Figure 6 shows the strip-chart and numerical read-outs of the peak and background readings for chips obtained at 700 fpm and 1000 fpm both at a 15 kV beam voltage. At 700 fpm (Fig. 6A) there is no significant difference between the tungsten peaks and background. However, the difference is significant between the cobalt peaks and the background. This indicates that at 700 fpm there is cobalt transfer, but no tungsten transfer. At 1000 fpm (Fig. 6B), however, the difference between the tungsten peak and background is significant and the cobalt peak to background ratio is also significant.

An example of the detection and identification of the film transfer of tool material is also shown in Fig. 7 (A and B). From a comparison of the outputs at 15 kV (Fig. 7A) and 30 kV (Fig. 7B) it can be seen that there is a larger amount of tungsten and cobalt recorded at 15 kV than at 30 kV, indicating that there was film transfer of tungsten and cobalt at 1000 fpm.

For better and direct comparison, the outputs at 7000 fpm and 1000 fpm are put together in Fig. 7 (C and D) It is also interesting to note that between the 700 and 1000 fpm, there was no increase in the cobalt transfer, but a significant increase in the tungsten transfer. This indicates that more tool material transfers at high speed.

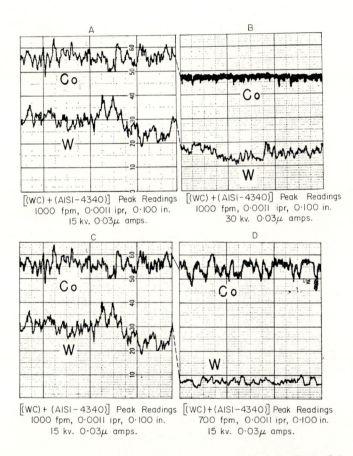

FIG. 7. Comparison of electron microscope analysis outputs (peak readings) of the machined chips of [Work (AISI-4340) + Tool (WC + Co)] system at different settings for the analyzer and cutting conditions.

From the foregoing discussions of the use of electron microprobe analysis, it should be noted that the measurements of the X-ray intensities of chips taken at different time intervals can be related to the actual flank and crater wear rates. These relations can be used to indicate tool wear rates. The tool wear data, measured by direct methods (Fig. 8A) and then X-ray intensity versus cutting time (Fig. 8B), obtained with the (AISI-4340) + (WC + Co) system at 300 fpm, 0.012 ipr and 0.100 in cut, are compared and correlated in Fig. 8C. From this figure it can be seen that the X-ray intensity measurements are linearly proportional to the actual wear measurements of both flank and crater wear (Fig. 8C). This shows that it is possible to use this method and technique as a measure of tool life.

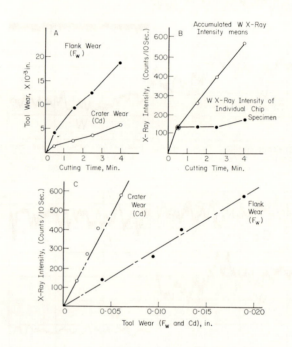

FIG. 8. Comparison of tool wear data by direct measurements and tungsten X-ray intensity data by electron microprobe analysis of chips of [(AISI-4340) + (WC + Co)] system, at 200 fpm, 0.0102 ipr, 0.100 in.

CONCLUSIONS

The results of this investigation indicate that the tool material transferred to the chip can be accurately examined by using electron microprobe analysis. It is possible to detect, identify and measure the size of tool material particles and the thickness of film transferred to the chip. It is also feasible to use tool wear rates measured by X-ray intensity counts for tool life tests.

Electron microprobe analysis is particularly useful in studies of minute wear of cutting tools since it provides sensitive and accurate measurements of elements transferred.

ACKNOWLEDGEMENTS

The experimental work reported in this paper was carried out as a part of the manufacturing method project (the Carborundum Co., Contract AF33(615)3250), under the technical direction of Mr. Floyd L. Whitney of the Advanced Fabrication Techniques Branch, MATF, Air Force Materials Laboratory, Wright-Patterson AF Base, Ohio.

REFERENCES

1. T. W. Loladze, *Wear of Cutting Tool*, Mashgiz, Moscow, 1968.
2. H. Opitz, Unsolved Problems Associated with Metal Removal Operations, *Proc. 2nd Buhl International Conference on Material*, Pittsburgh, 1966.
3. V. C. Venkatesh, Diffusion Wear of High-Speed Steel Tools, *Proc. 7th International M.T.D.R. Conference*, Birmingham, September 1966.
4. A. O. Schmidt, Spectrographic Measurements of Chips and their Correlation with Tool Wear, Paper, M.T.D.R., 1968.
5. M. E. Merchant, H. Ernst and E. J. Krabacher, Radioactive Cutting Tools for Rapid Tool-life Testing, *ASME Transactions*, **75**, 1958.
6. G. F. Wilson, Evaluation of the Radiometer Method of Tool Wear Determination, Rock Island Arsenal Lab. Tech. Report, 63-3598, October 1963.
7. N. H. Cook and A. B. Lang, Criticism of Radioactive Tool Life Testing, *ASME Trans.* **85**, 1965.
8. A. O. Schmidt and I. Ham, Evaluation of Cutting Performance of Newly Developed Tool Materials with Aero-space Materials, and Study on Correlations with Fabrication Parameters and Properties, Interim Progress Reports No. II to the Carborundum Co. Also "Research and Development of New and Improved Cutting Tool Materials," IR9-701 (VI), Carborundum Co., U.S.A.F. Contract AF33-(615)3250.
9. R. J. Babcock, A Study of the Basic Wear Mechanism of Cutting Tools Using Electron Microprobe Analysis, M.Sc. Thesis, Dept. of Ind. Eng., The Pennsylvania State University, June 1967.
10. I. Ham, Fundamentals of Tool Wear, Paper presented at the ASTME Seminar on "Reduction of Metal Cutting Theory to Practice," Cincinnati, Ohio, April 9–10, 1968.

THE EFFECT OF CUTTING CONDITIONS ON BEARING AREA PARAMETERS

D. J. Grieve, H. Kaliszer and G. W. Rowe

Department of Mechanical Engineering, The University of Birmingham

1. INTRODUCTION

In modern engineering the exacting requirements to ensure that a machine component will perform without failure during its life cycle are of paramount importance. Amongst the new requirements which have arisen with the development of more efficient manufacturing processes are the precision and high surface quality to provide sufficient strength, joint stiffness, wear resistance, etc, of machine parts. The precision and high quality of surfaces are directly related to the actual contact area which can be determined by a detailed analysis of bearing area.

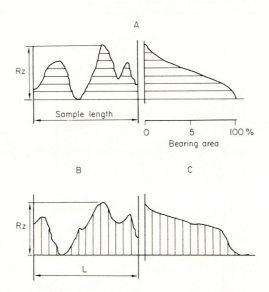

FIG. 1. Construction of the bearing area curve from the surface profile.

The bearing area (bearing fraction), or cumulative height distribution as it is sometimes called (Fig. 1A), represents the relationship of the total length of material intercepts at or above a specified level to a reference length L. It is usually expressed as a percentage. The bearing area may be illustrated by considering the profile to consist of a number of sticks resting on a reference plane passing through the bottom of the deepest valley (Fig. 1B). If the sticks are rearranged so that they are located in order of length then the top of the sticks will form the required bearing area curve (Fig. 1c). Common surface parameters such as

Ra (CLA), *Rz* (PVH) and RMS height reveal very little about a surface and can be misleading (Fig. 2, surfaces A and B). Since different surfaces may also be represented by the same bearing area (Fig. 3) the full description of a surface profile may require additional information about the slopes of the irregularities or the number of peaks per unit length and also curvatures of peaks and valleys. The bearing area alone will enable the surfaces A and B (Fig. 2) to be differentiated between in some ways although they have the same

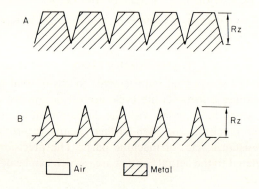

FIG. 2. Different surfaces with equal *Ra*, *Rz*, and RMS.

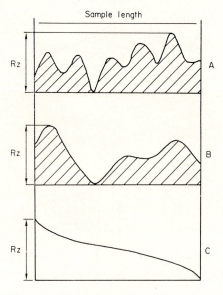

FIG. 3. Different surface profiles with equal bearing area curves.

Ra, *Rz*, and RMS values but differ in properties such as wear resistance, joint stiffness and lubricant retention. It should be noted that the vertical scale in Figs. 2 and 4 is exaggerated.

The purpose of this paper is to establish a relationship between cutting conditions (cutting speed and feed rate) and bearing area and other related parameters describing surfaces and machining operations and compare two methods of surface examination which are listed below.

(i) Electrical integrating instruments based on the Taylor Hobson Talysurf Mark 3 equipment.
(ii) Digital computation utilizing the digitized output of the amplified profile signal.

2. PREVIOUS STUDIES

Considerable work has already been done on how cutting conditions (speed, depth of cut and feed rate), tool geometry (tool nose radius, rake and flank angle) and tool wear affect the surface finish as measured by Ra.[1]

Other workers have investigated how tool geometry affects the actual shearing processes that occur in single-point metal cutting.

R. E. Reason[2] has suggested that the bearing area is a useful parameter, commenting nevertheless that the fraction determined from a profile is basically one of length not of area.

According to D'yachenko[3] the bearing area may be calculated as follows:

$$l = l_0 + by^v$$

where l = horizontal length (intercepts) of the material at or above a specific level,
  l_0 = horizontal length for $h^1 = h_{max}$ ($y = 0$),
  y = vertical distance between the highest and the specified level ($y = h_{max} - h$),
v and b = parameters determined experimentally.

The same author and other Russian scientists[4-6] have suggested that two perpendicular profiles should be taken and multiplied together to get an area relationship.

According to Rhyzhkov[6] the height of surface irregularities in two mutually perpendicular profile sections may differ considerably depending on the type of machining process. He analysed many types of machining processes and found the relationship between roughness in two perpendicular directions. He found that for decreasing roughness the ratio of Ra in the cutting direction to the Ra in the feed direction increased. The ratio lies between 0·10 and 0·30 for turning and 0·50 and 1·00 for grinding. Thus the roughness measured in the conventional feed direction has a dominant effect, and will govern the three-dimensional case.

According to Beletski[7] depending upon the dynamic rigidity of the machine–workpiece–tool system, and also when cutting with a built up edge the roughness in the cutting direction can exceed the roughness in the feed direction.

Greenwood and Williamson[8] have studied the relationship between two-dimensional surface representation, that is by a single profile, and three-dimensional representation utilizing adjacent profiles. They concluded that for most surfaces a single profile is typical of the whole surface when considering most surface parameters, i.e. Ra, Rz and the bearing area. They refute the assumption of D'yachenko[4] and stated that the surface asperities are randomly distributed, and uniform (the standard deviation of the asperity heights on a given surface is constant). From this it follows that the bearing area at any given level in any profile on the surface will be a constant irrespective of the profile position and direction (within statistical limits). In their experimental work they have found that it is difficult to get a profile that is not typical of the entire surface. This may be deduced by considering a regular triangular surface profile (Fig. 4). Profiles a and b have the same Ra, Rz and RMS unless 0 is very close to 90°, which is unlikely. The majority of surfaces are randomly distributed or nearly so, e.g. bead, or sand blasting or grinding, or else they are regular or a combination of regular and random.[9] In any of these cases a single profile is in some way representative of the whole surface when considering simple parameters such as *Ra*, *Rz* or bearing area.

According to R. E. Reason[2] the bearing area is generally related only to a small sample of the surface and ignores the variations resulting from waviness and errors of form. This difficulty occurs whatever parameter is being measured when using a stylus instrument.

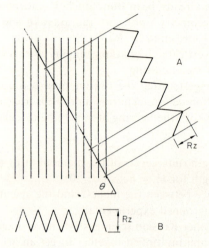

FIG. 4. Distribution of a regular triangular surface.

Several papers have been published containing details of the design of equipment suitable for finding the bearing area and amplitude density curves (the bearing area curve is a cumulative amplitude density curve).[10–12] Taylor Hobson[1] have produced a prototype CLA bearing area–peak count meter to operate in conjunction with a Talysurf III. This as well as showing the bearing area also shows the number of peaks crossed per traverse at a given level.

According to Dunin-Barkovsky[9] the profile of a machined surface can be described as the sum of two functions, i.e.

$$y(x) = g(x) + y'(x)$$

where $g(x)$ = non-random (non-stochastic) function representing the periodic component
of the profile,

$y'(x)$ = stochastic (random) function representing only random excitations superimposed upon the controllable factors of the process:

The surface profile obtained with a single point cutting tool is strongly periodic depending largely upon the shape of the tool cutting edge. In such a case the surface profile may be represented by a Fourier series as shown in Fig. 5. The theoretical surface roughness can be described as a function of four parameters:

$$Rz = \phi(\chi, \chi_1, f, r)$$

where χ, χ_1 = major and minor cutting edge angles,

f = feed,

r = nose radius.

The theoretical (ideal) surface roughness considered only geometrically can be classified into four cases as shown in Fig. 6.

Case 1. The surface irregularities are formed by the major and minor straight cutting edges connected by a rounded nose; the nose radius is less than the feed and depth of cut.

Cases 2 and 3. The surface irregularities are formed by one straight cutting edge and a rounded cutting edge.

Case 4. The nose radius is larger than the feed and the depth of cut and the surface irregularities are formed by the rounded cutting edge only.

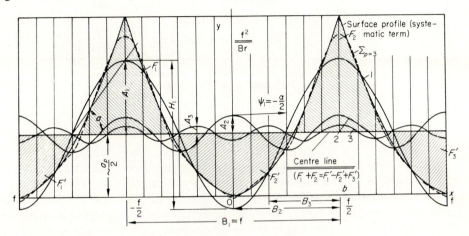

FIG. 5.

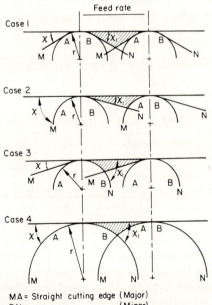

MA = Straight cutting edge (Major)
BN =   ,,   ,,   ,, (Minor)
AB = Round nose (Radius r)

FIG. 6. Profiles from various tool geometries.

3. ANALYSIS OF SURFACE PROFILES

If the equation $Rz = f^2/8r$ (see Appendix I) is put in dimensionless form

$$\frac{Rz}{r} = \frac{1}{8}\left(\frac{f}{r}\right)^2$$

then this equation will hold as long as

$$0 < \frac{f}{r} < 2\sin\chi$$

In the experiments described later the tool geometry was selected as follows:

$$\chi = \chi_1 = 45° \quad r = 0.4 \text{ mm } (1/64 \text{ in})$$

and the maximum feed rate

$$f = 0.25 \text{ mm/rev } (0.010 \text{ in/rev})$$

Therefore

$$0 < \frac{0.25}{0.4} < 2 \times 0.707$$

From the above it may be concluded that in all experiments the surface irregularities were formed by rounded tip of the tool only as shown schematically in Case 4 (Fig. 6).

In reality it is not usually possible to achieve conditions described in Fig. 6, and normally the actual surface roughness differs from the theoretical (ideal) considered geometrically. One of the main factors contributing to the deviations from the theoretical roughness is the build-up edge on the rake face which increases the height of surface irregularities considerably. Other factors which commonly contribute to the actual roughness are plastic and elastic deformations and also the friction on the minor cutting edge.

In general the actual roughness can be described as follows:

$$Rz = Rz_t + Rz$$

Rz_t = calculated theoretical surface roughness, which depends upon the four cases described above.

The residual factors grouped together under Rz consist of

$$Rz = Rz_{pl} + Rz_{el} + Rf$$

Rz_{pl} = deformation due to plastic stresses,
Rz_{el} = deformation due to elastic stresses,
Rf = deformation due to friction on the tool flank.

As experiments and measurements show, Rz_{el} is very small and seldom exceeds 3% of the sum of $Rz_{pl} + Rf$. In most cases the friction on the tool flank does not affect to a large extent the surface roughness in the feed direction. Only for very small feed rate is the effect of friction more pronounced.

Since the plastic deformations in the chip forming zone and also at the tool–workpiece interface largely depend upon the cutting speed and feed rate both these parameters have a very distinctive effect upon the actual surface roughness. Besides, the effect of cutting speed can also be analysed by considering its effect on tool geometry. Firstly at relatively low cutting speed a build-up develops altering the rake angle and in this way increasing the surface roughness. Secondly with increase in cutting speed the tool wear increases exponentially increasing the nose radius. This also affects the surface roughness by moving the bottom of the surface profile up towards the top of the profile leaving the sides comparatively unaffected. As a result the profile centre line moves up and the bearing area at mean line decreases. The higher the cutting speed the sooner and more rapidly will tool wear occur, increasing the probability of decreasing the bearing area at the centre line as compared with a bearing area produced by an unworn tool.

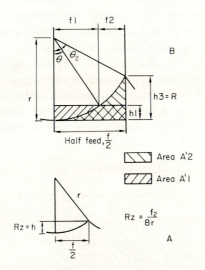

FIG. 7. The relationship between tool geometry and surface finish (Appendices I and II).

The effect of feed rate is analysed in Appendix II.

As can be seen from Fig. 7B, an increase in the feed rate will result in a greater arc of a circle being produced in the profile, decreasing the bearing at the centre line. Using a larger radius of the tool nose at the same feed rate will produce a smaller arc of a circle in the profile, increasing the mean line bearing area.

4. EXPERIMENTAL PROCEDURE AND RESULTS

About 100 En 8 steel components 4 in. diameter and 2 in. long were machined on a Harrison lathe. The cutting conditions were as follows:

Cutting speed, 30 to 160 m/min (90–500 ft/min).

Feed rate 0·08 to 0·25 mm/rev (0·0032 to 0·010 in.)

The depth of cut was constant for all components at 0·25 mm or 0·010 in.

The tungsten carbide tool tip insert had a radius of 0·4 mm (1/64 in.) and a major cutting edge angle $\chi = 45°$.

The components were examined by Taylor Hobson Talysurf III equipment for surface measurement. The multiparameter meter gave the value of Ra, peak height, valley height and Rz. The Ra, bearing area–peak count meter indicating the Ra, also the bearing area and the number of peaks crossed at the given level of traverse on the track. The level of examination on the track being within ± 3 times the Ra value of the track.

The components were initially examined for waviness. A component profile was recorded at low horizontal magnification ($\times 5$ usually), and vertical magnifications of between

1000 and 5000. Surfaces with little or no waviness were examined further. The bearing area at the mean line and at three levels above and below the mean line, the levels being spaced at intervals of half of the Ra value, was measured using the bearing area peak count meter. The stylus was moved along the same track again and the amplified signal was converted to digital form by a Solartron device adapted for use with Talysurf equipment, the digital signal being punched on five track paper tape (Ferranti Mercury Code). The readings were operated upon by computer programmes which evaluate the Ra, Rz and the bearing areas at 71 levels. The paper tape was printed by teleprinter and the ordinates shown punched on IBM data cards, put with a Fortran IV programme into the departmental IBM 1130 computer. This programme plotted a curve of bearing area against level and gave Ra and Rz values. However, as the card punching was done by hand it was laborious to use more than about 300 ordinates per component (unfortunately there is no tape reader for the IBM). Frequent checks on accuracy were made using the university's main computer, an English Electric KDF 9 with a 'K' autocode programme which read 5-track paper tape directly. About 1000 ordinates per component were used in these calculations.

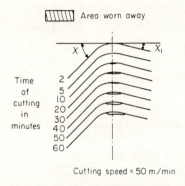

Fig. 8. Change in the cutting edge shape during turning ($\times$ 50).

To ensure consistent results the same equipment and workpiece dimensions were used for all experiments. The cutting conditions were set to avoid appreciable deterioration during a single cut, the carbide tool inserts being changed as soon as they showed any signs of deterioration. No cutting lubricant or coolant was used. All components were examined by means of the same stylus skid unit on the Talysurf, the sample examined being from near the middle of the component to decrease the scattering of results due to any tool wear that may occur during a single cut.

The effect of waviness on micro roughness parameters is difficult to analyse quantitatively except for a very simple model. As our experiments were intended to measure micro roughness only, waviness was removed as far as possible. BS 1134[13] suggests that wavelengths longer than 0·1 in. be removed when examining turned surfaces. Electrical filters do not have sharp cut-offs[13] so the effect of waves near the cut-off length is indeterminate. This will not affect results appreciably unless there are large components of waviness near the cut-off valve. To eliminate the effects of this phenomenon profiles with considerable waviness, of length near to the cut-off valve, were not included in the comparison. The limit of waviness was set so that the maximum amplitude of any waviness in the sample was less than half of the primary roughness amplitude. Surfaces that had high-frequency waviness or chatter, when examined, may be expected to have larger *Ra* and *Rz* values than the same surface without waviness.

For digital computation it is possible to set a finite limit on the wavelength, which was done when appropriate. For surfaces where less than the cut-off length was recorded it was not necessary to filter the signal. However, it is probable that the *Ra, Rz* are slightly smaller due to a larger probability that extremes of the surface have been omitted from the smaller sample.

The parameter chosen for observation was the bearing area at the centre line, the bearing area at this level is affected less than at any other level by un-typical high peaks or deep valleys. This may be considered as a guide to wear resistance as large centre line bearing areas mean that comparatively larger amounts of material must be worn away to wear the extremes of the surface down to the original centre line position. From the theory (Appendix II) the centre line bearing area is expected to decrease with increasing feed and cutting speed. To establish the validity of the results some components were examined several times on the same track, and also at different generators. The effect of sample size was also briefly examined. In a sample size of about 300 ordinates, which was used for most of the computations, the *Ra* at three different generators, taken at random on a component, varied from 75 to 95 μin but the bearing area at the centre line varied only from 44% to 46% (Fig. 9). This figure also shows that by excluding the top and bottom 10% of the surfaces, the bearing area was within a band of 8%, or for the entire surface, in a 10% band, despite an *Ra* variation of over 20%. The component was excluded from the general results due to excessive waviness.

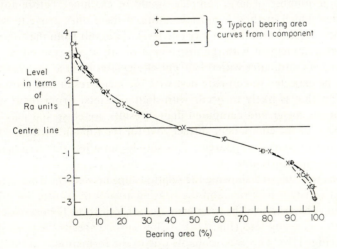

FIG. 9. Typical bearing area curves.

One generator of this surface was further examined to study the effect of varying the sample size. This generator was first examined by the bearing area meter, the length of the sample being about 7/16 in. The generator profile was next examined by digitizing 300 and then 900 ordinates, representing sample lengths of 0·06 and 0·18 in. respectively. The short sample was examined without filtration, the long sample being examined first without, then with, filtration to remove wavelengths longer than 0·1 in. The bearing area curves from these samples are given in Fig. 10. From this it is seen that either the instrument

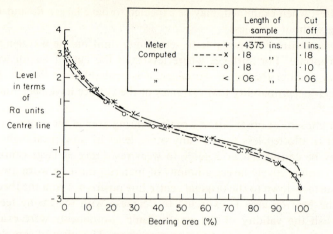

FIG. 10. Comparison between computed and measured bearing areas.

results underestimate the surface extremities or the digital computations overestimate the extremities. The former is possible to some extent, due to the electrical operation; while the stylus is travelling, the instrument reading falls off rapidly at low bearing areas, the reading changing further as soon as the traverse stops. The computer program fits a different centre line from that fitted by the instrument, this will account for some of the difference in the results. A large number of large samples should be examined before any definite conclusion concerning which method is most accurate, if there does prove to be a difference. The digital results are all within 7% at any one level, the result from the computation using 300 ordinates being coincident with the instrument result, at the centre line. However, the digital result using 900 ordinates with 0·1 in. cut off deviates most at the centre line by some 7%. This would be expected to compare best with the meter result and probably shows the range of variation that is likely to occur with digital results. From these observations it may be said that the meter and computed digital results are close and may be compared, for the middle 50% of the component surface, but for the top and bottom 20% of the surface some variation in results is apparent. This should not affect our results as we consider the bearing area at the centre line.

Series of graphs were plotted showing the relationships between bearing area at the centre line and speed at constant feed rate and the bearing area at the centre line to the feed rate at constant speed. The graphs of bearing area against feed show the decrease of centre line bearing area with increasing feed (Fig. 11a). At low cutting speeds the effect did not occur, or was reversed (Fig. 11b). This was probably due to the formation of a built up edge, and poor cutting at this lower speed. At higher speeds the centre line bearing area decreases even more rapidly with increasing feed (Fig. 11c) but the results were scattered much more due to comparatively rapid tool wear. Although tool inserts were changed whenever the wear became noticeable, it often happened that at extreme conditions no wear was noticed before a cut, but appreciable wear had occurred during the following cut. Figure 8 shows a typical example of tool wear. There is little difference between instrument and computed results, only about 3%, although some individual results are more scattered, as can be seen from Fig. 11. Both the curves for computed and instrument results show almost identical trends.

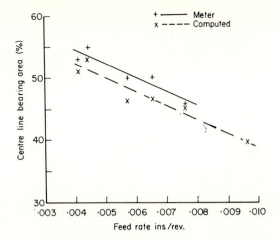

FIG. 11A. Bearing areas vs. feed rate; cutting speed = 220 ft/min.

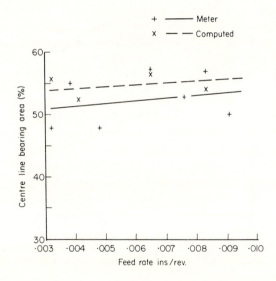

FIG. 11B. Bearing area vs. feed rate; cutting speed = 110 ft/min.

The results of centre line bearing area against cutting speed are given in Fig. 12. A theory predicts the centre line bearing area decreases with increasing the cutting speed Fig. 12A. However, as the tool tip wears continuously during a single cut the results are scattered, depending partly on whether the track of the component examined was nearer to the start or finish of the cut. At low feeds the effect was not noticeable (Fig. 12B) due to the fact that tool wear was negligible during single cuts; the individual results were also less scattered. The digital and instrument results were within 5%, but individual results were scattered more than this.

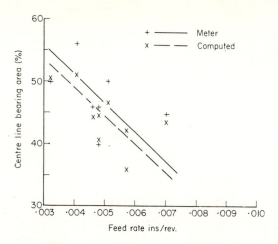

Fig. 11c. Bearing area vs. feed rate; cutting speed = 500 ft/min.

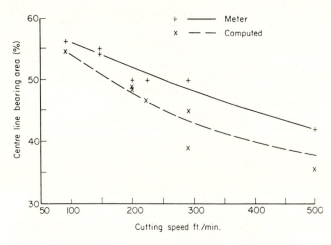

Fig. 12a. Bearing area vs. cutting speed; feed = 0·0057 in./rev.

5. CONCLUSIONS

The theory predicts that an increase in the feed rate or in the effective worn tool radius, and hence in cutting speed, will reduce the mean-line bearing area, other factors remaining constant. These relationships are observed.

The comparison between the electrical instrument results and digital computation show that the two methods compare favourably for the mid portion of the surface away from the extremities, but variation in results occurred close to the extremities. Large samples would be necessary to examine accurately the extremities of the profiles, due to the random nature of high individual peaks and valleys.

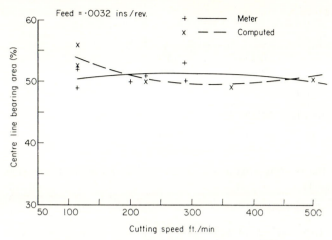

FIG. 12B. Bearing area vs. cutting speed; feed = 0·0032 in./rev.

ACKNOWLEDGEMENTS

The authors wish to thank Professor S. A. Tobias, Head of the Department of Mechanical Engineering, for providing facilities to carry out the above research.

Rank Taylor Hobson is gratefully acknowledged for lending the necessary instrumentation.

Our thanks also to Mr. R. A. E. Aston for assisting with the experiments.

REFERENCES

1. D. F. GALLOWAY. *Proc. Inst. Mech. Eng.* **153**, 113 (1945).
2. R. E. REASON. The bearing parameters of surface topography. *5th Int. M.T.D.R. Conf.*, Sept. 1964.
3. P. E. D'YACHENKO. *Criteria for Establishing Surface Irregularities.* Academy of Sciences U.S.S.R., Moscow, 1942 (Russian).
4. P. E. D'YACHENKO. *The Actual Contact Area between Touching Surfaces.* Consultants Bureau, New York, (1964).
5. N. B. DENNKIN. *Actual Contact Area between Hard Surfaces.* Academy of Sciences U.S.S.R., Moscow, 1962 (Russian).
6. E. V. RYZHKOV. *Contact Rigidity of Machine Members.* Moscow, 1966 (Russian).
7. D. G. BELETSKI. *Microgeometry during Turning.* Moscow, 1947 (Russian).
8. J. A. GREENWOOD and J. B. P. WILLIAMSON. *Proc. Roy. Soc.* **A295**, 300 (1966).
9. I. V. DUNIN-BARKOVSKY. Analysis of surface irregularities by the spectral method. Properties and Metrology of Surfaces Conference, Oxford, 1968.
10. Y. OONISHI. *Bull. Jap. Soc. Prec. Eng.* Vol. I, No. 4, March 1966, Tokyo.
11. M. PESANTE, *Annals of the C.I.R.P.* Band XII, Heft 2.
12. R. I. SALKIN and J. R. FREDERICK. *Annals of C.I.R.P.* Sept. 1966.
13. BS 1134–1961. Centre line average height method for assessment of surface texture.

APPENDIX I

The Peak-to-valley Height Rz *for a Round-nosed Cutting Tool*

Assume R large compared with feed and depth of cut as in Fig. 7A and Fig. 7B, Case 4:

$$r^2 = \left(\frac{f}{2}\right)^2 + (r - h)^2$$

$$r^2 = \frac{f^2}{4} + r^2 + h^2 - 2rh$$

h is small $\therefore$ ignore h^2

$$\therefore h = \frac{f^2}{8r} \simeq Rs$$

where f = feed,

 r = tool nose radius,

 h = peak-to-valley height Rz.

APPENDIX II

To show that the bearing area at the centre line decreases with increasing feed (Fig. 7A and Fig. 7B, Case 4),

where f = half of the feed,

 h_3 = peak to valley height Rz,

 r = cutting tool radius,

 f_1 = half length of profile below centre line,

 f_2 = half length of profile above centre line,

 $f = f_1 + f_2$

 f_1' = half length of profile below the centre line when the feed is increased by γf.

From geometry

$$f'_1 = r \sin \theta; \quad f = r \sin \theta_2$$

$$\frac{df_1}{d\theta} = r \cos \theta; \quad \frac{df}{d\theta_2} = r \cos \theta_2$$

$$A'_1 = fh_1, \, h_1 = r\,(1 - \cos \theta) = \text{valley depth}$$

$$A'_1 = fr - fr \cos \theta$$

$$A'_2 = fr - f\frac{(r - h_3)}{2} - \frac{r^2 \theta_2}{2}$$

$$h_3 = r\,(1 - \cos \theta_2)$$

$$A'_2 = fr - \frac{fr \cos \theta_2}{2} - \frac{r^2 \theta_2}{2}$$

By definition of the centre line $A'_1 = A'_2$

$$\cos \theta = \frac{\cos \theta_2}{2} + \frac{r\,\theta_2}{2f} = \frac{\cos \theta_2}{2} + \frac{\theta_2}{2 \sin \theta_2}$$

differentiating with respect to θ_2

$$- \sin \theta \, \frac{d\theta}{d\theta_2} = - \frac{\sin \theta_2}{2} + \frac{2 \sin \theta_2 - 2\,\theta_2 \cos \theta_2}{4 \sin^2 \theta_2}$$

$$\frac{d\theta}{d\theta_2} = \frac{1}{\sin \theta} \left[\frac{\sin \theta_2}{2} - \frac{1}{2 \sin \theta_2} + \frac{\theta_2 \cos \theta_2}{2 \sin^2 \theta_2} \right]$$

$$\frac{df_1}{df} = \frac{df_1}{d\theta} \frac{d\theta}{d\theta_2} \frac{d\theta_2}{df} = \frac{1}{\sin \theta} \frac{\cos \theta}{\cos \theta_2} \left[\qquad \right]$$

for decreasing bearing area with increasing feed we require f_2 to be decreasing, i.e. f_1 increasing, i.e. the fraction f_1/f is to be increasing, i.e.

$$\frac{f_1 + \gamma f \, (df_1/df)}{f + \gamma f} \times \frac{f}{f_1} > 1$$

i.e.

$$\frac{f}{f_1} \frac{df_1}{df} > 1$$

i.e.

$$\frac{\sin \theta_2 \cos \theta}{\sin^2 \theta \cos \theta_2} \left[\frac{\sin \theta_2}{2} - \frac{1}{2 \sin \theta_2} + \frac{\theta_2 \cos \theta_2}{2 \sin^2 \theta_2} \right] > 1$$

This was computed and found to be monotonically increasing from 1 upwards at $\theta = 0$ to $1 \cdot 54$ at $\theta = 88°$.

NOMENCLATURE

Bearing Area of Bearing Fraction
The bearing area, or fraction as it is sometimes called, is the ratio of the total horizontal lengths of the material surface, at or above a specified level, to a reference length, Fig. 1.
Centre line average height *Ra*.
This is defined as

$$Ra = \frac{1}{L} \int_0^L y \, dl$$

Centre Line or Mean Line
This is the straight centre line defined by $\int_0^L (y \,+) \, dl$ and $\int_0^L (y \,-) \, dl$, being equal and a minimum where
(*y* +) are ordinates above the centre line,
(*y* −) are ordinates below the centre line.

Peak-to-valley Height Rz
A measure of the vertical distance between the highest peak and lowest valley.

Amplitude Density—amplitude density curve
This is the ratio of the total horizontal length of the material surface between two specified levels, that are close together, to the sample length. The bearing area curve is a cumulative amplitude density curve.

Component
A turned surface about 4 in. diameter and 2 in. long.

M.T.C.—2K

Track

A surface profile along the tool feed direction parallel to the axis of the cylindrical component.

Sample Size

This is the length of track observed. For the Taylor Hobson equipment the sample size is 7/16 in. long. For computations, ordinates were spaced at 0·0002 in. intervals, therefore the sample size is 0·0002 × number of ordinates used.

Filtration

A method, electrical or digital of removing unwanted frequencies. Usually given in terms of "cut off", meaning that wavelengths above the cut off are removed from the calculation. Electrical filtration operates by means of conventional resistance, capacitance, inductance circuits. Digital filtration by fitting a continuously moving centre line extending equal amounts (equal to half of the cut off) on either side of each successive ordinate.

Top and Bottom Parts of the Surface

The top 10% is defined as the surface above the level at which the bearing area is 10%. The bottom 20% is the surface below the 80% bearing area level.

CUTTING PERFORMANCE COMPARISON OF TITANIUM AND TUNGSTEN CARBIDE TOOLS IN MACHINING OF GRAY IRON

INYONG HAM and DONALD S. ERMER

Department of Industrial Engineering, The Pennsylvania State University,
University Park, Pennsylvania

INTRODUCTION

An investigation was conducted to compare cutting performance, in terms of tool life, of a titanium carbide tool with that of cast-iron grade tungsten carbides. A preliminary test, at a given set of cutting conditions, using a titanium carbide tool and three commercially available cast-iron grade tungsten carbide tools indicated little variation in the tungsten carbides; therefore, only one was used for the study.

The cutting conditions for the tool-life tests were based on a statistical experimental design which allowed the determination of the possible effect of different foundry heats. An evaluation of the tool-life performance of both types of tools, as well as an economic comparison for an assumed set of cost, time, and workpiece values, was carried out.

TEST EQUIPMENT, MATERIALS AND TOOLS

The experimental tests were conducted mainly on a VDF machinability lathe with a 15 hp variable-speed drive. The test material was SAE-111 cast iron with the following nominal composition:

3.25 C min.,   2.40 Si max.,   4.25 C equiv.,
0.60–1.00 Mn,   0.12 P max.,   0.15 S max., 0.40 Cr max.,
Hardness 170–229 BHN.

The test billets were from two different heats and were 20 in long × 5 in O.D. × 3 in I.D. The actual chemical compositions of the test billets of both heats were analyzed as follows:

Heat No.	Total carbon	Combined carbon	Mn	P	S	Si
I	3.70/3.72	0.64	0.78	0.031	0.103	2.10
II	3.46/3.47	0.61	0.74	0.035	0.109	2.24

An experimental titanium carbide tool (tool A) with the following composition and properties was tested in comparison with three commercial tungsten carbide tools of cast iron cutting grade.

1005

(a) Tool A:[1]

   Nominal chemical composition (wt%):
   75.8 TiC, 12.5 Ni, 11.0 Mo, 0.7 C
   Nominal properties:
   Trans. rupture strength: 200,000 psi
   Hardness (Rockwell A): 92.7
   Density: 5.5 g/cc
   Young's Modulus: 67,000,000 psi
   Electrical resistivity: 95 micro-ohm-cm

(b) Tool B:

   3.7% Co, 0.3% Ta, 96% WC, $R_A = 92.5$, (C-3)

(c) Tool C:

   3.0% Co, 97.0% WC, $R_A = 92.7$, (C-4)

(d) Tool D:

   2.7% Co, 97.3% WC, $R_A = 93.0$, (C-4)

All tool inserts used had a tool geometry of $-5, -5, 5, 5, 15, 15, 1/32$, along with a standard tool holder of SBTR-16.

PRELIMINARY TESTS

A set of tests (cutting conditions: 750 fpm, 0.0053 ipr, and 0.050 in depth of cut) was made to compare the tool-life performances of three commercial cast-iron grade tungsten carbide tools (B, C and D) with that of the titanium carbide tool. The four tools were run in the following random order: C, B, A, D. Billets of the same heat (#II) were used to minimize structure variations. The significance of this was not known initially and was checked in the main test. The experimental results, flank wear rates and tool life values are shown in Fig. 1. From these data it is clear that the titanium carbide tool gives a distinctively longer tool life (about 5 times; 40 min tool life vs. 8 min) in comparison with the tungsten carbide tools, while tools B, C and D have almost identical wear rates and tool life values. After approximately 8 min of cut the tungsten carbide tools reached 0.015 in flank wear height while the

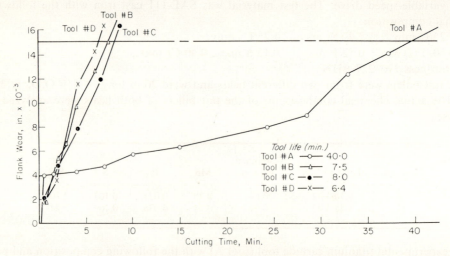

FIG. 1. Comparison of flank wear rates of titanium carbide tool (A) and three commercial cast iron grade tungsten carbide tools (B, C and D) in machining gray iron at 750 fpm, 0.0053 ipr, and 0.050 in.

titanium carbide tool showed only 0.005 in flank wear. The tungsten carbide tools had a very rapid wear rate, but showed a uniform flank wear pattern. The titanium carbide tool wear was gradual, but tended to form grooves, chip on the flank, and eventually failed by gross-chipping or fracture. All tools had some amount of cratering but this was not critical enough for tool failure. Another investigation on the comparative performances of tungsten and titanium carbide tools in finish machining reported similar results.[2]

The difference in tool life between the three tungsten carbide tools was very small and the wear characteristics have very similar patterns. Therefore, tool B was randomly chosen from the three tungsten carbide tools for a detailed comparison of performance with the titanium carbide tool.

COMPARATIVE TOOL-LIFE TESTS (MAIN TESTS)

To save time and material, but still obtain reliable data, the main tests were based on a 2^2 factorial design[4,5] for a given tool material, with two repetitions at the center points, as schematically illustrated in Fig. 2(a). Such replication gives an estimate of the experimental error during testing. The intervals between the levels of the variables cutting speed and feed

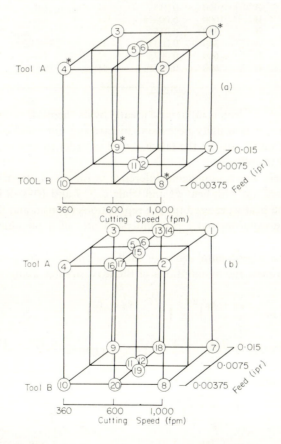

FIG. 2. Experimental design for the experimental tests at three different cutting speeds and feeds with tools A and B. (The asterisks indicate the trials which were made with heat #I. The other trials were performed with heat #II.) (a) Initial test design. (b) Augmented test design.

were made equal on a logarithmic scale to simplify the determination of the tool-life equation.

The twelve trials were carried out in a random order and the observed tool life data are given in Table 1a.

TABLE 1a. EXPERIMENTAL TEST CONDITION AND TOOL-LIFE DATA
(refer Fig. 2a)

| Test no. | Tool | Cutting condition | | | Tool-life | |
		Speed (fpm)	Feed (ipr)	Test material	T (min)	$y \, (= \ln T)$
1	A	1000	0.015	I	13.1	2.573
2	A	1000	0.00375	II	26.5	3.277
3	A	360	0.015	II	41.7	3.731
4	A	360	0.00375	I	122.0	4.804
5	A	600	0.0075	II	55.0	4.007
6	A	600	0.0075	II	55.6	4.020
7	B	1000	0.015	II	1.0	0.000
8	B	1000	0.00375	I	1.6	0.438
9	B	360	0.015	I	13.0	2.505
10	B	360	0.00375	II	77.3	4.348
11	B	600	0.0075	II	5.8	1.758
12	B	600	0.0075	II	10.1	2.313

The machinability of cast iron is influenced greatly by its chemical composition and microstructure.[2] Therefore, it is usually necessary, in carrying out machinability tests of cast iron, to ascertain the uniformity of the test material in order to minimize these effects and to check their degree of influence.

To test whether there was a significant difference in the results from the different heats, trials 1, 4, 8 and 9 were run with heat #I and trials 2, 3, 7 and 10 with heat #II. However, test billets of the same heat were used for the replications (trials 5 and 6, and 11 and 12) in order to estimate the experimental errors independently of a possible effect of different heats.

The estimates of the experimental error for each tool material are given by (the ln values of T are used to help meet the assumptions of the least squares analysis):

$$S_A^2 = \left(y_5 - \frac{y_5 + y_6}{2} \right)^2 + \left(y_6 - \frac{y_5 + y_6}{2} \right)^2$$

$$= \left(4.007 - \frac{4.007 + 4.020}{2} \right)^2 + \left(4.020 - \frac{4.007 + 4.020}{2} \right)^2$$

$$= 0.0043$$

$$S_B^2 = \left(y_{11} - \frac{y_{11} + y_{12}}{2} \right)^2 + \left(y_{11} - \frac{y_{11} + y_{12}}{2} \right)^2$$

$$= \left(1.758 - \frac{1.758 + 2.313}{2} \right)^2 + \left(2.313 - \frac{1.758 + 2.313}{2} \right)^2$$

$$= 0.0101$$

These estimates can be pooled into a common estimate of the experimental error as follows:

$$S^2_{(pooled)} = \frac{S^2_A + S^2_B}{2} = \frac{0.0043 + 0.0101}{2}$$

$$= \frac{0.0144}{2} = 0.0072$$

The estimate of the effect of the different heats on tool life is given by the difference between:

$$\frac{y_2 + y_3 + y_7 + y_{10}}{4} = \frac{3.277 + 3.731 + 0 + 4.348}{4} = \frac{11.356}{4} = 2.839$$

and

$$\frac{y_3 + y_4 + y_8 + y_9}{4} = \frac{2.573 + 4.804 + 0.438 + 2.565}{4} = \frac{10.380}{4} = 2.595$$

i.e., the difference is $(2.839 - 2.595) = 0.244$.

Under the null hypothesis that this difference is zero, i.e., the different heats have no effect a significant test can be performed using the following "t-test":

$$t = \frac{0.244 - 0}{\sqrt{[(0.0072)(1/4 + 1/4)]}} = \frac{0.244}{0.060} = 4.06$$

This t-value is smaller than $t_{0.025} = 4.303$ for two degrees of freedom, and the test results support the assumption that the different heats produce no statistically significant effect on the tool-life.

With this assumption the initial twelve trials were augmented to furnish additional information about the effect of feed. One more repetition at the center point (600 fpm, 0.0075 ipr) was carried out for each tool material and additional tests at the cutting conditions (600 fpm, 0.015 ipr) and (600 fpm, 0.00375 ipr) were also run for each tool. Additional tests were made at 600 fpm and 0.00375 and 0.015 ipr feeds for tool A. The final experimental design for the main tests is shown in Fig. 2b and the complete test results for all twenty trials are summarized in Table 1b.

The primary object of these tests was to examine and compare the effects of cutting speed and feed on tool life for both tools A and B. The experimental tests results are plotted in Figs. 3, 4 and 5 for cutting speeds of 360, 600 and 1000 fpm, respectively. In general, from these figures it can be said that the titanium carbide tool out-performed the commercially available C-3 tungsten carbide tool over the cutting conditions tested, and at a ratio of one to five or nine in terms of tool life. For example, at a cutting speed of 600 fpm tool B lasted about 6 min at all feeds tested, whereas tool A lasted 32 min at 0.015 ipr, 42 min at 0.00375 ipr, and 55 min at 0.0075 ipr, as shown in Fig. 4.

In most cases the titanium carbide tool (A) tended to form wear grooves on its flank at the side cutting edge as shown in Fig. 5. These grooves are detrimental and eventually lead to catastrophic failure as they grow. At the same time, tool A wore uniformly and progressively on both the tool face and flank. The tungsten carbide tool (B) wore in most cases uniformly, but tended to have severe cratering and maximum wear at the nose of the cutting

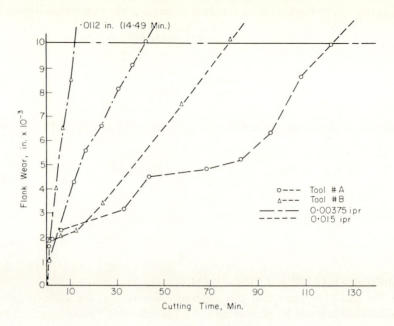

FIG. 3. Flank wear rates of tools A and B in machining gray iron at two different feeds (0.00375 and 0.015 ipr), 360 fpm, and 0.060 in cut.

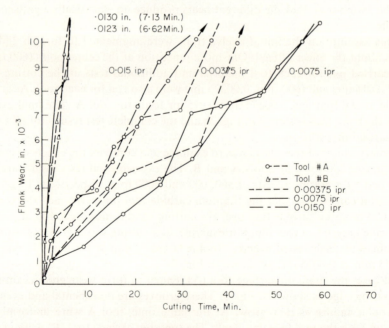

FIG. 4. Flank wear rates of tools A and B in machining gray iron at three different feeds (0.00375, 0.0075, and 0.015 ipr), 600 fpm, and 0.060 in cut.

TABLE 1b. EXPERIMENTAL TEST CONDITIONS AND TOOL-LIFE DATA
(refer Fig. 2b)

| Tool | Test no. | Cutting condition | | Tool-life | |
		Speed (fpm)	Feed (ipr)	T (min)	y ($= \ln T$)
	1	1000	0.015	13.1	2.573
	2	1000	0.00375	26.5	3.277
Tool #A	3	360	0.015	41.7	3.731
	4	360	0.00375	122.0	4.804
	5	600	0.0075	55.0	4.007
	6	600	0.0075	55.6	4.020
	7	1000	0.015	1.0	0.000
	8	1000	0.00375	1.6	0.438
Tool #B	9	360	0.015	13.0	2.565
	10	360	0.00375	77.3	4.348
	11	600	0.0075	5.8	1.758
	12	600	0.0075	10.1	2.313
	13	600	0.015	32.7	3.487
	14	600	0.015	29.8	3.395
Tool #A	15	600	0.0075	57.2	4.047
	16	600	0.00375	36.0	3.584
	17	600	0.00375	41.5	3.726
	18	600	0.015	5.78	1.754
Tool #B	19	600	0.0075	8.00	2.079
	20	600	0.00375	8.73	2.167

edge. These wear patterns of tool B are also shown in Fig. 5. In general, as feed increases the tool life decreases and the test results indicated this trend except for two cases with tool A at 600 fpm and feeds of 0.00375 and 0.0075 ipr. With a titanium carbide tool, in some cases, the tool wears by brittle chipping or breakage at the break-in zone at the beginning of the cut. Also, the tool wear patterns and tool life performance are distinctly different at small feed ranges which represent fine finish-machining conditions, i.e. in most cases the tool-life is shorter.[2] This may explain some of the discrepancy in the test results at low feed rates.

TOOL-LIFE RELATIONSHIPS

The functional relationship between the independent variables investigated, cutting speed (V) and feed (f), and the response, tool-life (T), is conventionally postulated as:

$$VT^n f^a = k$$

where n, a and k are empirically determined parameters. This equation is written in a more convenient form by taking logarithms of both sides:

$$\ln T = \left(\frac{1}{n}\right) \ln k - \left(\frac{1}{n}\right) \ln V - \left(\frac{a}{n}\right) \ln f$$

or

$$y = \ln T = B_0 + B_1 X_v + B_2 X_f \qquad (1)$$

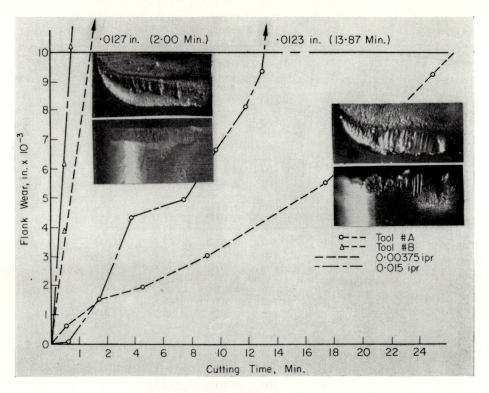

FIG. 5. Flank wear rates of tools A and B in machining gray iron at two different feeds (0.00375 and 0.015 ipr), 1000 fpm, and 0.060 in cut.

where y is the response tool-life on a logarithmic scale, X_v and X_f are the logarithmic transformations of cutting speed and feed, respectively, and where:

$$B_0 = \frac{1}{n} \ln k \qquad\qquad n = -\left(\frac{1}{B_1}\right)$$

$$B_1 = -\left(\frac{1}{n}\right) \qquad\qquad a = \frac{B_2}{B_1}$$

$$B_2 = -\left(\frac{a}{n}\right) \qquad\qquad k = \exp -\left(\frac{B_0}{B_1}\right)$$

The transformation equations to standardize the cutting conditions are:

$$X_v = 1 - 2\left[\frac{\ln 1000 - \ln V}{\ln 1000 - \ln 360}\right]$$

$$X_f = 1 - 2\left[\frac{\ln 0.015 - \ln f}{\ln 0.015 - \ln 0.00375}\right]$$

The levels of the independent variables and their corresponding standardized values are given in Table 2.

TABLE 2. LEVELS OF THE INDEPENDENT VARIABLES AND THEIR STANDARDIZED VALUES

Level	Independent variables		Standardized variables	
	Speed (V), fpm	Feed (f), ipr	X_v	X_f
High	1000	0.015	+1	+1
Center	600	0.0075	0	0
Low	360	0.00375	−1	−1

The three coefficients (B_0, B_1 and B_2) in the postulated model, eq. (1), are estimated by the method of least squares, and the predicting equation is:

$$\hat{y} = b_0 + b_1 X_v + b_2 X_f$$

where b_0, b_1 and b_2 are the least squares estimates of B_0, B_1 and B_2 respectively, and $\hat{y}$ is the predicted tool life on a logarithmic scale. The least square estimates are found from:

$$\mathbf{b} = (\mathbf{X'X})^{-1}\mathbf{X'y}$$

where $\mathbf{b}$ is the coefficient vector

$$\mathbf{b} = \begin{bmatrix} b_0 \\ b_1 \\ b_2 \end{bmatrix}$$

and the matrix of independent variables $\mathbf{X}$ for the 11 trials with tool A is:

$$
\begin{array}{cccc}
X_0 & X_v & X_f & \textit{Trial No.} \\
\end{array}
$$

$$
\mathbf{X} = \begin{bmatrix}
1 & 1 & 1 \\
1 & 1 & -1 \\
1 & -1 & 1 \\
1 & -1 & -1 \\
1 & 0 & 0 \\
1 & 0 & 0 \\
1 & 0 & 1 \\
1 & 0 & 1 \\
1 & 0 & 0 \\
1 & 0 & -1 \\
1 & 0 & -1 \\
\end{bmatrix}
\begin{array}{c}
1 \\ 2 \\ 3 \\ 4 \\ 5 \\ 6 \\ 13 \\ 14 \\ 15 \\ 16 \\ 17 \\
\end{array}
$$

Therefore:

$$
(X'X) = \begin{bmatrix} 11 & 0 & 0 \\ 0 & 4 & 0 \\ 0 & 0 & 8 \end{bmatrix}
\quad \text{and} \quad
(X'X)^{-1} = \begin{bmatrix} 1/11 & 0 & 0 \\ 0 & 1/4 & 0 \\ 0 & 0 & 1/8 \end{bmatrix}
$$

The vector $\mathbf{y}$ is called the response vector and is the column vector whose elements are the log-transformed tool-life values, i.e.,

$$\mathbf{y} = \begin{bmatrix} y_1 \\ y_2 \\ y_3 \\ y_4 \\ y_5 \\ y_6 \\ y_{13} \\ y_{14} \\ y_{15} \\ y_{16} \\ y_{17} \end{bmatrix} = \begin{bmatrix} 2.573 \\ 3.277 \\ 3.731 \\ 4.804 \\ 4.007 \\ 4.020 \\ 3.487 \\ 3.395 \\ 4.047 \\ 3.584 \\ 3.726 \end{bmatrix}$$

Thus, the estimated least squares coefficients for tool A are:

$$b_0 = 3.695386$$
$$b_1 = -0.671191$$
$$b_2 = -0.275673$$

and

$$\hat{y}_A = 3.6954 - 0.6712X_v - 0.2757X_f$$

The estimated least square coefficients for tool B are:

$$b_0 = 1.935767$$
$$b_1 = -1.618597$$
$$b_2 = -0.438893$$

and

$$\hat{y}_B = 1.9357 - 1.6186X_v - 0.4389X_f$$

The corresponding conventional equations, using the transformation equation and eq. (1), are given as follows for each tool for a depth of cut 0.060 in:

$$\text{Tool A: } VT^{0.761}f^{0.203} = e^{7.728} \tag{2}$$

$$\text{Tool B: } VT^{0.316}f^{0.200} = e^{6.030} \tag{3}$$

Although the value of n for tool A is rather high relative to previously reported values for carbide tools, it is the best estimate from the observed data in this experiment. The observed tool-life values and the tool-life curves based on eqs. (2) and (3) are given in Fig. 6, illustrating the longer tool life for tool A at a given cutting condition within the experimental range tested.

ECONOMIC ANALYSIS

The importance of being able to select economically optimum machining conditions ha long been recognized in the metal cutting field, and tools of type A will have an important effect on such a selection. The basic mathematical model generally used in the analysis of machining economics is a unit-cost model, or an analogous unit-time model if costs are neglected. In conjunction with these models two criteria are used in the determination of the optimum cutting conditions, minimum cost and maximum production rate. In this analysis the two types of tools are compared by these two conventional criteria for a single rough turning operation, although a criterion of maximum profit could also be used.[6]

Feed (ipr)	TOOL			
	Tool A		Tool B	
·0150	$VT^{·761}$	= 8107	$VT^{·316}$	= 967
·00750	$VT^{·761}$	= 10,102	$VT^{·316}$	= 1,110
·00375	$VT^{·761}$	= 12,339	$VT^{·316}$	= 1,276

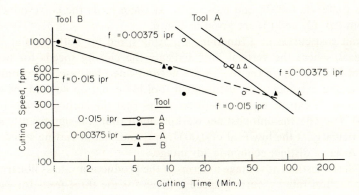

FIG. 6. Tool life data and curves of tools A and B in machining gray iron (0.06 in cut, 0.010 in flank wear limit).

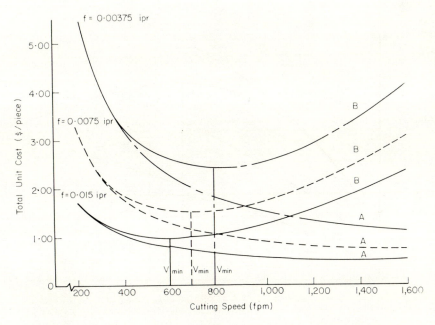

FIG. 7. Total unit cost vs. cutting speed (tool A: $n = 0.761$, tool B: $n = 0.316$).

For an economic comparison between tools A and B the following cost, time and work-piece values were assumed:

cost of operating time: $0.15/min (or $9/hr)
tool cost: $0.25/cutting edge (for both tools)
tool changing time: 0.50 min/cutting edge
handling time: 2.0 min/workpiece
diameter of workpiece: 6.0 in
axial length of cut: 18.0 in

It was also assumed that the depth of cut is a constant at 0.060 in, and that there are three possible feeds (0.00375, 0.0075 and 0.015 ipr). At these feeds the corresponding tool life equations from eqs. (2) and (3) are given in Fig. 6.

The total unit cost curves as a function of cutting speed for both tools at each feed are shown in Fig. 7, where the vertical lines indicate the cutting speeds for minimum cost (V_{min}) for tool B. The V_{min}'s for tool A are not indicated because they would have to be based on an invalid extrapolation of the fitted tool life equation beyond the experimental cutting conditions. It can be seen from these curves that the cutting speeds for minimum cost for tool B, $V_{min}(B)$, the unit cost per workpiece for tool A is significantly less than that of tool B, particularly at the low feed of 0.00375 ipr, and that at cutting speeds higher than $V_{min}(B)$ tool A has a distinct economic advantage.

Tool A also has a distinct advantage in terms of the production rate as illustrated in Fig. 8, where the production rate vs. cutting speed is given for the three feeds for each tool. The vertical lines here indicate the cutting speeds for maximum production rate for tool B, $V_{max}(B)$. As before, the V_{max} for tool A is not indicated because the derived tool-life equation is not valid at such high cutting speeds. It can be seen from these curves that tool A

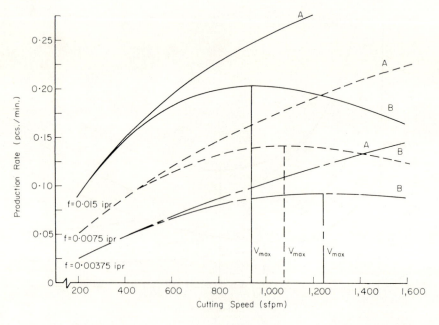

FIG. 8. Production rate vs. cutting speed (tool A: $n = 0.761$; tool B: $n = 0.316$).

gives a higher production rate than that of tool B at $V_{max}(B)$ and that this advantage increases significantly for cutting speeds above $V_{max}(B)$.

From these figures it can be seen that the titanium carbide tool (A) has significant economic and productive advantages as compared with the tungsten carbide tool (B). These advantages could be especially significant if the titanium carbide tool would be used at maximum allowable speeds consistent with the constraints of horsepower, rigidity, surface finish, etc.

CONCLUSIONS

1. A titanium carbide tool in comparison with a tungsten carbide tool gave significantly better tool life performance in machining SAE 111 cast iron.
2. An adequate tool-life relationship, including feed as well as cutting speed as a variable, can be developed from a limited number of tests with proper experimental design.
3. The economic potential of titanium carbide tools is very promising.
4. The wear pattern of both tools are different; the titanium carbide tool tends to form groove wear with less cratering while the tungsten carbide tool wears uniformly on the tool flank but with more cratering.

ACKNOWLEDGEMENTS

Acknowledgement is made to Dr. A. O. Schmidt, Professor of Industrial Engineering, Messrs. R. Babcock and T. Tanahashi, both graduate assistants, and Mr. R. Faria-Gonzalez, graduate student, the Department of Industrial Engineering, The Pennsylvania State University, for their participation in the test program in various capacities. The authors would also like to thank the Ford Motor Company for providing financial support as well as work and tool materials for the investigation.

REFERENCES

1. M. HUMANICK Jr., D. MOSKOWITZ and J. E. MAYER Jr., Titanium Carbide Cutting Tools, *Proc. Int. Conf. on Mfg. Tech.*, ASTME-CIRP meeting, Sept. 1967.
2. I. HAM, T. HOSHI and G. L. THUERING, Performance of Tungsten Carbide, Titanium Carbide, and Oxide Tools in Finish Turning of C-30 Gray Iron, *ASME Trans.* **87**, 1965.
3. I. HAM, J. R. ROUBIK and J. P. BUNCE, Machinability of High Strength Gray Cast Irons, ASME paper #57-A-239, Dec. 1957.
4. S. M. WU, Tool-life Testing by Response Surface Methodology—Parts I & II, *ASME Trans.*, Series B, **86**, No. 2, 1964.
5. R. LEVI, Evaluation of Machining Conditions by Factorial Experiments, *Int. Jr. Machine Tool Des. Res.*, **5**, 1965.
6. S. M. WU and D. S. ERMER, Maximum Profit as the Criterion in the Determination of the Optimum Cutting Conditions, *ASME Trans.*, Series B, **88**, No. 4, 1966.

SIMPLE EXPERIMENTS DEMONSTRATING MATERIAL BEHAVIOUR DURING CUTTING ON A LATHE

D. R. MILNER, E. F. SMART and G. W. ROWE

Department of Industrial Metallurgy, University of Birmingham

INTRODUCTION

These experiments have been selected on the basis of recent theoretical and practical research[1,2] to provide a concise approach to the understanding of metal behaviour during cutting. They are intended to be performed as a practical exercise, primarily for students and trainees. It is also hoped that they may be of value to those who are more experienced in machining but who wish to extend their interpretation of the details of the process and the influence of the material properties.

Three materials, Armco iron, structural mild steel, and a free-cutting steel are chosen to represent major categories of workpiece stock. The equipment used should be available in any modern laboratory engaged in machining research, but the auxiliary lathe equipment can be constructed easily and cheaply if necessary. The whole exercise occupies ten afternoons.

OBJECTIVES

The principal objective is to provide insight into the behaviour of different materials during ordinary lathe cutting.

The experimental results to be obtained are cutting forces, chip thickness, width and hardness. All are measured as functions of cutting speed over a wide range (5–800 ft/min). At a selected low speed (100 ft/min) and high speed (800 ft/min), quick-stop sections are prepared, showing the metal flow. The area of contact between tool and chip is also determined at these speeds, and the surface topography of the machined workpiece is recorded.

The energy expended in primary shear and in shear at the rake face of the tool is calculated for use in the theory. A relationship between shear plane angle and the rate of performing work at the rake face can be established for mild steel and the free-cutting steel. For the pure iron the theory must be modified to take account of the extensive lateral spread which occurs with annealed pure metals.

THEORETICAL BACKGROUND

Metallurgical comprehension of metal cutting is facilitated by considering the energy involved in shearing the metal. The primary process of chip formation is an intense shear as the metal changes direction to flow parallel to the rake face of the tool (Fig. 1).

The shear actually occurs over a narrow zone, but to a first approximation this can be represented by a single plane, shown as AB in Fig. 1. The inclination ϕ of this plane to the direction of cutting is the parameter which characterises the process. In most metalworking operations the geometry is pre-determined, but in cutting the shear plane is not dependent upon external constraint. The chip form, the cutting forces and the energy of chip formation all depend upon the shear plane angle.

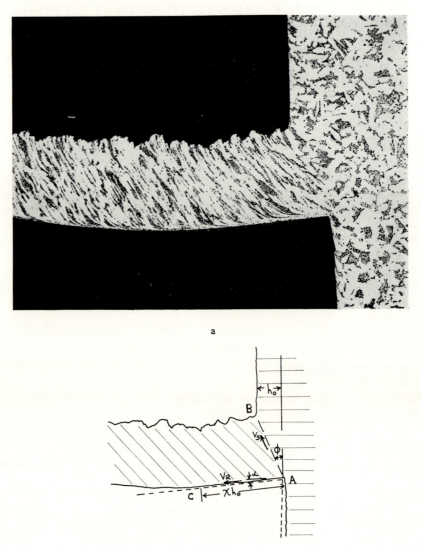

a

b

Fig. 1. (a) A "quick-stop" microsection showing the primary shear plane AB and the secondary shear at the tool face AC. (b) A diagram of the process.

In addition to this primary shear there is usually a region of secondary shear at the rake face of the tool, shown as AC in Fig. 1. This exerts a strong influence on the shear plane angle; if the drag on the tool is increased, the shear plane penetrates further into the work-piece and a thicker chip is produced. A quantitative description can be given by evaluating the energy and applying the principle of least action.

The rate of performing work W_S on the shear plane is simply the product of shearing force F_S and the velocity V_S in the shear direction. If the undeformed chip thickness/(feed rate) is h_0 and the width of the chip is w, the area of the shear plane is $w h_0$ cosec ϕ. Thus for a material of shear strength k_S,

$$\frac{dW_S}{dt} = F_S V_S = (w h_0 \text{ cosec } \phi) k_S V_S \tag{1}$$

To determine the rate of performing work W_R at the rake face it is necessary to know the contact area A_R and the shear stress k_R. The factors influencing the area of contact are complex and are not yet understood. Analytically it is convenient to express the contact length in terms of the undeformed chip thickness, introducing a multiplying factor χ. The area A_R is then given by $w (\chi h_0 \sec \alpha)$. In the experiments, χ is measured from the area of contact, which is easily visible on the tool. The shear strength can assume any value between zero and the full strength of the chip material, so a parameter β is introduced. When there is appreciable secondary shear, as in Fig. 1, β is equal to unity, but when, for example in the free-cutting steel, a soft film of manganese sulphide is present, β will be lower. With these parameters the rate of performing work at the rake face is

$$\frac{dW_R}{dt} = F_R V_R = w(\chi h_0 \sec \alpha) \beta k_R V_R \tag{2}$$

The velocities are geometrically determined and can be found from triangles of velocity.

Typical results for dW_S/dt and dW_R/dt are shown in Fig. 2.

In accordance with the principle of least action, it is postulated that the shear plane will assume an angle ϕ such that the overall rate of performing work is a minimum. It can clearly be seen from Fig. 2 that as $\beta\chi$ is increased, the minimum becomes sharper and moves to a lower value of ϕ. The total energy for chip formation also increases rapidly with $\beta\chi$. The analysis (2) shows that the equation for ϕ at the minimum is

$$\beta\chi \, k_R \sin^2\phi = k_S \cos \alpha \cos (2\phi - \alpha) \tag{3}$$

This can be solved easily by plotting the expressions on each side of the equation in terms of ϕ and the known parameters β, χ and α. The intersection of the curves predicts the value of the shear plane angle. In the absence of strain hardening, $k_R = k_S$.

If there is no drag on the tool face, $\beta\chi = 0$ and ϕ is given by

$$\cos (2\phi - \alpha) = 0; \quad 2\phi = 90 + \alpha \tag{4}$$

This sets an upper limit to ϕ, for frictionless cutting, with the shear plane bisecting the angle between the rake face and the direction of cutting (as seen in Fig. 2). The same prediction is made by the well-known "Minimum Energy" theory.[3] When there is finite tool-face drag the latter theory uses a "coefficient of friction". It is found experimentally that the ratio of tangential to normal force does not remain constant, and it can be misleading to quote a coefficient of friction, because the shear stress at the tool face is limited to the shear strength of the chip while the normal stress can increase indefinitely. At the other extreme, the shear stress can reach its limiting value even when the normal stress is relatively low, giving a very high apparent coefficient. The present theory emphasises the shear strength of the material and of the interfacial layer on the tool face, which makes it possible to interpret the results in terms of measurable metallurgical properties.

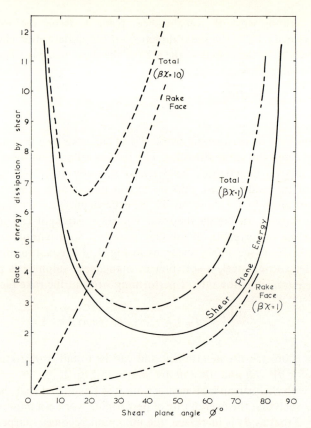

FIG. 2. The variation of total energy dissipation with shear plane angle ϕ, and the division of energy between the primary and secondary shear (rake angle $\alpha = 0$).

The theory is, however, still incomplete. The factors determining the length of contact on the rake face are not known, and in addition there can be appreciable increase in chip width w. This spread is important in annealed materials, where it increases the tool contact area A_R and in so doing increases the tool–face drag and reduces the shear plane angle. The iron used in these experiments demonstrates this feature, which for steels at high speeds is negligible.

EXPERIMENTAL MATERIALS AND PROCEDURE

Three materials were chosen to cover extremes of cutting performance. These were: Armco iron (very low C content), a plain mild steel (0·15% C), and a free-cutting mild steel (C·11% C, 0·26% S). All were used as $2\frac{1}{2}$–4 in diameter cylindrical stock, but the diameter is not critical. They were cut, as is usual, in the soft condition resulting from hot working and the hardness values were respectively 123, 176 and 179 VPN.

The tool material selected was tungsten-titanium carbide bonded with cobalt (Wickman Grade XL3). This is representative of tool materials used in industry for ferrous machining and is capable of cutting at the maximum speed used here.

The following tool angles were chosen as suitable for this exercise and reasonably representative of industrial practice:

Rake angle 6°; clearance angle 6°; approach angle 0°; trail angle 12°, front clearance angle 2°.

A $\frac{1}{32}$ in nose radius was ground onto the tool.

A medium feed (0·0063 in/rev) and depth of cut (0·050 in) were used, but these factors can be adjusted to suit the lathe available.

Ten speeds were chosen, ranging from 5 ft/min to 800 ft/min.

During cutting the forces were measured by a suitable two-component force dynamometer, calibrated in this instance up to 400 lb. The force component in the direction of cutting is related to the total energy consumed, while that in the radial direction is mainly due to the drag of the chip passing over the rake face of the tool, for a 6° rake angle tool. The chips were collected and their mean thickness was determined for each cut, using a micrometer and checking the results by weighing. The shear-plane angles were calculated from these measurements and the feed rate.

At two speeds, 100 ft/min and 800 ft/min, quick-stop sections were obtained. The device used has been described elsewhere.[4] It is essentially a humane-killer gun mounted over a pivoted cutting tool which is held in contact with a shear pin. When the gun is fired, the pin is fractured and the tool flies very quickly away from the workpiece, leaving the chip as formed at the cutting speed. The chip and a small amount of material round its root were sawn out and mounted for polishing and etching and examination in a metallurgical microscope.

The tools used at these two speeds were examined optically to determine the length of contact of the chip on the rake face.

Finally the surface topography was recorded, using a surface profilometer.

RESULTS

There will always be some scatter of individual points but the shapes of the graphs are usually reproducible by students. It is at once evident from the representative curves in Fig. 3a that Armco iron requires a much greater cutting force F_c than either mild steel or the free-cutting steel, particularly at low speeds, despite the fact that its inherent strength is no greater.

The cutting force for Armco iron decreases rapidly with increasing cutting speed but always remains higher than that for the steels. At low speeds the plain mild steel requires a greater cutting force than the free-cutting steel. This force also falls with increasing speed, but passes through an inflexion in the region of 100–200 ft/min. The cutting force for the re-sulphurised steel varies little, remaining low throughout the speed range. The variation of the force F_R along the tool face follows a similar pattern (Fig. 3b) showing a qualitative correlation with the theoretical prediction that a high cutting energy is associated with a high tool-face force. Figure 4 shows that the correlation between high cutting forces and low shear-plane angles is also generally substantiated, but the mild steel exhibits an anomalously high shear-plane angle at speeds between about 50 and 200 ft/min.

The high cutting force with Armco iron is also associated with the data on chip width shown in Fig. 5.

At low speeds the width of the chip exceeds the depth of cut by a factor 2·5. This is due primarily to the formation of a "collar" with Armco iron, and with other pure annealed materials. Some of the workpiece material, instead of forming a chip, is thrust ahead of the cut in the form of a radial collar, the size of which increases with each cut until an equilibrium

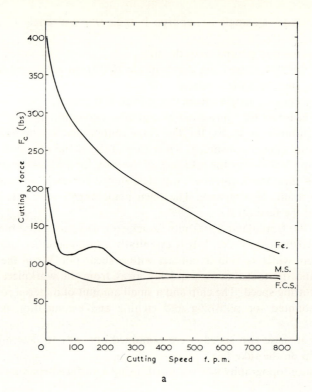

a

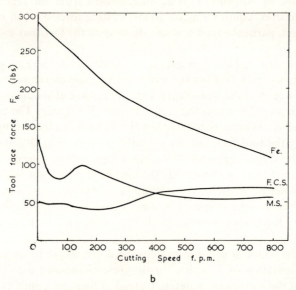

b

FIG. 3. The cutting force F_C (a) and the Force F_R resolved along the tool rake face (b) shown as functions of cutting speed for iron, mild steel and free cutting steel.

is reached. As the cutting speed is increased the size of the collar becomes smaller; this is associated with a reduction in the hardness of the chip (Fig. 6), which decreases from about 350 VPH at 5 ft/min to 230 VPH, roughly twice the workpiece hardness, at 200 ft/min, thereafter remaining constant. The marked reduction in cutting force with increasing speed arises from these two factors.

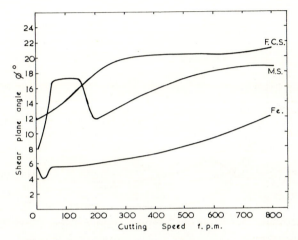

FIG. 4. The relationship between shear-plane angle and cutting speed.

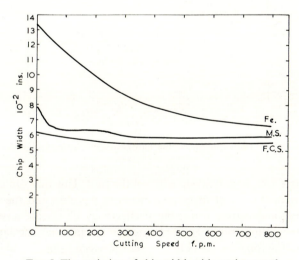

FIG. 5. The variation of chip width with cutting speed.

Both mild steel and the free-cutting steel exhibit comparatively little collar formation and can be cut with correspondingly lower energy consumption. With these materials there is less variation in hardness (Fig. 6) and in cutting force (Fig. 3) over the whole speed range.

The photomicrographs (Fig. 7) demonstrate the basic processes taking place in the metal, which result in the characteristic forms of the force curves.

At 100 ft/min the Armco iron shows a very low shear-plane angle, with a large amount of primary shear, and also extensive shearing of the metal adjacent to the tool face, associated with the high tool-face drag (Fig. 3b). The same pattern of behaviour is evident at 800 ft/min, but there is less shearing in both zones. The free-cutting steel is much less deformed (Fig. 7c) at both 100 and 800 ft/min, as might be anticipated from the low forces (Figs. 3a and 3b). At 800 ft/min the deformation in the mild steel is similar to that in the free-cutting steel,

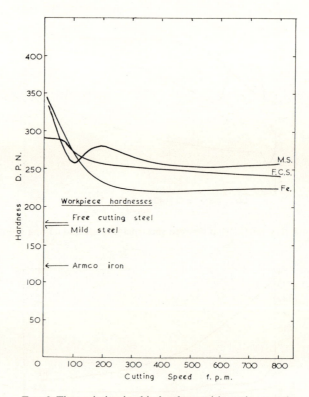

FIG. 6. The variation in chip hardness with cutting speed.

but at 100 ft/min there is a large built-up edge on the tool. This built-up edge has two contrary effects; it reduces the extent of contact between the chip and the tool and thereby reduces tool-face drag and cutting force, but it effectively blunts the nose of the tool and thereby increases the cutting force. The reduction in contact has the larger influence, as shown by the drop in the force curves (Fig. 3) over the speed range 5–150 ft/min. The built-up edge decreases in size with increased cutting speed and is eliminated at speeds above about 200 ft/min. It is responsible for the apparently anomalous behaviour shown in the cutting force, tool-face drag, shear-plane angle and hardness curves in this speed range for mild steel (Figs. 3–6).

The re-sulphurised steel gives a low–tool–face drag, and consequently a low cutting force, at low speeds. This can be attributed to the lubrication of the tool surface by the sulphide inclusions, which become elongated and tend to form a film on the rake face of the tool. The deformation and distribution of the sulphide inclusions in the free cutting steel can be

a

FIG. 7. Photomicrographs of quick-stop sections formed at 100 ft/min (upper row) and 800 ft/min (lower row). Magnification 50 ×. (a) Armco iron. (b) Mild steel. (c) Free-cutting steel.

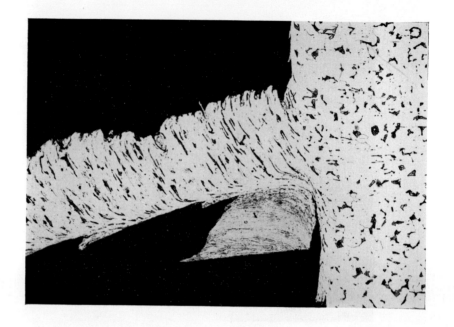

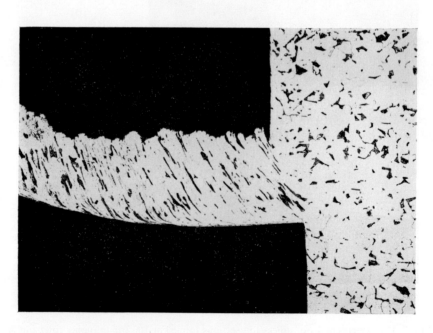

b

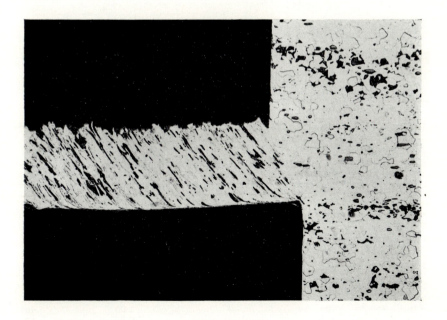

c

seen in Figs. 7c and 8. The secondary shear in these figures is less regular than that in the Armco iron arising from irregular deposition and removal of the sulphide film.

Examination of the tools after cutting shows that the contact between tool and chip is not uniform (Fig. 9). There are two divisions of the contact, one obviously metallic and complete, the other showing small fragments distributed over a larger zone. This distinction between a "sticking-friction" zone and a sliding friction zone has been discussed by several workers, particularly by Zorev.[5] Unless there are special circumstances, such as the presence of a soft sulphide film, the major part of the tool-face drag is provided by the former. It is important to recognise that the sticking friction implies only that the interfacial shear strength is greater than that of the substrate so that shear occurs beneath the surface. There may or may not be a strong adhesion or welding to the tool. Usually there is not, as can be seen from figures such as Fig. 7b, where the large built-up edge has been readily detached from the tool though there is only a relatively small coherent region at the root of the chip.

FIG. 8. A polished but unetched quick-stop section showing the deformation of the sulphide inclusion in free-cutting steel.

TABLE 1. TOOL CONTACT AREAS (STICKING FRICTION)
Cutting conditions: 6° rake XL3 tools, 0·0063 in/rev feed, 0·050 in depth of cut, dry cutting.

Workpiece material	Speed ft/min	Tool face force lb	Contact area mm²	Shear stress 10^3 lb/in²
Armco iron	100	255	6·2	26
	800	108	4·1	17
Mild steel	100	85	0·63	
	800	55	1·95	18
Free-cutting steel	100	47	1·06	46
	800	68	0·70	62

Table 1 shows the measured areas of sticking-friction contact, together with the deduced shear stresses at the tool face. It will be seen that the shear stress for Armco iron is approximately equal to the shear strength (22 tons/in² in the ordinary work-hardened condition) when the speed is low. At the higher speed it is softer. The result for mild steel is the same as for Armco iron at 800 ft/min, but at 100 ft min the built-up edge is formed and an anomalous result is obtained. The free-cutting steel appears to give excessively high shear

FIG. 9. A photograph of the contact area on a tool after cutting mild steel at 800 ft/min. (Magnification 25 ×.)

strengths, because the area of sticking friction is greatly reduced. This method is consequently inappropriate for assessing the shear stress at the rake face for free-cutting steels, which is actually lower than for iron or steel because the sulphide film is formed. The total area of contact is then difficult to assess because the tool is little damaged.

The surface topography left on the workpiece is also closely related to the conditions of shear and the chip form.

Figure 10 shows that the surface roughness is practically the same for all the materials at high speed, the main feature being the feed marks, which in practice could be reduced by using a tool with a larger radius at the tip. At 100 ft/min the free-cutting steel shows a similar result but Armco iron has a poorer surface. The mild steel shows a very rough surface, which can be attributed to severe shearing at the flank of the built-up edge. Some fragments of built-up edge are detached occasionally and remain adhering to the surface.

DISCUSSION

These experiments show in a qualitative manner that the cutting force, rake face force, shear plane angle ϕ, chip width and chip hardness are all related, and the theory gives a model from which the interrelationships can discussed. Thus, as the energy expended on the rake face of the tool is increased, the optimum shear plane angle is thrown further into the workpiece, as shown by Fig. 2. This increases the area of the shear plane and the cutting force. The strain imparted by shear is directly related to ϕ, so the chip is further hardened. The compressive force on the shear plane is also increased, producing greater lateral flow with increased collar formation and chip width. These features can also be deduced from minimum-energy theories depending on a coefficient of friction, but the present theory provides a basis of measurable properties of the material.

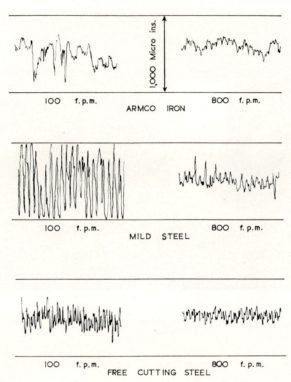

Fig. 10. Surface profilometer records for the three materials at low and high speeds.

In particular it highlights the importance of the behaviour of the material at the chip–tool interface. This is a complex factor that requires further elucidation, but the present exercise shows that the tool–face force F_R is very different for the three materials although their shear strengths differ little. The mild steel is strongly influenced by the built-up edge on the tool face in the lower part of the speed range. The striking feature of Armco iron is the steep fall in cutting force with increasing speed. This can readily be understood in terms of the softening of the iron as the chip temperature is increased at high speeds. The stress at the rake face is reduced, giving an increase in the shear-plane angle so that the area of the shear-plane as well as the shear stress acting on it are reduced. The lower force also produces less collar formation. With the free-cutting steel the high-stress sticking friction region on the tool is so limited that large changes in tool face force do not occur.

These experiments thus illustrate a number of important characteristics of the cutting behaviour of different materials. If considered in greater depth, they also suggest several lines of research which could usefully be pursued.

ACKNOWLEDGEMENTS

We thank Professor E. C. Rollason, head of the Department of Industrial Metallurgy, for his interest in setting up this exercise and in the research on which it is based.

REFERENCES

1. G. W. ROWE and P. T. SPICK, A New Approach to Determination of the Shear-plane Angle in Machining. *A.S.M.E. Jnl. Engg. for Industry*, Aug. 1967, pp. 272–282.
2. J. E. WILLIAMS, E. F. SMART and D. R. MILNER, unpublished.
3. M. E. MERCHANT, Basic Mechanics of the Metal-cutting Process. *Jnl. App. Phys.*, **16**, 1945, 267–318.
4. E. C. ROLLASON, Metallurgical Research on Machining at the University of Birmingham. Iron and Steel Inst. and Inst. Metals Conf. London, 1965.
5. N. N. ZOREV, *Int. Res. Prod. Engg. Conf. ASME*, p. 42, 1963.

SOME ASPECTS OF DRILL PERFORMANCE AND TESTING

D. J. Billau

Manufacturing Technology Centre, University of Technology, Loughborough

and

W. B. Heginbotham

Department of Production Engineering and Production Management,
University of Nottingham

THE performance of standard twist drills with various combinations of point angle, helix angle and web condition has been the subject of many researches in the past. There are also other forms of drill, e.g. spiral point, gun drills, etc., which also have been the subject of other investigations. It is not the object of this paper to compare the relative performance of drills of different types but to enumerate a method for the adequate control of general factors which affect drill performance in practice so as to make drill testing, evaluation and operation more predictable.

General Experimental Procedure

All the experiments reported here were carried out on a standard pillar drill manufactured by James Archdale & Co. Ltd. The spindle speeds were checked using a stroboscope and the regularity of the feed motions were checked by using a precision clock gauge.

Drill Mounting Technique

It was essential that the projecting length of the drill could be conveniently controlled. Conventional drills chucks are designed to grip on a cylindrical surface and, therefore, are only effective when gripping on the cylindrical shank of the drill itself. This means that variations in drill length can only be catered for by preparing special drills in each case. The chuck used for the experiments was an Erickson collet chuck shown in Fig. 1. This chuck was designed to grip and locate directly on the drill lands. The design of this chuck takes into account the back taper on the drill and is capable of producing concentric location of the drill about a mean axis of rotation to within a few ten thousandths of an inch with good reliability. By utilizing this chuck, therefore, the overhang of the drill in the chuck could be chosen conveniently whilst preserving similar gripping conditions for all projecting lengths.

Drill Material and Specification

One batch of quarter inch diameter high-speed steel standard drills were produced from a single cast by the International Twist Drill Co. Ltd. who produced the batch under specially controlled conditions throughout their manufacture and heat treatment. The drill material had the following composition: carbon 0.45%, molybdenum 5%, tungsten 6%, chromium 4%, vanadium 2%. The drills were heat treated to a hardness of between 63 and 65 Rockwell "C" scale, this figure being checked on the drill cutting edge after grinding each drill so as to ensure consistency. A typical micro-section of the drill material is shown in Fig. 2. Some $\frac{3}{8}$ in diameter and $\frac{1}{2}$ in diameter drills were also used during the course of the experiments and these were selected directly from manufacturers stock at random so

1035

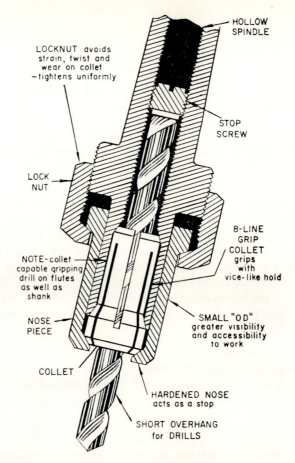

FIG. 1. Erickson drill chuck showing special double angle collet.

as to be representative of average quality twist drills. The point angle used was 118° with a 10° angle of lip relief, the relative lip height error was limited to 0·001 in. The web eccentricity was measured by rotating each drill in a "V" block and subtracting the readings taken at the bottom of each groove using a clock gauge thus measuring the eccentricity of the drill web. A maximum error of 0·0015 in was allowed for this eccentricity.

Chisel Edge Eccentricity

This error is a consequence of web eccentricity and can result in changes in drill life which are significant and to test this a number of drills were selected from a batch to give different amounts of chisel edge eccentricity. These were tested and the results are as in Fig. 3, and from this can be seen that an error of 0·0015 in in chisel edge eccentricity would result in approximately 5% error in average numbers of holes to produce failure for a given set of conditions. This tolerance of 0·0015 in was, therefore, deemed satisfactory.

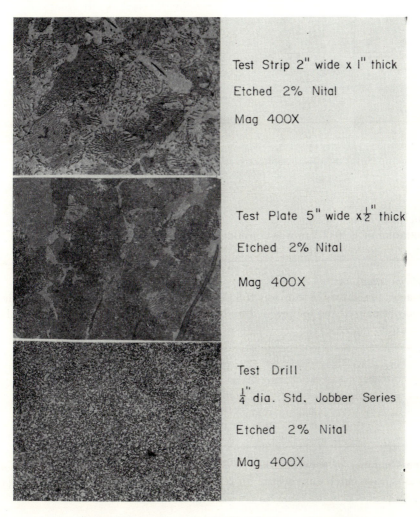

FIG. 2. Typical photomicrographs of drill and test material structures.

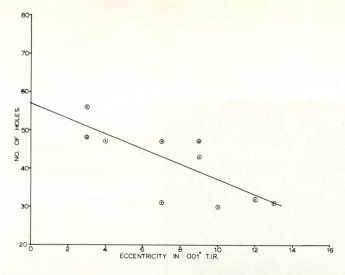

FIG. 3.

Web Thickness Error

A maximum of $\pm$ 0·003 in was considered a reasonable limit on the control of the web thickness.

Material Specification

The material used for the tests was B.S. 328 drill testing steel and this was supplied black rolled and normalized and tempered. Two batches of steel were supplied which had a specification as follows:

Test sample	Chemical composition %						
	C	Si	Mn	S	P	Cr	Ni
2″ × 1″	0·61	0·30	0·73	0·047	0·027	0·05	0·07
5″ × ½″	0·55	0·20	0·73	0·041	0·035	0·12	0·11

A hardness number of between 200 and 215 Brinell was maintained throughout the two batches used, each batch was checked for consistency in this way and also by taking random micro-sections.

Cutting Fluid

The cutting fluid used was Shell Dromus "B" used at a dilution of one part oil to ten parts distilled water by volume. The rate of flow of cutting fluid to the drill was approximately one-third of a gallon per minute kept constant for the whole of the tests.

Criterion for Failure when Drilling

An established method for deciding "failure" of a twist drill is the "screech" test. This technique relies upon the fact that a drill when worn to a certain state is excited into a state of vibration which results in a audible "screech". The amount of wear required to attain this condition is quite large and therefore in this state considerable damage has been done to the drill and consequently much work is required if the drill is to be re-conditioned for further work. Tests are normally carried out under accelerated conditions as a "short" test and therefore extrapolation of such results to practical conditions are subject to a certain degree of doubt.

An examination of worn drills reveals classification of wear patterns as follows:

1. Outer lip corner wear.
2. Wear at positions other than 1.
3. Flute cratering.
4. Body land wear.

These wear conditions produced effects similar to those found on the clearance face and rake face of lathe tools but of course there is a velocity variation across the cutting edges of a drill and the position of edge failure will be determined by the particular conditions imposed, i.e. failure can occur at the centre due to crushing or at the outer lip corner due to the higher rate of wear at higher cutting speeds and therefore higher temperatures.

The Effect of Drill Projecting Length upon Drill Life

This has been the subject of a limited amount of experimental work, i.e. Oxford (1954)[1] and Galloway (1956)[3], both of these authors expressing conflicting opinions, Oxford finding some significant variation but Galloway reporting that in his tests the effect was negligible. It must be pointed out, however, that these latter experiments were carried out using drills with larger L/d ratios. A preliminary set of experiments was carried out to evaluate the significance of the effect of projecting length on drill life using the "screech" test as a measure of failure for $\frac{1}{4}$ in diameter drills. It was also noted at this stage that there were no expressions for calculating drill life which took into account the projecting length of the drill. A complete factorial 3 factor 2 level 4 replication block of experiments were therefore designed with variables using drill speed, feed and projecting length as the independent variables with numbers of holes drilled as the dependent variable. The results obtained are shown in Table 1 and plotted in graphical form in Fig. 4, and the significance of the projected length is apparent.

TABLE 1

PRELIMINARY   INVESTIGATION   (TESTS RESULTS)

TEST CONDITION. R.P.M. x THOU./REV.	LENGTH OF DRILL PROJECTING FROM COLLET.					
	1"		2"		3"	
[I] 1600 x 10	1000*	881	73	87	23	23
	610	360	29	19	23	14
	$\bar{X}$ = 750		$\bar{X}$ = 52		MOD. VAL. = 23	
[II] 1750 x 10	278	295	26	31	19	28
	285	161	25	22	13	17
	$\bar{X}$ = 245		$\bar{X}$ = 26		$\bar{X}$ = 18	
[III] 1750 x 12	37	32	12	12	8	9
	48	54	12	5	9	5
	$\bar{X}$ = 43		MOD. VAL. = 12		MOD. VAL. = 9	

LUBRICANT :- SHELL DROMUS "E"

B.S. 328 [199 B.H.N.]   THROUGH HOLES. $\frac{1}{4}$" DIA. DRILLS (NEW)

118° POINT ANGLE. - 10° ANGLE OF LIP RELIEF

RELATIVE LIP HEIGHT LESS THAN ·001"

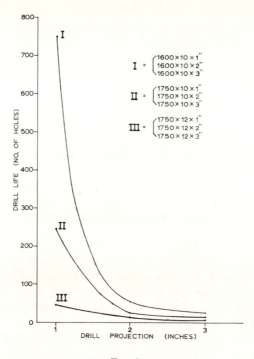

I = $\begin{cases} 1600 \times 10 \times 1'' \\ 1600 \times 10 \times 2'' \\ 1600 \times 10 \times 3'' \end{cases}$

II = $\begin{cases} 1750 \times 10 \times 1'' \\ 1750 \times 10 \times 2'' \\ 1750 \times 10 \times 3'' \end{cases}$

III = $\begin{cases} 1750 \times 12 \times 1'' \\ 1750 \times 12 \times 2'' \\ 1750 \times 12 \times 3'' \end{cases}$

Fig. 4.

Fig. 5. Four facet type drill lip relief surface.

After carrying out the preliminary experiments, however, they were not considered satisfactory from a practical point of view. The "screech" test is rather wasteful on material and drills due to the relatively large numbers of holes which have to be drilled before failure can be induced. It was, therefore, decided to look a little closer into the wear situation on a twist drill and to develop a more adequate method of representing drill wear.

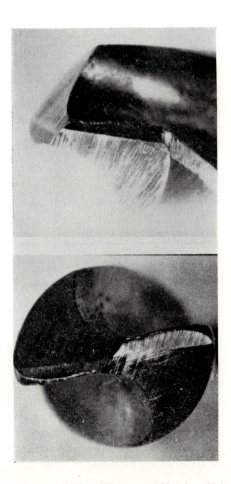

FIG. 6. Cutting condition and life. 1500 rpm × 103 cpi × 2½ in proj., 59 holes.

Assessment of Drill Performance by measuring Outer Lip Corner Wear

The width of wear band produced on the lip relief surfaces of a twist drill is not quantifiable in the same way as the wear band on the clearance face of a single point metal-cutting tool because the rate of wear varies along the cutting edge. For twist drills the following method is proposed: each cutting edge was prepared with primary and secondary clearances and a view of a typical edge is shown in Fig. 5. Using an edge of this type a typical wear situation is as shown in Fig. 6, and it can be seen that the trailing edge of the primary clearance band is undamaged and can be used as a datum for measurements. Total drill

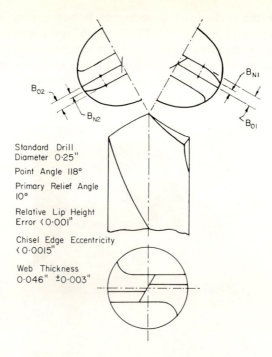

Standard Drill
Diameter 0·25"

Point Angle 118°

Primary Relief Angle
10°

Relative Lip Height
Error ⟨0·001"

Chisel Edge Eccentricity
⟨0·0015"

Web Thickness
0·046" ±0·003"

FIG. 7. Drill wear measurement.

lip wear W is therefore defined as the total loss of the dimension B (Fig. 7) on the two cutting edges.

$$W = (B_{o1} + B_{o2}) - (B_{N1} + B_{N2})$$

where B_{N1} and B_{N2} are as shown in Fig. 7 and can be positive or negative.

Measurement of the above quantities was carried out on a specially adapted Watts microscope with a jig to enable measurements to be taken normal to the plane of the primary clearance band. Typical measurements taken for two conditions of drill testing are as shown in Fig. 8, and the wear relationships follow a pattern which is identical to the classical wear relationships for lathe tools where this is:

1. A transient non-steady state with a high rate of wear until the initially sharp cutting edge is removed.
2. A "steady" state of cutting with a fairly uniform rate of wear.
3. An unstable region with a continuously increasing wear rate due to a continuous rise in temperature caused by the increasing friction between the worn cutting edge and the workpiece.

It is at some stage in this latter state that the "screech" and final failure occurs.

Under normal conditions it is considered that the drill lip wear is the more significant, and that other types of wear will be less significant, crater wear in the flutes does not assume significant proportions before lip wear causes failure of the drill and body land wear does not, in general, occur until a substantial wear band on the relief surface has developed.

It is possible to allocate some chosen wear condition W as the limit of life instead of relying on the "screech" test. This method also opens up the possibility of carrying out shorter

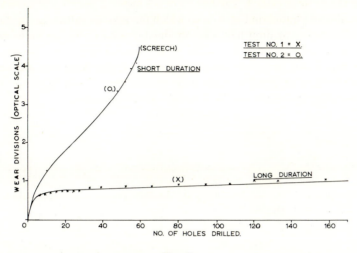

FIG. 8.

life tests by conducting tests so that the linear part of the wear characteristic is well established and then by extrapolation the performance for a pre-decided amount of wear could be determined. This is in line with current lathe tool testing techniques.

The Application of Wear Tests to Drill Life Forecasting

Having established a reliable method of measuring drill wear it is proposed that the "screech" test can be replaced by a "reliability wear test" which can be applied to any cutting conditions. With the "R.W." test the actual cutting conditions—about which information is required—can be used, even though the expected life may be of the order 2000–3000 holes. In order to obtain more reliable predictions from this shorter test it is combined with a factorial experimental layout and it is therefore suggested that the use of the "R.W." test can produce a more confident basis for drill life forecasting.

An accurate understanding of reliability is of importance when using twist drills in practice. It will be appreciated that if the average life of a set of twist drills under constant conditions is measured and one assumes that the variation in life and wear could be represented by a normal distribution, then it is clear that half the drills will fail before reaching this value. If a failure limit is set at two standard deviations below the average value then approximately one in twenty drills will fail prematurely whereas if the limit is set at three standard deviations below the mean then only one in two hundred drills should fail to reach the expected life. This is important for the control of drills in practice because very often drills fail catastrophically and it is vital to make sure that this does not happen otherwise down time and scrap can be increased due to premature failure of a high proportion of the drills in use. Hence a wear limit lower than the average to produce failure for a drill could be arbitrarily chosen or if sufficient experimental data was available on the progress of wear it would be possible to compute a statistically reliable value for the wear limit to virtually exclude the possibility of a drill failing prematurely.

A suitable method of statistically evaluating the reliability wear limit is to plot the wear value results on normal probability paper from which the mean, standard deviation and confidence limits can be obtained. The results of a 3 factor 2 level 4 replication full factorial experiment are shown plotted on Fig. 9 from which the various values can be obtained.

When the reliability wear limit has been obtained (from Fig. 9), this value can be used as the maximum wear value permissible on the steady state wear portion on the curves of the type shown in Fig. 8. As, therefore, should be apparent, if at least two life wear values are plotted on the steady state portion of the wear curves, then, either by simple extrapolation or simple arithmetic, the reliability life forecast can be estimated.

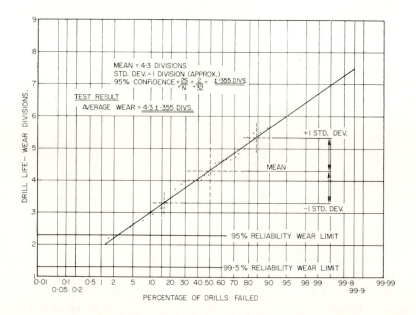

FIG. 9. Graphical solution of reliability value.

The Design of Drill Life Experiments

Assuming that the reliability wear limit has previously been determined, the following procedure can be used for the statistical analysis of the results in order to determine the effects and interaction of the variables upon drill life. The method of estimating the reliable drill life is also given.

In order to reduce the effect of the test material heterogeneity, the randomized block technique was used, the test also being arranged so that the second order interactions of the variables are confounded with the between block differences. Prior experimental data indicated that this particular interaction was non-significant.

A suitable factorial arrangement to meet these conditions is demonstrated as follows:

Using the Notation

	Factors	High level	Low level
Speed	A	a	No letter
Feed	B	b	No letter
Projection	C	c	No letter

The combination of all the factors at the low level indicated by (1).

Two blocks of four plots each containing the conditions as shown will confound the $\times B \times C$ interaction with the between blocks differences.

$$
\begin{array}{c|c}
abc & a \\
\hline
b & c
\end{array}
\qquad\qquad
\begin{array}{c|c}
ab & ac \\
\hline
bc & (1)
\end{array}
$$

The conditions are assigned to each plot within the blocks by random selection.

The between blocks bias is shown by the addition of the value q to one of the blocks, so that rewriting:

$$
\begin{array}{c|c}
abc & a \\
\hline
b & c
\end{array}
\qquad\qquad
\begin{array}{c|c}
ab + q & ac + q \\
\hline
bc + q & 1 + q
\end{array}
$$

The method of establishing the effects and interactions of the various factors can be seen as follows:

The A Effect

$$= (a - 1)(b + 1)(c + 1)$$

$$= (abc + ab + ac + a) - (bc + b + c + 1)$$

This between blocks differences is shown by the addition of q

$$\therefore = (abc + (ab + q) + (ac + q) + a) - ((bc + q) + b + c + (1 + q))$$

When the brackets are subtracted, the q effect is eliminated showing that the A effect is unaffected by the between blocks differences. This is also true for all the main effects and first order interactions.

In order to demonstrate the procedures so far discussed, the following results from a single replication 3 factor 2 level test are used as an example of the analytical procedure for processing the experimental results.

EXAMPLE OF FACTORIAL LAYOUT TO CONFOUND THE SECOND ORDER INTERACTIONS

Drill Wear Reliability Life Forecast Experiment

The 99·5% reliability wear limit is taken as 1·3 optical divisions based upon previous experimental data.

3 factors, 2 levels, single replication, utilizing two blocks.

Notation

			High level	Low level
A	Speed	(a)	1140 rpm	1500 rpm
B	Feed	(b)	103 rpi	240 rpi
C	Projection	(c)	1 in	$2\frac{1}{2}$ in

Block I

(1) 1140 × 103 × 1 in Drill No. 7	(2) 1500 × 103 × 2½ in Drill No. 3
(3) 1500 × 240 × 1 in Drill No. 6	(4) 1140 × 240 × 2½ in Drill No. 4

Block II

(5) 1500 × 103 × 1 in Drill No. 8	(6) 1140 × 240 × 1 in Drill No. 5
(7) 1500 × 240 × 2½ in Drill No. 2	(8) 1140 × 103 × 2½ in Drill No. 1

Results   99·5% *"Reliability Wear Tests"*

Block I	No. holes	Land 1	Land 2	Total	Loss	
Plot 1	0	0·75	0·75	1·50	—	
Drill 7	50	0·53	0·64	1·17	0·33	
1140 × 103 × 1 in	100	0·49	0·62	1·11	0·39	
	150	0·40	0·62	1·02	0·48	
	200	0·35	0·59	0·94	0·56	0·23
Plot 2	0	0·88	0·83	1·71	—	
Drill 3	50	0·50	0·49	0·99	0·72	
1500 × 103 × 2½ in	100	0·12	0·23	0·35	1·36	
	150	0·07	0·18	0·25	1·46	
	200	−0·13	−0·05	−0·18	1·89	1·17
Plot 3	0	1·04	0·88	1·92	—	
Drill 6	50	0·85	0·64	1·49	0·43	
1500 × 240 × 1 in	100	0·81	0·57	1·38	0·54	
	150	0·80	0·55	1·35	0·57	
	200	0·78	0·51	1·29	0·63	0·20
Plot 4	0	0·82	0·72	1·54	—	
Drill 4	50	0·42	0·26	0·68	0·86	
1140 × 240 × 2½ in	100	0·36	0·21	0·57	0·97	
	150	0·30	0·17	0·47	1·07	
	200	0·24	0·15	0·39	1·15	0·29

Block II	No. holes	Land 1	Land 2	Total	Loss	
Plot 5	0	0·67	0·87	1·54	—	
Drill 8	50	0·51	0·72	1·23	0·31	
1500 × 103 × 1 in	100	0·49	0·70	1·19	0·35	
	150	0·44	0·68	1·12	0·42	
	200	0·40	0·63	1·03	0·51	0·20
Plot 6	0	0·7	0·95	1·65		
Drill 5	50	0·40	0·66	1·06	0·59	
1140 × 240 × 1 in	100	0·38	0·60	0·98	0·67	
	150	0·35	0·56	0·91	0·74	
	200	0·34	0·54	0·88	0·77	0·18
Plot 7	0	0·76	0·67	1·43	—	
Drill 2	50	0·51	0·39	0·90	0·53	
1500 × 240 × 2½ in	100	0·47	0·34	0·81	0·62	
	150	0·40	0·3	0·70	0·73	
	200	0·33	0·27	0·60	0·83	0·30
Plot 8	0	0·69	0·76	1·45	—	
Drill 1	50	0·25	0·35	0·60	0·85	
1140 × 103 × 2½ in	100	0·20	0·25	0·45	1·00	
	150	0	0	0	1·45	
	200	−0·06	0·08	−0·14	1·59	0·74

99·5% *Reliability Life Forecast*

Condition	Initial wear	Loss
1500 × 240 × 1	0·43	0·20
	1·3 − 0·43 = 0·87	
Life forecast $= \left(\dfrac{0·87}{0·20} \times 150 \right) + 50$		= 702 holes
1500 × 240 × 2½	0·53	0·30
	1·3 − 0·53 = 0·77	
Life forecast $= \left(\dfrac{0·77}{0·30} \times 150 \right) + 50$		= 435 holes
1500 × 103 × 1	0·31	0·20
	1·3 − 0·31 = 0·99	
Life forecast $= \left(\dfrac{0·99}{0·20} \times 150 \right) + 50$		= 795 holes
1500 × 103 × 2½	0·72	1·17
	1·3 − 0·72 = 0·58	
Life forecast $= \left(\dfrac{0·58}{1·17} \times 150 \right) + 50$		= 125 holes
1140 × 240 × 1	0·59	0·18
	1·3 − 0·59 = 0·71	
Life forecast $= \left(\dfrac{0·71}{0·18} \times 150 \right) + 50$		= 641 holes

99·5% *Reliability Life Forecast (cont.)*

Condition	Initial wear	Loss
1140 × 240 × 2½	0·86	0·29
	1·3 − 0·86 = 0·44	

$$\text{Life forecast} = \left(\frac{0.44}{0.29} \times 150 \right) + 50 \qquad = 278 \text{ holes}$$

1140 × 103 × 1   0·33   0·23

1·3 − 0·33 = 0·97

$$\text{Life forecast} = \left(\frac{0.97}{0.23} \times 150 \right) + 50 \qquad = 685 \text{ holes}$$

1140 × 103 × 2½   0·85   0·74

1·3 − 0·85 = 0·45

$$\text{Life forecast} = \left(\frac{0.45}{0.74} \times 150 \right) + 50 \qquad = 141 \text{ holes}$$

Significance Tests

Block I

1. *abc* 1140 × 103 × 1 in Forecast = 685 holes	2. *b* 1500 × 103 × 2½ in Forecast = 125 holes
3. *c* 1500 × 240 × 1 in Forecast = 702 holes	4. *a* 1140 × 240 × 2½ in Forecast = 278 holes

Block II

5. *bc* 1500 × 103 × 1 in Forecast = 795 holes	6. *ab* 1140 × 240 × 1 in Forecast = 641 holes
7. (1) 1500 × 240 × 2½ in Forecast = 435 holes	8. *ab* 1140 × 103 × 2½ in Forecast = 141 holes

Coded form—Divide by 100

Block I Total $T = 17\cdot9$
Block II Total $T = 20\cdot12$

Grand total $= 38\cdot02$

$$\text{Grand correction factor} = \frac{T^2}{N} = \frac{38\cdot02^2}{8} = 180\cdot7.$$

Total Sum of Squares

Block I

$$6.85^2 + 2.78^2 + 1.25^2 + 7.02^2 = 46.92 + 7.728 + 1.562 + 49.28 \qquad = 105.49$$

Block II

$$4.35^2 + 1.41^2 + 6.41^2 + 7.95^2 = 18.923 + 1.988 + 41.088 + 63.203 \quad = 125.202$$

$$230.692$$

Total sum squares $= 230.692 - 180.7 = 49.992$

With 7 d.f.

Effect	abc	ab	ac	bc	a	b	c	(1)	S.O.S.
A	6·85	+1·41	+6·41	−7·95	+2·78	−1·25	−7·02	−4·35	1·21
B	6·85	+1·41	−6·41	+7·95	−2·78	+1·25	−7·02	−4·35	1·20
C	6·85	−1·41	+6·41	+7·95	−2·78	−1·25	+7·02	−4·35	42·5
$A \times B$	6·85	+1·41	−6·41	−7·95	−2·78	−1·25	+7·02	+4·35	0·192
$A \times C$	6·85	−1·41	+6·41	−7·95	−2·78	+1·25	−7·02	+4·35	0·01
$B \times C$	6·85	−1·41	−6·41	+7·95	+2·78	−1·25	−7·02	+4·35	4·26

Total sum treatments 49·372

Analysis of Variance

Source	Sum of squares	d.f.	Var. Est.
A	1·21	1	1·21
B	1·20	1	1·20
C	42·5	1	42·5
$A \times B$	0·192	1	0·192
$A \times C$	0·01	1	0·01
$B \times C$	4·26	1	4·26
Errors	0·62	1	0·62
Totals	49·992	7	

$B \times C = \dfrac{4.26}{0.62} = 6.9$ 1 d.f. for greater   1 d.f. for smaller

Not significant at 5% level

$A \times C = \dfrac{0.62}{0.01} = 62$ 1 d.f. for greater   1 d.f. for smaller

Not significant at 5% level

$A \times B = \dfrac{0.192}{0.62} = 3.1$ 1 d.f. for greater   1 d.f. for smaller

Not significant at 5% level

Revised Var. Est.	Sum of squares	d.f.
	0·62	1
	4·62	1
	0·01	1
	0·192	1
	5·442	4 d.f

$$\therefore \text{Revised Var. Est.} = \frac{5\cdot442}{4} = 1\cdot38 \text{ with 4 d.f.}$$

$$C = \frac{42\cdot5}{1\cdot38} = 31 \text{ for 1 d.f. for greater   4 d.f. for lesser}$$

Significant at 01 level

$$B = \frac{1\cdot38}{1\cdot2} = 1\cdot15 \quad 4 \text{ d.f. for greater   1 d.f. for lesser}$$

Not significant at 5% level

$$A = \frac{1\cdot38}{1\cdot21} = 1\cdot14 \quad 4 \text{ d.f. for greater   1 d.f. for lesser}$$

Not significant at 5% level

Naturally this single replication does little more than display the procedure for processing the experimental results and complete test requires repeated replication before definite conclusions can be drawn.

ACKNOWLEDGEMENTS

Acknowledgements are due to the Board of Governors of the Wolverhampton College of Technology and the Head of Department for providing the facilities to undertake this research and the laboratory staff for their valuable assistance.

The following firms also made significant contributions of material and provided information.

Erickson Tool Co., Solon. Ohio, U.S.A.

Shell Research Dept., Thornton, Cheshire.

International Twist Drill Ltd., Sheffield.

English Steel Corporation, Manchester.

Cleveland Twist Drill Co., Ohio, U.S.A.

P.E.R.A., Melton Mowbray.

Zwicky Ltd., London.

REFERENCES

1. CARL OXFORD, Short Drills v. Long Drills, 1954.
2. Drilling Very High Strength and Thermal Resistive Materials, Paper No. 258, A.S.T.M.E. Collected Papers.
3. GALLOWAY, Some Experiments on the influence of Various Factors on Drill Performance, 1956.
4. CARL OXFORD, Some Economic and Performance Factors Affecting Drill Selection, 1962.
5. On the Art of Cutting Metals. A.S.M.E. 1907.
6. S. PATKAY, Bearbeitbarkeit Bohrarbeit und Spiralbohren. Werkstattstechnik. 23, 1929, 3.
7. A. WALLICHS and W. MENDELSON, Zerspanbarkeitsuntersuchungen mit Spiralbohren. V.D.I. Verlag, 1932.
8. M. J. MORONEY, Facts from Figures.

SOME APPLICATIONS OF PHYSICAL METALLURGY IN METAL CUTTING

B. von Turkovich and S. Calvo

Istituto di Tecnologia Meccanica Politecnico di Torino, Italy

SUMMARY

In machining, large straining which takes place in a small zone at high strain rates produces a notable temperature rise in the deforming metal. The analysis of the process, based upon the theory of crystal imperfections, shows that the dynamic recovery in the flow zone is intimately related to mechanical energy conversion into heat. The principal mechanism of heat generation is the creation —annihilation of point defects and dislocations. It is shown also that the adiabatic instability in the deformation zone can take place only at cryogenic temperatures.

The influence of impurities and precipitates is briefly discussed.

1. INTRODUCTION

Metal cutting exhibits several interesting features of plastic deformation at high strains and strain rates. We believe these features merit a separate study, even though the experiments in machining are usually undertaken to improve the manufacturing operations.

On one hand the process of metal cutting encompasses a broad combination of stress, strain and rate levels, all within a small volume of test material. On the other hand, as is well known, the chip formation mechanism depends strongly upon the cutting speed, the initial workpiece temperature, the type and geometry of the tool in addition to basic metallurgical properties of the workpiece material. All these parameters are independently controllable; thus a link is anticipated which could permit the study of dynamic properties of metals. The connection is not a straightforward one, however. In establishing such a link one must remember that there are now several solutions of the chip formation problem obtained by the methods of mathematical theory of plasticity. Actually the number of correct solutions may be very large, indicating that there is probably no unique general solution. Some of these solutions or models are not interesting from a metallurgical point of view because the adopted simplifications mask almost completely the details of deformation process.

Others, which belong to the group of deformation zone models, are overly complex to permit, at the present time at least, any simple application of the physical theory of plasticity. The present work is based upon a rather simple version of the flow zone model so that it is possible to study the effects of strain rate, temperature, and stress distribution from a microscopic viewpoint. In such a sense it is connected with the theory of property testing. This idea is by no means a new one; in fact, Lira and Thomsen[1] have argued that metal cutting could be effectively employed in testing the material properties. A simple computation shows that this is true for a certain range of cutting speed, feed, etc., and for a rather narrow band of initial parameters, such as the workpiece temperature. Whether these restrictions would permit us, nevertheless, to obtain a consistent set of data over a wide range of stress and strain has not as yet been established. The purpose of the present work is to show the kind of physical and metallurgical effects to be expected when a given metal or alloy undergoes the deformation in conditions similar to those taking place during metal cutting. Conversely also, the peculiar combination of deformation and deformation rate in chip formation puts a severe restriction upon the choice of an appropriate microscale mechanism.

2. SOME PROPERTIES OF THE FLOW ZONE

A general discussion of the properties of deformation zone is necessary in order to specify its salient features. It would be very valuable to consider not only the polycrystals but also the deformation, i.e. the machining of single crystals. However, the dearth of experimental data on single crystals forces us to analyze the polycrystalline aggregates only. The presence of grain boundaries and the variety of crystal orientations introduce considerable difficulties, which under appropriate circumstances may be averaged over the deformation zone. Of course, such approximation ceases to be valid when the grain size is of the same order of magnitude as the volume of flow zone. It should be emphasized that very sensitive, high frequency response dynamometer measurements actually permit the resolution of force variation when the grain size of workpiece material is approximately equal (or somewhat larger) to the deformation zone volume.[2] However, under such circumstances the unique definition of the flow zone disappears. Thus, only the limiting cases of the single crystal on the one hand and of the fine grain polycrystals on the other, have sufficiently well-defined boundary conditions for a detailed analytical study. A critical set of parameters for the overall problem of chip formation is associated with the chip–tool interaction. These are the tool geometry, the presence (or its absence) of built-up-edge and the stress distribution along the contact surface. For any particular combination of these parameters one should expect a definable effect upon the flow zone geometry and, consequently, upon the strain and the strain rate. If the stress (the shear stress particularly) is considered to be a unique function of the strain, strain rate and the local temperature, i.e., the assumption for the existence of an equation of state is made, then the stress distribution can be computed. However, the existence of an equation of state has been denied experimentally for the face-centred-cubic metals, but it seems to be an adequate approximation for the real process in the case of body-centred-cubic metals and alloys. The role of local temperature needs emphasis. This temperature is the sum of the initial workpiece temperature and the temperature rise due to the conversion of mechanical work into heat. The basic properties of the crystals such as the elastic constants, the thermal conductivity and the specific heat are functions of the temperature. To establish, within some limits of precision, the local temperature an iterative procedure must be employed. In the case of orthogonal cutting, the total supplied energy per unit volume of metal removed is obtained from the relation:

$$W_T = F_C V_C / 12 b_1 t_1 V_C \quad [\text{lbf-in}/\text{in}^3] \tag{1}$$

where F_C = the cutting force, lbf;
             V_C = the cutting speed, ft/min;
             b_1 = the width of cut, in;
             t_1 = the feed, in.

It is assumed that the total specific energy, W_T, is composed only of the specific shear energy, W_S, and the specific frictional energy, W_F, all other energy components being either small or zero, i.e.:

$$W_T = W_S + W_F \tag{2}$$

For the Type 2 chip, i.e. the b.u.e. is absent, the frictional energy is simply:

$$W_F = Fr/A_0, \tag{3}$$

where F = the frictional force (the resultant cutting force component parallel to the rake face of tool), lbf;

   r = the chip thickness ratio, $r = t_1/t_2$ (t_2 = the thickness of the chip, in);

   $A_0 = t_1 b_1$, in^2.

When the b.u.e. is present, the frictional energy, W_F, cannot be easily determined from the force measurements because the effective rake angle is not known. The chip thickness ratio is also somewhat ambiguous in this case, but the relation (2) remains, however, valid.

The specific shearing energy, W_S, is the work, per unit volume of metal removed, spent in forming the chip in the deformation zone.

If this zone is represented by a very thin plane layer (shear plane), and if the shear stress is constant on this surface, then:

$$W_S = \tau\gamma \tag{4}$$

where τ = the shear stress on the shear plane, lbf/in^2;

   γ = the shear strain, in/in.

The shear energy, W_S, is the larger of the two parts of W_T. It varies from $0 \cdot 6$ W_T to $0 \cdot 9$ W_T. The energy, W_S, is the essential factor in metal cutting, and although its reduction is always desirable, it is not always possible to alter it substantially in normal cutting operations. W_F is, on the contrary, not necessary for the chip formation which can be effectively performed even with $W_F = 0$. When the chip formation process is approximated by a flow zone rather than a simple thin shear layer, expression (4) has to be modified to take into account the variation of the stress and strain throughout the zone.

$$W_S = \int \tau_{ij} d\gamma_{ij}, \tag{5}$$

where τ_{ij} = the deviatoric stress tensor;

   γ_{ij} = the strain tensor.

The integration is performed over the actual strain path from the initial state of metal in the workpiece to the final state in the chip. Unless the stress and strain distribution in the flow zone is known, the evaluation of the integral (5) is not feasible.

Since, in addition, the stress components depend on the strain, strain rate and the local temperature, a direct evaluation of the integral W_S is always difficult. The flow or deformation zone extends below the plane defined by the cutting velocity vector emanating from the cutting edge and the cutting edge itself. The depth to which the deformation zone penetrates into the workpiece has not, as yet, been determinated analytically. There is also a lateral bulging of the material as it crosses the deformation zone, which is very pronounced at low cutting speeds, i.e. when the zone has the appearance of a rather large wedge. The concept of deformation zone is incompatible with the assumption of rigid ideally plastic material, not only from the equilibrium of forces requirements but also from the impossibility to construct a kinematically admissible velocity field.[3] Even though the evaluation of the integral (5) presupposes the knowledge of the stress and strain distribution in the deformation zone, its numerical value is equal to $W_T - W_F$.

Since both W_T and W_F can be measured (in the absence of b.u.e.) without making any assumptions about the shape of the deformation zone, the energy W_S is always obtainable experimentally.

Whether one chooses to express this energy as $W_S = \tau\gamma$, i.e. assuming a thin shear layer, constant shear stress etc., or as in a zone, $W_S = \int \tau_{ij}\, d\gamma_{ij}$, these expressions are equivalent, viz.

$$W_S = \tau\gamma = \int \tau_{ij} d\gamma_{ij}, \tag{6}$$

whatever may be the actual shape of the zone with its stress and strain distributions. At the present time we do not have any mechanical criterion which would permit us to establish whether under a given set of cutting conditions, a wide or a narrow zone will develop and to determine the boundaries of the zone, so we are forced to postulate that the shape of the zone is governed by the physical properties of the workpiece material and the deformation velocity.

3. APPROXIMATE BOUNDARIES OF THE FLOW ZONE

The boundaries of the flow zone can be determined from the photographs of chip formation during the cutting operation itself or by photographing the chip-root after a suddenly interrupted cut.

Unfortunately even under the optimum conditions the photographs are not very precise so that there is always some uncertainty regarding the location of the boundaries. Guided

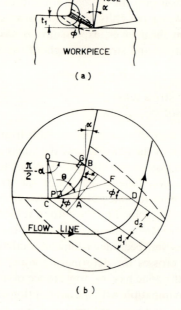

(a)

(b)

Fig. 1.

by the analysis of the typical photographs of the flow zone, several auxiliary assumptions are made regarding the position of the boundaries:

(a) The central layer of a roughly wedge-shaped flow zone is parallel to the shear plane and has a thickness, d, given by the formula (Fig. 1)

$$d = d_1 + d_2 = \eta t_2 \left[\cos (\phi - \alpha) + \sin \phi\right], \text{ [in]} \qquad (7)$$

where t_2 = the chip thickness, in;
　　　　　η = a coefficient, $0 < \eta \ll 1$;
　　　　　α, ϕ, have the usual meaning.

The portion of a thickness $d_1 = \eta t_2 \cos (\phi - \alpha)$ is above the shear plane defined by the angle ϕ, and that of a thickness $d_2 = \eta t_2 \sin \phi$ is below this plane.

The radius of curvature of the flow line in the layer is:

$$r = \eta t_2 \tan \left(\frac{\pi}{4} + \frac{\alpha}{2}\right), \text{ [in]} \qquad (8)$$

or

$$r = \frac{d \tan (\pi/4 + \alpha/2)}{\cos (\phi - \alpha) + \sin \phi}.$$

The flow lines in the layer are, in this case, circular arcs. The strain at the point P on the arc CB is, according to Lira and Thomsen[1] (Fig. 1b):

$$\gamma = \frac{DF}{GP} = \frac{DG - FG}{GP} = \cot \phi - \tan \theta; \qquad (9)$$

$$- (\phi - \alpha) \leqslant \theta \leqslant (90 - \phi),$$

where θ is the normal angle formed by the tangent on the arc CB at P and the normal to shear plane.

The strain rate is simply:[1]

$$\dot{\gamma} = \frac{V_C}{r} \frac{\sin \phi}{\cos^3 \theta} \frac{12}{60} = \frac{0 \cdot 2 \, V_C}{\eta \, t_2} \frac{\sin \phi}{\tan (\pi/4 + \alpha/2)} \frac{1}{\cos^3 \theta}, \text{ [sec}^{-1}\text{]} \qquad (10)$$

(b) The upper boundary has a maximum inclination angle: $\phi = \pi/4 - (\beta\text{-}\alpha)/2$, i.e. it corresponds to Merchant's solution of the chip formation problem. Obviously, it can, in certain cases, coincide with the central layer.

(c) The lower boundary can be determined only by solving the problem of stress distribution in the workpiece. As a first approximation it is assumed that the shear stress in the upper and lower boundaries has the same value.

(d) The details in the immediate vicinity of the cutting edge are entirely ignored.

(e) The strain change along any flow line is considered small within the zone except for the central layer, so that the total strain is given by Merchant's formula:

$$\gamma_{\text{tot}} = \tan (\phi - \alpha) + \cot \phi.$$

These auxiliary assumptions permit one to concentrate all the significant deformation processes in the central layer. The present scheme of the flow zone has several disadvantages.

First, there is a sharp discontinuity in the strain rate at the entry and the exit of the parallel sides central layer.

Second, the strain rate has a minimum value inside the central layer which does not appear to agree with the experimental data of Goriani and Kobayashi.[4] The first disadvantage can be remedied by stipulating that the entire flow line be a curve which has at least first and second derivatives continuous within the entire flow zone and its vicinity. A hyperbola, a parabola or a catenary would satisfy this requirement. The second disadvantage is much more difficult to remedy. As can be easily seen, the values of $1/\cos^3 \theta$ can vary by a factor of 50 within the zone of interest. If, nevertheless, it is desired to have a strain rate distribution within the zone which has a maximum value in the central layer, i.e. $V_C/r \sin \phi \ll 1$ at the upper and lower boundaries of the central zone, and $V_C/r \sin \phi \gg 1$, somewhere within the zone, then r would have to behave as $1/\cos^4 \theta$.

There are no simple functions which have a curvature of the type

$$\rho = \text{const}/\cos^4 \theta$$

as can be easily seen from the equation:

$$\frac{\text{const}}{\cos^4 \theta} = \frac{[r^2 + (dr/d\theta)^2]^{\frac{3}{2}}}{[r^2 + 2(dr/d\theta)^2 - r(d^2r)/d\theta^2]} ; r = f(\theta),$$

the solution of which must also satisfy the subsidiary condition that $r = f(\theta)$ have as the asymptotes the original workpiece surface and the chip parallel to the tool face, i.e. the asymptotes forming an angle $\psi = 90 + \alpha$. If the curved position of the flow line is a parabolic arc, then the strain rate remains constant in the layer, because radius of curvature for the parabola has the form:

$$\rho = p/\cos^3 \theta,$$

where p is the parameter of the parabola $y^2 = 2px$, with the axis parallel to the shear plane. If the parabola is referred to oblique axes $x'y'$ with included angle ψ then $y'^2 = (2p/\sin^2 \psi)x'$. At the points B and C (Fig. 1b) the strain rate is discontinuous. The strain rate within the parallel layer is given for this case, with a sufficient degree of approximation, by the expression:

$$\dot{\gamma} = \frac{0 \cdot 2 \, V_C \sin \phi}{\eta t_2} \cos \left(\frac{\pi}{4} + \frac{\alpha}{2} \right), \ [\text{sec}^{-1}] \tag{11}$$

which depends upon the rake angle α only by a factor of 3 for the practical values of the angle α. Since the strain rate must vary by orders of magnitude to have a notable influence upon the shear stress in the flow zone, expression (11) gives an adequate estimate of the strain rate.

In fact, virtually identical results are obtained by the original formula due to Shaw:[5]

$$\dot{\gamma} = \gamma \frac{0 \cdot 2 \, V_C \sin \phi}{d},$$

where d is the thickness of the deformation layer, and
             γ the shear strain.

In metal cutting γ can vary from 1·5 to 8, i.e. by a factor of 5. The cutting velocity can be varied rather easily over at least three to five orders of magnitude. Thus, in principle, the cutting operation provides a suitable means for a property test so far as the strain rate effect is concerned. The fact that these high strain rates are achieved only at rather high strains is however a serious limitation, because it is not possible to compare metal cutting results with those obtained by shock loading, which usually have only a relatively small strain.

4. HEATING THE DEFORMATION ZONE

The temperature at which the plastic deformation takes place has a fundamental influence upon the microscopic phenomena of straining. The temperature influences not only the elastic constants but also all thermally activated processes. An estimate of the temperature in the deformation zone is obtained from the expression:

$$T_S = \frac{W_S}{J\rho c} + T_0,$$

where J is the mechanical equivalent of heat,
   ρ the density of material,
   c the specific heat, and
   T_0 the initial workpiece temperature.

At high cutting speed W_S is virtually independent from the speed, feed, and to some extent also the rake angle; thus the temperature T_S cannot be changed unless T_0 is variable.

The corrections of T_S, due to heat transfer, moving coordinates, etc., do not change the level of T_S appreciably except at very slow cutting speeds.

Two problems present themselves at this point:

1. What is the primary process of energy conversion?
2. Can the phenomenon of adiabatic instability take place in the central layer of the flow zone?

These two problems are intimately connected with work-hardening and recovery, and therefore with the theoretical estimates of the effective shear stress in the flow zone. From the metallurgical or physical viewpoint, the plastic deformation is the consequence of dislocation movement and creation of new dislocations and assorted point defects. For discussion purposes the processes in the pure metals are analyzed first; the impure metals and alloys have a more complicated mechanism and can be only briefly reviewed.

All metals and alloys possess an initial dislocation density ρ_0, which is generally of the order of 10^5 to 10^8 cm^{-2}, for materials considered as workpiece. Higher dislocation densities are due to previous plastic working and may reach levels of 10^9 to 10^{11} cm^{-2}.

The dislocations are not simply arranged in the crystalline grains forming the workpiece. In some metals such as copper and aluminium they are arranged in densely packed groups forming walls of the cells. The interiors of the cells are sparsely populated. In other materials, such as stainless steel, the arrangement is quite different; there are no cells and the dislocations are distributed randomly in the glide planes. The conservative motion of the dislocations is possible only in the glide planes, even though these may not be the planes of maximum shear stress.

The dislocations can move nonconservatively by climbing out of their glide planes, but such motion involves diffusion of point defects to or from the dislocation line. Obviously, such motion is thermally activated and hence the importance of the flow zone temperature which influences the diffusion process. The dislocations move when the shear stress appears on a glide plane, but since the dislocations have a stress field of a rather long range the original distribution and quasi-equilibrium of dislocations is altered and the interactions between dislocations requires an increase of the stress if the movement is to continue.

Such long-range interactions lead to a shear stress which is proportional to the square root of dislocation density:

$$\tau = \text{const. } Gb\rho^{\frac{1}{2}}, \qquad (12)\dagger$$

as shown by G. I. Taylor.[6]

In addition the crossing of dislocations has a contribution to the shear stress necessary to maintain the deformation rate. The flow stress given by eq. (12) is a process which does depend upon the temperature of the medium (not only for the elastic constant, G) but also because the mechanism of crossing, cutting and recombination of dislocations depends very strongly upon the temperature. Thus the total shear stress is a sum of the thermal and athermal parts:

$$\tau = \tau_T + \tau_E \qquad (13)$$

The influence of the grain boundaries can under certain conditions be assimilated into the component τ_E. In others it contributes a part which is a function of the grain diameter. For the metal containing a substantial concentration of impurities and for the alloys the terms τ_E and τ_T are more complicated because the interaction of dislocations with impurities and alloying elements and phases has also an elastic as well as a thermal part. The athermal portion of the flow stress, τ_E, does not depend upon the velocity of deformation and is usually the predominant part of the flow stress, τ. The thermally activated part depends not only upon the temperature but also upon the strain rate.

Only at high temperatures does its influence begin to compete with the athermal part in the contribution to the overall flow stress.

In metal cutting the contribution of the thermal part has not as yet been estimated since there are very few experimental data covering large variations of the initial workpiece temperature, T_0.

The dislocation activity in the workpiece material increases as the stress ahead of the flow (or deformation) zone increases.

The workpiece material is very likely to be in a certain state of work-hardening just before it enters the flow zone. Let this state be characterized by a shear stress τ_W. The material advances into the flow zone and the stress increases by $d\tau_W$ giving rise to a change in the shear strain, $d\gamma$. During this process of deformation many dislocation sources become activated within each crystalline grain so that a great number of dislocation loops is formed. These loops spread out in the glide planes, travelling a mean distance $\bar{l}$ before reaching an obstacle. The increase in shear strain is then:

$$d\gamma \simeq b\bar{l}^2 dN, \qquad (14)\ddagger$$

where b = the Burgers vector of the dislocation,

   dN = the increase of the number of dislocation loops per unit volume.

† G = the shear modulus, b = the Burgers vector of a dislocation.
‡ All equations are in the c.g.s. system of units unless otherwise specified.

The increase of the dislocation density is simply,

$$d\rho \simeq \bar{l}dN \simeq d\gamma/b\bar{l}. \tag{15}$$

If $\bar{l}$ remains constant, then, by combining eqs. (12) and (15), Taylor's formula for stress–strain relationship results,

$$\tau_W = \text{const. } G\,(b\gamma/\bar{l})^{\frac{1}{2}}, \tag{16}$$

which is the well-known parabolic law of deformation. If the mean distance $\bar{l}$ changes proportionally to the dislocation density then the combination of (12) and (15) leads to a very rapid work-hardening slope (of the order of G = the shear modulus), i.e.:

$$\tau_W = \text{const } G\gamma, \tag{17}$$

where the constant is of the order of one.

According to eq. (17), even a relatively small strain can give rise to a large shear stress. This situation is not inconsistent with the concept of elastic-plastic or rigid-plastic solid.

It has the same characteristcs of a shock-wave appearing at the boundary of the flow zone. Since the dislocations can accelerate very rapidly, i.e. within 10^{-9} sec (a dislocation can reach $0.9c$, where c is the speed of sound in the material), the traverse of a distance of the order of 10^{-6} cm takes place in a time interval of the order of 10^{-9} sec. Since the deformation progresses until the flow zone is fully developed, and all experimental evidence points out that the shear stress is constant or virtually constant (with, perhaps, a slight decrease in the vicinity of the upper boundary of the central layer), a strong process of dynamic recovery must take place in the deformation layer. The recovery process is always accompanied by a release of thermal energy. We proceed now to examine the nature of the shear stress in the flow zone and the details of the recovery process. In heavily deformed metals the cell size reaches a minimum value and does not change with further deformation.[7] The cell size is about 1μ under these conditions and it grows somewhat upon the increase of deformation temperature. The new dislocation loops are generated by the sources in the interior of the cells and the dislocations move until they reach the vicinity of the dense dislocation tangles or arrays of the cell walls.

The elastic interactions with the dense groups impedes further movement of new dislocations, but the edge dislocations can climb, and the screw dislocations cross-slip in the interior of the cells so that the dislocations of opposite sign and coming out of neighbouring cells can meet and mutually annihilate. The screw dislocations can easily cross-slip at the stress levels prevailing within the flow zone. The edge dislocations, by contrast, climb only when enough point defects have been generated by the plastic deformation and these defects have diffused into the neighbourhoods of edge dislocations.

For the metals which do not form cells, the flow stress in the deformation zone is estimated on the basis of the following model. As a dislocation glides in a slip plane it cuts a number of forest dislocations which pierce the glide surface.

It is assumed, now, that the force required to move the dislocation has to overcome only this obstacle. In addition, we assume that the mean free path $\bar{l}$ travelled by a dislocation is proportional to the average distance between the forest dislocation and that the stress required to force the dislocation through the forest is inversely proportional to $\bar{l}$, i.e.

$$\tau \simeq \frac{Gb}{\bar{l}}. \tag{18}$$

It is to be remembered that the model implies that the mean free path of the dislocations is not constant as in Taylor's model but varies with dislocation density. (We consider, however, Taylor's model as a proper representation for the metals which form the cells during heavy deformation.)

Since $b\bar{l}d\rho \simeq d\gamma$, as shown in eq. (15), eliminating $\bar{l} \simeq d\gamma/bd\rho$ and substituting in (18) one obtains:

$$\tau = \xi Gb^2 \frac{d\rho}{d\gamma}. \tag{19}$$

The constant $\xi = \tau d\gamma/Gb^2 d\rho$ is the ratio of incremental plastic work, $\tau d\gamma$, to the energy supplied to form a $d\rho$ density of dislocations. (The energy per unit length of dislocation line is, approximately, Gb^2.) Since $Gb^2 d\rho$ is essentially the stored energy, which represents about 2 to 5% of the total energy, the value of the constant ξ is 20 to 50.

As the dislocations intersect one another a jog is produced on them. If the jog is dragged along with the dislocation it may produce a point defect in each successive step or if it is firmly anchored a dipole may form.

The increase of the number of point defects is proportional to the number of jogs and to the number of lattice spacings within the distance, $\bar{l}$, i.e.

$$dn \simeq \left(\frac{\bar{l}^2 dN}{\bar{l}^2} \right) \frac{\bar{l}}{b},$$

or upon substitution of (15) and (18)

$$dn = \zeta \frac{\tau d\gamma}{Gb^3}, \tag{20}$$

where ζ is a constant .[8] Clearly, it means that the number of point defects and the number of dislocations produced in the flow zone are proportional to each other. The ratio $\zeta = Gb^3 dn/\tau d\gamma$ relates the energy consumed in forming point defects to the incremental plastic work. Stuewe[9] has proposed the following mechanism for the recovery effect in heavily deformed metals. If the radius of the cell is r, then a dislocation loop sweeps an area $A = \pi r^2$ and the loop length is $2\pi r$. Let N be the number of produced loops so that the contribution of the edge dislocations to the density of dislocations is $\rho = \pi r N = AN/r$.

The shear strain is equal to $\gamma = NAb = r\rho b$ and the time rate of increase of the density is, then,

$$\dot{\rho} = \frac{\dot{\gamma}}{rb}. \tag{21}$$

ρ is a mean value. In the centre of a cell it is about 0 and in the walls approximately 2ρ.

As it moves, each dislocation loop cuts $A\rho$ other dislocations which pierce the glide plane. About 1/4 of these intersections produce jogs which upon further movement of the dislocation generate point defects. The number of jogs/second is then:

$$\frac{dN}{dt} \frac{A\rho}{4} = \frac{\dot{\gamma}\rho}{4b}. \tag{22}$$

Only half of these jogs are oriented in such a way that their dragging produces vacancies. The production of interstitial ions and dipoles is ignored. Only vacancies are considered. Each jog produces about $r/4b$ vacancies. Then the concentration rise of the vacancies is:

$$\frac{dc}{dt} = \frac{\dot{\gamma}\rho r}{32Mb^2},$$

(23)

where M is the number of lattice sites per unit volume (cm³). The vacancies can diffuse to the edge dislocations causing them to climb. Since the shortest diffusion distance of a vacancy is $x = (2\rho)^{-\frac{1}{2}}$, the time for the traverse is $t = x^2/D$† (which assumes a concentration gradient). If we assume in addition that the adsorbed vacancies lead immediately to the climb of dislocation, then the number of annihilated vacancies is

$$M\frac{dc}{dt} = -Mc2\rho D.$$

(24)

When a dislocation climbs a lattice distance it can collide in the new glide plane with another dislocation of opposite sign and mutually annihilate. Then for two dislocations of length i to annihilate at least lq/b vacancies must be available. Under such circumstances the reduction of dislocation density is given, using eq. (24), by

$$\frac{d\rho}{dt} = -\frac{4MDc\rho b}{q}.$$

(25)

Let $\pi r^2 \rho$ dislocations be distributed in cell walls, so that their separation one from another is about $2/\rho r$. At least half of their distance must be traversed by climb in order to meet the dislocations of the opposite sign in the neighbouring cell. Then $q = 1/\rho rb$ and eq. (25) reads:

$$\frac{d\rho}{dt} = -4MDb^2 r\rho^2 c.$$

(26)

We obtain, finally two equations, one for the concentration of vacancies change and the other for the dislocation density change by combining eqs. (23) and (24) and eqs. (21) and (26):

$$\frac{dc}{dt} = \frac{\dot{\gamma}r\rho}{32Mb^2} - 2D\rho c$$

(27)

$$\frac{d\rho}{dt} = \frac{\dot{\gamma}}{rb} - 4MDb^2 r\rho^2 c.$$

(28)

In the steady state $dc/dt = d\rho/dt = 0$ and we obtain:

$$C^* = \frac{\dot{\gamma}r}{64Mb^2 D}$$

(29)

$$\rho^* = \left(\frac{16}{r^3 b}\right)^{\frac{1}{2}}$$

(30)

† D is the diffusion coefficient, $D = D_0 \exp(-U_D/kT)$.

Upon substitution of $x = c/c^*$; $y = \rho/\rho^*$ and $\theta = \dot{\gamma}/4\,(r/b)^{\frac{1}{2}}t$ eqs. (27) and (28) transform into:

$$\frac{dx}{d\theta} = py\,(1 - x),\qquad(31)$$

$$\frac{dy}{d\theta} = 1 - xy^2,\qquad(32)$$

where

$$p = \frac{32D}{\dot{\gamma}r^2}.\qquad(33)$$

Stuewe has solved these equations numerically and has shown that for the values of $p \leqslant 3\cdot4$ the curves $\sqrt{y}(\theta)$ possess a maximum.

Since $\sqrt{y}$ is directly related to $\sqrt{\rho}$ and $\sqrt{\rho}$ is proportional to τ, the shear stress, we expect that the stress–strain curve has also a maximum, before reaching a steady value for large strains.

Such results have been obtained by several investigators.[10–11] For the case of chip formation we do not have as yet such experimentally established curves and it is for this reason that the study of flow zone models is important. It is only in flow zone models where the stress–strain curve has a definite physical meaning and can be effectively correlated with other type of tests.

Since the shear stress is given as $\tau = $ const $Gb\rho^{\frac{1}{2}}$, by setting the constant equal to $0\cdot5$ the density of dislocations is:

$$\rho^* = \left(\frac{2\tau^*}{bG}\right)^2.$$

From eq. (30) we obtain:

$$\left(\frac{16}{r^3b}\right)^{\frac{1}{2}} = \left(\frac{2\tau^*}{bG}\right)^2$$

or

$$r = b\left(\frac{G}{\tau^*}\right)^{\frac{4}{3}}.\qquad(34)$$

A. B. Draper[12] has obtained electron microscope transmission photographs of the chips of stainless steel and copper at low cutting speed. The values of G, τ, and b are known in his experiments from direct measurements. The measured values of the average cell diameter fit well with the prediction obtained by eq. (34). The stationary vacancy concentration is:

$$c^* = \frac{\dot{\gamma}}{64MbD}\left(\frac{G}{\tau^*}\right)^{\frac{4}{3}}.\qquad(35)$$

c^* is obviously a non-equilibrium concentration. In the flow zone, c^* can be computed if the estimate of $\dot\gamma$ is accurate. Using some of Stuewe's data for copper the following results are obtained:

$$T = 350°C,\ \dot\gamma = 0\cdot77\ \text{sec}^{-1},\ \tau^*/G = 1\cdot64 \times 10^{-3},$$
$$p = 1\cdot3 \times 10^{-2},\ c^* = 7\cdot1 \times 10^{-3},\ \text{and}\ \rho^* = 1\cdot6 \times 10^{10}\ \text{cm}^{-2}.$$

The ratio τ^*/G in the flow zone is higher than 5×10^{-3}, and the temperature $T \approx 200°C$. Since the strain rate is of the order $10^3\ \text{sec}^{-1}$, the value of c^* is of the order of 10^{-2}. For higher strain rates, i.e. $\dot\gamma \approx 10^5$ to $10^6\ \text{sec}^{-1}$, the value of c^* is prohibitively high. The recovery is then governed very probably by the annihilation of dipoles because if two dislocation loops L_1 and L_2 are very close to each other, i.e. in the glide planes separated by only a few lattice parameters, then the attractive force between them is equal to the climb force. According to Friedel,[13] the climb condition is given by:

$$\frac{h}{b} \leqslant \frac{Gb^3}{2\pi\,(1-\nu)\,U_f.\sin\psi};$$

h is the separation of the planes where the dislocation segments L_1 and L_2 are located, $\nu = $ Poisson's ratio, $U_f = $ the energy for defect formation, $\psi = $ the angle between the Burgers vector, b, and the tangent on the dislocation line. A very strong super-saturation of defects (vacancies and interstitial ions) is possible with this mechanism.

A substantial reduction of density of dislocations is achieved by cross-slip of screw dislocations. Cross-slip is not controlled by diffusion of point defects, and can occur rather readily for metals with high stacking fault energies, such as the b.c.c. group and of course aluminium.

Due to the complexity of the recovery process and its dependence upon the details of the metallic structure, the precise investigation of the shear stress in metal cutting could certainly provide the much needed information.

One of the first problems is to investigate the cell size and type in the chips of various metals and alloys. Equation (34) provides a formula to correlate the shear stress with the cell diameter. Since the deformation process leaves the chip with a super-saturation of point defects, rapid quench experiments can be effectively extended to materials deformed at substantially high strain rate than hitherto possible. We suggest, therefore, that the problem of cross-slip and climb of dislocations which is an important question in the theory of work-hardening is also a key problem in the theory of the flow stress in metal cutting. Metal cutting chips can be easily prepared into samples for transmission electron microscopy studies.[12]

The observation of surface effects under the electron microscope, at lower magnification than that required by transmission technique, have been carried out successfully by S. Ramalingam.[2]

In metal cutting one observes heating in flow zone which under appropriate circumstances may be considered as an adiabatic process.

The heating is definitely due to the creation and annihilation of point defects and dislocations. From the previous discussion of the recovery process it is apparent that both processes are concurrent. Which one predominates in the energy dissipation in metal cutting cannot at the present time be decided. It is quite possible that in some metals for a given temperature and a range of cutting speed one may predominate over the other and that the reverse may be true when the speed and temperature are changed.

Let us now examine various processes of creation of point defects which are compatible with the deformation in the flow zone.

Three principal mechanisms are as follows:[14]

(a) A moving dislocation may emit vacancies if it gives rise to intense local heating. Seitz[15] has shown that a uniformly moving dislocation in copper at room temperature would move about 10 meters before it evaporated one vacancy on each atomic length of its line. However, if a dislocation is held up by a localized obstacle which bends it into a curve of radius less than $10b$ and then breaks through, the energy released when the curved segment suddenly straightens may be sufficient to evaporate vacancies.

(b) Point defects can be created when two edge dislocation segments of opposite sign approach each other on adjacent glide planes or when these two dislocations meet on the same glide plane. In the first case if the separation is a lattice spacing, the configuration is equivalent to a row of vacancies or a row of interstitial ions. In the second case, the annihilation of two dislocations by collision, the energy release per atomic plane is of the order required to produce a vacancy-interstitial ion pair. It is very likely that this pair will recombine giving away its energy to the lattice in the form of heat. The mechanism of intersection of two screw dislocations with different Burgers vector produces a similar effect when the dislocations meet and are unable to intersect one another. One then sweeps around the other, forming edge segments which are separated by the Burgers vector.

(c) When a moving screw dislocation cuts another and acquires a jog which is then dragged behind in a non-conservative motion, the result is the formation of a row of vacancies or of interstitial ions. This is probably the most frequent process in metal cutting. We have made use of it in the discussion of the recovery. An estimate of the vacancy concentration in the flow zone as a function of deformation is obtainable by assuming that all the work per unit volume $\tau d\gamma$ goes into the formation of vacancies, i.e.

$$\tau d\gamma = U_v dN_v, \tag{36}$$

where U_v is the energy to form a vacancy and N_v the number of vacancies per unit volume. If the active glide planes are intersected by dislocations with a density ρ_D and the expanding loops in the glide planes are circular in shape, then as each loop expands it acquires an increasing number of jogs per unit length. Thus, it requires an increasing stress to keep expanding. Assuming that there are ρ_L loops per unit volume each of radius R, then each loop cuts $\pi R^2 \rho_L$ dislocations. As indicated earlier, only $\frac{1}{4}$ to $\frac{1}{6}$ of these intersections form jogs which produce vacancies. As the loop expands from R to $R + dR$ it creates $R^2 \pi \rho_D dR/4b$ vacancies. The change of the number of vacancies/unit volume is then:

$$dN_v = \pi R^2 \rho_D \rho_L dR/4b, \tag{37}$$

corresponding to a strain increase of:

$$d\gamma = 2\pi R b \rho_L dR. \tag{38}$$

If ρ_D and ρ_L are constant, we have:

$$N_v = \frac{\pi R^3 \rho_D \rho_L}{12b}, \tag{39}$$

$$\gamma = \pi R^2 \rho_L b, \tag{40}$$

and finally:[16]

$$N_v = \frac{\rho_D}{12b^2} \left(\frac{1}{\rho_L \pi b}\right)^{\frac{1}{2}} \gamma^{\frac{3}{2}}.$$ (41)

A modification is necessary in the above computation to make it more compatible with cell structure.

We suppose that the loops spread from Frank-Read sources in the interior of the cell. The number of loops, n, which have spread from a single source is obtained according to Eshelby, Frank and Nabarro[17] from the expression:

$$n = \frac{\tau R \pi (1 - \nu)}{Gb},$$ (42)

where ν is Poisson's ratio. If the mean radius of the loops is approximately equal to the radius R of the outermost loop, then eqs. (37) and (38) are replaced by

$$dN_v = n\pi R^2 \rho_D \rho_L dR/4b,$$ (43)

and

$$d\gamma = 2n\pi R \rho_L b dR.$$ (44)

Using eqs. (42) and (36) we obtain:

$$n = \frac{U_v R \pi (1 - \nu)}{Gb} \frac{dN_v}{d\gamma}.$$ (45)

Substituting the values dN_v and $d\gamma$ from (43) and (44) leads to:

$$n = \frac{U_v R^2 \pi (1 - \nu)}{8Gb^3} \rho_D.$$ (46)

The value n from eq. (46) can be substituted in eqs. (43) and (44) which are integrable if we assume that ρ_D and ρ_L are constant:

$$N_v = \frac{R^5 \pi^2 (1 - \nu) \rho_D^2 \rho_L U_v}{160 Gb^4},$$ (47)

$$\gamma = \frac{R^4 \pi^2 (1 - \nu) \rho_D \rho_L U_v}{32 Gb^2}.$$ (48)

The relationship between N_v and γ is now

$$N_v = \text{const. } \gamma^{\frac{5}{4}}.$$ (49)[18]

Since the dislocations in the forest piercing the primary glide planes multiply as fast as the primary dislocations loops, i.e. $\rho_D \approx n\pi R \rho_L$, the number of vacancies per unit volume is then (using $\frac{1}{6}$ for the intersections ratio):

$$N_v = \frac{R^2 \rho_D^2}{30b}.$$ (50)

The corresponding strain is:

$$\gamma = \tfrac{1}{2}\rho_D Rb.\tag{51}$$

The vacancy concentration produced under such conditions is of the order of:

$$C_v = 10^{-1}\gamma^2.\tag{52}$$

Of course, this is not an equilibrium concentration. The maximum equilibrium concentration at the melting point temperature is 10^{-4} for vacancies and 10^{-8} for the interstitial ions.[19] A transient concentration of the order of magnitude of $10^{-2}\gamma^2$ may be physically unattainable even in very rapid cutting.

On the basis of the above analysis it is apparent that the vacancy creation-annihilation mechanism cannot account for all energy dissipation in metal cutting. Even if such high concentrations are not probable and somewhat lower are perhaps not unreasonable, the electron microscopic examination of chip which has been obtained at high speed and then immediately quenched should reveal typical structural features associated with metal quenched from the temperatures close to the melting point. This is an experiment at high strain rate and high strain which cannot be performed so easily with other testing techniques.

The transient concentration of point defects in shear zone can be estimated rather directly also. Again we assume that the specific shear energy W_S is completely converted into energy necessary to form point defects. Then:

$$W_S = N_d U_d = c_d N U_d = c_d \frac{m}{a^3} U_d,\tag{53}$$

where W_S = the specific shear energy, ergs/cm³;
  N_d = the number of defects/cm³;
  N = the number of atomic sites in a perfect crystal/cm³;
  U_d = the energy of formation of defects, ev;
  a = the lattice constant, cm;
  m = the number of atoms in a cube whose side is lattice parameter.
In the case of copper:

$$W_S \approx 10^{10} \text{ ergs/cm}^3, \ U_d = 4 \text{ ev}, \ m = 4, \ a = 3 \cdot 6 \times 10^{-8} \text{ cm}.$$

The concentration c_d is then:

$$c_d = \frac{W_S a^3}{m U_d} \approx 1\cdot8 \times 10^{-2}.$$

The annihilation rate of the point defect is:

$$\frac{dN_d}{dt} = \frac{\dot{W}_S}{U_d}.\tag{54}$$

Substituting, for copper, $\dot{W}_S = 1 \cdot 58 \times 10^4$ erg cm⁻³ sec⁻¹ at the initial workpiece temperature of 300°K and the cutting speed $V_C = 1500$ cm sec⁻¹ (the shear layer temperature is about $0\cdot5\ T_m$), we have:

$$\frac{dN_d}{dt} = \frac{1\cdot58 \times 10^{14}}{1\cdot6 \times 4 \times 10^{-12}} \approx 2\cdot5 \times 10^{25}, \ \left(\frac{\text{annihilations}}{\text{cm}^3 \cdot \text{sec}}\right).$$

The average lifetime of a defect, $\bar{t} = c_d N/(dN_d/dt)$, is about:

$$\bar{t} = \frac{1\cdot8 \times 10^{-2} \times 0\cdot85 \times 10^{23}}{2\cdot5 \times 10^{25}} \approx 6 \times 10^{-5} \text{ sec.}$$

It is very difficult to estimate how far these defects would travel before annihilation since the expression $x^2 = D\bar{t}$ is clearly not applicable in our case. Some of the difficulties concerning the large point defect concentration in the previous discussion can be partially overcome by adoption of an analysis of dissipation mechanism due to Nicholas.[20] We have not paid particular attention earlier to the details of point defect generation, i.e. what should be equilibrium concentration of jogs in a dislocation line which moves though a forest of dislocations cutting its glide plane. The assumption was made that either $\frac{1}{4}$ or $\frac{1}{6}$ of the intersections produces jogs which in turn produce vacancies. To estimate the number of defects produced per intersection we consider that the incremental work in the shear zone is $dW_S = \tau dy$. In terms of dislocations on a set of parallel slip planes of equal area we obtain $dy = L_D b dx$, where L_D is the length of these dislocations per unit volume. Then the energy spent when unit length of dislocation moves through unit distance is $dW_S/L_D dx = \tau b$.

The moving dislocation can loop around the interesting dislocation and produce a row of defects when the loop reunites, after which the jogs may glide conservatively a certain distance, produce new defects or get annihilated. The energy dW_S dissipated in producing n_d defects per intersection is

$$dW_S = \rho L_D dx n_d U_d, \tag{55}$$

which is equal to $L_D dx \tau b$, so that

$$n_d = \frac{\tau b}{\rho U_d}. \tag{56}$$

Since $U_d = 11Gb^3/15$ and $\tau = \frac{1}{2}Gb\sqrt{\rho}$, we obtain

$$n_d = \frac{15}{22} \frac{\sqrt{\rho}}{\rho b}. \tag{57}$$

If $\rho = 10^{12}$ cm^{-2} and $b = 2\cdot5 \times 10^{-8}$ cm, $n_d = 27$. This estimate is in good agreement with overall arguments presented by Nicholas. The further refinement would consist in the computation of the mean speed of jog oscillation along the dislocation line. We have not solved this problem as yet.

Following Nicholas it can be easily shown that a dislocation would travel under the stress corresponding to that normally found in the shear zone an average distance $\bar{l} = KGb/\tau$ before dissipating its energy of formation. For copper setting $K = 1$, $G = 7\cdot5 \times 10^3$ kg/mm^3, $\tau = 29$ kg/mm^3, $b = 2\cdot5 \times 10^{-8}$ cm, we obtain $\bar{l} \approx 200$Å $= 2 \times 10^{-6}$ cm. Since the cell diameter is about 10^{-4} cm it is quite obvious that such mechanism may dissipate enough energy in the annihilation of two independent dislocations as seen in the process

$$\tfrac{1}{2}a\,[01\bar{1}] + \tfrac{1}{2}a\,[0\bar{1}1] = 0,$$

or its equivalent

$$\tfrac{1}{2}a\,[01\bar{1}] + \tfrac{1}{2}a\,[\bar{1}01] = \tfrac{1}{2}a\,[\bar{1}10],$$

as suggested by Seitz[15] (*a* is the lattice constant). The trouble with this mechanism is that it is not general enough. If the dislocation can glide longer distance than $\bar{l}$ it could gain from the applied stress an energy ten or more times that which is released when it combines with another dislocation.

As indicated earlier, we believe that the real mechanism is very likely a combination of the point defects and dislocations annihilation.

The importance of metal cutting as a test of various hypotheses of energy dissipation is now obvious. It is the only test which can be performed with relative ease at very high speeds, strains, strain rates, and at a variety of initial temperatures. Since all processes which enter into play are also very strongly dependent upon the crystalline structure, a careful microscopic examination of the chip and chip roots of different materials would supply the necessary data.

Nabarro, Basinski and Holt[21] have shown the relationship between the linear work-hardening theories and the amount of stored energy. Any theory which is based upon the concept that the flow stress is due to the interaction between dislocations and which yields the correct value for the inverse work-hardening coefficient $K = Gd\gamma/d\tau$ automatically predicts that the fraction of the deformation work stored in the crystalline structure is small and constant. Their argument is as follows: the work done per unit volume is $dW_S = \tau d\gamma = (K/G)\tau d\tau$. The increase in stored energy is the energy of dislocations which are added to the structure, i.e. $dW_d = \frac{1}{2}b^2 Gd\rho_D$. For any work-hardening mechanism for which there are no large pile-ups of dislocations, the stress and the dislocation density are related: $\tau = \frac{1}{4}Gb\sqrt{\rho_D}$. Therefore, $\tau d\tau = \frac{1}{32}G^2 b^2 d\rho\rho$. It follows that $dW_d/dW_S = 16/K$. The experimental value of K is 300, which leads to a stored energy of 5% of W_S.

For polycrystals and for single crystals deformed deeply in the third stage the amount of stored energy decreases steadily to about 1%.

5. ADIABATIC INSTABILITY

The problem of adiabatic instability is associated with the question whether the plastic deformation can lead to melting of metal. Cottrell[22] gave a general argument that it will not melt, but if the slip occurs suddenly in a few closely spaced glide planes, the temperature rise may weaken the metal substantially. The adiabatic instability can be defined as the condition at which the deformation process becomes unstable due to the adiabatic conversion of mechanical work into heat which produces a temperature rise at which the actual flow stress has a negative slope as a function of strain. Neglecting the strain rate effect, the condition of adiabatic instability for a tensile test is given by the inequality[23]

$$\rho c < \frac{-\left(\dfrac{\partial \sigma}{\partial T}\right)_\epsilon}{\dfrac{1}{\sigma}\left(\dfrac{\partial \sigma}{\partial \epsilon}\right)_T - 1}, \tag{58}$$

where ρc is the thermal capacity per unit volume, and ϵ is the linear strain.

Pomey[24] has repeatedly drawn attention to the possibility that in metal working one might expect, under certain conditions, the onset of adiabatic instability. In the following we examine the possibility of adiabatic instability in the deformation zone of chip formation process. The shear stress in the deformation zone is not known precisely except when the process of chip formation is imagined to be represented by a single plane shear surface.

As is well known, the shear stress determined on this basis is independent of the strain and the strain rate, but depends upon the temperature of the flow layer. If we follow Pomey's arguments and assume an equation of state for the deforming material, the change of shear stress is given as:

$$d\tau = \left(\frac{\partial \tau}{\partial \gamma} \right)_{T,\dot{\gamma}} d\gamma + \left(\frac{\partial \tau}{\partial \dot{\gamma}} \right)_{T,\gamma} d\dot{\gamma} + \left(\frac{\partial \tau}{\partial T} \right)_{\gamma,\dot{\gamma}} dT. \tag{59}$$

This equation leads to the same instability condition as that of Basinski (eq. 58) if we neglect the influence of strain rate.

Recalling that $dT = \tau d\gamma / J\rho c$, eq. (59) becomes:

$$\frac{d\tau}{d\gamma} = \left(\frac{\partial \tau}{\partial \gamma} \right)_{T,\dot{\gamma}} + \left(\frac{\partial \tau}{\partial \dot{\gamma}} \right)_{T,\gamma} \frac{d\dot{\gamma}}{d\gamma} + \left(\frac{\partial \tau}{\partial T} \right)_{\gamma,\dot{\gamma}} \frac{\tau}{J\rho c}, \tag{60}$$

$\left(\dfrac{\partial \tau}{\partial \gamma} \right)_{T,\dot{\gamma}}$ is the work-hardening coefficient at given constant temperature and strain rate.

$\left(\dfrac{\partial \tau}{\partial \dot{\gamma}} \right)_{T,\gamma}$ is the strain rate sensitivity of the metal at given constant, strain and temperature.

$\left(\dfrac{\partial \tau}{\partial T} \right)_{\gamma,\dot{\gamma}}$ is the thermal coefficient of the stress at given constant, strain and strain rate (it is a negative quantity).

We assume that each of these coefficients is independent of the others. This may be an adequate approximation for a small region of $\{\gamma,\dot{\gamma},T\}$ space, but it is manifestly wrong for the entire space. Also the auxiliary effects of the hydrostatic component of the stress tensor are neglected.

A velocity–temperature compensation of the stress τ implies:

$$\left(\frac{\partial \tau}{\partial \dot{\gamma}} \right)_{T,\gamma} d\dot{\gamma} = - \left(\frac{\partial \tau}{\partial T} \right)_{\gamma,\dot{\gamma}} dT. \tag{61}$$

The ratio $d\dot{\gamma}/d\gamma$ is according to Pomey equal to $\dot{\gamma}/\gamma$, implying the proportionality between strain and strain rates. In terms of dislocations mechanics we write $d\gamma = \rho b dx$, and $d\dot{\gamma} = \rho b dv$, because $\gamma = \rho b x$, and $\dot{\gamma} = \rho b v$ at any given instant of deformation. Thus the change of average velocity of the dislocation is proportional to the average length of the travel. Equation (60) becomes:

$$\frac{d\tau}{d\gamma} = \left(\frac{\partial \tau}{\partial \gamma} \right)_{T,\dot{\gamma}} + \left(\frac{\partial \tau}{\partial \dot{\gamma}} \right) \frac{\dot{\gamma}}{\gamma} - \left| \left(\frac{\partial \tau}{\partial T} \right)_{\gamma,\dot{\gamma}} \right| \frac{\tau}{J\rho c} \tag{62}$$

The condition of adiabatic instability is

$$\frac{\tau}{J\rho c} \left| \left(\frac{\partial \tau}{\partial T} \right)_{\gamma,\dot{\gamma}} \right| > \left(\frac{\partial \tau}{\partial \dot{\gamma}} \right)_{T,\gamma} \frac{\dot{\gamma}}{\gamma} + \left(\frac{\partial \tau}{\partial \gamma} \right).$$

Since $\dot{\gamma}/\gamma$ is always a large number in metal cutting, $(\partial \tau / \partial \dot{\gamma})$ and $(J\tau / \partial \gamma)$ are also positive no matter how small, and since $|(\partial \tau / \partial T)|$ has a small value, $J\rho c$ will govern the onset of the instability.

Since the specific heat, c, decreases with decreasing temperature according to Debye's law, $c_v = \text{const.} (T/\Theta)^3$, we can expect that for the shear zone, the instability can occur only at cryogenic temperatures and very high cutting velocity. Actually, even with moderate lowering of workpiece temperature (to about $100°K$) substantial effects are already noticeable on the quality of surface finish and chip appearance (thin deformation zone), although the onset of instability is obviously not reached.

6. EFFECTS OF IMPURITIES, PRECIPITATES AND ALLOYING ELEMENTS AND SURFACE DAMAGE

Almost all materials currently employed in the metal working industry contain for one purpose or another a substantial density of impurities and precipitates. Others, being alloys, frequently have structures which are quite different from the parent metals.

The deformation characteristics of such materials are strongly dependent upon the presence of immobile impurities or precipitates along the dislocations. The impurity clouds can move by diffusion when dislocations escape from their midst and re-form around dislocations when these stop. Since at low strains these effects are very pronounced, one is tempted to believe that these effects may persist even at high strains and strain rate. It is therefore rather startling to find that the macroscopic mechanism of chip formation has only a few features which can be immediately ascribed to the impurity effects.

In fact, the chip formation mechanism does not even recognize the existence of particular structures of materials. Both pure metals as well as the impure ones exhibit the constant, although not the same, shear stress and deform at about the same strain rate and strain.

From the physical metallurgy viewpoint, this state of affairs is of particular interest.

One can study the effects which lead to the suppression of the impurity and precipitate hardening, its consequences upon recovery and annealing and arrangement of dislocations. The technologically important effect, however, is connected with the physical state of the newly formed surface. From the purely industrial viewpoint, the surface state of pure metals and alloys is seldom a problem. The state of the surface however is very important in high-strength alloys, precipitation hardened alloys and some fatigue-resistant materials. In all such materials the residual effects of machining are much more complex than in the simple structures where we are essentially concerned with dislocation–dislocation interactions and the dislocation–orientation change interactions (by decreasing the mean free path of dislocations motion through reduction of grain size).[2] In impure metals and alloys we have to consider, in addition, the dislocation–solute interactions as well as dislocation–second phase mechanism.

Since all these processes are sensitive to temperature and presence of adsorbed ions from the surface, the damaged layer, which is the newly machined surface, has mechanical and thermal properties which are considerably different from those of the material in the interior of the piece.

The thickness of the damaged layer is a function of two sets of variables: one set comprises the tool geometry, feed and depth of cut, and the other the cutting speed, initial workpiece temperature and the physical state of the material, i.e. the dislocation density, type and concentration of impurities, type of crystalline structure, grain size and chemical affinity for the fluids and gases coming in the contact with the new surface during the machining process.

As a general rule the stronger and therefore more complex is the material, the thicker is the damaged layer. This is to be expected since almost all high-strength materials are structurally metastable systems, so that the severe perturbation of the surface layer leads to exchange processes with the adjacent sub-layer with the resulting increase in damaged layer thickness. It is, therefore, in the study of surface damage where the dislocation and point defects physics should play an important technological role.

7. CONCLUSIONS

We have shown that a microscale analysis of the chip formation mechanism permits us to use metal cutting as a property test. In order to allow for the gradual increase of the shear strain and to obtain a reasonable computation of the strain rate, a modification of the flow zone model has been suggested.

The flow stress theory in deformation zone has been developed on the basis of dislocation mechanism. The theory is in general agreement with the experimental results. Particular emphasis has been placed upon the dynamic recovery process.

The heating of the deformation zone has been analysed in detail, showing that both point defects and dislocation annihilation do indeed account for the energy dissipated. In addition we have shown that the adiabatic instability in the flow zone can take place only in the workpiece initially at cryogenic temperature.

The effects of impurities and solutes upon the surface damage has been emphasized.

The main conclusion is simply that it is possible and fruitful to apply dislocation theory to the metal cutting analysis.

ACKNOWLEDGEMENTS

The authors are grateful to Prof. G. F. Micheletti, Director of the Institute for Mechanical Technology at Politecnico di Torino, for his support during the course of this study and C.N.R. under whose grant one of the authors (B. v. T.) is the visiting professor at Politecnico di Torino, Italy.

REFERENCES

1. F. LIRA and E. G. THOMSEN, *J. Eng. for Industry, Trans. A.S.M.E.*, Ser. B, **89**, 1967, 489.
2. S. RAMALINGAM, Plastic Deformation in Metal Cutting. Ph.D. Dissertation, University of Illinois, Urbana, 1967.
3. H. KUDO, *Int. J. Mech. Sci.* **7**, 1965, 43.
4. V. L. GORIANI and S. KOBAYASHI, *Ann. CIRP*, **15**, 1967, 425.
5. M. C. SHAW, *J. Appl. Physics.* **21**, 1950, 599.
6. G. I. TAYLOR, *Proc. Roy. Soc.* **A145**, 1934, 362, 388.
7. G. H. GREWE and E. KAPLAN, *Phys. Stat. Sol.*, **6**, 1964, 699.
8. G. SAADA, Thesis, Paris, 1960.
9. H. P. STUEWE, *Acta Metall.* **13**, 1965, 1337.
10. C. ROSSARD and P. BLAIN, *Rev. Mét.* **55**, 1958, 573.
11. D. HARDWICK and W. J. McG. TEGART, *J. Inst. Met.* **90**, 1961–62, 17.
12. A. B. DRAPER, Sub-grain structure of Machined Copper and Stainless Steel. Ph.D. Dissertation, University of Illinois, Urbana, 1967.
13. J. FRIEDEL, *Dislocations*, p. 117, Pergamon Press, London, 1964.
14. F. R. N. NABARRO, *Theory of Crystal Dislocations*, p. 382, Oxford University Press, London, 1967.
15. F. SEITZ, *Adv. Phys.* **1**, 1952, 43.
16. G. H. VAN BUEREN, *Acta Metall.* **1**, 1953, 469, 607.
17. J. D. ESHELBY, F. C. FRANK and F. R. N. NABARRO, *Phil. Mag.* **42**, 1951, 351.
18. G. H. VAN BUEREN, *Acta Metal.* **3**, 1955, 519.
19. J. FRIEDEL, *op. cit.*, p. 82.
20. J. F. NICHOLAS, *Acta Metal.* **7**, 1959, 544.
21. F. R. N. NABARRO, Z. S. BASINSKI and D. B. HOLT, *Adv. Phys.* **13**, 1964, 192.
22. A. H. COTTRELL, *Dislocations and Plastic Flow of Crystals*, Oxford University Press, 1953.
23. Z. S. BASINSKI, *Proc. Roy. Soc.* **A240**, 1959, 229.
24. J. POMEY, *Ann. CIRP*, **13**, 1965/66, 93.

FLOW ZONE MODELS IN METAL CUTTING

B. von Turkovich and G. F. Micheletti

Istituto di Tecnologia Meccanica, Politecnico di Torino, Italy

SUMMARY

This paper is concerned with the examination of properties of the flow zone models in comparison with the singular (shear) plane constructions.

It is pointed out that the physical properties of the materials can be effectively included in a flow zone model and that the allowance for the shear stress variation shows the possibility of tensile hydrostatic stress in the vicinity of the cutting edge. The recognition of the randomness of cutting forces and the correlation study of these open a new avenue in metal-cutting research. Surface damage still remains a neglected problem in metal-cutting theory.

1. INTRODUCTION

It is paradoxical that such a seemingly simple deformation process as metal cutting has hitherto resisted a complete solution even though the amount of technical literature dealing with this technologically and economically fundamental process is, as it is to be expected, rather staggering. In this paper we are concerned with the models of the process and the consequences we can draw from various models. We also wish to draw attention to some new, until now, not very much exploited avenues of research. It is our hope that by pointing out the less promising approaches, and also those methods of analysis which have reached the point of diminishing return, that new, more successful, attacks on the problem can be eventually devised.

2. SINGULAR SURFACE MODELS

In the group of singular surface models belong all those schemes of chip formation which assume a single surface on which the deformation takes place. The curvature of the shear surface is the consequence of particular stress distribution on the tool–chip contact area, or it may be due to the geometry of cut. The four principal types are illustrated in Figs. 1 to 4 with the traces OA. The input data are: F_C, F_T, the components of cutting force R, the angle β (so-called friction angle on the tool face), t_1 and t_2, the feed and chip thickness, the depth of cut, b_1, the rake angle α, for the orthogonal cutting. In the case of oblique cutting, additional angles are necessary. The role of cutting speed is nominal, i.e. the model works for any speed greater than zero.

The problem is to determine the shear angle for the shear plane model and for the curved surfaces cases, the shape of the trace OA and the stress distribution. The solution of the problem can be obtained by the following three procedures:

(a) the plasticity method based upon the rigid–ideally plastic medium hypothesis;[1-6]
(b) minimum cutting force concept; [7-9]
(c) the elastic–plastic method, which, in our particular case, is an application of Muskhelishvili's complex variable procedure. [10]

There is a large number of exact solutions, which attests for the power of these procedures.[11] Except for some slip-line solutions, the results of these calculations yield a functional relationship

$$\Phi = f' \, (a, \beta, n\pi),$$ (1)

where n is a number smaller than 1.

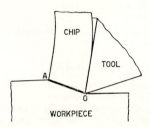

FIG. 1.

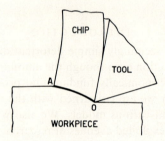

FIG. 2.

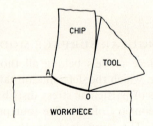

FIG. 3.

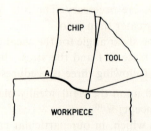

FIG. 4.

When the actual chip formation process is *closely approximated* by one of the geometric constructions in Figs. 1–4, measured values of the angle Φ or the chip thickness t_2 are in good agreement with those predicted from eq. (1). Since the physical properties of the workpiece and tool materials play no role in the theory, any predictions *ab initio* are of course impossible, but a great deal of useful information can be obtained, however, by the introduction of modifications based upon the experimental data.[12,13] Most experimental data relating Φ, β, and α to the various solutions are, unfortunately, in very poor agreement with eq. (1).

The principal reason, in our opinion, resides in the fact that the simple models in Figs. 1–4 do not approximate the real process.

Since many of the repair efforts introduced into the theory to bridge the disagreement between eq. (1) and experimental results usually violate some of the basic assumptions in (a), (b) and (c), they can hardly be considered as better solutions. A practical attitude toward this problem would be simply not to ask for more than the simple models can effectively yield.

To make our position clear, we want to state that if these simple models are used for the systematic treatment of experimental data, their function is amply fulfilled. If, however, one expects that further research along such lines would eventually yield a complete solution of the metal cutting mechanics problem, we feel that such expectations are not justified in view of the past performance. We don't believe that such solution is obtainable and we fail to see in what sense one would enhance the art and technology of metal machining by insisting upon more or less naïve models.

3. FLOW ZONE MODELS

By abandoning the severe restrictions of singular surface models, the concept of flow zone in metal cutting purports a closer agreement with actual mechanism of chip formation. Unfortunately the heuristic features of the flow zone models have received only a limited experimental and theoretical attention in comparison with the well-established singular surface models.

Reasons for such relative neglect stems perhaps from the fact that the flow zone concept is a much more complex idea than any of the previous schemes. The flow zone models show that the deformation process by which the chip is formed takes place within a rather large roughly dihedral space in front of the tool edge (Fig. 5).

We will not discuss any effects in the secondary deformation zone along the tool face, realizing, however, that they may have an influence upon the phenomena in the flow zone. The concept of flow zone is independent of friction along the tool face, and the mechanism will work even with zero friction.

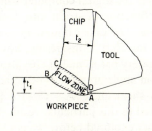

FIG. 5.

There are at least three obvious advantages of the flow zone model over the singular surface models. The first and perhaps the foremost one is the connection it affords with the machined surface damage.

The second feature is the recognition that different materials machine in unlike fashion, although many of them produce continuous chips.

Specifically, the microstructure of materials has a recognized role in the process and a host of technological and theoretical problems stem from the considerations of micro-scale mechanism.

The observation that the cutting speed and the initial workpiece temperature are indeed non-arbitrary parameters in chip formation, because each of them influences the geometry and the stresses in the flow zone, is the third advantage. Upon a little reflection some of the lesser advantages become apparent also, such as the treatment of the initial engagement of the tool and workpiece and the exit of tool from the work.

The position and shape of the boundaries of the flow zone are difficult to determine. The boundary which separates the flow zone from the workpiece is defined as the surface where the shear stress has reached the yielding level. The yield stress is a function of temperature, strain rate and initial state of the workpiece material.

At low cutting speed the strain rate effect is very likely small. Experiments have indeed shown that the boundary, at low cutting speed, extends very far ahead of the cutting edge (at least several times the feed) and penetrates deeply below the cutting edge into the work-piece.[14]

When the cutting speed is increased the length of the lower boundary is reduced, as well as its penetration into the workpiece, showing clearly the influence of the cutting speed upon the yield stress. If the workpiece material is in annealed state the boundary is larger than that for the strain hardened workpiece at the same cutting speed and the initial temperature but, for materials with large concentration of impurities and precipitates the lower boundary of the flow zone is always shorter than that corresponding to a purer sample. These effects are the consequence of the different strain rate sensitivity of yield point in various types of materials. The effects are much more pronounced at small strain, i.e. immediately after the elastic limit has been reached. The mathematical determination of the shape and position of the lower boundary is a complicated task and it will be discussed in a later paper. The analytical solution gives a stress and displacement distribution which depend upon the cutting velocity V_C through the moving coordinates:

$$X_1' = (X_1 - V_C t)/\left(1 - \frac{V_C^2}{C_t^2}\right)^{\frac{1}{2}}; \quad X_2' = X_2; \quad X_3' = X_3,$$

(C_t = the transverse sound velocity) for the shear deformation. Specifying a yield criterion a surface is obtained for the shear stress $\tau_{ij} = k$. This becomes the first approximation of the lower boundary. Next step involves the digital computer solution of the plasticity equations considering workhardening as a function of strain rate. Unfortunately the results hitherto obtained are not very encouraging. The upper boundary is almost undetectable because of the severe distortion of metal in the chip. Considering, however, only a few photographic data, it seems that the upper boundary has a smaller curvature than the corresponding lower boundary. The average inclination angle appears to be equal, for the majority of examined photographs, to the Merchant's shear angle $\Phi = (\pi/4) - [(\beta - \alpha)/2.]$ We are not certain what significance is to be attributed to this particular result. Its frequency of occurrence in a larger set of data should be further explored before we could generalize these observations.

Nevertheless, in view of these observations a rather curious geometric construction seems to fit rather nicely the boundaries of the flow zone (Fig. 6). Traces OD and AB are parallel, the same is true for OB and AC' (AC' is slightly adjusted so that the trace exits under an angle $\pi/4$ across the free surface at C). This construction has several intriguing properties. Its included angle at $P, \{\pi/4 - [(\beta - \alpha)/2] - \Phi\}$ varies with the cutting speed because Φ increases with the speed V_C. No flow zone is obtained except in the case when Merchant's shear formula is satisfied, i.e. $\Phi = (\pi/2) - [(\beta - \alpha)/2]$. Very small shear angles imply a huge flow zone. The triangle OAA' represents the portion of the deformation zone remaining

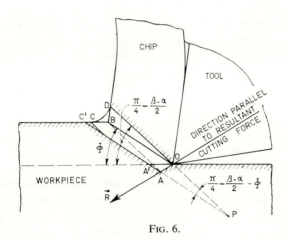

FIG. 6.

in the workpiece. Its depth depends upon the orientation of the resultant cutting force, i.e. upon the friction and rake angles. For $\beta - \alpha = 0$, the surface damage should be zero. Damage is thus always present for the negative rake angles but could be zero for large positive $\alpha - \beta$. The same scheme can be constructed even when built-up edge is present, or for a rounded edge tool.

There are several flow zone models discussed and analysed in the literature.[15-16] The most complete case is due to Palmer and Oxley.[15] Although Oxley's model requires extensive computer calculations, its superiority to all others derives from its direct relationship with the experimental observations, i.e. no *a priori* statement about the shape and nature of the flow zone are necessary. Simplified versions of the model introduced somewhat later by Oxley himself or his workers, are less instructive from our viewpoint than the original one. This opinion is reached by stipulating that the principal problems which a correctly constructed orthogonal metal-cutting model must solve are: (a) given cutting speed, V_C, feed, t_1, depth of cut, b_1, rake angle, α: predict the chip thickness t_2 as a function of the friction angle β (β may vary along the tool face and be a function of V_C also) for a material with known stress–strain, strain-rate and temperature characteristics; (b) given the same data as in (a): determine the cutting force, F_C, and the thrust force, F_T. (We emphasize that the problem of tool wear, although an essential ingredient of the overall machining process, has nothing to do with the chip formation mechanism pure and simple.) These conditions provide a test for comparative evaluation of various models. It is not necessary therefore to ask whether a given model provides a complete solution, but it is sufficient to check how close it approaches the experimental data by solving the problems (a) and (b).

The important contribution of Palmer and Oxley[15] for the analysis of stress distribution in the flow zone is the inclusion of strain-hardening effects in the slip-line fields. Thus by writing their equations in the integrated form

$$p + 2 \int k \, \frac{\partial \phi}{\partial s_\alpha} \, ds_\alpha - \int \frac{\partial k}{\partial s_\beta} \, ds_\alpha = \text{constant along } \alpha\text{-line}$$

$$p - 2 \int k \, \frac{\partial \phi}{\partial s_\beta} \, ds_\beta - \int \frac{\partial k}{\partial s_\alpha} \, ds_\beta = \text{constant along } \beta\text{-line}$$

where p is the hydrostatic stress and k, the maximum shear stress, is now variable.

It is seen from these equations that the intersection of slip-lines of different families is not necessarily orthogonal as is the case in the rigid-plastic case. This angle is now a variable dependent upon the distribution of the shear stress through the deformation zone. The terms $(\partial k / \partial s_\alpha)$, $(\partial k / \partial s_\beta)$ represent the rate of change of the shear stress k normal to the considered slip-line and measured along ds_β and ds_α respectively. Of course, the shear stress k varies also as a function of strain rate and temperature but this variation need not be included in the first trial of field construction.

The consequence of variable terms $(\partial k / \partial s_\alpha)$, and $(\partial k / \partial s_\beta)$ in eqs. (2) is the possibility that the hydrostatic stress in the vicinity of the tool edge may become tensile, thus promoting the formation of cracks in the newly generated surface. It is also to be expected that the rate of hardening would influence the thickness of the deformation zone. Oxley, Humphreyt and Lazarideh[19] have found that the materials with high rate of hardening produce a thicker flow zone than those having a much lower hardening rate. These authors pointed out that the cutting forces, friction angle and chip thickness ratio are much less dependent upon the hardening rate than the shape of the flow zone.

The applications of eqs. (2) to the model shown in Fig. 6 are not very difficult if the sharp corner at A is slightly rounded, the trace OA moved a small distance to right and the trace OD displaced parallel to itself upward just enough to avoid the strong singularity at O. Zorew, Kufarew, Del and Goldschmidt[20] have also performed an extensive investigation of the stress distribution in the flow zone and reached the conclusion that the stress distribution is very non-uniform and found the sign change in the hydrostatic stress.

These considerations indicate that the potential of the flow zone model for the analysis of surface damage and study of actual chip formation mechanism outweights the disadvantage due to its time-consuming computation.

It is our suggestion that in the research on metal cutting a greater emphasis should be given to the flow zone models even at the expense of the singular surface types.

CUTTING FORCES VARIATION

If the cutting forces are measured by means of a dynamometer with a very high natural frequency and high sensitivity and the signals exhibited in a C.R.O. screen, a strong fluctuation of the cutting forces about a D.C. value is noticed. In some cases the D.C. component cannot be identified. Peklenik and Kwiatkowski[21] have pointed out the random character of such force variations. The random variation of cutting force components F_C and F_T can be imagined as a resultant of random processes of deformation in the flow zone and frictional forces along tool face and flank.

It is known that all polycrystalline aggregates possess a random variation of principal crystalline orientations changing from grain to grain.

The grain size is also randomly variable. For ordinary, large volume test specimens, it is possible to establish an average yield strength and its variance. In the case of metal cutting the size of deformation zone is frequently of the same order as the grain diameters and sometimes even smaller. Thus, we know that the mean values as well as their variances will be different from material to material and for the same material depending on the grain size and the physical state of the grains.

Peklenik and Kwiatkowski have also shown that the correlation functions for the cutting forces change from material to material. These authors state: "In a number of force and temperature measurements it has been shown that the fluctuation of these parameters are random and should be strongly considered in cutting mechanics investigations." We agree with this statement and believe that the study of random variation of cutting forces is an essential factor in the overall theory of machining.

It is premature at this time to engage in a long discussion of eventualities, but in view of limited experience we came to the conclusion that the correlation function shapes of cutting forces can indeed identify various metals according to the ease of machining. In some instances the power spectra of cutting forces appear even more instructive. Obviously, the force and surface roughness profiles can be correlated, and these in turn with the machine response. Therefore this approach to metal-cutting study appears to us as extremely promising and worthwhile.

SURFACE DAMAGE

The introduction of new high strength materials into modern mechanical technology puts new emphasis upon the problem of surface damage. It is really disappointing that metal-cutting research has not as yet produced a consistent theory for the evaluation of the surface damage phenomena. For instance, it is not known to what an extent is the final surface state dependent upon the flow zone deformation and how large is the contribution of the flank surface rubbing. In practical three-dimensional machining the effects of the nose radius upon the roughness have been investigated, but the physical state of the surface layer is unknown. A promising avenue of research is anticipated from the application of moving flat indenters over the edge of a plate. Some estimates of the surface damage may also be obtained from the study of flat die indenting a semi-infinite plate or semi-infinite medium. This problem appears to be connected with the problem of expansion of a spherical cavity in an infinite medium.[22]

CONCLUSIONS

In this review we have indicated that the singular surface models of metal cutting do not promise a great deal of new information for the machining problems. By an examination of the properties of flow zone models we reached the conclusion that further effort in this particular direction would yield very important and instructive information.

The application of random processes method to the study of cutting forces is a new and most promising development in the recent metal-cutting theory. Although we realize that the effective study along these lines requires sophisticated instrumentation and availability of a large computer, the cost of data acquisition is amply compensated by the value of the obtained information. The problem of the surface damage is again brought to foreground. Its importance is increasing in view of the need for high integrity of machined surface of high strength materials.

REFERENCES

1. E. H. Lee and B. W. Shaffer. *J. Appl. Mech.*, **18**, (1951) 405.
2. M. C. Shaw, N. H. Cook and J. Finnie. *Trans. ASME*, **15**, (1953) 273.
3. H. Ll. D. Pugh. *Proc. Conf. on Tech. Eng. Manuf.* I.M.E., London, 1958.
4. E. Usui, K. Kikuchi and K. Hoshi. *J. Eng. Ind., Trans. ASME*, Ser. B, **86**, (1964) 95.
5. R. Hill. *J. Mech. Phys. Solids*, **3**, (1954) 47.
6. W. Johnson. *Annals CIRP*, **14**, (1967) 315.
7. M. E. Merchant, *J. Appl. Phys.*, **16**, (1945) 267, 318.
8. V. Piispanen. *J. Appl. Phys.*, **19**, (1948) 876.
9. H. Ernst and M. E. Merchant. *Trans. Am. Soc. Metals*, **29**, (1941) 229.
10. U. Wegner. *Zeit. angew. Math. Mech.*, **38**, (1958) 200.
11. H. Kudo. *Int. J. Mech. Sci.*, **7**, (1965) 43.
12. S. Kobayashi and E. G. Thomsen. *J. Eng. Ind., Trans. ASME*, Ser. B, **84**, (1962) 63.
13. J. D. Cumming, S. Kobayashi and E. G. Thomsen. *J. Eng. Ind., Trans. ASME*, Ser. B, **87**, (1965) 480.
14. S. S. Chang and W. B. Heginbotham. *J. Eng. Ind., Trans. ASME*, Ser. B, **83**, (1961) 351.
15. W. B. Palmer and P. L. B. Oxley. *Proc. Inst. Mech. Engrs.*, **173**, (1959) 623.
16. P. L. B. Oxley. *Int. J. Mech. Sci.*, **3**, (1961) 68.
17. K. Okushima and K. Hitomi. *J. Eng. Ind., Trans. ASME*, Ser. B, **83**, (1961) 545.
18. F. Lira and E. G. Thomsen. *J. Eng. Ind., Trans. ASME*, Ser. B, **89**, (1967) 489.
19. P. L. B. Oxley, A. G. Humphreys and A. Lazarideh. *Proc. Instn. Mech. Engrs.*, **175**, 18, (1961) 881.
20. N. N. Zorew, G. L. Kufarew, G. D. Del and G. Goldschmidt. *Annals CIRP*, **14**, (1967) 337.
21. J. Peklenik and A. W. Kwiatkowski. *Annals CIRP*, **15**, (1967) 67.
22. R. F. Bishop, R. Hill and N. F. Mott. *Proc. Phys. Soc.*, **57**, (1945) 147.

AN EXPERIMENTAL STUDY OF PLASTIC DEFORMATION DUE TO OBLIQUE CUTTING

C. Y. CHOI

University of Aston in Birmingham

INTRODUCTION

As in the case of orthogonal cutting, Merchant assumed that the chip is formed by a process of shear on a plane which makes an angle with the surface generated.[1] Thus the geometry involved in oblique cuttings, as Merchant suggested, can be represented by a model such as shown in Figs. 1 and 2. In Fig. 1 the primary rake angle (normal rake angle) is designated by a_n and the normal shear angle is represented by ϕ_n. The thickness of the chip is denoted by t_2 and the thickness of the chip before removal is designated by t_1. Knowing t_1, t_2 and a_n, ϕ_n can be determined from the following equation readily derivable from the geometry of Fig. 1,

$$\tan\phi_n = \frac{t_1/t_2 \cos a_n}{1 - t_1/t_2 \sin a_n} \tag{1}$$

In Fig. 2, the direction of motion of tool relative to work is indicated by vector L_1, and it makes an angle β with the X axis (i.e. the cutting edge). Plane $OBCy$ is an extension of the shear plane, while plane $ACBx$ is parallel to the tool face and passes through the terminal point A of vector L_1. If the cutting tool advances along the work piece a distance L_1, then a point which was originally at O on the chip will have advanced up the tool face a distance L_2 in the direction of chip flow as given by F, and will have advanced relative to the workpiece a distance L_3 in the plane of shear, thus bringing it to point D. Therefore L_3 represents the direction of shear of the metal on the shear plane. It is obvious that L_3 does not, in general, coincide with OB, which is lying on the tool face and perpendicular to the cutting edge. The angle between L_3 and OB is designated by δ. From the geometry of Fig. 2, it is possible to derive the following relation between δ and F

$$\tan\delta = \frac{\tan\beta \cos(\phi_n - a_n) - \tan F \sin\phi_n}{\cos a_n} \tag{2}$$

The shear strain in the chip is therefore determined by

$$\epsilon_s = \frac{L_3}{Ex}$$

or

$$\epsilon_s = \frac{\cot\phi_n + \tan(\phi_n - a_n)}{\cot\delta} \tag{3}$$

Merchant's single shear plane model implies that the plastic deformation due to oblique cutting can be represented by a two-dimensional simple shear and is a function of ϕ_3 and δ or t_2/t_1 and F only.

While Merchant considers that the chip flow angle is a variable, Stabler reports that experimentally the angle F is equal to the oblique angle β, for all test conditions he investiated, including tool and work materials, rake angles and speeds, and proposes the Chip Flow Law.[2,3]

$$F = \beta \qquad\qquad (4)$$

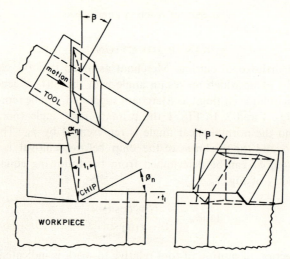

FIG. 1.

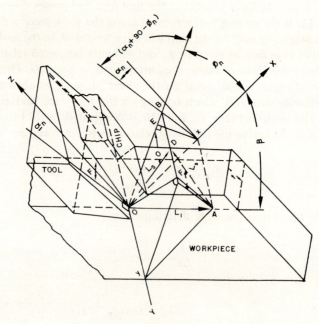

FIG. 2.

Applying the Chip Flow Law to Merchant's simple shear plane model, eqs. (2) and (3) become

$$\tan\delta = \frac{\tan\beta \cos(\phi_n - a_n) - \tan\beta \sin\phi_n}{\cos\delta} \tag{5}$$

$$\epsilon_s = \frac{\cot\phi_n + \tan(\phi_n - a_n)}{\cos\delta} \tag{6}$$

Equations (5) and (6) indicate that the combination of the Merchant's model and Stabler's Chip Flow Law lead to the conclusion that the plastic deformation due to oblique cutting contains only one parameter, i.e. t_2/t_1 or ϕ_n. If Merchant's model and Stabler's Chip Flow Law are true, the whole geometrical problem of oblique cutting will be reduced to the finding of the shear angle ϕ_n only. Unfortunately, further researches[4] revealed that not all experimental results are in agreement with Merchant's model or Stabler's Chip Flow Law.

It seems, as in the case of orthogonal cutting, the main shortcomings of the single shear plane model are that the deformation does not take place in a single plane and that the deformation can not be represented by a simple shear along the shear plane.

Therefore, it is the main object of this paper to investigate the actual plastic deformation due to oblique cutting.

PLASTIC DEFORMATION DUE TO OBLIQUE CUTTING

Actually the deformation due to oblique cutting is three-dimensional though it is subjected to several restrictions due to symmetry and the streamlined flow (which produces the continuous chip), and hence is less complicated than the general three-dimensional deformation.

Hsü[5] suggests that the deformation due to oblique cutting may be divided into five steps (Fig. 3): (a) a pure shear in the x_1x_2-plane, (b) a simple shear in the x_1x_3-plane, (c) a simple shear in the x_1x_2-plane, (d) a simple shear in the x_2x_3-plane, and (e) a rotation about the x_3 axis through an angle equal to $(90° - a)$ (not shown in Fig. 3). Each of the first four steps is represented in Fig. 3 as a deformation of a unit cube (dotted). Each step may be represented by a matrix and the overall deformation is the product of the five matrices, as follows:

$$
\begin{vmatrix}
\cos(90°-a_n) & -\sin(90°-a_n) & 0 \\
\sin(90°-a_n) & \cos(90°-a_n) & 0 \\
0 & 0 & 1
\end{vmatrix}
\begin{vmatrix}
1 & 0 & 0 \\
0 & 1 & 0 \\
0 & \gamma_2 & 1
\end{vmatrix}
$$

$$
\times
\begin{vmatrix}
1 & \gamma_1 & 0 \\
0 & 1 & 0 \\
0 & 0 & 1
\end{vmatrix}
\begin{vmatrix}
1 & 0 & 0 \\
0 & 1 & 0 \\
\gamma_3 & 0 & 1
\end{vmatrix}
\begin{vmatrix}
t_1/t_2 & 0 & 0 \\
0 & t_2/t_1 & 0 \\
0 & 0 & 1
\end{vmatrix}
\tag{7}
$$

$$
=
\begin{vmatrix}
t_1/t_2 \sin a_n & \gamma_1 t_2/t_1 \sin a_n - t_2/t_1 \cos a_n & 0 \\
t_1/t_2 \cos a_n & t_2/t_1 (\gamma_1 \cos a_n + \sin a_n) & 0 \\
\gamma_3 t_1/t_2 & \gamma_2 t_2/t_1 & 1
\end{vmatrix}
$$

where γ_1, γ_2 and γ_3 are as defined in Fig. 3 and a_n is the primary rake angle (normal rake angle).

MEASUREMENT OF THE PLASTIC DEFORMATION IN THE CHIP

As can be seen in eq. (7), for a given tool (fixed a_n) four quantities (γ_1, γ_2, γ_3 and t_2/t_1) are required to represent the deformation in the chip.

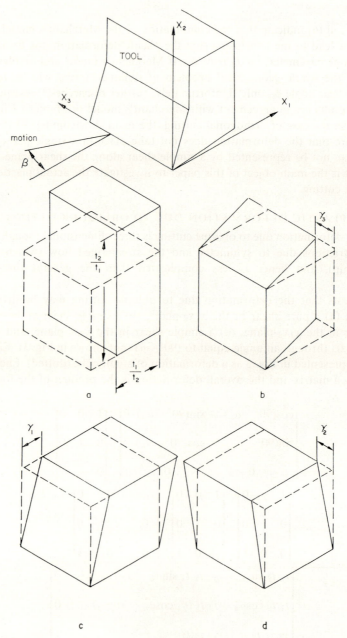

Fig. 3.

The ratio t_2/t_1 can be easily measured, since the chip thickness t_2 can be measured with a micrometer.

To determine the simple shear γ_3, let us first consider a unit cube of the undeformed material with one of its edges lying on the cutting edge. After deformation it becomes a parallelepiped in the chip, which is shown in its three orthogonal projections in Fig. 4b; that in the tool face (top left) that in the plane normal to the cutting edge (top right) and

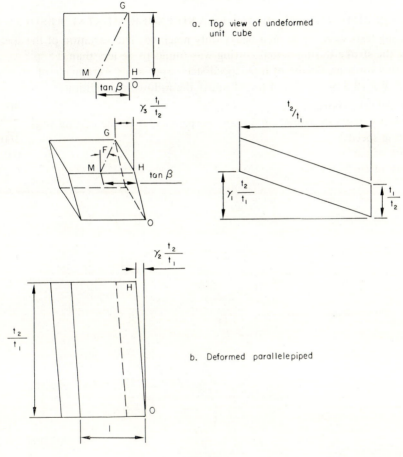

Fig. 4.

that in the plane through the cutting edge normal to the tool face. A line element MG which was originally in the direction of the motion of the tool and on the top of unit cube (Fig. 4a) will become MG at the deformed parallelepiped (Fig. 4b.) Some reflection will show that the angle F is generally defined as the chip flow angle. From the geometry of Fig. 4, we have

$$\gamma_3 = t_2/t_1 \tan\beta - \tan F \qquad (8)$$

Therefore, the simple shear γ_3 can be determined by the cutting ratio t_2/t_1 and the chip flow angle F.

The measurement of the simple shears γ_2 and ν_1 is more difficult than those of the t_2/t_1

and γ_3, because the location of the line element OH at the deformed parallelepiped is required. Suppose the position of OH is known and is represented by the coordinates of points O and H (x_{01}, x_{02}, x_{03} and x_{H1}, x_{H2}, x_{H3}) then simple shears γ_1 and γ_2 can be determined by

$$\gamma_1 = \frac{x_{H2} - x_{02}}{t_2 \cos a_n} \tan a_n \tag{9}$$

$$\gamma_2 = (x_{H3} - x_{03}) t_2 \tag{10}$$

EXPERIMENTAL TECHNIQUES AND EXPERIMENTAL RESULTS

The cutting tests were made on a heavy-duty machine. The variation of the speed of the ram along the stroke during actual cutting was found to be less than 0.5%.

The cutting tools used were of high speed steel. Specimens were made of an aluminium alloy HE-9-WP (B.S.S. 1476) and tested under the following conditions:

(1) Primary rake (normal rake) $30°$
(2) Clearance angle measured on a plane perpendicular to the cutting edge $7°$
(3) Cutting speed   160 fpm
(4) Depth of cut   0·015 in
(5) Width of cut   0·25 in

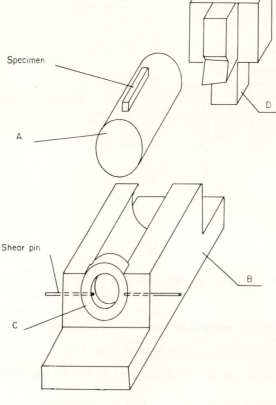

Fig. 5.

Strips of the work material, $\frac{1}{4}$ in wide and $1\frac{1}{2}$ in long, were held in the sliding cylinder A of the quick stopping mechanism (Fig. 5). The mechanism was in turn clamped on the table. During the cutting process, A is prevented from sliding through B by a ring C, held in place by shear pins. After the tool had cut about $\frac{3}{4}$ in through the specimen, the tongue D hit the end of A and the pins sheared. Thus the cutting action was arrested abruptly. The specimen was then taken down for taking measurement on the MU214B Universal Measuring Machine.

In order to measure the simple shears γ_1 and γ_2, a small hole was drilled on the specimen such as shown in Fig. 6a, then a copper wire was fitted in the hole. During the cutting process, part of the copper wire was also deformed and became part of the chip (Fig. 6b). For the convenience of measurement, the position of points O and H were measured with

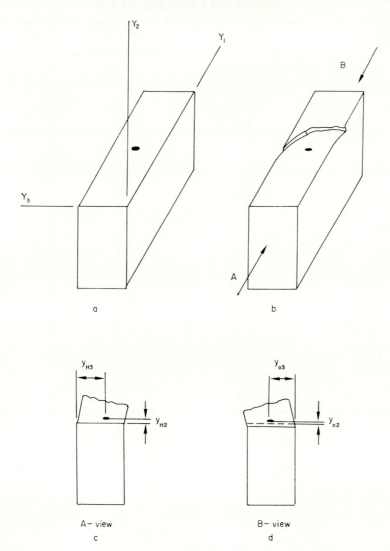

FIG. 6.

reference to coordinate system γ_1, γ_2, γ_3 such as shown in Fig. 6a. Consequently, equations 9 and 10 for determining the simple shears γ_1 and γ_3 were modified and became

$$\gamma_1 = \frac{y_{H2} - y_{02}}{t_2 \cos\alpha_n} - \tan\alpha_n \tag{11}$$

$$\gamma_2 = \frac{y_{H3} - y_{03}}{t_2 \cos\beta} - (\gamma_1 \sin\alpha_n - \cos\alpha_n) \tan\beta \tag{12}$$

The experimental results obtained are shown in Figs. 7 to 10. It can be seen that

(1) The chip flow angle F obtained experimentally is not in agreement with Stabler's Chip Flow Law; and the deviation of F from β is not linear.

(2) The cutting ratio t_2/t_1 and the shear angle $\tan^{-1}\gamma_1$ decrease with increase in oblique angle β.

(3) The value of γ_2, the chip flow angle F and consequently the simple shear γ_3 increase when the oblique angle β is increased.

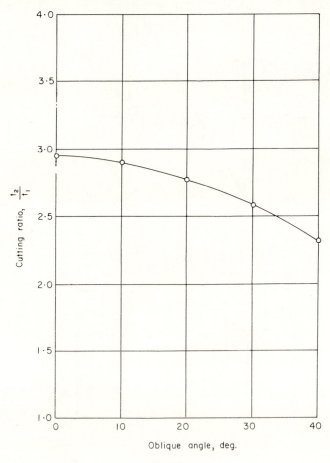

FIG. 7.

REPRESENTATION OF FINITE DEFORMATION
IN TERMS OF PRINCIPAL STRAINS

A finite deformation, for the convenience of investigation, can be resolved into several steps in different ways. For example Hsü suggested that the plastic deformation due to oblique cutting can be represented by a pure shear (t_2/t_1), three simple shears $(\gamma_1, \gamma_2$ and $\gamma_3)$ and a rotation $(90°-a_n)$.[5] However, for the purpose of comparing two different finite deformations, it is desirable to represent them by principal strains in terms of pure shears, because it is known that simple shears occur along a noncoaxial strain path, or a strain

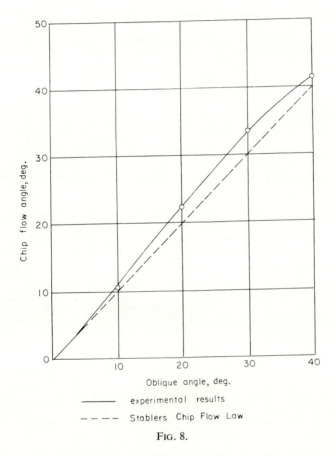

FIG. 8.

path in which not all the incremental strains have the same principal axis with respect to the material.[6,7] It is, therefore, desirable to introduce briefly the analysis of finite deformation analysis.[8,9]

The position of a material point P in the undeformed body is denoted by a rectangular coordinate system X^K $(K = 1, 2, 3)$. After deformation the position of the corresponding spatial point p is designated by a new set of rectangular cartesian coordinates x^k $(k = 1, 2, 3)$. By employing coordinate systems X^K and x^k (Fig. 11), the motion, which carries various material points through various spatial points, can be represented by

$$x^k = x^k(X^K) \tag{13}$$

or

$$X^K = X^K(x^k) \tag{14}$$

A homogeneous deformation can be represented by the affine transformation

$$x^k = D_{kK}X^K \tag{15}$$

where D_{kK} is a tensor of the second rank in three dimensions.

For continuous media, the inverse transformation of (15) exists and can be expressed by

$$X^K = D_{Kk}x^k \tag{16}$$

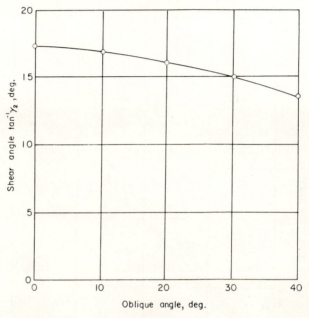

Fig. 9.

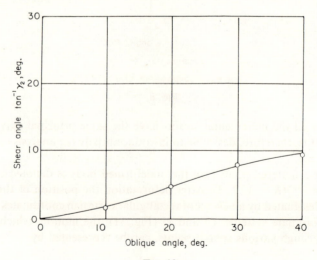

Fig. 10.

where D_{Kk} is defined by

$$D_{Kk}\, D_{kK} = 1 \tag{17}$$

Through eqs. (13) and (14), we have

$$\mathrm{d}x^k = \frac{\partial x^k}{\partial X^K}\, \mathrm{d}X^K \quad \mathrm{d}X^K = \frac{\partial X^K}{\partial x^k}\, \mathrm{d}x^k \tag{18}$$

Meanwhile, from eqs. (15) and (16) we get

$$\frac{\partial x^k}{\partial X^K} = D_{kK} \quad \frac{\partial X^K}{\partial x^k} = D_{Kk} \tag{19}$$

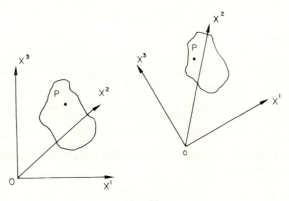

FIG. 11.

Hence, (18) becomes

$$\mathrm{d}x^k = D_{kK}\mathrm{d}X \quad \mathrm{d}X^K = D_{Kk}\mathrm{d}x^k \tag{20}$$

The square of the lengths of the infinitesimal line elements in the undeformed and deformed body may be expressed as

$$\mathrm{d}S^2 = \delta_{KL}\mathrm{d}X^K\mathrm{d}X^L$$
$$\mathrm{d}s^2 = \delta_{kl}\mathrm{d}x^k\mathrm{d}x^l \tag{21}$$

By substituting (20) into (21), we obtain

$$\mathrm{d}S^2 = c_{kl}\mathrm{d}x^k\mathrm{d}x^l \quad \mathrm{d}s^2 = C_{KL}\mathrm{d}X^K\mathrm{d}X^L \tag{22}$$

where

$$c_{kl} = \delta_{KL}D_{Kk}D_{L1} \quad C_{KL} = \delta_{kl}D_{kK}D_{1L} \tag{23}$$

which are called respectively Cauchy's deformation tensor and Green's deformation tensor. Both are symmetrical, i.e. $c_{kl} = c_{1k}$ and $C_{KL} = C_{LK}$, and both are positive-definite.

The ratio $\mathrm{d}s/\mathrm{d}S$ of the lengths of infinitesimal line elements $\mathrm{d}x^k$ and $\mathrm{d}X^K$ is a function of the direction of either $\mathrm{d}X^k$ or $\mathrm{d}x^k$ and is called the stretch. Let N^K and n^k be respectively the unit vectors along dX^K and dx^k, then the stretch $\Lambda_{(N)} = \lambda_{(n)}$ is defined by

$$\Lambda_{(N)} = \frac{ds}{dS} = \sqrt{(C_{KL}N^K N^L)}$$

(24)

$$\lambda_{(n)} = \frac{ds}{dS} = \frac{1}{\sqrt{(c_{kl}\, n^k n^l)}}$$

where

$$N^K \equiv \frac{dX^K}{dS} \quad n^k \equiv \frac{dx^k}{ds}$$

(25)

If we introduce the inverse of the Cauchy's deformation tensor, c_{k1}, defined by

$$c_{k1}^{-1}\, c_{k1} = 1$$

(26)

eq. (24_2) may be written as

$$\lambda_{(n)} = \frac{ds}{dS} = \frac{\sqrt{(c_{k1}^{-1})}}{\sqrt{(n^k n^l)}}$$

(27)

An infinitesimal sphere at X^K swept by a vector dX^K is given by

$$\delta_{KL}dX^K dX^L = dS^2 = K^2$$

(28)

Through the motion (13), the material points of this sphere are carried into a ellipsoid at the point x^k of the deformed body, i.e.

$$c_{k1}dx^k dx^l = dS^2 = K^2$$

(29)

This ellipsoid is called the material strain ellipsoid. Similarly, by inverse mapping, the infinitesimal sphere

$$\delta_{k1}dx^k dx^l = ds^2 = k^2$$

(30)

at x^k is carried into an ellipsoid at X^K

$$C_{KL}dX^K dX^L = ds^2 = k^2$$

(31)

called the spatial strain ellipsoid.

Now let dX_1^K and dX_2^L be two orthogonal vectors at X^K, i.e.

$$\delta_{KL}dX_1^K dX_2^L = 0$$

(32)

According to (21_1) and (22_1) eq. (32) becomes

$$c_{k1}dx_1^k dx_2^l = 0$$

(33)

This indicates that perpendicular diameters of an infinitesimal sphere at X^K are deformed into conjugate diameters of the material strain ellipsoid at x^k.

An ellipsoid has three diameters which are each perpendicular to its conjugate diameters. Actually these axes are the mappings of a set of orthogonal axes of the sphere at X^K. Therefore, we may conclude that at a point $P(X^K)$ of the undeformed body there exist at least three orthogonal diameters which remain orthogonal after deformation and constitute the principal axes of the material strain ellipsoid at $p(x^k)$ (Fig. 12a).

It is obvious that what was said in the last paragraph applies to the inverse deformation, i.e. at the point $p(x^k)$ of the deformed body there exist at least three orthogonal diameters which remain orthogonal after the inverse deformation takes place, and constitute the principal axes of the spatial strain ellipsoid at $P(X^K)$ (Fig. 12b).

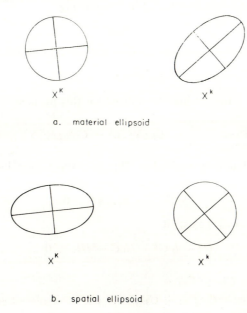

a.   material ellipsoid

b.   spatial ellipsoid

FIG. 12.

A little reflection shows that, owing to the fact that the materials are continuous media and if it is further assumed that the state of strain in question is neither spherical nor cylindrical, the principal axes of the strain ellipsoids must coincide with the corresponding diameters of the undeformed spheroid (Fig. 12).

This leads to the following conclusions:

(1) A homogeneous finite deformation can be resolved as follows:

(a) An infinitesimal spheroid with its centre at X^K is stretched along the principal axes of the spatial strain ellipsoid. The principal stretches are characterized by
$$\Lambda_{(\alpha)} = \sqrt{C}_{(\alpha)}.$$

(b) The ellipsoid (i.e. the deformed spheroid) is rigidly translated in parallel transport to x^k.

(c) The deformed ellipsoid is then rigidly rotated about x^k until its principal axes coincide with those of the material strain ellipsoid. However, it should be noted that the deformation in question can actually be resolved in many other ways.

(2) In order to describe a homogeneous finite deformation in terms of principal stretches (or principal strains), it is necessary and sufficient to obtain the following data:

(a) The principal stretches (or principal strains).

(b) The principal directions of the material strain ellipsoid.

(c) The principal directions of the spatial strain ellipsoid.

(3) The difference between the principal directions of the material strain ellipsoid and those of the spatial strain ellipsoid represents the rotation of the principal axes.

DETERMINATION OF THE PRINCIPAL STRAINS
AND THE PRINCIPAL DIRECTIONS

Since the stretches $\Lambda_{(N)}$ along the three orthogonal diameters, which are carried into the principal axes of an ellipsoid at x^k, take extremum values, the analytical determination of these three directions may thus be effected by minimizing

$$\Lambda^2_{(N)} = C_{KL}N^KN^L$$

with respect to N^L subject to the condition that N^K is a unit vector, i.e.

$$\delta_{KL}N^KN^L = 1$$

Lagrange's method of multipliers may be utilized for this purpose. Thus

$$\frac{\partial}{\partial N^M}\, \Lambda^2_{(N)} = \frac{\partial}{\partial N^M}\left[C_{KL}N^KN^L - C(\delta_{KL}N^KN^L - 1)\right] = 0$$

where C is an unknown Lagrange multiplier. This gives three linear homogeneous equations for N^L

$$(C_{KL} - C\delta_{KL})N^L = 0 \tag{34}$$

A nontrivial solution of (34) exists if

$$C^3 - IcC^2 + IIcC - IIIc = 0 \tag{35}$$

where

$$Ic = C_{11} + C_{22} + C_{33} \tag{36}$$
$$IIc = C_{11}C_{22} + C_{22}C_{33} + C_{33}C_{11} - C_{12}{}^2 - C_{23}{}^2 - C_{31}{}^2$$
$$IIIc = C_{11}C_{22}C_{33} + 2C_{12}C_{23}C_{31} - C_{11}C_{23}{}^2 - C_{22}C_{31}{}^2 - C_{33}C_{21}{}^2$$

If we assume that the material in question is incompressible and that the strain is not equal to zero, the characteristic equation (35) always possesses three real distinct roots $C_{(\alpha)}$, $(\alpha = 1, 2, 3)$, called eigenvalues.

If the reference frame X^K chosen coincides with the principal directions of the material strain ellipsoid, we have

$$N^L_{(\alpha)} = \delta^L_{(\alpha)}$$

Hence through (34) we obtain

$$C^K_{(\alpha)} = C_{(\alpha)}\delta^K_{(\alpha)} \tag{37}$$

and

$$\Lambda_{(\alpha)} = \sqrt{(C_{(\alpha)})} \tag{38}$$

It is clear that
$$\Lambda_{(\alpha)} = e^{\epsilon(\alpha)}$$

where $\epsilon(\alpha)$ are the natural strains. For $\alpha = 1, 2, 3$, we have

$$C_{11} = C_1; C_{22} = C_2; C_{33} = C_3; C_{12} = C_{23} = C_{31} = 0$$

This indicates that the eigenvalues are the normal components of the deformation tensor in the principal axes, and the shear components in this frame of reference are zero. The square roots of the eigenvalues are, therefore, the principal stretches.

It is obvious that the same arguments apply to the inverse deformation.

Thus, the determination of the directions of the principal axes of the material and spatial strain ellipsoids and the determination of the principal stretches (or principal strains) are simply equivalent to the diagonalization of the Green's deformation tensor and the inverse of the Cauchy's deformation tensor.

DISCUSSION

Examples of plastic deformation due to oblique cutting represented by principal strains in terms of pure shears are shown in Figs. 13 and 14. In Figs. 13b and 14a, an undeformed unit cube is illustrated by chain lines and its edges are originally parallel to the spatial principal axes N_1, N_2 and N_3 respectively. During the cutting process, the cube is first rotated rigidly to the directions of the material principal axes n_1, n_2 and n_3. Then it is stretched

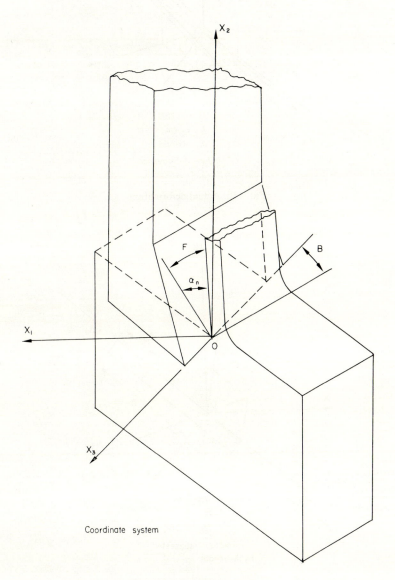

Fig. 13a. Coordinate system.

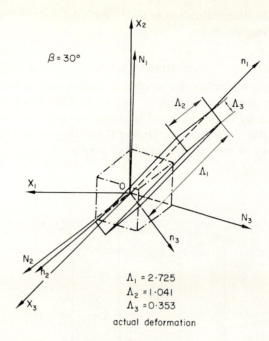

$\Lambda_1 = 2\cdot725$
$\Lambda_2 = 1\cdot041$
$\Lambda_3 = 0\cdot353$

actual deformation

FIG. 13b. Actual deformation.

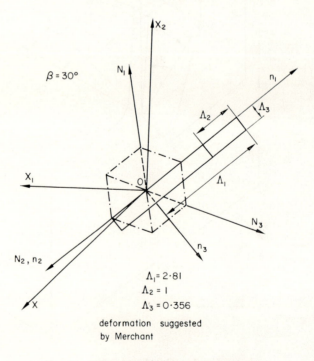

$\Lambda_1 = 2\cdot81$
$\Lambda_2 = 1$
$\Lambda_3 = 0\cdot356$

deformation suggested
by Merchant

FIG. 13c. Deformation suggested by Merchant.

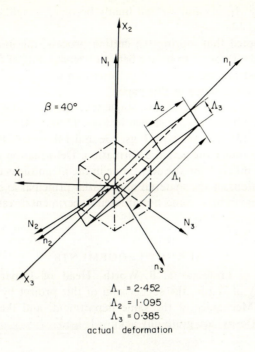

$\Lambda_1 = 2.452$
$\Lambda_2 = 1.095$
$\Lambda_3 = 0.385$
actual deformation

FIG. 14a. Actual deformation.

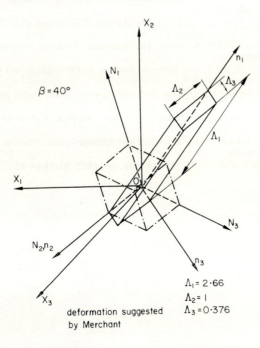

$\Lambda_1 = 2.66$
$\Lambda_2 = 1$
$\Lambda_3 = 0.376$

deformation suggested
by Merchant

FIG. 14b. Deformation suggested by Merchant.

along n_1, n_2 and n_3 by Λ_1, Λ_2 and Λ_3 and finally becomes a right parallelepiped such as shown in Figs. 13b and 14a.

Here it should be noted that, during the cutting process, the unit cube is rotated and stretched simultaneously. In order to have a better understanding of the cutting process it is also necessary to investigate the history of deformation (e.g. strain path and strain rate). However, this is beyond the scope of this paper.

For the purpose of evaluating Merchant's single shear plane model, corresponding results obtained from eqs. (1), (2) and (3) are shown in Figs. 13 and 14.

Comparison of Figs. 13b and 14a with Figs. 13c and 14b reveals that Merchant's model does not and can not produce the actual deformation. Deformation due to oblique cutting proposed by Merchant differs from the actual one in the magnitudes of the principal strains as well as in the directions of the material and spatial principal axes. Therefore, it is not surprising that other researchers found not all their experimental results in agreement with Merchant's analysis.

ACKNOWLEDGEMENTS

The encouragement of Professor T. B. Worth, Head of Department of Production Engineering, University of Aston, the supervision of this project by Professor T. C. Hsü, Professor of Applied Mechanics of the same department, and the financial support of Rolls-Royce Limited, Derby, are gratefully acknowledged.

REFERENCES

1. M. E. MERCHANT, Basic Mechanics of The Metal Cutting Process, *Trans. A.S.M.E.*, **66**, 1944, A168.
2. G. V. STABLER, The Fundamental Geometry of Cutting Tools, *Proceedings of The Institution of Mechanical Engineers*, **165**, 1951, 14.
3. G. V. STABLER, The Chip Flow Law and Its Consequences, *Proceedings of the 7th International M.T.D.R. Conference*, 1966, 243.
4. M. C. SHAW, N. H. COOK and P. A. SMITH, The Mechanics of Three Dimensional Cutting Operation, *Trans. A.S.M.E.*, **74**, 1952, 1055.
5. T. C. HSÜ, An Analysis of the Plastic Deformation due to Orthogonal and Oblique Cutting, *Journal of Strain Analysis*, **1**, 1966, No. 5, 375.
6. T. C. HSÜ, The Characteristics of Coaxial and Non-coaxial Strain Paths, *Journal of Strain Analysis*, **1**, 1966, No. 3, 216.
7. T. C. HSÜ, A Study of Large Deformations by Matrix Algebra, *Journal of Strain Analysis*, **1**, 1966, No. 4, 313.
8. C. TRUESDELL and R. TOUPIN, The Classical Field Theories 1960. *Handbuch der Physik*, Vol. III/1, 1960, Springer-Verlag, Berlin.
9. A. C. ERINGEN, *Nonlinear Theory of Continuous Media*, 1962, McGraw-Hill, New York.

ON THE METAL-CUTTING MECHANISM
WITH THE BUILT-UP EDGE

Koichi Hoshi

Hokkaido University, Sapporo, Japan

and Tetsutaro Hoshi

Kyoto University, Kyoto, Japan

NOMENCLATURE

Cutting conditions

a rake angle of orthogonal cutting tool, deg
v cutting speed, mpm (meters per minute)
t_1 depth of cut in orthogonal cutting, mm
H_v micro-Vickers hardness number
     Metal-cutting geometries (referring to Fig. 5)

u thickness of built-up edge measured in the cutting direction at point D, m m
L contact length between built-up edge and chip measured parallel to tool rakeface, mm
Φ angle of the start boundary of primary plastic flow to the cutting direction, deg
ψ over-cut depth, mm
h depth of deformed layer, mm
T distance between points H and G, mm
t_2 chip thickness, mm
     Stress assumptons (referring to Fig. 12)

τ_s shear stress on the start boundary of primary flow
σ_s normal stress to the start boundary of primary flow
τ_f shear stress on the built-up edge boundary
σ_f normal stress to the built-up edge boundary
$S_1 = \sigma_f/\tau_f,\ S_2 = \sigma_s/\tau_f,\ S_3 = \tau_s/\tau_f$

1. INTRODUCTION

In the history of metal-cutting research, the built-up edge has been one of the most interesting subjects,[1–6] and is still producing diversified discussions. One of the authors has continued to work on[7, 8] this problematic point; recently, he finished a series of metal-cutting tests from which many formerly uncovered features came to light.

In this study, orthogonal cutting of carbon steels was investigated for the purpose of obtaining advanced knowledge on the formation of the built-up edge and the metal-cutting mechanism when it was present.

In particular, the following points were explored:

(1) the type of physical process by which a portion of the work metal was deformed to form the built-up edge;

(2) the type of adhesion of the built-up edge material onto the tool face;

(3) the metal removal model by which the worked metal plastically transformed into chips when the built-up edge was present;

(4) the manner in which the size and shape of the built-up edge varied according to the change in cutting speed and tool rake angle; also periodical fluctuations of the built-up edge in a cutting at a given condition were observed; and the cause of such variations were studied.

Experimental data were presented in the form of micro-photographs of the middle sections of the partly formed chips obtained by quick-stopping the orthogonal turning tests of the steel tube ends, at cutting speeds ranging from 8 mpm (meters per minute) to 165 mpm. By those micro-photographs, the size and shape of the built-up edge were analyzed, its formation and cutting mechanism were estimated, and a simplified metal flow model was proposed, which was further followed by a stimulation of mechanical force equilibrium in cutting with the built-up edge.

2. MATERIAL AND EQUIPMENT FOR THE STUDY

For the quick-stop tests to obtain the partly formed orthogonal chips having the built-up edge, a tubular work-piece was machined beforehand out of a solid block of 0.25% carbon steel with mechanical properties and compositions as listed below.

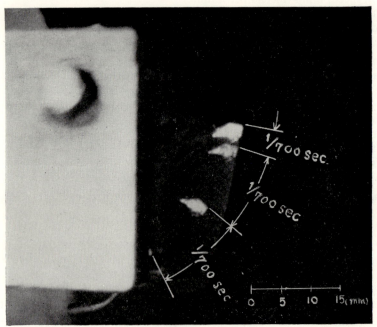

FIG. 1. High-speed flashlight photograph to show the speed of the escaping tool when the cutting process is quick-stopped. The average speed of the escaping tool is found to be 170 mpm during the initial 1/700 sec, followed by a constant speed of 470 mpm.

Mechanical properties of material

Yielding point kg/mm²	Tensile strength kg/mm²	Total elongation %	Reduction of area %	Hardness H_B	Impact vale kg-m/cm²
30.7	52.6	34.7	58.0	145	5.1

Chemical composition of material (wt. %)

C	Si	Mn	P	S	Ni	Cr	Mo	Cu
0.24	0.30	0.68	0.015	0.012	0.14	0.13	0.04	0.28

The quick-stop orthogonal turning tests employed P20 carbide tools and also 5% Co and 10% Co type high-speed steel tools. The tool was mounted on a quick-stop device fixed in such a way that the tool could be suddenly removed along the direction of cutting, from the cuttng position.

The speed of the tool escape was preliminarily calibrated using high-speed flashlight photography as shown in Fig. 1. From the 170 mpm initial average speed calculated from the photograph, the test rig was found to be capable of stopping cutting without interfering with the chip formation from a cutting speed of 65 mpm or below.

The quick-stop tests were performed on an engine lathe Shoun-Cazeneuve, having a 500 mm swing over bed, a 1000 mm distance between centers and a variable speed spindle drive motor rated at 11 kW.

3. EXPERIMENTAL RESULTS AND INTERPRETATIONS

3.1. *Formation of Built-up Edge*

3.1.1. *Secondary flow, built-up layer, and built-up edge.* Partly formed chip quick-stopped from various cutting speeds contain built-up edges of different sizes. In a series of micro-photographs shown in Fig. 2, the specimen of a relatively high cutting speed, top picture (a) at 188 mpm, involves no built-up edge, but only a flow of metal along the tool face which is conventionally termed as "the secondary flow" is observed. A 20 to 30 μ ($\mu = 10^{-3}$ mm) thick secondary flow zone is observed at cutting speeds of 130 mpm and above.

At slower cutting, a layer of stationary metal is observed between the secondary flow and the tool face as is seen in the one taken at 41 mpm, Fig. 2(b). Micro-hardness tests indicated that this kind of layer (297 to 322 HV) was as hard as the typically formed built-up edge (297 to 439 HV) and was work-hardened more than the original work-piece (153 to 159 HV) or the formed chip (236–256 HV). The stationary layer is the bulk of metal deformed by the secondary flow, thus work-hardened and becoming stationary. The removed chip is separated from this layer by the secondary flow inside the chip. The thickness of the layer increases when the cutting speed is reduced, as shown in Fig. 3, until the layer attains the size of a typical built-up edge like the one at 35 mpm seen in Fig. 2(c). It is proposed, therefore, that such a layer be given the name "built-up layer" and treated as a premature body of the built-up edge according to the definition listed below.

Proposed definition of built-up layer and built-up edge

	Thickness u as indicated in Fig. 3 (mm)	Typical cutting speed range in steel cutting (mpm)
Built-up layer	0 to 0.1	40 to 130
Built-up edge	0.1 and above	under 40

Above the speed range of built-up layer, only secondary flow zone exists.

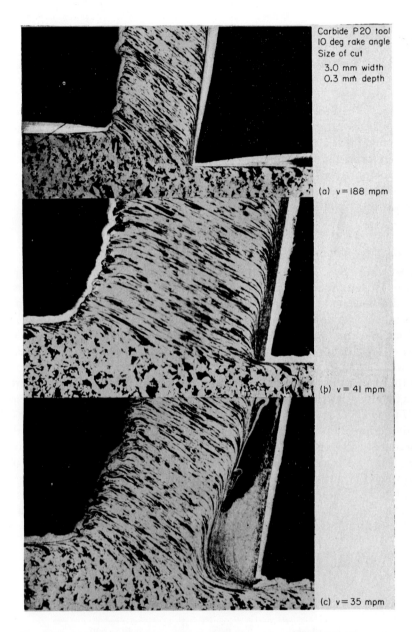

FIG. 2. Typical photographs showing the effect of cutting speed on built-up edge.

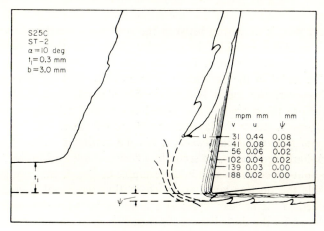

FIG. 3. Effect of cutting speed on the thickness u and the over-cut depth ψ of the built-up edge, depicted from the quick-stopped photographs.

FI . 4. Magnified view of the built-up edge formed on a 10 deg rake angle carbide P20 tool at a cutting of 10 mpm speed and 0.3 mm depth. Numbers in the left-hand photograph indicate the reading of Micro-Vickers hardness test at room temperature.

In a magnified view of the typically formed built-up edge (Fig. 4), there appears no clear boundary between the built-up edge and the removed chip. The flow of structure indicates that the secondary flow is occurring between the stationary metal and the removed chip. Since the metal is work-hardened by the secondary flow, a portion of metal joins with the built-up edge as a new addition to it. Thus, the built-up edge constantly tends to grow in size as long as the metal in the secondary flow zone is work-hardenable. At a higher cutting speed, however, the metal is less work-hardenable due to the higher cutting temperature; therefore, the secondary flow zone loses its hardenability as soon as a thinner built-up layer is formed. Hence the layer does not grow further.

3.1.2. *Growth and fracture of built-up edge.* When a cutting is started, a layer of initially cut metal is work-hardened due to the secondary flow and adhered to the tool face to form a nucleus of the built-up edge. As seen in the example of Fig. 5, the continuous secondary flow occurs outside the nuclear layer, so mnch so that the metal is continuously work-hardened and added to the neclear layer; as a result, a built-up edge is brought up to its typical size, as shown by the dotted curve *DEG* in Fig. 5, in an extremely short time afer the start of cutting.

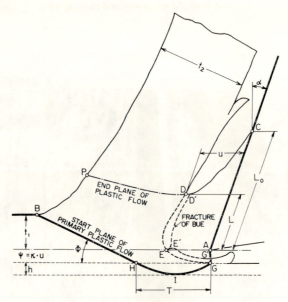

FIG. 5. Proposed model of metal cutting with the built-up edge.

In its formation process, the built-up edge grows not only in the cutting direction but also toward the generated surface. The growth in the latter direction results in metal removal in excess of the given depth of cut. The over-cut depth is empirically proportional to the thickness u of the built-up edge, so that approximately it holds that

$$\psi = k \cdot u,$$

where k is a constant whose value lies around 0.283.

At a low cutting speed, the metal in the secondary flow zone is always work-hardenable even after the built-up edge attains its typical size. Thus the built-up edge continually grows in width and depth.

When it grows, however, to such a size that the shearing stress along an inner boundary such as $D'E'G'$ (Fig. 5) exceeds a certain limit, fracture occurs along the inner boundary and the upper part of the fractured layer $DEE'D'$ slides up with the chip while the lower part $EGG'E'$ slides down to remain over the generated surface. The built-up edge continues another growth after the fracture so that the growth and fracture alternates periodically, maintaining an average size of the built-up edge representative of the given set of cutting conditions.

3.1.3. *Adhesion of built-up edge nucleus.* Past knowledge explains that the built-up edge adheres to the tool face by a kind of welding process resulting from the high stress and temperature of cutting. In addition to the theory above described, the present study suggests that the mechanical inlaying of the deformed metal into the asperities of the tool surface, as depicted in Fig. 6, is one of the primary causes of the adhesion. This finding is based on the following observations.

First, the built-up edge occurs at an extremely slow cut of small depth. Figure 7 shows a built-up edge formed in a 0.1 mm deep cut at 28 mm per minute speed. Cutting temperature measured even at a higher cutting speed (1081 mm per min; work metal, brass) indicated that the maximum temperature was only 105°C registered near the tool–chip interface as seen in Fig. 8. According to the conventional welding theory, adhesion is not likely to occur at such a low cutting temperature.

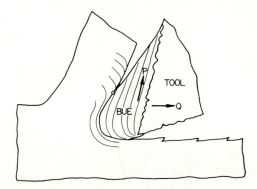

FIG. 6. Inlaying of the built-up edge onto the tool surface.

Second, all of the quick-stopped partly formed chips are left with the built-up edge, while the built-up edge is always left on the tool face when the cutting is finished in the conventional manner instead of quick-stopping. This indicates that the built-up edge is tied to the tool face strongly against the force acting parallel to the tool face as the force P in Fig. 6, but weakly against the horizontal force Q which exerts itself at the instant of quick-stopping.

3.2. *Metal-cutting Model with Built-up Edge*

Although the hardness of the built-up edge is measured at room temperature about three times higher than that of the original work metal, this must not lead to an interpretation that the built-up edge is acting as a substitute for the cutting tool on which the chip slides. Instead, the chip and the built-up edge are actually a continuous body and are separated by the shear of the secondary plastic flow.

A model as depicted in Fig. 5 is proposed to explain the chip formation with a built-up edge. Tool–chip contact length is L_0 at the start of a cutting; though, as the built-up edge

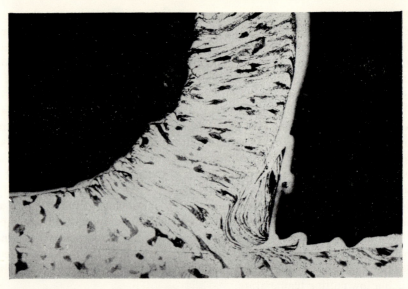

FIG. 7. Example of the built-up edge formed at an extremely low cutting speed. 0.15C steel was being cut at 28 mm per min speed and 0.1 mm depth by a 8 deg rake angle high-speed steel tool (10% Co type).

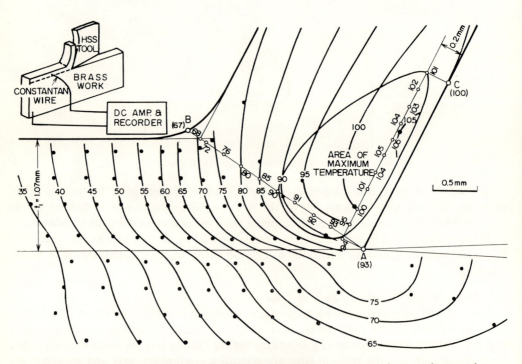

FIG. 8. Cutting temperature measured by the embedded constantan wire at a low-speed cutting. Numbers indicate temperatures in deg C. Work material: 60% Cu brass, Pb under 0.3%, T.S. above 36 kg/mm², Tool material: 10% Co type HSS, Room temperature: 18.3 °C, Tool rake angle $a = 25$ deg, Cutting speed $v = 1081$ mm per min, Depth of cut $t = 1.07$ mm. 0.17 mm diameter constantan wire put in 0.2 mm diameter hole.

grows, the contact length is reduced to L and an over-cut of a depth ψ occurs. The boundary where the original work metal first undergoes plastic flow is indicated by the curve $BHIG$. The plastic flow in the region of $GHIG$ produces a deformed layer of a depth h left over from the generated surface. The largest portion of the primary plastic flow occurs right after the metal passes the boundary curve $BHIG$; then the rest of plastic flow slowly continues until the metal reaches an end boundary curve somewhat like the curve PD. Although no clear-cut boundary of built-up edge exists, it may be well surmised that the stationary pile of work-hardened metal lies between the dotted curve passing through the point E and the tool face, where the point E is the intersection of a line of cutting direction passing through the tool tip point A and a line parallel to the tool face passing through the point D where the chip escapes from the contact with the built-up edge.

It is commonly known that the presence of a built-up edge on a tool tip reduces the cutting force. The reason for this was formerly sought based on the assumption that the face slope of the built-up edge served as a tool face with an increased rake angle so that the chip was formed in a shear plane with an increased shear angle. However, the chip formations observed in the present study did not agree with the above theory. First of all, the chip departs from the built-up edge at the point D (Fig. 5) and does not slide over the face slope DC. Real contact length L between the chip and the built-up edge is only a small portion of the surface of the built-up edge. Second, the built-up edge does not have a sharp edge but has a rounded tip, so that it is impossible to define a simple shear plane in which most of the chip formation takes place.

When the metal cutting involves the built-up edge, the chip formation should be discussed on a more complex model than the conventional shear plane concept. The following section will present results of an experimental study on the geometries of the metal-cutting model as proposed in Fig. 5 under variations in cutting parameters.

3.3. Metal-cutting Model Geometries under Various Cutting Parameters

3.3.1. *Test procedure.* A total of 44 quick-stop cutting tests were carefully planned to

TABLE 1. RESPONSE OF THE METAL-CUTTING MODEL GEOMETRIES TO THE CHANGE IN CUTTING CONDITIONS

Item	When built-up edge grows in a continuous cut so that the thickness u is increased	When cutting speed v is increased	When tool rake angle a is increased
Thickness of built-up edge u is		decreased	decreased
Contact length L is	increased	increased	decreased
Chip thickness t_2 is	unaffected	unaffected	decreased
Inclination of start plane of primary flow Φ is	unaffected	unaffected	increased
Distance T between G and H is	increased	decreased	decreased
Over-cut depth ψ is	unaffected	decreased	decreased
Thickness of distorted layer over finished surface h is	increased	decreased	decreased

investigate the effect of the size fluctuation of the built-up edge, the cutting speed, and the tool rake angle on the geometries of the metal-cutting model. On every micro-photograph of the partly formed chip, characteristic geometries such as the thickness of the built-up edge u, contact length L, etc., were readily obtained from the pattern of the metal flow. Statistical tests were applied on those data to identify the response of those geometries to the investigated effects, and the conclusions were drawn as summarized in Table 1.

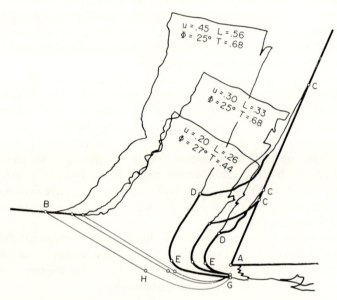

FIG. 9. Fluctuation of built-up edge in a continuous cutting. Depth of cut $t_1 = 0.3$ mm, cutting speed $v = 17$ mm. 5% Co type high-speed steel tool.

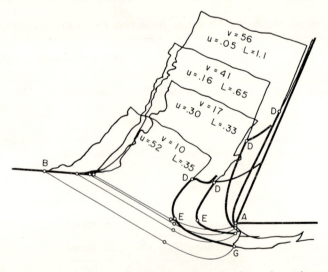

FIG. 10. Variation of built-up edge and cutting geometries by change in cutting speed v mpm. Depth of cut $t_1 = 0.3$ mm. u indicates thickness of built-up edge, L contact length, both in mm. 5% Co type high-speed steel tool.

3.3.2. *Fluctuation of built-up edge in a continuous cutting.* In cutting at an identical condition, growth and fracture of the built-up edge alternates so that its size fluctuates continually. The test data shows that a thicker built-up edge (greater *u*) is accompanied by a longer contact with the chip (greater *L*), as illustrated by three selected test results in Fig. 9.

3.3.3. *Effect of cutting speed on built-up edge and cutting geometries.* Increased cutting speed brings about a thinner built-up edge (less *u*), as is commonly known, but a longer contact with the chip (greater *L*), as illustrated by four test results in Fig. 10. It is not apparent that the chip thickness t_2 is effected from those variations.

3.3.4. *Effect of tool rake angle on built-up edge and cutting geometries.* Cutting by a tool with a greater rake angle results in a smaller built-up edge (less *u*), a shorter contact (less *L*), and a thinner chip (less t_2) as illustrated in Fig. 11.

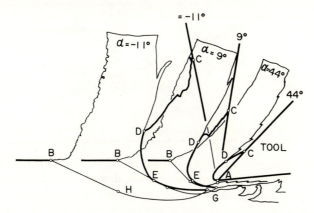

FIG. 11. Variation of built-up edge and cutting geometries by change in tool rake angle α. Depth of cut $t_1 = 0.3$ mm, cutting speed $v = 17$ mpm, 5% Co type high-speed steel tool.

4. ANALYTICAL SIMULATION OF THE METAL-CUTTING MODEL WITH BUILT-UP EDGE

4.1. *Purpose of Simulation*

In order to simulate the principle which ruled the variations of the built-up edge and the metal-cutting model geometries as observed in the preceding section, the equilibrium of cutting force was computed for the 44 test conditions, such that the normal to shear stress ratios were obtained on the start boundary of primary flow and on the built-up edge boundary, based on the observed metal-cutting geometries.

4.2. *Method of Simulation*

(Assumptions on the start boundary of primary flow.) As proposed in the metal-cutting model, the work metal started a primary plastic flow when it crossed the boundary surface *BHIG* (Fig. 12). A simplified assumption was made in which the shear strain and therefore the maximum shear stress on the start boundary was acting in the direction along the boundary itself. As illustrated in Fig. 12, the magnitude of the maximum shear stress was denoted by $-\tau_s$ and it was uniformly distributed. The normal stress to the start boundary was assumed constant and was denoted by $-\sigma_s$ (compression) along the straight part of the start boundary *BH*, while it varied linearly along the curved part *HIG* in such a way that a tensile stress of $+\sigma_s$ was attained at the end point *G*.

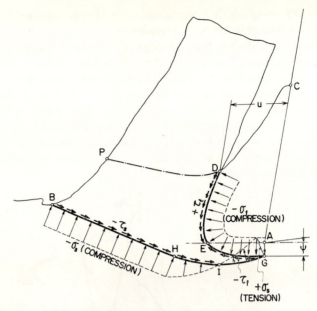

Fig. 12. Simplified stress distribution assumed on the start plane of primary flow *BHIG* and the built-up edge boundary *DEG*.

(Assumptions on the built-up edge boundary.) Along the built-up edge boundary, a uniform shear stress, of which the magnitude was denoted by τ_f, was assumed since the stationary built-up edge adjoined the deforming secondary flow region over this boundary. Opposite to the shear on the upper part of the built-up edge boundary *DE*, the lower part *EG* slid over the underlying metal, so that the sign of the shear stress was negative on the latter part. Normal stress to the built-up edge boundary was simply assumed to be uniform and was designated by $-\sigma_f$. Among the stress ratios: $S_1 = \sigma_f/\tau_f$, $S_2 = \sigma_s/\tau_f$, and $S_3 = \tau_s/\tau_f$, defined from the above stress assumptions, an equation $(S_1 + S_2)^2 + S_3{}^2 = 1$ should hold according to the Mohr's stress circle at the point *G*.

(Equilibrium equation of the force.) With the curved portions of the two boundary surfaces further approximated by straight planes as shown in Fig. 12, (1) the magnitude of the resultant cutting force, (2) the direction of the resultant cutting force, and (3) the resultant momentum of the cutting force were analytically derived on both of the two boundaries and were equated to each other according to the equilibrium theory.

(Computation of the force equilibrium.) The derived equation contained in its variable the cutting geometries ψ, u, L, Φ and T, the cutting parameters a and t_1 and the stress ratios S_1, S_2 and S_3. The first seven variables were given their values from experimental data, and the unknown stress ratios were determined by a digital computer in such a way that the equilibrium equation was best satisfied.

4.3. Result of Simulation

For various cutting speeds and tool rake angles, the simulation presented the stress ratio $S_1 = \sigma_f/\tau_f$: the normal to shear stress ratio which most probably existed on the built-up edge boundary, and that on the start boundary of the primary flow. Also the magnitude of the resultant cutting force, its direction, and the resultant momentum of the cutting force were predicted.

The contact length between the built-up edge and the chip, in particular, was found to correlate well with the resultant force magnitude, as shown in Fig. 13, irrespective of the cutting speed and the tool rake angle. This prediction agrees with the well-known fact that the cutting force reduces with the built-up edge, so that this fact is possibly attributed to the reduced contact length, instead of the increased effective rake angle as was formerly conceived.

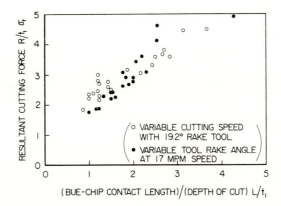

Fig. 13. Predicted correlation of built-up edge–chip contact length L to resultant cutting force R. (Result of computer simulation.)

5. SUMMARY

An experimental study was undertaken in order to study the formation and the cutting mechanism of the built-up edge.

Investigations of the quick-stopped partly formed chip produced at various cutting parameters clearly demonstrated that the secondary plastic flow in chip formation was the direct source of the work-hardened metal which forms the built-up layer or the built-up edge. Adhesion of the built-up edge nucleus onto the tool face was discussed from the mechanical inlaying principle.

Growth and fracture of the typical built-up edge and the metal-cutting model were described. Variation of the built-up edge and the metal-cutting model geometries was explored by a series of quick-stop tests, and the test results were processed by a computer simulation of the cutting force equilibrium, which results indicated that the variation of the cutting force was mainly controlled by the contact length between the built-up edge and the chip.

REFERENCES

1. W. ROSENHAIN and A. C. STURNEY, Flow and Rupture of Metals During Cutting, *Proc. I.M.E.*, Part 1, 1925.
2. DIGGS and HERBERT, *Proc. I.M.E.*, 1928.
3. H. ERNST and M. MARTELOTTI, Die Bildung und Wirkung der "Aufbauschneide", *Werkzeugmaschine*, Heft 18, 1936.
4. V. D. PRIANISHNIKOFF, *Am. Machinist*, **180**, 1936.
5. RICKICHI MURANAKA, Research Paper of Toyama University.
6. M. E. MERCHANT, *Tool Engineers Handbook*, A.S.T.E., 1949.
7. KOICHI HOSHI, On the Built-up Edge and Counter-plot for it, *J.S.M.E.*, Abst., 1937.
8. KOICHI HOSHI, On the Characteristics in the Cutting Action of Steel and Cast Iron, *J.S.M.E.*, 1938.

SONIC TESTING OF GRINDING WHEELS

J. Peters, R. Snoeys and A. Decneut

Department of Mechanical Engineering, University of Louvain
Centre de Recherches Scientifiques et Techniques de l'Industrie des Fabrications
Métalliques (C.R.I.F.)

SUMMARY

The "grade or hardness" of a grinding wheel does not have a real physical definition. Grinding wheel hardness has a different meaning for the manufacturer and for the user. With this statement in mind, the modulus of elasticity of the grinding wheel material has been considered a physically defined grinding wheel criterion. The paper gives a detailed description of a precise and fast measuring method for the practical determination of the E-modulus by a sonic test.

It is shown that the E-modulus meets the manufacturers' as well as the users' needs. The "grade chart" relates the wheel composition with the empirically measured "hardness" and with the E-modulus value.

Finally the E-modulus is related to the practical grinding work.

NOMENCLATURE

a_{ns}	constant depending upon the mode shape and edge conditions, with n = number of nodal diameters, s = number of nodal circles
b	width of testbar, normal to direction of flexion, m
C_1, C_2	coefficients mainly dependent on mode of vibration
C_{Ddh}, C_{lbh}	coefficients mainly dependent on dimensions, m^{-1}
d	internal diameter, m
D	external diameter, m
E	modulus of elasticity, Nm^{-2}
F_T	tangential force, N
F'_T	tangential force per 1 mm width of cut, Nmm^{-1}
f	natural frequency, Hz
f_1	natural frequency of a perforated disc, Hz
G	shear modulus, Nm^{-2}
h	thickness of the circular plate or depth of testbar in direction of flexion, m
H	scratching force, N
I	moment of inertia of a cross section of the bar, m^4
l	length of testbar, m
m	mass, kg
N	$Eh^3/12(1 - \nu^2)$ bending stiffness coefficient (or modulus of flexural rigidity)
R	radius, m
Ra	surface roughness (C.L.A.) μm
T	period, sec
t_B	sand blast depth, mm
V	total apparent volume, m^3
$V_B(v_B)$	volume of bond material, $m^3(\%)$
$V_K(v_K)$	volume of grains, m^3 ($\%$)

1113

$V_P(v_P)$	volume of pores, $m^3 (\%)$
V_W	metal removal, mm^3
V'_W	metal removal per 1 mm width of cut, $mm^3\ mm^{-1}$
Z'	metal removal rate per 1 mm width of cut, $mm^3\ mm^{-1}\ sec^{-1}$
ϵ	relative error
κ	frequency ratio
λ	correction coefficient
μ	mass density, $kg\ m^{-3}$
ν	Poisson's ratio
ϕ_S	wear ratio, $\%$

1. INTRODUCTION

A grinding wheel is a complex composition of abrasive grains and bonding material. The general performance of the grinding wheel depends on the nature of these grains, on the characteristics of the bonding material and on the concentration of grain and bonding material in the apparent volume of the grinding wheel. The practical performance of a grinding wheel is also a function of the dressing conditions which determine the actual state of the cutting surface, the grinding wheel speed, the grinding fluid, etc. These parameters are also very important, but they do not depend directly on the grinding wheel material as such and their influence will not be discussed here.

2. THE GRADE OF A GRINDING WHEEL

One of the standard grinding wheel characteristics which has been discussed to a great extent during the last decennium is the grade of the grinding wheel.[1-7, 17]

The basic reason for misunderstanding in this field is the lack of a valid definition of the grade or "hardness" of the grinding wheel. The most scientific sounding definition refers to the force required to pull a grain out of the bond. However, a practical measurement of this grade criterion seems to be rather difficult. Although Peklenik[6] developed an interesting device performing a measurement which approximates the mentioned definition almost completely, Pahlitzsch[1] and Colwell[2, 3] enumerated more than twenty other methods developed in order to approximate this "hardness" criterion. Apart from the scientific definition of the hardness there are the points of view of both manufacturers and users which are briefly discussed in the following paragraphs.

2.1. The hardness indication of the manufacturer

The grade indication (K, L, M, etc.) of the manufacturer relates to the composition of the wheel. It is assumed that in so far as the ingredients are unchanged and the manufacturing process is under control, a wheel will have well-defined characteristics of structure, porosity and grade. A variation of the composition will influence these characteristics and the grade scale is then related to the composition of the grinding wheel.

In order to represent the wheel composition, the ternary diagram has been used (Fig. 1). The volumetric composition of a grinding wheel is given by three volumes, i.e. the volume of grains (V_K), the volume of bond material (V_B) and the volume of the pores (V_P). The sum of the three volumes equals the total apparent volume (V):

$$V_K + V_B + V_P = V \tag{1}$$

Changing the volumes in eq. (1) to percentages of the total apparent volume:

$$\frac{V_K}{V} \times 100 = v_K$$

$$\frac{V_B}{V} \times 100 = v_B$$

$$\frac{V_P}{V} \times 100 = v_P$$

gives the following basic equation:

$$v_K + v_B + v_P = 100\% \tag{2}$$

This equation expresses that the sum of the percentage volume of grains (v_K), the percentage volume of bond (v_B) and the percentage volume of pores (v_P) equals 100%.

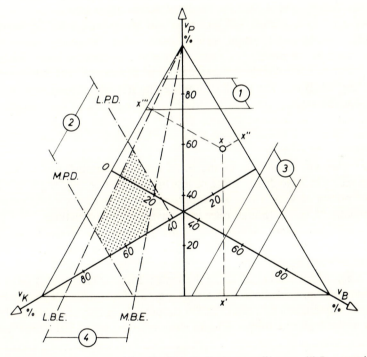

FIG. 1. Volumetric wheel composition. Iso lines in the ternary diagram: (1) Iso porosity lines. (2) Iso grain volume lines (Iso structure lines). (3) Iso bond volume lines. (4) Lines of constant ratio of grain volume and bond volume.

These three quantities can be represented on the axis of an equilateral triangle. Each corner represents 100% of the considered quantity and zero of the two others. Any point (see Fig. 1, point X) in the triangle represents a definite wheel composition since the sum of the distances from any point to the sides of an equilateral triangle equals the height of the triangle which is said to be 100%.

In such a diagram some interesting geometric loci can be found. The lines which are perpendicular to one of the three axes are iso lines:

Iso porosity lines (1)†

Iso grain volume lines (2)

Iso bond volume lines (3)

Iso grain volume lines have the same percentage of grain volume or the same "packing". Most grinding wheel manufacturers consider these lines as "Iso structure lines". The realizable structures however do have an upper and lower limit. The maximum packing density (M.P.D.) which can be obtained by shaking and by the application of an adequate pressure level. Higher packing densities will always be accompanied by breaking down a certain quantity of grains. The lower packing density (L.P.D.) limit is imposed by the need for mutual contact of the grains during manufacturing in order to avoid untolerable deformations of the grinding wheel shape during firing. Since the maximum and the lower packing density depend on the grain size, different values for the M.P.D. and the L.P.D. line are possible. The lower packing density limit can be influenced by means of the addition of some porosifying additive, disappearing in the course of firing. The maximum packing density can also be shifted by mixing different grain sizes.

Some other interesting geometrical loci that can be indicated in the ternary diagram are lines corresponding to a constant *ratio of grain volume to bond volume* (4). These are lines passing through the point P of 100% porosity. The extreme values for this ratio are indicated in Fig. 1 by means of the "maximum bond equivalent" (M.B.E.) and the "lower bond equivalent" (L.B.E.). The L.B.E. is imposed by strength requirements of the grinding wheel body. The maximum value, the M.B.E. limit, is imposed by practical manufacturing experience. These two limits together with the L.P.D. and the M.P.D. lines determine the practical or useful area in the ternary diagram which corresponds to approximately 10% of the total surface of the triangle.

The ternary diagram forms one of the bases for the manufacturer's grade indication. Each of them seems to have drawn more or less logical lines of "equal grade" or "iso hardness".[5, 8] All points of such a line are supposed to represent the composition of wheels considered equally "hard" although they may have different structures.

Ljubomudrov[9] and some manufacturers use iso porosity lines (1) as lines of iso hardness. Manufacturers using this kind of grade indication assume the hardness to be constant with unchanging porosity.

Not all grinding wheel makers are using the same iso hardness lines as illustrated in Fig. 2. (Fig. 2 is a detail of Fig. 1, identical indications are used for similar iso lines in both figures.) Here the lines are inclined with respect to the iso porosity lines; in addition, the lines show break-points in the region corresponding to close grinding wheel structures (large value of v_K).

From the above it is clear that the classic alphabetic grade indication is a conventional one, depending on each manufacturer. Further investigations will show that the described grade significance does not cover the given definition of the grinding wheel "hardness", i.e. the force required to pull out grains.

2.2. *Hardness from the point of view of the user*

The user simply wants an indication enabling him to make a proper choice of a grinding wheel for an optimum grinding process.

† The numbers refer to the iso line indications in Fig. 1.

The performance of a grinding wheel, however, is not only a complicated function of all *grinding wheel parameters* such as grain nature and grain dimensions, nature of bond material, volumetric composition of the wheel, etc., but, as said before, depends on the *grinding conditions also*. The workpiece material, the geometric relationship of the contact area, the grinding wheel speed, the dressing conditions, etc., affect the grinding wheel performance. This means that the wheel characteristic the user wants covers more than what one physical property of the grinding wheel material can provide.

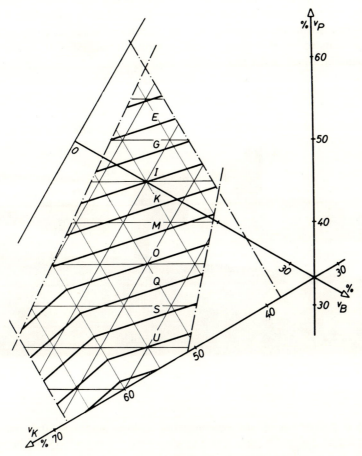

Fig. 2. Iso grade lines. Manufacturer's grade indication in the useful area of the ternary diagram. The three axes correspond to those of Fig. 1.

Since there are a great number of parameters determining the practical grinding performance, it is essential that those parameters be at least well defined. Presently, however, most of the grinding wheel characteristics do not have a true physical definition. This situation leads to difficulties in comparing practical grinding results as it creates a serious risk of inter-mixing different influences. Quite often the user does have to conclude that after he has found the "good" wheel, it is rather difficult to buy exactly the same ones.

With this statement in mind, the modulus of elasticity of the grinding wheel material has been considered a physically well defined grinding wheel criterion. The Young's modulus of elasticity is a characteristic which is connected with the rigidity of the grinding wheel material. As will be shown later, the *E*-modulus meets the manufacturer's as well as the user's needs and it gives a reliable indication on the behaviour of the wheel at work. Obviously the first question remains how to measure the *E*-modulus.

3. THE SONIC TESTING

3.1. *Test equipment*

The measurement of the *E*-modulus is easy and *reliable with a sonic testing device*.[10] Figure 3 shows the test equipment GRINDO-SONIC. Figure 4 gives a simplified block diagram of the electronic instrument. Actuating the "TEST" button causes all counters to be cleared and control circuits to be set in their initial positions. Immediately thereafter the hammer strikes the grinding wheel under test.

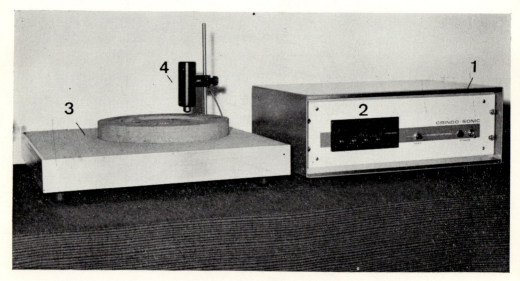

FIG. 3. GRINDO-SONIC test equipment. 1. Electronic instrument. 2. Display. 3. Table with grinding wheel. 4. Electromagnetic shock exciter.

The signal is picked up by the piezo-electric sensor, amplified and correctly shaped before being fed to the main gate which is closed at the beginning of the measurement cycle.

At the same time the amplified signal is also subjected to detection and integration resulting in an "envelope" waveshape. This signal is used to control the hysteresis circuit which imposes the following conditions on the measurement:

(a) The signal level must exceed a minimum amplitude before it can be taken into consideration.

(b) Once this amplitude has been reached the moment of sampling will be delayed to allow the interfering overtones to die out insuring, however, that the amplitude is still well above the instrument sensitivity threshold.

At this moment the main gate is opened, allowing eight cycles of the input signal to trip the clock gate. A pulse train from the crystal clock is counted and the numerical result is displayed on the front panel. It will be shown that this number, actually $2T$ in microseconds, constitutes a measure of the relative hardness of the grinding wheel. The interlock controls protect the reading from being disturbed before another cycle is initiated.

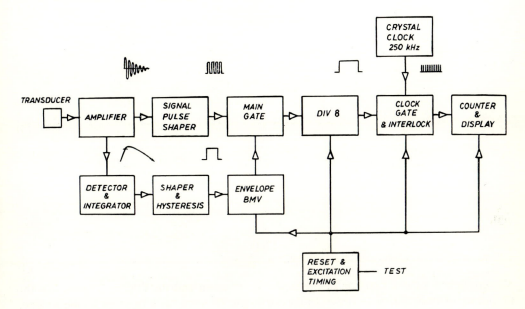

FIG. 4. Simplified block diagram. Main sub-units of the GRINDO-SONIC test equipment.

Special attention has been paid to the method of excitation and to the detection of the signal in view of designing a dependable instrument. Experience has shown that the smaller wheels up to 50 cm (20 in) diameter are conveniently measured on the table.

The main advantages of the GRINDO-SONIC equipment are:

—The extreme rapidity of the measurement, resulting in numerical readings with very little spread, hardly ever reaching 1%.

—The ability to measure both resinoid and vitrified grinding wheels as well as test bars.

—No critical attention either to overtones or to the damping introduced by the supports.

3.2. *Relation between the natural frequency and the modulus of elasticity*

The natural frequency of the mechanical vibration of a body is determined by the shape factor and several physical constants. The shape factor includes both the geometry and the dimensions of the body. The physical constants involved are the modulus of elasticity, the density and Poisson's ratio.[11, 12, 13] Two examples have been considered, first the disc and afterwards the bar.

3.2.1. *The disc.* For the circular plate, the general form of the relation between the natural frequency and the modulus of elasticity is:

$$f = \frac{a_{ns}}{2\pi R^2} \sqrt{\left(\frac{N}{\mu h}\right)} \tag{3}$$

where $a_{ns} = a_{2.0} = 5.25$ for the two-nodal diameter mode of a free plate.

Substituting

$$N = \frac{Eh^3}{12(1 - \nu^2)} \quad \text{and solving for } E$$

$$E = \frac{3\pi^2(1 - \nu^2)\mu D^4 f^2}{a_{ns}^2 \cdot h^2} \tag{4}$$

Substituting

$$\mu = \frac{4m}{\pi D^2 h} \quad \text{in eq. (4)}$$

$$E = \frac{12\pi(1 - \nu^2)m D^2 f^2}{a_{ns}^2 \cdot h^3} \tag{5}$$

or

$$E = C_1 \frac{m D^2 f^2}{h^3} \tag{6}$$

with

$$C_1 = \frac{12\pi(1 - \nu^2)}{a_{ns}^2} \tag{7}$$

The coefficient C_1 is a function of the mode of vibration and the Poisson's ratio. For the first mode of vibration ($n = 2$, $s = 0$) the coefficient C_1 varies as a function of the ν value.

ν =	0.0	0.1	0.15	0.20	0.25	0.30
C_1 =	1.362	1.3536	1.3365	1.3126	1.2818	1.2442

Poisson's ratio has been measured for both ceramic and resinoid test bars, using the well known formula giving Poisson's ratio as a function of Young's modulus E and the shear modulus G.

$$\nu = \frac{E}{2G} - 1 \tag{8}$$

E has been measured by a bending test and G has been measured by a torsion test. Poisson's ratio ν was ranging from 0.039 to 0.053. The corresponding values of C_1 differs only 0.2%. A good value of C_1 for both ceramic and resinoid materials is 1.36.

A special case is the disc with an axial hole, such as an abrasive wheel. If f is the natural frequency of the disc without hole and f_1 the natural frequency of the perforated disc, one will have

$$\kappa = \frac{f}{f_1} > 1 \tag{9}$$

This means that the natural frequency decreases when the disc is perforated. The ratio κ for the two-nodal diameter mode is given by McMaster[12]

$$\kappa = \frac{1}{1 - (d/D)^2} \tag{10}$$

with d = internal diameter, D = external diameter.

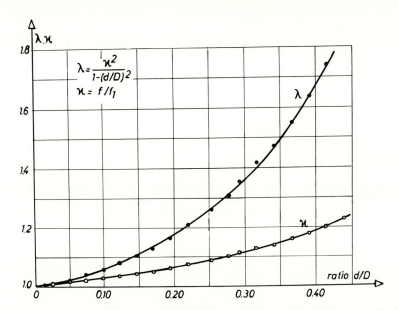

FIG. 5. Correction factor λ for wheels with a central hole. The experimental correction factor as a function of d/D.

κ has also been checked empirically;[4] the results being represented in Fig. 5. The maximum deviation of McMaster's equation is 1.5% for values of $d/D < 0.25$. For values of $d/D > 0.25$, however, McMaster's values lead to important errors. For perforated discs, the E-modulus will be found by substituting

$$\mu = \frac{m}{V} = \frac{4m}{\pi h D^2 (1 - d^2/D^2)} \tag{11}$$

and $f = \kappa f_1$ from eq. (9) in eq. (4)

$$E = C_1 \frac{\kappa^2}{1 - d^2/D^2} \frac{m D^2 f_1^2}{h^3} \tag{12}$$

defining

$$\frac{\kappa^2}{1 - d^2/D^2} = \lambda \tag{13}$$

Equation (12) leads to

$$E = \frac{C_1 \cdot \lambda \cdot m \cdot D^2 \cdot f_1}{h^3} \tag{14}$$

λ is a dimensionless correction factor depending essentially on the ratio d/D. The value of λ is plotted in Fig. 5.

A mathematical equation for λ can be found using McMaster's formula (eq. 6)

$$\lambda = \left[\frac{1}{1 - (d/D)^2} \right]^3 \tag{15}$$

If $d/D < 0.25$ the error will be less than 3%.

In order to compare the modulus of elasticity of a number of similar wheels, eq. (14) has been rewritten as

$$E = C_{D\hat{d}h} \cdot mf_1{}^2 \tag{16}$$

with

$$C_{D\hat{d}h} = C_1 \, \lambda \, D^2/h^3. \tag{17}$$

The coefficient $C_{D\hat{d}h}$ is constant for a given value of D, d and h. Equation (16) shows that for a set of grinding wheels with comparable dimensions the modulus of elasticity depends only on the natural frequency and the mass.

3.2.2. *Rectangular and square bars*

The equation giving the relation between E-modulus and natural frequency for bars is:

$$E = \frac{4\pi^2}{a_n{}^2} \frac{ml^3f^2}{I} \tag{18}$$

defining $C_2 = 4\pi^2/a_n{}^2$, eq. (18) leads to

$$E = C_2 \frac{ml^3 f^2}{I} \tag{19}$$

For the fundamental flexural mode of vibration: $a_n = 22$ or $C_2 = 0 \cdot 08156$ eq. (19) can be rewritten as

$$E = C_{lbh} \cdot mf^2 \tag{20}$$

with

$$C_{lbh} = \frac{C_2 \, l^3 \times 12}{bh^3} \tag{21}$$

This is an equation of the same form as formula (16).

The E-modulus given by eq. (20) and based on the natural frequency measurements (E_{Dyn}) has been compared with E-modulus values resulting from bending tests (E_{Stat}). The results obtained with twelve different test bars are given in Fig. 6. Within the range of experimental errors, similar values of the E-modulus have been found for the sonic and bending test.

The mean difference between E_{Dyn} and E_{Stat} was 2.2%. This spread takes into account the inaccuracy of the dimensional measurements and all other errors.

4. EVALUATION OF THE SONIC TESTING

4.1. *E-modulus and standard grade indication*

A series of grinding wheels with classical grade indications ranging from G to R were subjected to the sonic testing using a standard GRINDO-SONIC instrument. Figure 7 gives the natural frequency and the E-modulus versus the manufacturer's indication. Except for the grinding wheels with letter indication I and K, the manufacturer's indication are consistent with the E-modulus values. The deviations have been checked by comparing the test results obtained with those resulting from another measuring method due to Prof. Peklenik.[6]

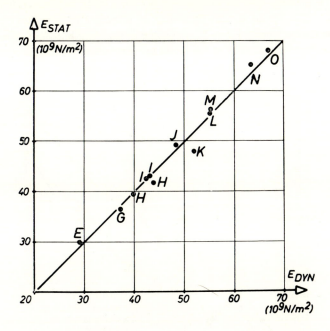

FIG. 6. Results of bending test. No systematic difference exists between the results of the two test series.

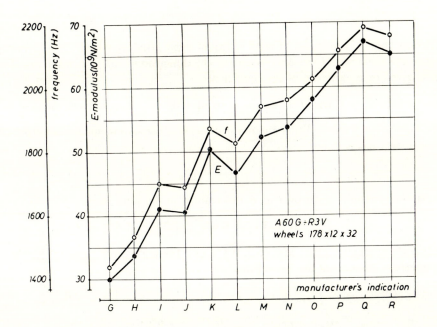

FIG. 7. Natural frequency and modulus of elasticity, plotted versus the classical "grade" indication, do have the same trend for a given shape of grinding wheel.

Basically this is a measurement of the average force required to scratch a groove into the wheel surface. The results are given in Fig. 8. Clearly the same deviations have been found in the case of the wheels I and K. The time consuming hardness determinations can thus conveniently be replaced by a quick sonic test procedure which furthermore is subject

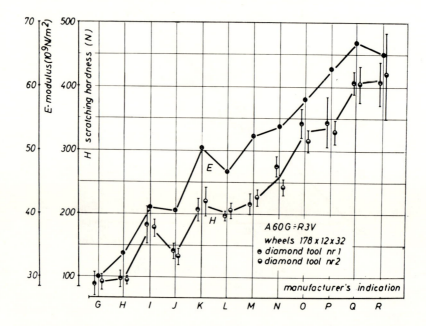

FIG. 8. Scratching hardness. The results of sonic test and scratching test are similar.

to much less spread. For series of tests on equal sized grinding wheels, it is important to investigate the influence of the distribution of the dimensions on the E-modulus. This influence is given by a variation of the coefficient:

C_{Ddh} in eq. (16) ($E = C_{Ddh} mf^2$)   or C_{lbh} in eq. (20) ($E = C_{lbh} mf^2$) in the case of bars.

By estimating the possible dispersion of the dimension measurements the resulting relative uncertainty on C_{Ddh} is

$$\epsilon_{C_{Ddh}} = \epsilon_\lambda + 2 \epsilon_D + 3\epsilon_h = 1\% + 0.2\% + 1.8\% = 3\%. \tag{22}$$

The maximum possible error on C_{lbh} is determined in the same way:

$$\epsilon_{C_{lbh}} = \epsilon_b + 3\epsilon_l + 3 \epsilon_h = 0{\cdot}4\% + 1.2\% + 1.8\% = 3.4\%. \tag{23}$$

Estimating the frequency error le ss than 0·8 % and the mass error equal 0.1 %, the corresponding errors influencing the E-modulus are

(a) grinding wheels: $\epsilon_E = \epsilon_{C_{Ddh}} + \epsilon_m + 2\epsilon_f = 3\% + 0.1\% + 1.6\% = 4.7\%$ $\qquad$ (24)

(b) bars: $\epsilon_E = \epsilon_{C_{lbh}} + \epsilon_m + 2\epsilon_f = 3.4\% + 0.1\% + 1.6\% = 5.1\%$ $\qquad$ (25)

This corresponds to approximately 20 to 25 % of a class width.

Figure 9 gives a practical example. For a series of wheels and a series of bars with comparable dimensions, the natural frequency has been plotted versus the *E*-modulus. The distribution of the dimensions of the bars and of the grinding wheels is mentioned in the figure. If a series of wheels have similar dimensions it may be stated that a determination of the "grade" of the grinding wheels can simply be performed by the sonic test.

This conclusion is very useful for checking the "grade" of a large number of equal sized grinding wheels.

Figure 9 further shows that the same relation is obtained for abrasive materials composed of grains with different friability (grains A and WA). This proves that the sonic measurement is not influenced by the friability of the grains. The manufacturer's grade indication is also shown in Fig. 9, and illustrates that they are not always justified.

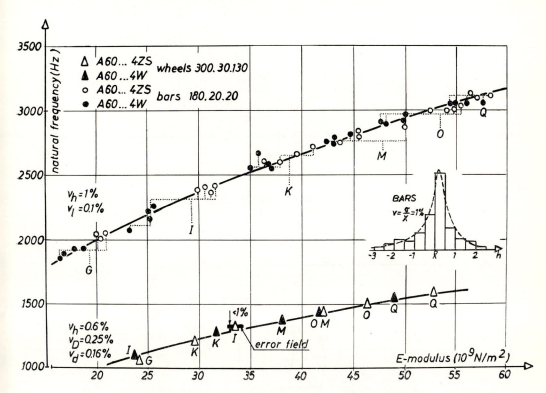

FIG. 9. A simple frequency measurement is sufficient for a grade classification in the case of series of wheels (or bars) with comparable dimensions.

4.2. Grade chart for grinding wheels

In order to check the manufacturer's letter indication three sets of empirically measured iso grade lines have been drawn in the ternary diagram. Conforming to the discussion in section 1, the classical iso grade lines have a somewhat different orientation in the ternary diagram depending on the manufacturer. It is interesting to compare them with empirically determined lines using different criteria, e.g. the scratching force measurement, the sand blast test (determination of the depth of a hole obtained by a jet of sand or Zeiss Manchensen test) and the sonic *E*-modulus measurement. For this purpose, investigations described in ref. 14

have been very useful. Daude measured the grade of 30 grinding wheels composed of the same grains (Al$_2$O$_3$ (99%) "Edelkorundweisz" $\neq$ 46: standard) identical bond material, all of which had been manufactured in the same production conditions. In Fig. 10 the results of the scratching test are plotted perpendicularly to the ternary diagram surface, in each of the volumetric location points of the grinding wheels. Figure 11 gives the results of the sand

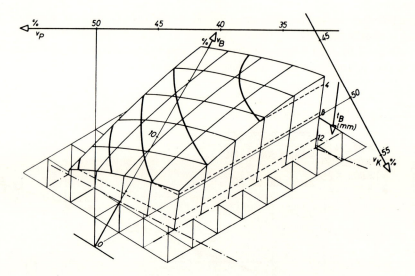

Fig. 10. Three-dimensional diagram showing the scratching force as a function of the wheel composition, which is given by the position in the ternary diagram.

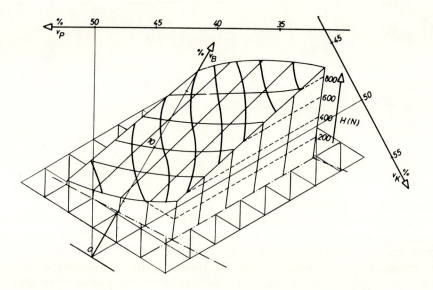

Fig. 11. Three-dimensional diagram showing the sand blast depth as a function of the wheel composition.

blast test. Into each of these diagrams horizontal planes parallel with the ternary diagram plane have been drawn. The construction leads to the evaluation of the iso grade lines which have been projected into the basic ternary diagrams. Figure 12 gives the superposition of each of the sets of iso grade lines derived from Figs. 10 and 11.

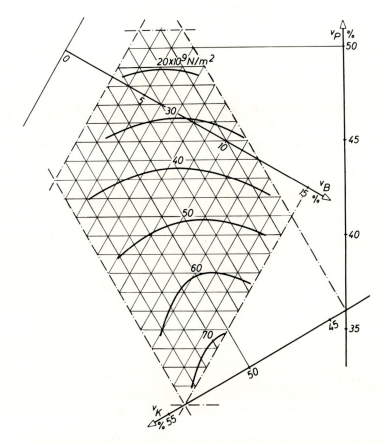

FIG. 12. The grade chart shows the trend of iso grade lines corresponding to two different criteria as a function of the wheel composition.

As a result, a somewhat different orientation of the iso grade lines in the ternary diagram has been noted. For "soft" and medium wheels, both criteria will lead to practically identical classifications of the grades. The iso hardness lines are approximately iso porosity lines. A better approximation, though, are lines with an inclination conforming to Fig. 2. For very hard wheels, the criteria may vary between iso porosity lines and lines with equal bond percentage. In this area the sand blast test is not so accurate. This conclusion is found by considering the mutual distance between the iso hardness lines.

Conforming to the former graphical representations relating the wheel composition with hardness measurements, the E-modulus values obtained by Daude and Snoeys can be plotted in the same fashion. Projecting the iso E-modulus lines in the ternary diagram leads to Fig. 13.

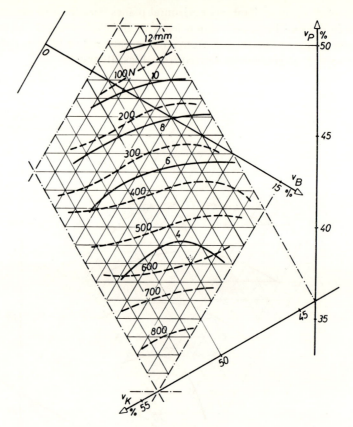

FIG. 13.

Considering the E-modulus as a fixed grinding wheel characteristic, being a material constant of the grinding wheel which is independent of the measurement method, it can be stated for example that a grinding wheel with $v_{K1} = 44\%$ and $v_{B1} = 11\%$ will have the same specifications as a wheel with $v_{K2} = 51\%$ and $v_{B2} = 6\%$, both grinding wheels having the same porosity ($v_P = 43\%$).

It should be of interest to compose such a set of E-modulus maps as shown in Fig. 13 for different grain dimensions, other types of grains and different qualities of bonding materials. Such information would give a more complete picture of the physical meaning of the grades of a grinding wheel. Figure 13 already indicates the trend towards a physically determined grade chart that relates the more classic definition of the grinding wheel hardness to the manufacturer's indication based upon a volumetric composition.

4 3. Practical grinding test results

Finally, some results of surface grinding[15] and cylindrical plunge grinding[16] are given in Figs. 14 and 15. The concerned grinding wheels did have the same structure (which means that they are situated in the ternary diagram on a line of constant v_K) and the same grain size was used. The E-modulus, however, varies over a very large range (10 classes).

Sonic Testing of Grinding Wheels

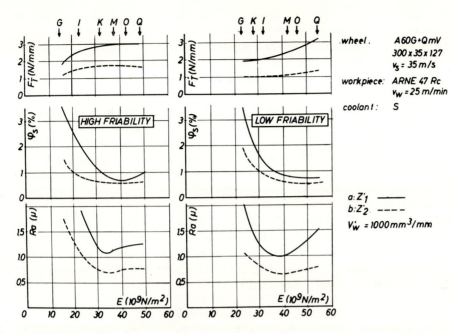

FIG. 14. Influence of modulus of elasticity and grain friability on tangential force (F_T), wear ratio (ϕ_s) and surface roughness (R_a).

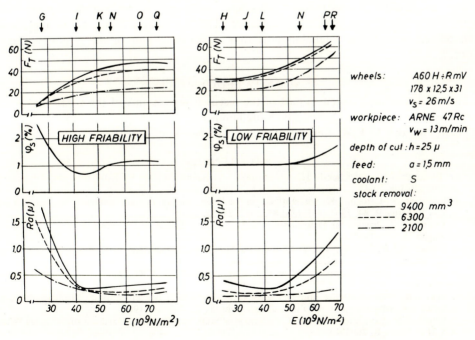

FIG. 15. Influence of modulus of elasticity and grain friability on tangential force per 1 mm width of cut (F'_T), wear ratio (ϕ_s) and surface roughness (R_a).

Similar tests have also been run using two different types of grinding wheels. The difference between the two series concerned the friability of the grains. For the series of grinding wheels with a high friability of the composing grains, the tangential force is increased by using "harder" grinding wheels. For increasing E-modulus the grain is held better by the bond, the tangential force increases and the wear ratio as well as the roughness decreases. For higher E-moduli the predominant factor, however, is the grain fracture. The tangential force tends toward a constant value even though the wheel grade is increased. The wear ratio and the roughness also follow the same trend. For this grade range and this kind of grain the wheel behavior is dominated by the grain friability and no longer by the bond characteristics. In the case of low friability, the tangential force is governed by the bond. It increases with increasing E-modulus. Instead of breaking the grains wear off and become dull. A tendency toward surface burning is observed.

It seems that the friability of the grain is as important as the "hardness" or better as the E-modulus of the grinding wheel. Both have an important influence on the grinding behaviour of the grinding wheel. A practical grinding test will always give the total influence and obviously the results of such a test are not suitable as a grade criterion of the grinding wheel material. A separate criterion for grade and friability is required in order to evaluate the relative importance of both characteristics.

5. CONCLUSIONS

From the various test series carried out the possibilities of the sonic testing of grinding wheels are brought to light.

The measuring method turned out to be very suitable for the determination of the grinding wheel hardness both for vitrified and resinoid wheels.

The "grade chart" shows that the E-modulus could replace the "standard grade indication" which has no real physical significance. The E-modulus, however, is a physically well-defined value giving a reliable indication of the behaviour of the grinding wheel at work.

6. ACKNOWLEDGEMENTS

This work was undertaken as part of the research programme of the CRIF and is published by courtesy of the Director R. Stinglhamber.

Thanks are due to J. W. Lemmens p.v.b.a. for their kind cooperation.

7. REFERENCES

1. G. Pahlitzsch, Vergleichende Harteprüfungen von Schleifkörpern Schleif-Polier und Oberflachentechnik, 20 June 1943.
2. L. Colwell, R. Lane and K. Soderburg, On the Determination of the Hardness of Grinding Wheels. *A.S.M.E.* Series B, **84**, 1962, 113–128; **85**, 1963, 27–33.
3. L. Colwell, A Process for Determining the Hardness or Grade of Grinding Wheels. University of Michigan, March 1962.
4. J. Peters and R. Snoeys, The E Modulus, a Suitable Characteristic of Grinding Wheels. Report CRIF-MC 9, University of Louvain, August 1965.
5. R. Snoeys, Recherche des propriétés des meules abrasives. Report CRIF-MC 3, University of Louvain, March 1964.
6. J. Peklenik, Neue Statische und Dynamische Prüfmethoden der Physicalisch-Mechanische Eigenschaften von Schleifkorpern. 18 Forschungsbericht, T. H. Aachen, 1960.
7. J. Peklenik, R. Lane and M. Shaw, Comparison of Static and Dynamic Hardness of Grinding Wheels. *A.S.M.E.* Series B, **86**, 1964.
8. R. Snoeys, Influence du volume de liant et de la structure sur la dureté des meules. Report CRIF-66R4, University of Louvain, August 1966.

9. V. LJUBOMUDROV, *Schleifwerkzeuge und ihre Herstellung*. Moscow, 1953.
10. A. DECNEUT, Classifying Grinding Wheels with the GRINDO-SONIC. Report CRIF-67R8, University of Louvain, Oct. 1967.
11. S. TIMOSHENKO, *Vibration Problems in Engineering*, 3rd ed. D. Van Nostrand, New York.
12. R. McMASTER, *Nondestructive Testing Handbook* II. Natural frequency vibration tests, p. (51.5). The Ronald Press Company, 1963.
13. W. FLÜGGE, D. YOUNG, *et al. Handbook of Engineering Mechanics*, p. (61.24). Mc-Graw Hill, 1962.
14. O. DAUDE, Untersuchung des Schleifprozesses. Ph.D. Thesis, T. H. Aachen, July 1966.
15. J. PETERS, R. SNOEYS and X. LOGÉ, Etude sur la rectification en plongée et les conditions de dressage. Report CRIF-MC 17, University of Louvain, Sept. 1966.
16. J. PETERS, A. DECNEUT and G. THUNUS, Recherche sur les conditions de travail optimales en rectification—Un critère d'utilisation économique. Report CRIF-MC 22, University of Louvain, Nov. 1967.
17. M. MOSER, Grundsätzliche und technische Fragen der Erzeugung von Schleifwerkzeugen aus keramisch gebundenen Korundkörnern. *Periodica Polytechnica*, **9**, No. 3–4, Dudapest.

ANALYSIS OF THE STATIC AND DYNAMIC STIFFNESSES OF THE GRINDING WHEEL SURFACE

Raymond Snoeys and I-Chih Wang

Department of Mechanical Engineering, University of Cincinnati, Cincinnati, Ohio 45221, U.S.A.

SUMMARY

The stability of the grinding operation is greatly influenced by the elastic deformation of the grinding wheel-workpiece contact area. It has been found that a conservative stability-criterion for the grinding operation depends on the ratio of contact area stiffness to machine stiffness. This study is an attempt to correlate the stiffness of the grinding wheel surface to important parameters such as wheel surface roughness and grinding forces. In turn, this provides an explanation of the influence of grain size, dressing conditions, infeed rate, and stock removal upon the stability of the grinding operation.

The Hertzian theory of elastic contact between two smooth elastic bodies has been applied to investigation of the static and dynamic characteristics of the grinding wheel and the workpiece. It has been found that the Hertzian results are valid at high loads but, at lower loads, the theoretical results analysis predicts much larger values than are given by the experimental stiffness curve of the grinding wheel. This difference can be eliminated by considering the surface of the grinding wheel to be rough instead of smooth.

INTRODUCTION

The stability of the grinding operation is influenced by the characteristics of the grinding wheel, especially the hardness and the grain size and by other grinding parameters such as the dressing conditions, wheel and work speeds, etc. Only recently a theoretically-based explanation has been formulated[1] that separates the various parameters governing the stability of the concerned operation.

As a result of these investigations, it turns out that the compliance of the wheel-work contact area is a basic parameter with respect to stability requirements in grinding. This contribution deals with this basic parameter by discussing the influence of the grinding wheel characteristics, the geometrical surface constellations of the contact area and the grinding conditions.

Starting from the closed loop representation of the grinding operation, the classical stability requirement has been derived for the feedback system. The resulting requirement indicates the interrelation of the basic parameters of the grinding process which influence the stability. Special attention has been paid to the contact stiffness which is studied in more detail. Therefore, theoretical investigations and experimental measurements have been performed in order to determine the general trends in grinding stability due to this important stiffness characteristic.

CLOSED LOOP REPRESENTATION OF THE GRINDING OPERATION

Following basic statements in refs. 1 and 2 led to the composition of the closed loop system of the grinding operation:

1. At any instant of time (t), the sum of the total wear of workpiece (δ_w) and grinding

wheel (δ_s) must be equal to the total infeed (u_0), minus the deformation amount of contact area (y_k) and machine structure (y_m):

$$\delta_w(t) + \delta_s(t) = u_0(t) - y_k(t) - y_m(t) \tag{1}$$

which gives in Laplacian form (s) as:

$$\delta_w(s) + \delta_s(s) = u_0(s) - y_k(s) - y_m(s) \tag{2}$$

where subscripts w and s indicate workpiece and grinding wheel, respectively.

2. The desired depth of cut is the increment in total feed during the previous revolution of the workpiece as

$$\Delta u_0(t) = u_0(t) - u_0(t - \tau_w) \tag{3}$$

where τ_w is the revolution time of the workpiece.

Applying the Laplace transform and rearranging yields:

$$\frac{u_0(s)}{\Delta u_0(s)} = \frac{1}{1 - e^{-\tau_w s}} \tag{4}$$

3. The workpiece shape at any instant of time is equal to the shape one revolution before plus the instantaneous depth of cut, or:

$$\delta_w(t) = \delta_w(t - \tau_w) + \Delta\delta_w(t) \tag{5}$$

and similarly for the grinding wheel shape:

$$\delta_s(t) = \delta_s(t - \tau_s) + \Delta\delta_s(t) \tag{6}$$

Equations (5) and (6) can be transformed to:

$$\frac{\delta_w(s)}{\Delta\delta_w(s)} = \frac{1}{1 - e^{-\tau_w s}} \tag{7}$$

$$\frac{\delta_s(s)}{\Delta\delta_s(s)} = \frac{1}{1 - e^{-\tau_w s}} \tag{8}$$

4. It is assumed that the instantaneous depth of cut or wear amount is proportional to the cutting force (F_c); this simplification corresponds to a similar assumption used in the stability of a single tool operation[3-6] for a large application range and has been proven practically by Hahn[7] or:

$$F_c = k_s\Delta\delta_s \tag{9}$$

$$F_c = k_w\Delta\delta_w \tag{10}$$

where k_s is the wear stiffness of grinding wheel and k_w is the cutting stiffness of workpiece.

5. The dynamic compliance (G_m) of the machine structure and of the contact area is given by:[1, 3]

$$\frac{y_m(s)}{F_c(s)} = \frac{1}{k_m} = G_m(s) \tag{11}$$

$$\frac{y_k(s)}{F_c(s)} = \frac{1}{K} \tag{12}$$

where k_m and K are the static stiffnesses of machine structure and contact area, respectively.

The interrelationship of the set of simultaneous equations (2), (4), (7), (8), (9), (10),

(11) and (12) are illustrated by Fig. 1. The physical meaning of this block diagram is easily seen: the total infeed amount u_0 causes a deformation of the contact area, y_k and a contact force, F_c, is built up. This force will deflect the machine structure an amount, y_m, causing an instantaneous wear of the wheel, $\Delta\delta_s$, and leads to an instantaneous depth of cut $\Delta\delta_w$. The instantaneous depth of cut or wear must be combined with the shape of the wheel and the form of the workpiece which are determined by the depth of previous cuts and the wear history of the grinding wheel. This, in its turn, leads to the total wear of the wheel and the workpiece.

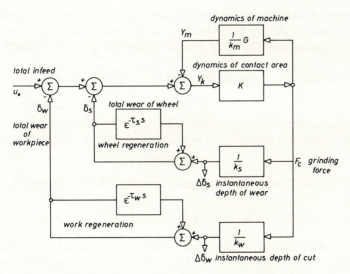

FIG. 1. Closed loop representation of plunge grinding operation.

STABILITY REQUIREMENTS

The transfer function of the grinding operation can be derived either from the fundamental simultaneous equations or from the block diagram (Fig. 1). It is the ratio of the instantaneous depth of cut to the instantaneous infeed.[1]

$$\frac{\Delta\delta_w(s)}{\Delta u_0(s)} = \frac{1}{\{1 + (k_w/k_s)\,[(1 - e^{-\tau_w s})/(1 - e^{-\tau_s s})]\} + [1 - e^{-\tau_w s}]\,[(1/K) + (1/k_m)\,G_m(s)]k_w}$$

The characteristic equations derived from eq. (13) can be written in the form of:

$$\tag{13}$$

$$-\frac{1}{1 - e^{-j2\pi\nu_w}} - \frac{K_v}{R}\frac{1}{1 - e^{-j2\pi\nu_s}} - \frac{k_w}{K} = \frac{k_w}{k_m}\,G_m(j\omega) \tag{14}$$

where $K_v/R = k_w/k_s$, $s = j\omega$, $\omega\tau_{s;w} = 2\pi(n_{s;w} + \nu_{s;w})$.

In eq. (14) ν_w and ν_s are defined as the decimal phase angles of the dynamical component of the cutting force for two consecutive revolutions of workpiece and grinding wheel; $n_{s;w}$ are integers corresponding with a certain complete number of revolutions of the wheel or workpiece; K_v is the speed ratio, and R is the grinding ratio. The right side of the equation depends on the compliance of the machine structure and the ratio of k_w and k_m. The geometrical loci corresponding to the left side of eq. (14) is a straight line parallel to the imaginary axis.[1] Indeed, it has been found that the real parts of each of the three terms of the left side of eq. (14) are constants, independent of v_s and v_w. No intersection of the polar plots of both sides of the equation can occur (in other words, the grinding operation is stable) if the negative real part, R_e, of the transfer function of the machine, G_m, multiplied by the ratio of cutting stiffness is smaller than the abscissa value of the straight line:

$$\frac{k_w}{k_m} \quad R_e < \tfrac{1}{2}\left(1 + \frac{K_v}{R}\right) + \frac{k_w}{K} \tag{15}$$

This condition is similar to the asymptotic stability requirement in regenerative chatter problems in single tool cutting.[3, 6] The K_v/R ratio usually can be neglected with respect to unity because, in common grinding practice, this term turns out to be 2 to 4 orders of magnitude smaller.

The cutting stiffness, k_w, is directly proportional to the width of cut, w:[3]

$$k_w = w . k_w{}^* \tag{16}$$

where $K_w{}^*$ is the cutting stiffness per unit contact width.

The stiffness of the contact area also is proportional to the width of contact

$$K = w . K^* \tag{17}$$

in which K^* is the stiffness per unit contact length. Equation (15) can be written as:

$$R_e < \frac{k_m}{w}\left(\frac{1}{2k_w{}^*} + \frac{1}{K^*}\right) \tag{18}$$

by neglecting K_v/R compared to unity and by introducing eqs. (16) and (17). The cutting stiffness per unit contact length, $k_w{}^*$, has been proven to be a frequency depending function.[1] This can easily be illustrated by Fig. 2.

If, for some reason, the grinding wheel moves periodically relative to the workpiece, an undulation on the workpiece surface will occur. The generation of this undulation is fully

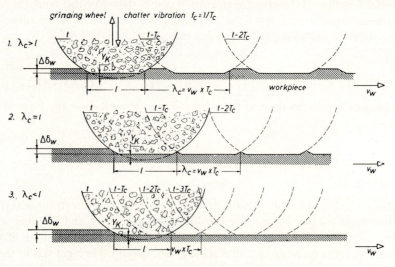

FIG. 2. Limitation of work undulation due to low work speed.

possible only if the contact length of the interaction zone workpiece-grinding wheel is smaller than the wave length on the workpiece surface. When the vibration frequency is very high or when the speed of the workpiece is too small, the surface cannot be modulated, even for large amplitudes of the applied dynamic force. For a given workpiece speed the cutting stiffness, which is defined as the ratio of the cutting force and the depth of cut, depends on the frequency of the vibration motion.

It has been shown[1] that the order of the minimum workspeed required for a generation of waves at a vibration frequency, f, is:

$$v_{w\min} = l \cdot f = 2(y_c \cdot D_{eq})^{0.5} \cdot f \tag{19}$$

in which l is the contact length, y_c is the radial contact deformation of the wheel and workpiece, and D_{eq} is the equivalent diameter of the workpiece. By putting some typical values† of the various parameters in eq. (19), it can be concluded that a significant number of practical grinding work is performed in a workspread range, where generation of waves in the workpiece surface is not possible or strongly attenuated. This is especially true for production grinding. The deformation of the contact area during finishing operations is much smaller, and, therefore, waves on the workpiece occur at considerably higher frequencies.

The dynamic cutting stiffness, k_w^*, becomes very large for workspeeds smaller than $v_{w\min}$. The stability equation 18 can consequently be simplified in such cases to:

$$R_e < \frac{1}{w} \frac{k_m}{K^*} \tag{20}$$

In other words, in much of the grinding work the operation is stable when the real part, R_e, of the machine transfer function remains smaller than the product of the inverse of the width of cut and the ratio of the machine stiffness, k_m, and the contact area stiffness, K^*. This is a conservative criterion because in eq. (18) some positive terms on the left side are neglected.

It is known that the grinding stability is favorably influenced by an increase of the machine stiffness, k_m, a decrease of the dynamic compliance of the machine structure (introduced in eq. (20) by R_e) and a reduction of the contact width, w. This is consistent with earlier conclusions[3-6] concerning single tool cutting operations. A new element, however, is the fact that the contact area stiffness, K^*, turns out to be one of the most significant factors governing the stability of grinding operation. It is important to know how stiff the contact area really is, and what parameters will affect this contact stiffness. Theoretical and experimental investigations have been performed in order to provide an answer to these questions.

CONTACT STIFFNESS MEASUREMENTS

It is quite difficult to measure the contact stiffness during grinding. However, to obtain an estimation of the order of magnitude of the compliance of the contact area, some stiffness tests have been run without any relative motion in the tangential direction.[1] This approximation is partially justified because it has been shown, theoretically, that the superposition of a tangential load, distributed in the contact area in proportion to the radial

† Typical values are: deformation $y_c = 130\,\mu$in; $D_w = 4$ in; $D_s = 12$ in; $D_{eq} = D_w \cdot D_s/(D_w + D_s) = 3$ in; common chatter frequency 500 cps. According to eq. (19), $v_{w\min} = 2 \times (130 \times 10^{-6} \times 3)^{0.5} \times 400 \times 60/12 = 100$ ft/min.

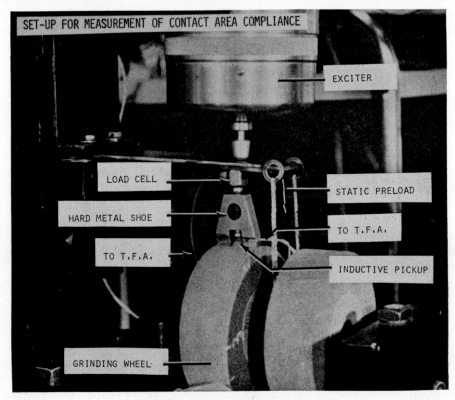

FIG. 3. Set-up for measurement of contact area compliance.

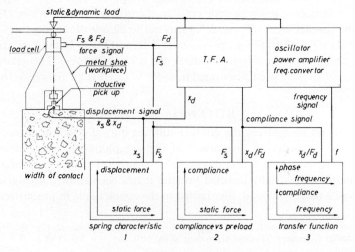

FIG. 4. Contact compliance measurement scheme.

load, does not drastically influence the total deflection of the contact area in the radial direction even for high values of the corresponding coefficient of friction.[3]

The test apparatuses are shown in Fig. 3 and the scheme of the test set-up is given in Fig. 4. The small hard steel shoe is pushed against the grinding wheel with a preload, a dynamic exciting force is superimposed, and both the static and the dynamic force components are measured by means of a load cell. The deformation of the contact area is detected by a noncontacting inductive type of probe.

The TFA (transfer function analyzer) measures the ratio of the dynamic excitation force and the A.C. component of the deformation while sweeping through the frequency range. This yields the transfer function of the contact area for a given preload. The set-up also gives the dynamic compliance versus preload at a constant frequency. Finally, plotting static load versus static deformation leads to the conventional spring characteristic of the contact area.

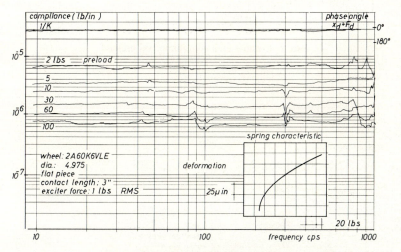

FIG. 5. Transfer function of contact area.

From Fig. 5 it is seen that the compliance is almost independent of the frequency in the considered frequency range. For a constant exciting frequency and a variable preload, the compliance decreases drastically in the range of 0–50 lb/in but remains almost constant in a higher preload range as shown in Fig. 6. The figure also shows the deformation or spring characteristic. The deformation of the contact area for a load of 50 lb/in is of the order of 150 μin. The results of stiffness measurements on the same wheel at different areas are plotted in Fig. 7. The stiffness, K^*, can vary by as much as a factor of almost 10 in the load range usually applied in grinding practice.

If the contact width, w, is long enough to neglect three-dimensional effects at both ends of the contact, it has been proven that the stiffness of the contact area is a linear function of the contact width, corresponding to eq. (16). The value of the contact stiffness, of course, is influenced by a number of parameters: the grinding wheel composition, the dressing conditions, the wheel and workpiece diameter, etc. Therefore, it is necessary to have an adequate model for the contact area in order to drive a mathematical expression relating these various parameters describing the contact compliance and, by means of eq. (20), predict the influence of those parameters upon the stability of the grinding operation.

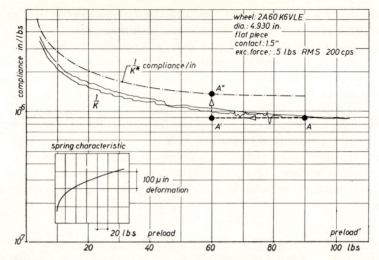

FIG. 6. Contact area compliance vs. preload.

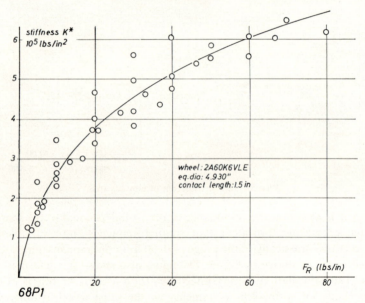

FIG. 7. Contact stiffness vs. radial force.

THEORETICAL APPROACH OF CONTACT COMPLIANCES

The theoretical derivation of the contact stiffness can lead to quite different solutions, depending on the assumptions or hypothesis concerning the nature of the grinding wheel surface.

1. Wheel Surface Considered as a System of Individual Springs

With this model it is hypothesized that each grain on the surface of the grinding wheel is supported by a single spring, and that the workpiece deformation is negligible. It is assumed that the radial positions of the grains with respect to the theoretical external circumference of the wheel are distributed linearly and each of the individual springs has the same stiffness constant, k (Fig. 8).

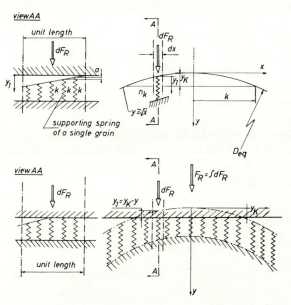

FIG. 8. Model of grinding wheel surface.

It is further assumed that the interval a between two consecutive grain locations measured in the radial direction is small with respect to the deformation, y_k, of the wheel surface. In that case the force, dF_R, needed to deflect a part of the surface of the wheel with dimensions equal to ldx is given by:

$$dF_R = \frac{k}{2a} (y_k - y)^2 \, dx \qquad (21)$$

The total radial force necessary to deflect the surface over y_k by a flat surface as shown in Fig. 8 is then:

$$F_R = 2 \int_0^{dF_{Rmax}} dF_R$$

or

$$F_R = 2 \int_0^{x_k} (k/2a) (y_k - y)^2 \, dx \qquad (22)$$

Integrating and rearranging yields:

$$y_k = \left(\frac{6}{5} \frac{a}{k} D_{eq} F_R \right)^{0.25} \qquad (23)$$

The definition of the contact stiffness, K^*, leads to

$$K^* = \frac{\partial F_R}{\partial y_k} \qquad (24)$$

or

$$K^* = 3.72 \left(\frac{k}{a} \right)^{0.25} . D_{eq}^{-0.25} . F_R^{0.75} \qquad (25)$$

The parameters influencing the contact stiffness of this model of the grinding wheel surfaces are the stiffness of the individual grains, k; the grains spacing distribution, a; the equivalent diameter, D_{eq}; and the preload, F_R.

The preload of the contact area is the most important because this parameter occurs in eq. (25) with the highest exponent. Equation (25) shows that the contact stiffness will be larger for small values of a. Theoretically speaking, the grinding stability will be influenced unfavorably by using fine grit sizes and gentle dressing conditions. The spring constant, k, reflects the influence of the wheel composition since higher grades of wheels with larger E moduli yield larger k values. Finally, eq. (25) indicates that the contact stiffness increases for smaller wheel (or work) diameters although the influence is reduced to the square root of the diameter variations.

This model appears to be adequate for describing the contact stiffness. Test results corresponding to the data of Fig. 7 yield to a force exponent of 0·67 in the static preload range of 2 to 20 lb/in instead of the 0·75 of eq. (25). This can be checked by the log–log representation of previous test data (Fig. 9).

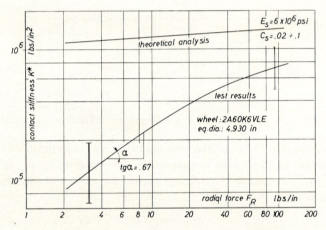

FIG. 9. Theoretical and measured contact stiffness.

A more elaborate approach would be required to match more precisely the exponents of each of the parameters with the trends determined by practical stiffness measurements. It would be possible to introduce a more precise distribution assumption of the grain location or to assume some nonlinear spring support of the individual grains. However, the advantage of the initial approach, in addition to its simplicity, is the formulation of some significant conclusions relating the dressing conditions, the grain size, the wheel hardness of the work and wheel diameters to the stability of the grinding operation. For further refinements, it is felt that a more fundamental theory ought to be developed starting from the classical Hertz theory in which special attention must be paid to the surface condition of grinding wheels and to the relative speed of the workpiece and the grinding wheel.

2. Stiffness of Smooth Surfaces using the Elastic Theory

Although the assumption of "smooth" surfaces is difficult to accept because of the wheel's surface roughness, some general trends can be derived using the classical approach derived from Hertz's theory. Recalling briefly the general assumptions of this theory:

(1) the local deformations are considered only;

(2) the contact surface is generally elliptical (in the case of two cylinders, the elliptical surface degenerates to a rectangular surface);

(3) the sum of the compression stresses on the contact surface is equal to the total applied force;

(4) the compression stresses on the surface outside the contact area are zero;

(5) in the contact area only compression stresses occur;

(6) the material of both bodies is isotropic and homogeneous, and the stresses are always within the proportional limit.

Obviously these initial assumptions have to be kept in mind, for applications or extrapolations of the theoretical results for the grinding wheel-workpiece contact stiffness.

Starting from Weber's[9] and Povitsky's work[10] the analytical determination of the contact stiffness is summarized in the Appendix. It leads to eq. (A.15), giving the compliance of the contact area, C, as a function of the radial force or preload, the diameter of workpiece, D_w, and grinding wheel, D_s, and material constants, p, including the E modulus and the Poisson ratio c:

$$C = 2p_s \left[\ln \frac{D_s}{(2F_R p_s D_{eq})} .5 - \frac{c_s}{2(1 - c_s)} \right] + 4 \frac{P_s{}^3 F_R{}^2}{D_s} \tag{26}$$

The contact area stiffness per unit contact length can be determined as

$$K^* = \frac{\partial F_R}{\partial y_k} = \frac{1}{C} \tag{27}$$

By applying actual values of the material constants, related to the previously discussed test conditions, the variation of the stiffness versus preload can be found by a numerical solution of eqs. (26) and (27). The results are plotted in Fig. 9. It can be observed that the stiffness values for the theoretical Hertzian curve are always higher than the actual measurements. Although both curves tend to the same asymptotic value, a significant difference exists in the low contact force range.

3. Stiffness of Rough Surfaces

The theory of the contact of two smooth elastic bodies, a problem originally solved by Hertz, has been extended to treat the case when one of the bodies is rough.[11] The model of roughness was chosen to represent certain features of actual physical surfaces of asperities whose heights were distributed according to a Gaussian probability law. Especially in the low contact-force range, contact occurs only at a finite number of these asperities. While contact never takes place over a continuous smooth surface, an "effective" contact length can be defined. For the case of contact of a smooth sphere on a rough plane, the variation of effective diameter of contact region with load is shown in Fig. 10.

This figure indicates that the deformation, characterized by the "effective" contact diameter, is always bigger for rough surfaces. The compliance of a rough surface is thus larger, especially in the lower load range. This is also consistent with the actual measurements on a grinding wheel (Fig. 9). The theoretical approach used in ref. 13 seems to be very useful for describing the stiffness of the contact area. Investigations are suggested in this area in order to find a good mathematical model describing the surface roughness of a grinding wheel, to develop the concerned theory for cylinders in contact, and to compare the theoretical values with test results. This theory can probably be extended to cases with a tangential load in order to introduce the relative tangential speed of both contact surfaces.

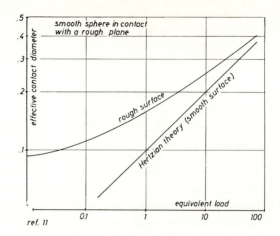

FIG. 10. Effective contact diameter vs. load.

INFLUENCE OF GRINDING CONDITIONS UPON STABILITY

1. Introduction

The influence of the grinding conditions upon the stability of the operations is indirectly indicated by the value of the contact stiffness, K^*, of eq. (20):

$$R_e < \frac{1}{w} \cdot \frac{k_m}{K^*} \qquad (28)$$

The contact area stiffness, K^*, introduces, on a double basis, the influence of the grinding wheel characteristics upon the stability of the operation. The material characteristics of the grinding wheel and the features of the wheel surface, of course, influence directly the value of the contact stiffness (parameter k/a in eq. (25) and parameters c_s and p_s in eq. (26)). Indirectly, however, the wheel characteristics influence the contact stiffness as follows: the static value of the grinding force, which can be considered as the preload of the contact area, also affects the contact stiffness as is suggested by Fig. 7 and by eqs. (25) and (26). The grinding force in its turn depends on the wheel characteristics; a softer wheel usually leads to smaller cutting force. For identical stock removal rates, the preloads on hard wheels are higher, the contact area is stiffer and the stability is influenced unfavorably.

2. Influence of Grinding Conditions

(a) *Stock removal rate*. The contact stiffness, K^*, also reflects the influence of the stock removal rate upon the stability of the grinding operation. For higher stock removal rates and all other work conditions unchanged, the grinding force increases almost directly with the variation of the stock removal. Thus, the static preload of the contact area is proportional to the stock removal rate; higher stock removal leads to an increase of the contact area stiffness which in its turn corresponds to a less favorable stability[†] (eq. 28).

[†] The stability theory is developed for a linear system; the contact stiffness, which is an important part of it, turns out to be a nonlinear element dependent on the preload. The theory can still be applied by taking an appropriate choice of the contact stiffness corresponding to the work conditions. The theory is then restricted to small amplitudes of vibration.

(b) *Wheel and workspeed.* Also, the wheel speed will affect the grinding force.[12] An increase of the wheel speed reduces the force and yields an improvement of the stability. The influence of the workspeed upon the stability cannot be clarified by eq. (28) only because the grinding forces (F_c) are almost independent of the workspeed. The cutting stiffness, however, defined as

$$k_w{}^* = F_c/\Delta\delta_w \tag{29}$$

is strongly influenced by the wheel speed because F_c remains constant and the depth of cut $\Delta\delta_w$ varies inversely with the workspeed. Using reported grinding test data[12–14] it is possible to show that the cutting stiffness, $k_w{}^*$, is basically a linear function of the wheel-speed ratio[1] or

$$k_w{}^* = k_c k_v = k_c(v_w/v_s) \tag{30}$$

where k_c is the cutting stiffness coefficient of workpiece ($k_c = k_w/K_v$). Referring to eq. (18) it can be concluded that the stability of the operation will improve for small values of $k_w{}^*$ requiring high wheel speeds and small workspeeds.

(c) *Grindability.* Obviously a difficult to grind material, corresponding to a large value of the cutting stiffness, or to a small with dull cutting grains, contributes to high grinding force levels, yielding a decrease in the stability of the operation.

(d) *Forced vibrations.* Finally it is interesting to note that the apparent spring constant of the contact area can be influenced by the magnitude of the dynamical deformation of the contact zone. Although the stiffness increases with the preload or with the static deformation, the apparent dynamic stiffness will decrease for increasing vibration amplitudes.[1] This phenomenon could be applied for stabilizing grinding machines by means of forced vibrations and it explains why the chatter amplitude sometimes stabilizes at a very small level.

3. Conclusions

As a result of this discussion, it must be emphasized that the influence of the work conditions is probably more significant in determining the actual stiffness of the contact area than the properties of the grinding wheel itself. In other words, the indirect influence of the resulting cutting force level dominates the influence related directly to the elastic properties of the abrasive material.

Using eq. (28), an estimation of the order of magnitudes of R_e, k_m and K^* indicates that a significant amount of grinding work is performed under unstable conditions:

(small damping)	$- 10 < R_e < -0.5$	(high damping)
(weak machine)	$5 \times 10^4 < k_m < 10^7$ lb/in	(rigid machine)
(small grinding force)	$10^5 < K^* < 10^6$ lb/in	(large grinding force)

The corresponding maximum widths of contact are found by:

$$w_{\max} = \frac{k_m}{K^*} \frac{1}{R_e} \tag{31}$$

For an average contact stiffness value of $K^* = 5 \times 10^5$ lb/in and a normal damped grinding machine structure ($R_e = 4$), the static stiffness of the machine (with workpiece) for a grinding wheel of width $w_{\max} = 1$ in must be: $k_m{}^* = w_{\max} \times K \times R_e$ (32) or $k_m = 1 \times 5 \times 10^5 \times 40 = 2 \times 10^6$ lb/in. Such high stiffness of grinders is quite difficult to realize.

Introducing the estimated maximum and minimum values of R_e, k_m and K^*, in eq. (31), the limits of the range of the maximum width of cut are 0·005 and 200 in.

Fortunately, it is quite possible to grind in unstable conditions for a considerable period of time. Especially in the case grinding chatter due to the regeneration of waves on the grinding wheel, regeneration occurring at small workspeeds ($v_w < v_{w\min}$, eq. (19)), develops very slowly. Therefore, a great deal of grinding can be accomplished before the amplitude of vibration becomes objectionable.

CONCLUSIONS

The model of the grinding operation based upon the basic equation allows a general formulation of a stability criterion. Using this theory it becomes possible to relate the grinding conditions to stability requirements. The basic parameter in this theory is the compliance of the contact area of the wheel workpiece. This contact stiffness has been studied theoretically and experimentally and yields the following conclusions:

1. The higher the stiffness of the contact area the worse the problem of grinding stability. The compliance of the contact area depends on the grinding wheel characteristics; in this respect fine grit sizes and gentle dressing conditions are a disadvantage and soft wheels with small E modulus values are preferable.

2. Higher grinding forces yield larger values of the contact stiffness and reduces the stability. This is the reason why the stability is improved for high wheel speed, small work speeds, sharp cutting grains, etc., and explains a decrease of stability due to dull cutting grains, high stock removal, etc.

3. Grinding work often is performed in unstable conditions because the building up of self-excited vibrations is sometimes very slow allowing the work to be accomplished before the amplitudes of the vibration becomes objectionable.

4. Forced vibrations are not always detrimental with respect to the stability of the grinding operation because in specific cases vibrations can yield a decrease of the contact stiffness and thus stabilize the cutting operations.

5. As an extremely simplified stability criterion the required static stiffness of a grinder may be estimated at approximately 2×10^6 lb/in for each inch of the width of the grinding wheel.

ACKNOWLEDGEMENTS

The authors acknowledge with thanks the assistance of Prof. J. Lemon, Messrs. T. Comstock, D. L. Brown, T. Nakajima, and S. Wang of the University of Cincinnati for their assistance. The paper is based on research performed under Air Force Contract AF33(615)–5412, which is supervised by Mr. F. L. Whitney, Wright–Patterson Air Force Base, Ohio.

APPENDIX

Analytical Determination of the Contact Stiffness

Half the length of contact between two cylinders is given by:[10]

$$\frac{l}{2} = b = \left[2F_R \left(p_s + p_w \right) \frac{D_s D_w}{D_s + D_w} \right]^{1/2} \tag{A.1}$$

where the material constants are

$$p_s = \frac{1 - c_s^2}{\pi E_s} \quad \text{and} \quad p_w = \frac{1 - c_w^2}{\pi E_w}.$$

The pressure distribution on the contact area is given:

$$P_c = \frac{2F_R}{\pi b^2} [b^2 - x^2]^{0.5} \tag{A.2}$$

In the derivation of the Hertzian equations, the contact bodies are considered as slightly curved semi-planes. Using the semi-plane $0 < y$, the equation for the stresses can be expressed in terms of a stress function, ϕ, as[9]

$$\sigma_y = \frac{\partial \phi}{\partial y} - y \frac{\partial^2 \phi}{\partial y^2}, \quad \tau_{xy} = - y \frac{\partial^2 \phi}{\partial x \partial y}, \quad \sigma_x = \frac{\partial \phi}{\partial y} + y \frac{\partial^2 \phi}{\partial y^2} \tag{A.3}$$

For the case of plane strain, we have $\sigma_z = c(\sigma_x + \sigma_y)$, then,

$$\sigma_x + \sigma_y + \sigma_z = (1 + c)(\sigma_x + \sigma_y) = 2(1 + c)(\partial \phi / \partial y) \tag{A.4}$$

These equations generally hold for a semi-plane with normal load. For the elliptical Hertzian load distribution, we can obtain

$$\frac{\partial \phi}{\partial y} = - \frac{2F_R}{\pi b^2} Re\, [(b^2 - z^2)^{0.5} + iz] \tag{A.5}$$

where $z = x + iy$. If $y = 0$ then

$$\left(\frac{\partial \phi}{\partial y} \right)_{y=0} = - \frac{2F_R}{\pi b^2} (b^2 - x^2)^{0.5} \tag{A.6}$$

which agrees with eq. (A.2). Hence, for the deflection v in the y-direction, we have

$$E \frac{\partial v}{\partial y} = \sigma_y - c(\sigma_x + \sigma_y) = (1 + c) \left[(1 - 2c) \frac{\partial \phi}{\partial y} - y \frac{\partial^2 \phi}{\partial y^2} \right] \tag{A.7}$$

By integrating with respect to y, eqs. (A.5) and (A.7) become

$$\phi = \frac{2F_R}{\pi b^2} Re\, i \left[\frac{z}{2} (b^2 - z^2)^{0.5} + \frac{b^2}{2} \sin^{-1} \frac{z}{b} + \frac{iz^2}{2} \right] \tag{A.8}$$

and

$$\frac{Ev}{1 + c} = 2(1 - c)\, \phi - y \frac{\partial \phi}{\partial y}$$

The deflection, v, is now evaluated at the center of the cylinder, v_c; with $x = 0$, $y = R_i$, then

$$\frac{Ev_c}{1 + c} = \frac{2F_R}{\pi b^2} \left\{ 2(1 - c)Re\, i \left[\frac{i\, R_i}{2} (b^2 + R_i^2)^{0.5} + \frac{b^2}{2} \sin^{-1} \frac{i\, R_i}{b} \right. \right.$$

$$\left. \left. - \frac{i\, R_i^2}{2} \right] + R_i\, Re\, [(b^2 + R_i^2 - R_i)^{0.5}] \right\} \tag{A.9}$$

Thus,

$$\frac{i\, b^2}{2} \sin^{-1} \frac{i\, R_i}{b} = - \frac{b^2}{2} \sin b^{-1} \frac{R_i}{b} = - \frac{b^2}{2} \ln \left[\frac{R_i}{b} + \left(\frac{R_i^2}{b^2} - 1 \right)^{0.5} \right] = - \frac{b^2}{2} \ln \frac{2\, R_i}{b} \tag{A.10}$$

since $b \ll R_i$,

$$(R_i{}^2 + b^2)^{0.5} = R_i + \frac{b^2}{2\,R_i} \tag{A.11}$$

then

$$v_c = \frac{1+c}{E}\,\frac{2\,F_R\,(1-c)}{\pi}\left[\ln\frac{2\,R_i}{b} - \frac{c}{2(1-c)}\right] \tag{A.12}$$

The relative deformation of the two cylinders, due to the Hertzian pressure, becomes $y_k = |v_{cs}| + |v_{cw}|$ and

$$y_k = 2F_R\left\{p_s\left[\ln\frac{D_s}{b} - \frac{c_s}{2(1-c_s)}\right] + p_w\left[\ln\frac{D_w}{b} - \frac{c_w}{2(1-c_w)}\right]\right\} \tag{A.13}$$

It must be emphasized that Hertzian deformation is not proportional to F_R, since b is a function F_R as shown in eq. (A.1). If the cylinder w is considered more rigid than the cylinder s, eq. (A.13) can be simplified as

$$y_k = 2F_R\,p_s\left[\ln\frac{D_s}{b} - \frac{c_s}{2(1-c_s)}\right], \tag{A.14}$$

since $p_w \ll p_s$.

The compliance of the contact area can be defined as

$$C = \frac{d\,y_k}{d\,F_R}.$$

Making use of eqs. (A.14) and (A.1) then,

$$C = 2p_s\left\{\ln\frac{D_s}{(2F_R\,p_s\,D_{\mathrm{eq}})^{0.5}} - \frac{c_s}{2(1-c_s)}\right\} + \frac{4p_s\sqrt[3]{F_R{}^2}}{D_s} \tag{A.15}$$

REFERENCES

1. J. Lemon, K. Okamura, R. Snoeys, et al., Study of Grinding Process as Applied to High-strength and Thermal-resistant Alloys, Interim Engineering Reports I thru V, Air Force Contract AF33 (615)–5412, University of Cincinnati, 1967.
2. R. Snoeys, Instabiliteit van het slijpproces, Ph.D. Thesis, University of Louvain, 1966.
3. G. W. Long and R. L. Kegg, Final Report on Effect and Control of Chatter Vibrations in Machine Tool Processes, Technical Report AFML-TR-65-177, June 1965, U.S. Air Force, Wright–Patterson Air Force Base, Ohio. Prepared under Contract AF33(657)-9143 in the Cincinnati Milling and Grinding Machines, Inc.
4. J. Tlusty, Selbsterregte Schwingungen an Werkzeugmaschinen. Berlin, Verlag Technik, 1962.
5. S. Tobias, Schwingungen an Werkzeugmaschinen, Hauser, Munchen, 1961.
6. J. Peters and P. Van Herck, Ein Kriterium for die dynamische Stabilitat von Werkzeugmaschinen, Industrie Anzeiger, February 1963.
7. R. S. Hahn, On the Mechanics of the Grinding Process under Plunge Cut Conditions, Trans. ASME, 72, Feb. 1966.
8. Chang Keng Lui, Stresses and Deformations due to Tangential and Normal Loads on an Elastic Solid with Applications to Contact Stresses, Ph.D. Thesis, University of Illinois.
9. C. Weber, The Deformation of Loaded Gears and the Effect on their Load-carrying Capacity, Institut fur Maschinenelemente, Einzelbericht No. 102, Braunschweig, 1949.
10. H. Povitsky, Stresses and Deflections of Cylindrical Bodies in Contact with Application to Contact of Gears and of Locomotive Wheels, Journal of Applied Mechanics, June 1950, pp. 191–201.
11. J. A. Greenwood, and J. H. Tripp, The Elastic Contact of Rough Spheres, Journal of Applied Mechanics, 66-WA/APM-28, 1966.
12. K. Guhring, Hochleistungsschleifen. Dissertation, T. H. Aachen, 1967.
13. K. Okamura, and T. Nakajima, Study on Ultra-high Speed Grinding, Research Report, Kyoto University, 1967.
14. K. Bruckner, Der Schleifvorgang und seine Bewertung durch die auftretenden Schnittkrafte. Dissertation, T. H. Aachen, 1962.

A COMPUTER METHOD FOR HYPSOMETRIC ANALYSIS OF ABRASIVE SURFACES

H. T. McAdams, L. A. Picciano and P. A. Reese

Cornell Aeronautical Laboratory, Inc., Buffalo, New York 14221

SUMMARY

Previous analyses of grinding as a stochastic process have employed a description of the abrasive surface in terms of profiles taken either parallel or transverse to the cutting direction. The approach taken in this paper employs a complete mathematical description of the abrasive surface and can provide contour maps delineating the spatial relation and geometry of cutting points as they are involved in the grinding process.

A computer method is presented for regenerating the abrasive surface, given a finite number of elevations relative to a reference plane. Employing general vector-space theory, the method develops an equation of the surface in terms of locally defined "patches" and can provide plan view contours at selected elevation levels. These contours are of interest in analyzing the chip generation process and in predicting abrasive wear lands. By virtue of its step-wise approach to the description of surfaces, the system overcomes the restrictions of classical response-surface and trend surface methodology. By the use of principal components analysis, the method can produce a description of the surface which is optimum in the sense that the vector space model is of minimum dimensionality. The analysis is believed to offer an advantage over power spectral density methods in that the basic platforms derived are interpretable in terms of abrasive particle geometry.

INTRODUCTION

The important role played by the topography of the abrasive surface in the cutting performance of grinding wheels, discs and belts is well recognized.[1] Certain sites on the abrasive tool, often referred to as active cutting points, engage the workpiece while other sites seldom, if ever, come into contact with the work material. Factors controlling the conditions for interaction between abrasive and workpiece surfaces include, in addition to the geometry of these surfaces, the kinematics of the grinding process.

Grinding, viewed as a time sequence of cutting events, is seen to be a material-removal process for both tool and workpiece, as shown in Fig. 1. Chips removed from the workpiece

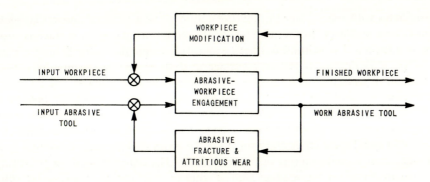

Fig. 1. Grinding as a stochastic process.

modify its surface and thus change the conditions under which it will subsequently make contact with the abrasive particles. Similarly, material is lost from the abrasive surface through abrasive fracture and attritious wear, and this loss in turn modifies the conditions for subsequent contact with the workpiece. To study a grinding process over a significant portion of its time history, therefore, requires a means for describing in some detail the sequence of cutting events as they affect both workpiece and abrasive tool.

Several attempts have been made to view grinding in the context of a stochastic process. McAdams[1,2] developed methods for describing the abrasive surface in terms of profiles taken either parallel or transverse to the cutting direction. Though it was emphasized that profiles taken parallel to the cutting direction are best suited to describing the sequence of cutting events, the registry of grit particles in a direction transverse to their motion was recognized to be of equal importance in assessing the forces felt by the grit particles. Peklenik[3] emphasized profiles taken in a direction transverse to the cutting direction but took into account the correlation of such profiles when separated by a given distance in the cutting direction. Both cases represent attempts to provide a three-dimensional model of the abrasive surface in terms of two-dimensional profiles.

A third approach to a three-dimensional visualization of the abrasive surface is one which delineates the intersection of the abrasive topography with planes passed through the abrasive surface at specific heights above a datum plane.[4] These intersections are elevation contours in the same sense as the contours on a topographic map of terrain. This approach, which can be called the hypsometric method, is the subject of this paper. A method is provided for mathematical description of the surface, and a computer method is presented for automatic construction of elevation contours. These contours provide a concise picture of the configuration of active cutting points in directions both parallel and transverse to the cutting direction and make evident the grits which interact in the generation of chips.

MATHEMATICAL DESCRIPTION OF ABRASIVE SURFACE

For purposes of developing a mathematical description of an abrasive surface, reference is made to Fig. 2. Though the coordinate system is applicable to any abrasive surface, it will be convenient to interpret the system as pertaining to a coated abrasive and to regard the backing sheet as the datum plane to which grit elevations are referenced. It is desired to develop an explicit description of the abrasive surface as a function

$$z = f(x,y)$$

which associates with every point (x,y) in the datum plane a height z above that plane.

A well-known approach to the mathematical description of such a surface is provided by regression theory. The method has been extensively employed in the field of experiment design, where it is known as response surface analysis,[5] and in the mining industry,[6] where it is known as trend surface analysis. In general, the response variable z is known at a finite number of points in the (x,y)-plane and, because error is usually present in the measurement of z, a least-squares technique is employed in developing the response-surface or trend-surface equation. Conventionally, the surface is developed by expanding the function $z = f(x,y)$ as a Taylor series and truncating the series at some degree believed to represent the significant variation in the response variable. The criterion used for truncating the series is usually based on the residual sum of squares of deviations between the observed data and the corresponding responses computed from the fitted equation.

The conceptual difficulties incident to the conventional application of response surface theory has been previously discussed.[7] The validity of the residual sum of squares as a criterion for goodness of fit was questioned, and the need for a criterion based on interpolative feature of the mathematical model was emphasized.

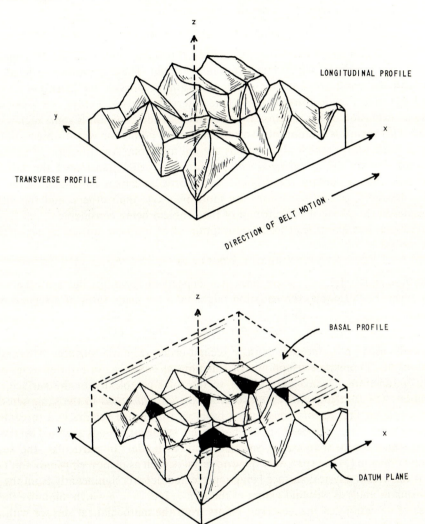

FIG. 2. Abrasive coordinate system.

Strictly speaking, a regression model is statistically supported only at points at which observations are available, yet inferences are drawn with regard to a continuum. When considered as a mathematical function, the model is observationally defined only on a domain consisting of a finite number of points. Extension of this domain to some continuous region of space must be based on physical rather than statistical argument. Vector space and function theory concepts are employed as a logical framework for a physically oriented analysis of the representation of an abrasive surface.

For the present purpose, the basic notion of a function is exemplified by a set of ordered triples (x,y,z). Thus

$$f = \{(x_1,y_1,z_1), (x_2,y_2,z_2), \ldots, (x_n,y_n,z_n)\}$$

denotes a set of n position coordinates in the (x,y)-plane and a corresponding set of elevations on the abrasive surface. The domain of the function is

$$D = \{(x_1,y_1), (x_2,y_2), \ldots, (x_n,y_n)\}$$

and interpolation consists of extending the domain D to some augmented set D^* containing points in the (x,y)-plane at which interpolation is desired. These additional ordered pairs, together with the corresponding interpolated elevations, provide an *extension f^** of the function f. In fitting a regression equation to the observed points, therefore, one is essentially providing an extension of the function in which the extended domain is some region of the real plane. This region is a union of two sets, the set of points for which observations are available and the set for which observations are not available. Any function f^* defined on the extended domain D^* is *mathematically* legitimate as an extension of the function f provided only that it contain the original set of ordered triples contained in the set f. Physically, however, certain extensions are more plausible than others, and the ultimate choice resides in the physical background of the problem being considered.

The considerations above lead to the conception of an abrasive surface as a *generalized response surface*

$$z = f(x,y) = b_1 f_1(x,y) + b_2 f_2(x,y) + \ldots + b_p f_p(x,y) \tag{1}$$

where the $f_i(x,y)$, $i = 1,2,\ldots,p$ are linearly independent functions but are otherwise of arbitrary form. For example, the equation might take the usual form of a second-degree polynomial

$$z = b_1 + b_2 x + b_3 y + b_4 xy + b_5 x^2 + b_6 y^2 \tag{2}$$

as commonly used in connection with a second-order response surface. Alternatively, the functions $f_i(x,y)$ might be trigonometric functions as employed in Fourier series or any other set of functions deemed appropriate to the total complexity of the surface under consideration. The functions $f_i(x,y)$ are regarded as *basis functions* in the same sense that three non-coplanar line vectors constitute a basis set for all line vectors in a three-dimensional vector space. Once the basis set has been fixed, stringent bounds are fixed on the class of surfaces which can be spanned by the mathematical model. In particular, the second-degree polynomial in (2) can span only quadric surfaces such as planes, ellipsoids and hyperboloids. If the actual physical surface being represented departs significantly from the form (2), least-squares analysis will find values of the f_i, $i = 1, 2, \ldots, p$ which minimize the sum of squares of deviations of the observed points from the mathematical surface within the constraints of the model.

It is at this point that the discrepancy between the complex surface of a coated abrasive and an idealized surface such as that represented by eq. (2) becomes evident. In order to follow the many convolutions of a real abrasive surface, the regression equation would need to be of very high degree or would otherwise need to have a large number of terms. Efforts to fit a complex surface by means of a high-degree polynomial based on a finite number of observations, however, may result in an oscillatory function which gives highly biased results in regions of data voids, even though it satisfies the observations quite well. Such oscillatory behavior is especially likely if the surface has steep gradients, as it is quite likely to have in view of the crystallographic nature of grit particles.

The dilemma can be avoided by a simple stratagem: instead of fitting a single equation to the entire set of observations, one can confine attention to small subregions within which the surface is relatively well-behaved. The method has been applied to a variety of subjects with considerable success, including the topography of terrain[8] and the mapping of forest canopies.[9]

Though a central problem is that of delineating the appropriate size of subregion to be considered, the choice is relatively straightforward in the case of describing an abrasive surface. Since grit size is a natural unit comprising the structure of the abrasive surface, it is logical to employ a subregion compatible with the size of this natural structural unit.

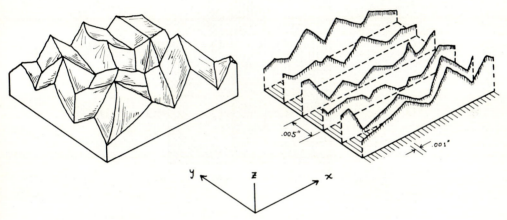

FIG. 3. Regular sampling of an abrasive belt.

Consider an abrasive belt sampled at regular intervals in the longitudinal and transverse directions, as shown in Fig. 3. In the (x,y)-plane, the sampling points constitute a regular grid and the entire set of grid points can be grouped into local subsets, called *design regions*, each of which can be treated as a separate problem in regression analysis. These local surfaces can then be blended together to produce a continuous function defining the entire surface of interest.

Figure 4 illustrates the way in which local surface fitting is applied to subsets of the sample points. In the example, the points were taken nine at a time. In the figure, it is presumed that the process has been in progress for $N - 1$ steps. At the Nth step, the operation involves sampling points 9, 10, 11, 16, 17, 18, 23, 24 and 25. A surface suitably adjusted to the complexity of the data is fitted to these points and is then used to compute or interpolate grit elevations within the region $ACKI$, called the *interpolation region*. In principle, a fine grid of points within $ACKI$ is set up, and the interpolated elevation computed at every point on this fine grid. The scales chosen for the x and y increments in this interpolation region are arbitrary, but in all cases they would be smaller than the original distances between sampling positions. At the $(N + 1)$-st step, a surface is fitted to sampling points 10, 11, 12, 17, 18, 19, 24, 25 and 26 and used to interpolate grit elevations within the interpolation region $EGOM$. Similar interpolation is carried out at steps $N + 2$, $N + 3$, and so on. In regions such as $EGKI$, two estimates of interpolated elevations are available and in regions such as $FGKJ$, four estimates are available. This multiplicity of estimates results from overlap of the interpolation regions, and the amount of overlap can be chosen at the discretion of the analyst. The basic consideration in this matter is the propagation of error in the least-squares process and is discussed in more detail in Appendix I. The redundant estimates are reconciled by the use of a concept called blending functions, as discussed in Appendix II.[10]

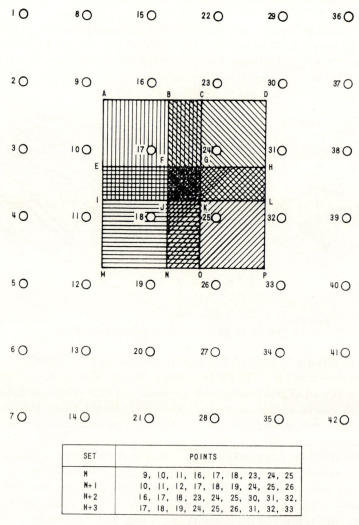

SET	POINTS
N	9, 10, 11, 16, 17, 18, 23, 24, 25
N+1	10, 11, 12, 17, 18, 19, 24, 25, 26
N+2	16, 17, 18, 23, 24, 25, 30, 31, 32,
N+3	17, 18, 19, 24, 25, 26, 31, 32, 33

Fig. 4. Sample grid.

If the size of local subregions is determined on the basis of some natural feature of the surface under consideration, the choice of basis functions is not particularly critical. For example, a second-degree polynomial of the form of eq. (2) can be postulated to fit, reasonably well, subregions of the abrasive surface if these regions are constrained to a size not greater than grit size. Equation (2) can be rewritten as

$$Z = b_1 + b_2x + b_3y + b_4xy + b_5(3x^2 - 2) + b_6(3y^2 - 2) \tag{3}$$

so that the basis functions are orthogonal on the set of points constituting the local array of the design region. Equation (3) has the advantage that calculations are simplified and error propagation is under more rigid control. If eq. (3) is assumed, the basis functions can be visualized as a set of component surfaces into which each of the local surface patches can be resolved (Fig. 5). Inasmuch as each of the basis functions can be expended in terms of trigonometric components, it is clear that trigonometric functions would serve as an alternate set of basis functions for the design regions and that choice of basis elements is not unique. There is, however, a means to select basis functions for the design regions and

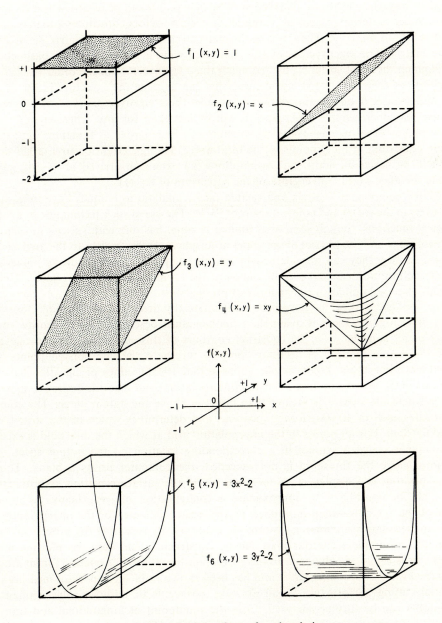

FIG. 5. Basic planforms for surface description.

that choice of basis elements is not unique. There is, however, a means to select basis functions in such a way that most rapid convergence of the expansion is assured.

By invoking principal component analysis, one can approach the choice of basis functions analytically. As the sample array sweeps over the entire set of sampling points, a local surface is generated for every design region. It is presumed that each of these local surfaces can be resolved as a linear combination of simpler surfaces which are common to all the design regions. However, if the nine points composing the local region are ordered in a consistent way, the elevations observed at these nine points can be regarded as a vector and, if the number of design regions in N, the set of N such vectors constitutes a matrix of order $9 \times N$. Call this matrix Z. Then, if Z' is the matrix transpose of Z, the matrix ZZ' is a 9×9 square matrix. The eigenvectors of this matrix provide a preferred set of basis elements for spanning the local surfaces. By arranging these eigenvectors in the order of decreasing magnitude of their associated eigenvalues, one can arrange these vectors in the order of their importance. If one retains only the first k of these eigenvectors ($k < 9$), the residual variance, based on sample observations, will be less than for any other set of k vectors chosen by any other means. These eigenvectors can be regarded as functions defined on a domain consisting of the 9 points in the local array. Though extension to a larger domain must still be based on additional assumptions, an extension arrived at in this way will provide the most rapid convergence of the expansion of eq. (1).

Principal component analysis and related factor analytic techniques have been applied extensively in the social and political sciences.[11, 12] The use of such techniques in analyzing empirical functions, though somewhat limited to date, has met with success in such areas as meteorology and the physics of the upper atmosphere.[13-16] Applied to the analysis of an abrasive surface, they can provide insight into the shape of potential cutting points in the abrasive surface and can serve as an analog of the correlation and power spectral density methods already familiar in the analysis of profiles.[1, 3]

Once the local subregions have been mathematically described and blended together at points of contiguity, there is available a mathematical description of the entire abrasive surface under consideration. To construct contours of this surface at constant elevations above the datum plane it is necessary only to threshold the interpolated points at the desired elevation levels. For example, suppose that 10,000 values of elevation have been generated in the process of data extension (interpolation) and that it is desired to map those regions where the surface is greater than 0.020 in above the datum plane. The computer interrogates each of 10,000 storage locations and determines which of the stored values exceed 0.020 in. For all points in the interpolation grid at which the threshold is exceeded, the computer prints a symbol in a corresponding location on the output sheet. At all positions where the threshold is not exceeded, the computer prints a blank. Thus the output printout is a display of all locations where the surface elevation exceeds 0.020 in and might be considered to approximate a configuration of wear lands. A numerical count of the symbols, when multiplied by the least count area of the interpolation grid, provides the total area contained within the contours. By repeating the process at various thresholds, a complete contour map can be constructed, and data for plotting a curve of area versus height are obtained. In the geographic literature, this relation for landforms is called the *hypersometric integral*[17] and has been shown to be an important measure of the potential cutting performance of an abrasive surface.[1] In addition, the configuration of contours can be further analyzed, from the standpoint of longitudinal and transverse separation, to obtain insight into the geometric factors affecting chip generation.

EXPERIMENTAL RESULTS

Sample elevation data from a 36-grit coated abrasive were provided by the Coated Abrasives Division of the Norton Company. By means of automatic mechanical profiling equipment, a stylus traces over the abrasive surface and digitally records heights of the surface at one-thousandth-inch intervals along the track. The digitized profile can be regarded as a profile in the cutting direction. For purposes of the study, twelve parallel tracks separated transversely by five-thousandth-inch intervals were made available. Thus a section of 36-grit coated abrasive sampled on a rectangular 0.001 in × 0.005 in grid

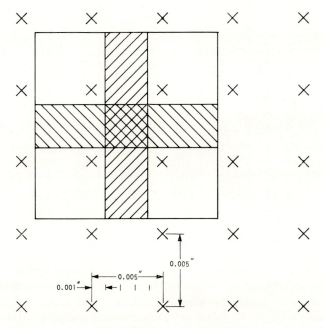

FIG. 6. Regions of overlap.

constituted the data for the study. Sampling intervals were chosen appreciably smaller than nominal grit particle size so that the contouring process would be capable of delineating the actual shapes of grit cross sections. Actual data used as input to the contouring program was on a square sampling grid in which the spacing of points was 0.005-in in both the longitudinal and transverse direction. Thus in the longitudinal direction only every fifth point was retained as input to the contouring program. The points which were skipped were, however, available for comparison with corresponding points interpolated by the mathematical model. Thus it was possible to evaluate the validity of the computation not only for the data used as inputs, as is normally done in regression analysis, but for interpolated points as well.

The local sampling array consisted of nine points in a square array and the model used was that of eq. (3). Elevations were interpolated on a square grid having dimensions of 0.001 in and the successive interpolation regions were overlapped as shown in Fig. 6. Linear blending functions were employed to resolve redundant estimates in the regions of overlap. Elevations were thresholded at 0.005-in intervals.

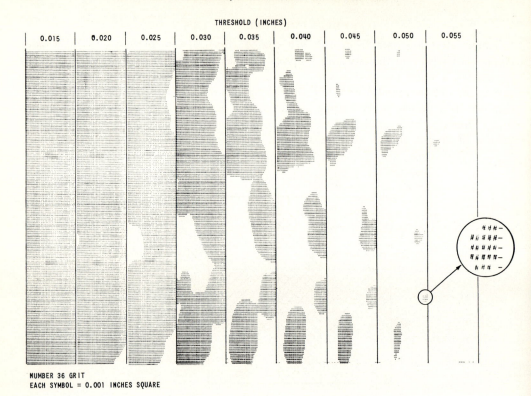

Fig. 7. Elevation contour maps of abrasive surface.

Figure 7 presents contour maps of a portion of the abrasive surface. The precision of these maps can be appreciated by comparing the interpolated points with the corresponding points actually recorded by the profiling equipment. Table 1 shows the residual sum of squares of deviations for (1) the observed sample points spaced at 0.005-in intervals and used as input to the computation and (2) the observed sample points spaced at 0.001-in intervals and not used as input to the computation. The corresponding standard deviations are 0.00086 in for the observed points used as inputs to the interpolation and 0.00090 in for the observed points not used as inputs to the calculation. The fact that these two quantities differ by only about 5% indicates that the basis functions chosen exhibited good ability to interpolate the abrasive surface.

TABLE 1. RESIDUAL SUMS OF SQUARES FOR ABRASIVE SURFACE

	Sum of squares† (in²)	Number of points	Mean squares	Standard deviation (in)
For observed points in input array	0·000592	800	0·00000074	0·00086
For observed points not in input array	0·000648	3200	0·00000081	0·00090

† Sum of squares of deviations between observed and computed grit elevations.

For each of the twelve tracks, the hypsometric functions $\beta(z)$ and $\mu(z)$ were computed. These quantities, introduced in an earlier paper,[1] are related to the corresponding terrain measurements of hypsometric integral and ridge-to-valley spacing. More specifically, $\beta(z)$ denotes that fraction of the length of the track in which elevations exceed z inches above the datum plane. The quantity $\mu(z)$ denotes the average number of times per unit track length that the track profile crosses a line drawn parallel to the datum plane and at height z above it.

Tables 2 and 3 summarize $\beta(z)$ and $\mu(z)$ for the twelve tracks. The composite relationships, as obtained by averaging the results for the twelve tracks, are plotted in Figs. 8 and 9.

TABLE 2. SUMMARY OF BETA FUNCTION FOR TWELVE TRACKS

Track no.	1	2	3	4	5	6	7	8	9	10	11	12	Average
z (in)							$\beta(z)$						
0.015	1.000	1.000	1.000	1.000	1.000	1.000	1.000	1.000	1.000	1.000	1.000	1.000	1.000
0.020	1.000	1.000	1.000	1.000	1.000	1.000	1.000	1.000	1.000	1.000	1.000	1.000	1.000
0.025	0.974	0.972	0.972	0.962	0.977	0.984	0.982	0.980	0.981	0.972	0.966	0.962	0.972
0.030	0.860	0.852	0.842	0.818	0.834	0.806	0.826	0.810	0.802	0.754	0.712	0.704	0.802
0.035	0.528	0.563	0.604	0.594	0.592	0.584	0.588	0.574	0.504	0.494	0.480	0.491	0.550
0.040	0.308	0.300	0.306	0.358	0.373	0.370	0.339	0.324	0.298	0.287	0.284	0.291	0.320
0.045	0.164	0.158	0.148	0.194	0.234	0.212	0.184	0.179	0.180	0.186	0.175	0.180	0.183
0.050	0.071	0.066	0.078	0.090	0.117	0.107	0.091	0.096	0.106	0.108	0.102	0.106	0.094
0.055	0.032	0.033	0.025	0.032	0.054	0.050	0.034	0.046	0.060	0.042	0.048	0.049	0.042
0.060	0.014	0.010	0.009	0.008	0.005	0.012	0.010	0.016	0.020	0.015	0 014	0.013	0.012
0.065	0.004	0.000	0.000	0.000	0.001	0.000	0.004	0.006	0.002	0.003	0.005	0.000	0.002

TABLE 3. SUMMARY OF MU FUNCTION FOR TWELVE TRACKS.

Track no.	1	2	3	4	5	6	7	8	9	10	11	12	Average
z (in)							$\mu(z)$ Crossings/inch						
0.015	0.5	0.5	0.5	0.5	0.5	0.5	0.5	0.5	0.5	0.5	0.5	0.5	0.5
0.020	4.5	4.5	3.5	4.5	3.5	3.5	3.5	3.5	3.5	3.5	4.5	6.5	4.1
0.025	13.5	13.5	14.5	15.5	12.5	15.5	14.5	17.5	19.5	23.5	19.5	20.5	16.7
0.030	29.0	29.0	29.0	25.0	22.0	22.0	26.0	27.0	26.5	24.5	27.5	25.5	26.0
0.035	25.0	23.0	25.0	24.0	24.0	24.0	25.0	27.0	22.0	20.2	25.5	25.5	24.1
0.040	20.0	17.0	17.0	20.0	19.0	22.0	20.0	15.0	15.0	14.0	14.0	18.0	17.6
0.045	8.0	7.0	10.0	13.0	13.0	11.0	11.0	11.0	9.0	13.0	11.0	11.0	10.7
0.050	4.0	5.0	4.0	5.0	9.0	9.0	7.0	7.0	6.0	6.0	8.0	11.0	6.8
0.055	3.0	2.0	2.0	2.0	2.0	3.0	2.0	3.0	4.0	3.0	2.0	3.0	2.6
0.060	1.0	0.0	0.0	0.0	1.0	0.0	1.0	1.0	1.0	1.0	1.0	0.0	0.8
0.065	0.0	0.0	0.0	0.0	0.0	0.0	0.0	0.0	0.0	0.0	0.0	0.0	0.0

DISCUSSION

The conditions for engagement of abrasive particles with the workpiece in a grinding process change with time as a consequence of the fact that both workpiece and abrasive suffer material loss. Because of the sequential nature of the individual cutting events, the depth to which a grit engages the workpiece is determined in part by the conditions under which one or more previous grit engagements occurred. In short, a chip has basically two

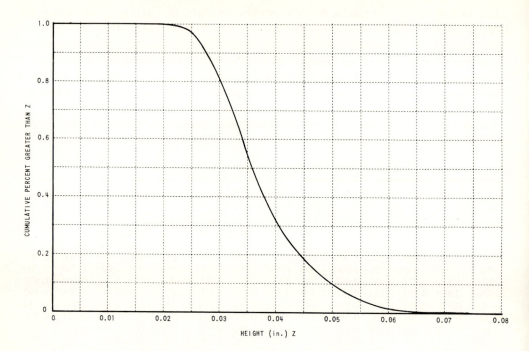

Fig. 8. Hypsometric β-curve for coated abrasive.

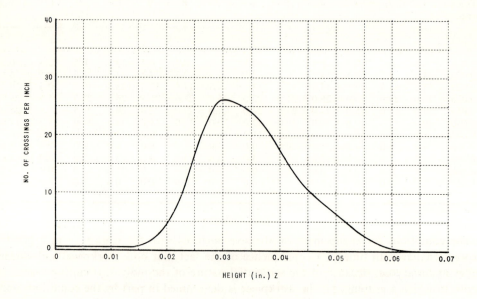

Fig. 9. μ-curve for coated abrasive.

surfaces of interest, one produced by the current cutting event, the other by prior cutting events.

A two-dimensional analysis performed on a longitudinal profile of the abrasive surface is not adequate to delineate the actual geometry of a chip. This fact can be appreciated by considering a typical contour map of the abrasive surface. If stock is being removed at a constant rate, the surface of the abrasive can be regarded as advancing into the workpiece in a manner which maintains a zone of interference of some average thickness T. If the abrasive surface is contoured on a basal plane corresponding to this thickness, the configuration of contours might appear as in Fig. 10.

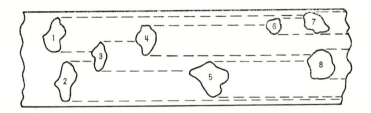

FIG. 10. Contour configuration in relation to chip removal.

A chip removed by grit No. 1 leaves on the workpiece a new surface which may be encountered not by a single grit but possibly by several, such as Nos. 3, 6 and 7. If the workpiece is advancing at a constant rate, the thickness of a chip or part of a chip resulting from the interaction between the grit pairs 1–3, 1–4, 1–6, and 1–7 will all be different; the forces experienced by grits Nos. 3, 4, 6 and 7 will be correspondingly affected.

Figure 10 may also be interpreted as a configuration of wear lands after the abrasive has been subjected to considerable usage. As the abrasive wears, the feed force must be increased to compensate for the development of wear lands on the grit particles, if constant cutting rate is to be maintained. Otherwise the cutting rate will decrease in response to the decreased penetration of the grit particles into the workpiece under the given feed force. Quite clearly the size of the wear lands and their rate of development over the use history of the abrasive product are important considerations affecting its performance. Hypsometric analysis of the product, in terms of such functions as $\beta(z)$ and $\mu(z)$, seems particularly suited to the study of this relationship, both for design and quality control purposes.

The description of abrasive profiles by such quantities as correlation functions and power spectral density finds an analogous concept in the resolution of the abrasive surface into basic surface planforms. A profile, either longitudinal or transverse, can be expanded in terms of trigonometric functions or, more generally, in terms of a set of basis functions peculiarly suited to its geometry. Similarly, it is possible to approximate the abrasive surface as a linear combination of elementary surfaces chosen in such a way that they efficiently span the class of surfaces of interest. Inasmuch as it is the entire abrasive surface, rather than a particular profile, which interacts with the workpiece, such a characterization of abrasives is indicated to have advantages over profile descriptions.

The methods described and demonstrated are exploratory. The mathematical tools presented, however, are very general and should not be interpreted as being restricted to this particular demonstration. Adjustable constraints in the application of the contour method include, in addition to broad choice of basis functions, the number of points to be included in the local design regions, the scale of interpolation grid, the extent of overlap of local interpolation regions, and the form of the blending functions. By virtue of the fact that generalized response surface theory employs linear combinations of basis functions to represent the abrasive surface, the computer model can be readily adapted to present the spatial distribution of slopes, curvatures, and other derived quantities pertaining to the cutting performance of an abrasive surface.

Generalized response surface theory and the associated contouring methodology is applicable to a variety of problems, in addition to those previously alluded to, and is believed to offer a means for broadening the general scope of regression theory in the design and analysis of experiments.

ACKNOWLEDGEMENTS

The ideas on which this paper is based were gleaned from previous research performed under contracts with The Carborundum Company, The Abrasive Grain Association, and the U.S. Army Materiel Command. The paper represents a synthesis and an extension of these concepts under internal support by Cornell Aeronautical Laboratory, Inc.

For supplying the coated abrasive data for this study, the authors acknowledge with thanks the kind cooperation of Mr. Bruce Duke, Mr. Roger Story and Mr. Ed Keyes of the Coated Abrasives Division of The Norton Company, Troy, New York. Thanks are also due to Mr. Robert Coakley and Mr. Gregory Lewandowski of Cornell Aeronautical Laboratory, Inc. for their assistance respectively in computer processing of the data and in preparation of the manuscript. For their encouragement of the development of the methodology of the paper, the senior author also wishes to acknowledge the interest of Mr. Arthur Stein, Head, Weapons Research Department, Cornell Aeronautical Laboratory and Dr. R. J. Hader, Department of Experimental Statistics, North Carolina State University at Raleigh.

APPENDIX I

Mathematical and Statistical Foundations of Generalized Response Surface Methodology

Assume that a variable ζ, which shall be referred to as the *response variable*, depends on p experimental variables $x_1, x_2, \ldots, x_p$ and that a functional relation

$$\zeta = f(x_1, x_2, \ldots, x_p) \tag{1}$$

exists. Under certain conditions, the response equation (1) can be expanded as

$$\zeta = \beta_1 f_1(x_1, x_2, \ldots, x_p) + \beta_2 f_2(x_1, x_2, \ldots, x_p) + \beta_3 f_3(x_1, x_2, \ldots, x_p) + \ldots \tag{2}$$

where the β_i, $i = 1, 2, 3, \ldots$ are constants to be determined. The functions $f_1, f_2, f_3, \ldots$ are of arbitrary form provided only that they are linearly independent and do not involve the constants β_i. Equation (2) is thus a linear function of the β_i, although the f_i may be nonlinear in $x_1, x_2, \ldots, x_p$. Equation (2) is said to be a *linear model*, and the functions f_i may be regarded as *basis vectors* spanning a *vector space* comprising a certain class c of functions.

Suppose, now, that the function (1) is to be estimated from experimental observations. The variables $x_1, x_2, \ldots, x_p$ constitute a p-dimensional space called the x-space, and one may estimate (1) from observations taken at n points in this space. These n points constitute what shall subsequently be referred to as the experimental design. In general, ζ cannot be observed at these points because of error. Rather, it is possible only to observe a variable z related to ζ by

$$z = \zeta + \epsilon \tag{3}$$

where ϵ is a random error. Then (2) assumes the form

$$z = \beta_1 f_1(x_1, x_2, \ldots, x_p) + \beta_2 f_2(x_1, x_2, \ldots, x_p) + \ldots + \beta_r f_r(x_1, x_2, \ldots, x_p) + \epsilon \tag{4}$$

Since ϵ is a random variable, the responses observed at each design point also constitute a random variable. As a result, it is possible only to obtain from the observations an equation of the form

$$\hat{z} = b_1 f_1(x_1, x_2, \ldots, x_p) + b_2 f_2(x_1, x_2, \ldots, x_p) + \ldots + b_r f_r(x_1, x_2, \ldots, x_p) \tag{5}$$

where $\hat{z}$ is an estimate of z and b_i, $i = 1, 2, \ldots, r$ is an estimate of β_i.

Clearly, two types of errors can affect the approximation of the function (1). First, if (1) is to be approximated by the linear model (2), then (1) must belong to the class of functions c spanned by the basis functions $f_0, f_1, \ldots, f_r$. Second, some means must be found for minimizing or controlling the effect of the random errors ϵ, since these will affect the estimation of the β_i. Generally, the form of (1) is unknown at the outset, and the experimenter has the option of assuming a set of basis functions according to experience or prior knowledge concerning the system under study. For control of random errors, the theory of least squares is employed.

In view of eq. (5), each observation z_j obtained in the data-collecting phase of abrasive surface analysis can be represented as

$$z_j = b_1 f_1(x_{1j}, x_{2j}, \ldots, x_{pj}) + b_2 f_2(x_{1j}, x_{2j}, \ldots, x_{pj}) + \ldots + \epsilon_j \tag{6}$$

where the b_i are estimates of β_i and the ϵ_j are random errors as computed from

$$\epsilon_j = z_j - \hat{z}_j \tag{7}$$

At the outset, $\hat{z}$ is not known, and it is the object of the least-squares algorithm to estimate $\hat{z}$ in such a way that

$$\sum_{j=1}^{M} \epsilon_j^2 = \text{a minimum.} \tag{8}$$

We proceed to display the theory of this algorithm.

In matrix notation, eq. (6) can be written as

$$\mathbf{z} = x\mathbf{b} + \epsilon \tag{9}$$

where $\mathbf{z}$ and ϵ are n-rowed column vectors (or $n \times 1$ matrices), $\mathbf{b}$ is a r-rowed column vector (or $r \times 1$ matrix), and x is a matrix of dimension $n \times r$. The set of points at which observations are made will be referred to as the *design region*, and the set of functions $f_0, f_1, \ldots, f_r$ will be referred to as the *basis functions* or simply the *basis*. For the computer contouring program, only two position variables x and y are involved. If x is identified with x_1 and y is identified with x_2, then the basis functions are functions of only two variables.

For the two-variable case, the x matrix is generated as shown on p. 1164.

The columns of the matrix are identified with the basis functions, the rows with the design points. In the example shown, x_1 assumes K distinct values and x_2 assumes M distinct values, so that there are $n = KM$ points in the design. Each basis function is evaluated at every point in the design, and the resulting $n \times r$ array constitutes the X-matrix. If the array is rearranged, so that the rows become columns and the columns become rows, the

X - Matrix

	f_1	f_2	f_r
X_{11},X_{21}	$f_1(X_{11},X_{21})$	$f_2(X_{11},X_{21})$	$f_r(X_{11},X_{21})$
X_{11},X_{22}	$f_1(X_{11},X_{22})$	$f_2(X_{11},X_{22})$	$f_r(X_{11},X_{22})$
.	.	.	.
.	.	.	.
X_{11},X_{2M}	$f_1(X_{11},X_{2M})$	$f_2(X_{11},X_{2M})$	$f_r(X_{11},X_{2M})$
X_{12},X_{21}	$f_1(X_{12},X_{21})$	$f_2(X_{12},X_{21})$	$f_r(X_{12},X_{21})$
X_{12},X_{22}	$f_1(X_{12},X_{22})$	$f_2(X_{12},X_{22})$	$f_r(X_{12},X_{22})$
.	.	.	.
.	.	.	.
X_{12},X_{2M}	$f_1(X_{12},X_{2M})$	$f_2(X_{12},X_{2M})$	$f_r(X_{12},X_{2M})$
.	.	.	.
.	.	.	.
X_{1K},X_{21}	$f_1(X_{1K},X_{21})$	$f_2(X_{1K},X_{21})$	$f_r(X_{1K},X_{21})$
X_{1K},X_{22}	$f_1(X_{1K},X_{22})$	$f_2(X_{1K},X_{22})$	$f_r(X_{1K},X_{22})$
.	.	.	.
.	.	.	.
X_{1K},X_{2M}	$f_1(X_{1K},X_{2M})$	$f_2(X_{1K},X_{2M})$	$f_r(X_{1K},X_{2M})$

resulting matrix is the transpose of X. The matrix X and its transpose X' will be used extensively in the sequel.

Consider (9) and write the error vector **e** as

$$\mathbf{e} = \mathbf{z} - x\mathbf{b} \tag{10}$$

The sum of squares of the components $e_1, e_2, \ldots, e_n$ of the error vector **e** can be written as

$$Q = \mathbf{e}'\mathbf{e} = (\mathbf{z} - X\mathbf{b})' \, (\mathbf{z} - X\mathbf{b}) \tag{11}$$

where $\mathbf{e}'$ is the transpose of **e** and $(\mathbf{z} - X\mathbf{b})'$ is the transpose of $(\mathbf{z} - X\mathbf{b})$.

In extenso, (11) becomes (since $n = KM$)

$$Q = \sum_{l=1}^{n} e_i^2 = \sum_{k=1}^{K} \sum_{m=1}^{M} e^2{}_{km}$$

$$= \sum_{k=1}^{K} \sum_{m=1}^{M} [z_{km} - b_1 f_1(x_{1k},x_{2m}) - b_2 f_2(x_{1k},x_{2m}) - \ldots - b_r f_r(x_{1k}, x_{2m})]^2 \tag{12}$$

By differentiating (12) with respect to the b_j, one obtains a set of p equations of the form

$$\frac{\partial Q}{\partial b_j} = 0 \tag{13}$$

which can be solved for the b_j to minimize Q. The result can be summarized succinctly in the form

$$X'X\mathbf{b} = X'\mathbf{z} \tag{14}$$

where $X'X$ is a square matrix of order r. Equation (14) provides the so-called *normal equations* of least squares.

Then

$$\mathbf{b} = (X'X)^{-1}X'\mathbf{z} \tag{15}$$

is the formal solution minimizing the error sum of squares.

It is of interest to investigate the statistical properties of the least squares solution under certain assumptions. For the error vector ϵ, we assume that

$$E(\epsilon) = 0$$

$$E(\epsilon\epsilon') = I\sigma^2 \tag{16}$$

where E denotes expectation and I is the identity matrix of order η. Equations (16) are equivalent to the assumption that the errors are uncorrelated, with mean zero and constant variance σ^2.

From eq. (15), it is clear that

$$\mathbf{b} = CX'\mathbf{z} \tag{17}$$

where $C = (X'X)^{-1}$. Note that, for the design points, (4) can be written as

$$\mathbf{z} = X\boldsymbol{\beta} + \boldsymbol{\epsilon} \tag{18}$$

Substituting (18) into (17) one obtains

$$\mathbf{b} = CX'(X\boldsymbol{\beta} + \boldsymbol{\epsilon}) = CX'X\boldsymbol{\beta} + CX'\boldsymbol{\epsilon} = \boldsymbol{\beta} + CX'\boldsymbol{\epsilon} \tag{19}$$

Since $\boldsymbol{\epsilon}$ is a random vector, $\mathbf{b}$ is also a random vector.
Taking expectations in (19), one sees that

$$E(\mathbf{b}) = E(\boldsymbol{\beta} + CX'\boldsymbol{\epsilon}) = E(\boldsymbol{\beta}) + E(CX'\boldsymbol{\epsilon}) = \boldsymbol{\beta} + CX'E(\boldsymbol{\epsilon}) = \boldsymbol{\beta} \tag{20}$$

Thus, the estimates provided by (15) are *unbiased, provided the postulated form of the function ζ is correct*. In the event of an incorrect choice of model, the estimates of the coefficients will be biased to an extent depending on the degree of discrepancy between the postulated and true models. It is for this reason that the contouring program incorporates sufficient flexibility to allow changes of basis if it becomes evident that such a change is necessary to fulfil the requirements of the particular contouring problem under consideration.

Unbiasedness derives from the ability to substitute for $\mathbf{z}$ its equivalent $X\boldsymbol{\beta} + \boldsymbol{\epsilon}$. Suppose the model is inadequate and requires additional basis functions so that the true model is

$$\mathbf{z} = X\boldsymbol{\beta} + X_1\boldsymbol{\beta}_1 + \boldsymbol{\epsilon}$$

where $X_1\boldsymbol{\beta}_1$ denotes the additional terms in the expansion. Then

$$\mathbf{b} = CX'\mathbf{z} = CX'(X\boldsymbol{\beta} + X_1\boldsymbol{\beta}_1 + \boldsymbol{\epsilon})$$

$$= CX'X\boldsymbol{\beta} + CX'X_1\boldsymbol{\beta}_1 + CX'\boldsymbol{\epsilon}$$

and

$$E(\mathbf{b}) = \boldsymbol{\beta} + CX'X_1\boldsymbol{\beta}_1$$

$$= \boldsymbol{\beta} + A_1\boldsymbol{\beta}_1$$

The matrix $A_1 = CX'X_1 = (X'X)^{-1}X'X_1$ is called the *alias matrix*. Though it is useful in indicating the extent of confounding among various coefficients in the correct model, it is defined only in relation to an alternative hypothesis.

Consider, now, the covariance matrix of **b**. Denoted $V(\mathbf{b})$, the covariance matrix is of order $r \times r$ and is given by

$$V(\mathbf{b}) = E[\mathbf{b} - E(\mathbf{b})][\mathbf{b} - E(\mathbf{b})]' \tag{21}$$

But, from (20),

$$\mathbf{b} - E(\mathbf{b}) = \mathbf{b} - \beta \tag{22}$$

and applying the results of (19) gives

$$\mathbf{b} - E(\mathbf{b}) = \beta + CX'\epsilon - \beta = CX'\epsilon \tag{23}$$

Therefore,

$$V(\mathbf{b}) = E[CX'\epsilon][CX'\epsilon]'$$

$$= E[CX'\epsilon\,\epsilon'XC'] \tag{24}$$

or, since C is symmetric,

$$V(\mathbf{b}) = E[CX'\epsilon\,\epsilon'XC] \tag{25}$$

But, by the assumption of (16),

$$E(\epsilon\epsilon') = I\sigma^2$$

Therefore,

$$V(\mathbf{b}) = CX'XC\sigma^2 = C\sigma^2 \tag{26}$$

The results of (26) can be expressed as follows:

(a) The diagonal elements of the matrix $C = (X'X)^{-1}$, when multiplied by σ^2, the variance of the individual observations, provide the variances of the estimated coefficients in the model.

(b) The off-diagonal elements of C similarly provide the covariance between two estimated coefficients, b_i and b_j.

The variable z is often referred to as the *response*. We wish to study the variance to which the estimated response $\hat{z}$ is subject as we consider different points in the x-space. The necessary information can be obtained by an extension of the above reasoning.

Consider an arbitrary point (x_1, x_2) in the x-space and define a corresponding vector **x**′ as

$$\mathbf{x}' = [f_1(x_1, x_2)\, f_2(x_1, x_2) \ldots f_r(x_1, x_2)] \tag{27}$$

Then the estimated response at that point is

$$\hat{z} = \mathbf{x}'\mathbf{b} = b_1 f_1(x_1, x_2) + b_2 f_2(x_1, x_2) + \ldots + b_r f_r(x_1, x_2) \tag{28}$$

where **b** is a column vector of coefficients.

Then,

$$E(\hat{z}) + E(\mathbf{x}'\mathbf{b}) = \mathbf{x}'E(\mathbf{b}) = \mathbf{x}'\beta \tag{29}$$

and

$$\begin{aligned}
\mathrm{Var}\,(\hat{z}) &= E[\hat{z} - E(\hat{z})][\hat{z} - E(\hat{z})]' \\
&= E[\mathbf{x}'\mathbf{b} - \mathbf{x}'\beta][\mathbf{x}'\mathbf{b} - \mathbf{x}'\beta]' \\
&= E[\mathbf{x}'(\mathbf{b} - \beta)][\mathbf{x}'(\mathbf{b} - \beta)]' \\
&= E[\mathbf{x}'(\mathbf{b} - \beta)(\mathbf{b} - \beta)'\mathbf{x}] \\
&= \mathbf{x}'E[(\mathbf{b} - \beta)(\mathbf{b} - \beta)']\mathbf{x}
\end{aligned} \tag{30}$$

But $E(\mathbf{b} - \beta)'(\mathbf{b} - \beta) = V(\mathbf{b})$. Therefore,

$$\mathrm{Var}\,(\hat{z}) = \mathbf{x}'(X'X)^{-1}\mathbf{x}\sigma^2 \tag{31}$$

Equation (31) gives the variance of the estimated response at an arbitrary point (x_1, x_2)

in the sampling plane. Note that this variance depends strongly on the form of the X matrix, which is determined both by the location of the design points and the form of the basis functions.

Equation (31) theoretically provides an estimate of the variance of the estimated response at every point in the x-space. In the event that the x-space is two-dimensional, as it is in the present study, it is possible to display contour maps of this variance. The variance is computed at every point in an array of points in the x-space plane, and these values are then thresholded and displayed by the contouring program. The results for the 3×3 design array, using the basis functions of the example in the text, are displayed in Fig. 11. Compare this figure with Fig. 6. It is seen that, by constraining the interpolation region to be appreciably smaller than the local design region, one assures that the variance of the estimated response will be no greater than the variance inherent in the input measurements, provided there is no bias or systematic lack of fit in the model assumed for the surface.

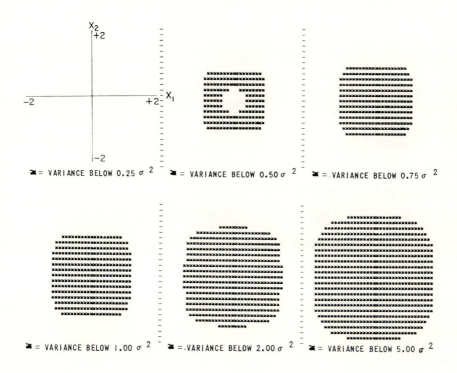

FIG. 11. Variance maps.

APPENDIX II

Theory of Blending Functions

The theory of blending functions, as used in computer contouring by means of generalized response surface theory, can be illustrated by a two-dimensional problem, as illustrated in Fig. 12.

Let us suppose that a function $f_1(x)$, defined on the real line, has been fit to the points P_1, P_2, P_3 and that a function $f_2(x)$, also defined on the real line, has been fit to the points

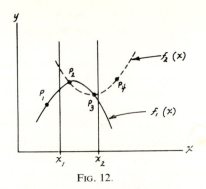

Fig. 12.

P_2, P_3, P_4. The problem is to "blend" these functions in some interval $x_1 \leqslant x \leqslant x_2$, to obtain a function $f(x)$ defined as

$$f(x) = \begin{cases} f_1(x) & x < x_1 \\ \phi[f_1(x), f_2(x)] & x_1 \leqslant x \leqslant x_2 \\ f_2(x) & x > x_2 \end{cases}$$

and to satisfy certain continuity requirements at x_1 and x_2.

Toward this end we define a blending function $g(x)$ as

$$g(x) = \begin{cases} 0 & x < x_2 \\ h(x) & x_1 \leqslant x \leqslant x_2 \\ 1 & x > x_2 \end{cases}$$

where the form of $h(x)$ is yet to be specified. Then

$$f(x) = [1 - g(x)] f_1(x) + g(x) f_2(x)$$

$$= f_1(x) - g(x) f_1(x) + g(x) f_2(x)$$

and it is seen that, if $h(x)$ is continuous in the interval $x_1 \leqslant x \leqslant x_2$, then $f(x)$ is continuous over the entire real line.

In the example reported in the text, $g(x)$ is defined as

$$g(x) \begin{cases} 0 & x < x_2 \\ \dfrac{x - x_1}{x_2 - x_1} & x_1 \leqslant x \leqslant x_2 \\ 1 & x > x_2 \end{cases}$$

where x_1 and x_2 denote the boundaries of the region of overlap between two contiguous interpolation regions.

In general, it might be desirable to require not only that $f(x)$ be continuous at x_1 and x_2, but that its derivatives up to some specified order be also continuous. We can examine this requirement by invoking the chain rule of differentiation. If

$$f(x) = f_1(x) - g(x) f_1(x) + g(x) f_2(x)$$

then

$$f'(x) = f'_1(x) - g(x) f'_1(x) - g'(x) f_1(x) + g(x) f'_2(x) + g'(x) f_2(x)$$

Since

$$g'(x) = \begin{cases} 0 & x < x_1 \\ h'(x) & x_1 \leqslant x \leqslant x_2 \\ 0 & x > x_2 \end{cases}$$

we see that

$$f'(x) = \begin{cases} f'_1(x) & x < x_1 \\ f'_2(x) & x > x_2 \end{cases}$$

but that continuity of $f'(x)$ is not necessarily assured as $x \to x_1$ from below and $x \to x_2$ from above. For this requirement to be satisfied, one must have

$$\lim_{x \to x_1} g'(x) = 0$$

and

$$\lim_{x \to x_2} g'(x) = 0$$

as well as $g(x_1) = 0$ and $g(x_2) = 1$. These constraints suggest, as an example, a third-degree polynomial

$$g(x) = b_0 + b_1 x + b_2 x^2 + b_3 x^3$$

with

$$g'(x) = b_1 + 2b_2 x + 3b_3 x^2$$

Imposing the boundary conditions, one can obtain the required function. For example if $x_1 = 0$ and $x_2 = 1$, then

$$g(x) = 3x^2 - 2x^3$$

is a function satisfying the boundary conditions or, in general,

$$g(x) = 3 \left(\frac{x - x_1}{x_2 - x_1} \right)^2 - 2 \left(\frac{x - x_1}{x_2 - x_1} \right)^3$$

A similar approach can be used to force continuity of second and higher derivatives if desired. Thus the contouring model can be developed in such a way that, in the event that slopes or curvatures of the surface are of interest, discontinuities in these derived quantities can be avoided.

Two-dimensional blending, as required in contouring an abrasive surface, is a straight-forward extension of the above theory. For example, consider the problem of blending the regions, 1, 2, 3, and 4 in the regions of overlap indicated in Fig. 13.

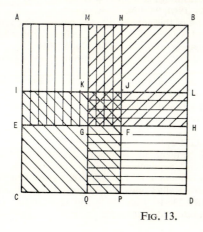

REG. 1
REG. 2
REG. 3
REG. 4

FIG. 13.

Let g = a blending function as described above. Then the blending between regions 1 and 2, f_{12}, is given by

$$f_{12}(x,y) = [1 - g(x)]f_1(x,y) + g(x)f_2(x,y).$$

This assures continuity within the area $AEHB$. Now we blend between regions 3 and 4 obtaining f_{34}.

$$f_{34}(x,y) = [1 - g(x)]f_3(x,y) + g(x)f_4(x,y)$$

f_{34} is continuous within region $ILDC$. Note that we have not produced a function which is continuous in the y direction.

Now blend f_{12} and f_{34} in the y direction obtaining $f_{1234}(f,y)$.

$$f_{1234}(x,y) = [1 - g(y)]f_{12}(x,y) + g(y)f_{34}(x,y).$$

This function is continuous throughout the entire region $ABCD$. In this process we have blended twice in the region $KJFG$, once in the x and once in the y direction. By blending regions 1 and 2 with regions 3 and 4, then composite region 1–2 with composite region 3–4, we have avoided having to explicitly write a two-directional blending function for the region $KJFG$. The procedure can, of course, be extended to more than four regions by blending first in the x direction obtaining continuous strips, then blending these strips in the y direction.

For the four-region case, we can combine the three blending steps into the single equation

$$\begin{aligned}
f_{1234}(x,y) &= [1 - g(y)]\{[1 - g(x)]f_1(x,y) + g(x)f_2(x,y)\} \\
&\quad + g(y)\{[1 - g(x)]f_3(x,y) + g(x)f_4(x,y)\} \\
&= [1 - g(y)][1 - g(x)]f_1(x,y) + [1 - g(y)]g(x)f_2(x,y) \\
&\quad + g(y)[1 - g(x)]f_3(x,y) + g(y)g(x)f_4(x,y).
\end{aligned}$$

By an extension of this type of argument, blending in higher dimensions can conceivably be achieved without difficulty using only one-dimensional blending functions.

REFERENCES

1. H. T. McAdams, The Role of Topography in the Cutting Performance of Abrasive Tools, *Journal of Engineering for Industry*, *Trans. ASME*, Series B, **86**, 1964, 75–81.
2. H. T. McAdams, Markov Chain Models of Grinding Profiles, *Journal of Engineering for Industry*, *Trans. ASME*, Series B, **86**, 1964, 383–387.
3. Janez Peklenik, Contributions to a Correlation Theory for the Grinding Process, *Journal of Engineering for Industry*, *Trans. ASME*, Series B, **86**, 1964, 97–106.
4. An interesting example of the application of this approach to the physical characterization of surfaces is found in J. B. P. Williamson, Physical Aspects of a Surface, Technical Paper EM68-513, American Society of Tool and Manufacturing Engineers, Dearborn, Michigan, 1968.
5. G. E. P. Box, The Exploration and Exploitation of Response Surfaces: Some General Considerations and Examples, *Biometrics*, **10**, 1954, 16.
6. W. C. Krumbein, Trend Surface Analysis of Contour-type Maps with Irregular Control-point Spacing, *Journal of Geophysical Research*, **64**, No. 7, 1959, 823–839.
7. H. T. McAdams, Regression Models: Facts and Fallacies, Paper presented at a meeting of the Society of Mining Engineers of AIME, New York, New York, 27 February to 3 March 1966.
8. Characteristics of Vegetation at Field Test Sites and Various Geographic Areas (U), Cornell Aeronautical Laboratory, Inc., Report No. GM-2338-G-1, September 1967.
9. Off-road Mobility Research, Second Semiannual Technical Report (U), Cornell Aeronautical Laboratory, Inc., Report No. VJ-2330-G-2, September 1967.
10. A similar concept of blending functions is found in D. T. Ross, S. A. Coons, and J. E. Ward, Investigations in Computer-aided Design for Numerically Controlled Production, Interim Engineering Progress Reports IR8-236-I, December 1964; IR8-236-II, June 1965; and IR8-236-III, March 1966, Massachusetts Institute of Technology, Cambridge, Massachusetts.

11. A. NEAL and S. RETTIG, Dimensions of Alienation Among Manual and Non-manual Workers, *American Sociological Review*, **28**, 1963, 599–608.
12. H. ALKER and B. RUSSETT, *World Politics in the General Assembly*, Yale University Press, New Haven, Conn., 1965.
13. C. K. STIDD, The Use of Eigenvectors for Climatic Estimates, *J. Appl. Meteor.* **6**, 1967, 255–264.
14. M. GRIMMER, Space Filtering of Monthly Surface Temperature Anomaly Data in Terms of Pattern, Using Empirical Orthogonal Functions, *Quarterly J. Roy. Meteor. Soc.*, **39**, 1963, 395–408.
15. C. L. MATEER, On the Information Content of Umkehr Observations, *Journal of Atmospheric Sciences*, **22**, 1965, 370–381.
16. E. N. LORENZ, Empirical Orthogonal Functions and Statistical Weather Prediction, Sci. Report No. 1, Statistical Forecasting Project, Department of Meteor., MIT, Cambridge, Massachusetts, 1956.
17. A. N. STRAHLER, Hypsometric (Area-Altitude) Analysis of Erosional Topography, *Geol. Soc. Am. Bull.*, **63**, 1952, 1117–1142.

A METHOD FOR EXTRACTING ABSOLUTE DIGITAL DISPLACEMENT INFORMATION FROM OPTICAL GRATINGS

A. RUSSELL

National Engineering Laboratory, East Kilbride,
Machinery Group: Machine Tools and Metrology Division

SUMMARY

In this new system absolute positional information is derived from an optical grating without any pulse-counting techniques. Signals from the grating are simply amplified and logically selected to provide a decimal read-out of the least significant digit. More significant digits can be provided either by coarser gratings or by a form of switching, such as a shaft digitizer.

The transducer can be used in any application where a grating is used at present and is suitable for direct digital control. The transducer is not prone to electrical interference.

The optical grating is at present used as a displacement measuring element in systems applied to measurement and control of multi-axis machines. The accuracy of the device has been established but the measuring systems are incremental in nature which recessitates the addition, subtraction and storing of pulsedin formation. Electrical interference or brief discontinuity of supplies, introduces error into the system and reference to datum must be re-established to remove this error. The systems are also wasteful in that they do not take full advantage of the detailed positional information present in the grating signals. Coded systems using various forms of radial encoders are absolute in nature and they restore correct information when interference or discontinuity has been removed. The resolution and accuracy of the mechanical contact and magnetic types of encoder are, however, inferior to those of optical gratings. The optical encoder is better than its mechanical and magnetic counterparts but it is very wasteful in that each channel is to the radix 2. These encoders measure shaft rotation only and this, for many reasons, may not bear a stable and accurate relationship to a linear motion which it is desired to measure.

To overcome these objections to the coded systems and to give the optical grating the advantages of an absolute transducer the two techniques have been "married" to produce a hybrid system in which an optical grating measures the linear motion of the tool-holder and the shaft encoder measures the rotary motion of the lead screw. The accuracy of the complete system is determined solely by the finest track of the optical grating. The number of tracks necessary on the grating for any particular application depends on (a) the resolution required of the finest bit size and (b) the limit of accuracy one can expect from the shaft digitizer due to inherent backlash and wind up, etc., of the lead screw.

If we first consider the grating. To establish a decimal absolute system it is necessary to relate ten units of the finest grating to one unit of the next coarser grating or one unit of the finest track of the shaft encoder where only a single track grating is used. One could count the units of the grating by counting techniques but there are two major objections to this.

1. The device would not be absolute, but partially incremental in operation.
2. The grating would have to be very accurately related to the shaft encoder throughout its length.

A better alternative is to interpolate or divide the grating units by some absolute method and to relate each unit to a unit of the next track of intelligence (Fig. 1a). This produces an absolute relationship between the two sets of information and overcomes objection 1 stated. A high related accuracy is, however, still required at the transition points of the second channel of intelligence. By modifying the relationship between two sub-divisions as

(a)

(b)

FIG. 1. Relating an optical grating and shaft encoder.

in Fig. 1b, the tolerances are greatly increased as, theoretically, the second track need only be related within $\pm \frac{1}{2}$ unit of the first track. The second track is strobed, or interrogated, only during the transitions 0 to 9 and 9 to 0 of the first track. Other coarser tracks of information can be strobed likewise. This system has, however, three serious failings.

1. The transition 9 to 0 of the grating calls for a number N to be registered from the second track while transition 0 to 9 requires that $N - 1$ be registered.
2. The information from the coarser tracks has to be stored, after an interruption or interference false information may be stored until all coarser tracks have again been interrogated.
3. All coarser tracks require a related accuracy tolerance of $\pm \frac{1}{2}$ unit of the grating if they are all to be interrogated from the grating.

A system which overcomes the above objections is detailed in the next figure (Fig. 2). Row A represents cycles of information from a grating and Row B represents ten sub-divisions of each cycle produced by an interpolator. Row C is the "lead" line of a two-line interrogate system, this is "on" for the duration of counts 0 to 4 and "off" for the duration of counts 5 to 9. The "lag" interrogate line of the pair is shown in Row D and is the inverse of Row C. Rows G and H represent the switch points of the second track of intelligence. The switch points in each row are 180° out of phase with each other. The "lead" line of Row C is made to interrogate or strobe the "lead" configuration of Row G while the "lag' line of Row D is made to interrogate the "lag" switch configuration of Row H. The resultant actual interrogations of boths Rows G and H can be seen in Row F and the transitions of Row F will be seen to be accurately related to Rows C and D. It will be observed that the accuracy of Row F will be maintained provided that the switch points of the second track is within a relative tolerance of $\pm \frac{1}{4}$ cycle of one unit of the first track. The "lead" and "lag" line of the second track appears in Rows J and K and are formed, as before, from 0 to 4 and 5 to 9 of Row F. These are used to interrogate the third track of information, and so forth. Again the tolerance of the third track of information is $\pm \frac{1}{4}$ of the period of the second track of information. It is important to realize the following advantages of the described system.

1. The information displayed at each decade of the register (Rows D, F and L) is a direct indication of the information present at its own source, i.e. grating or switches which has undergone only amplification and selection. There is no need for pulse techniques of any kind or storage involving "active" electronics such as flip-flops, etc.
2. The accuracy and resolution of the complete system is dependent only on the qualities of the grating used in the first or finest track.
3. The tolerances required of the switch points of the coarser tracks is wide enough for practical manufacturing techniques of shaft digitizers ($\pm 9°$ on a ten turns per inch lead screw).
4. A system comprising one grating and a four track decimal digitizer will be capable of generating absolute decimal information over a 'stroke' of 10 inches with a digital resolution of 0·0001 in (1 part in 100,000).

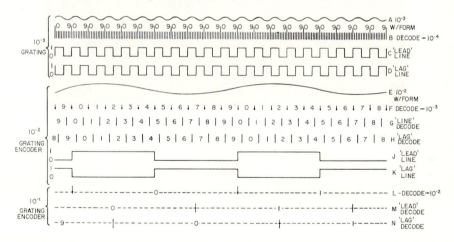

FIG. 2. System of interrogation for absolute indication from grating/encoder combination.

The practical capabilities of the system for manual and automatic control are obvious. We will now describe the method of interpolation used in order to sub-divide the cycles emitted from the optical gratings. By a suitable arrangement of indices and photocells in the optical assembly we can extract two waveforms from the gratings which form a sine and cosine relationship, i.e. there is a 90° phase shift between the two signals. If we invert the sine waveform by means of a suitable amplifier we can generate a further waveform which we shall call the inverted sine waveform. It has been long known that if these three waveforms are applied to a resistance chain further waveforms can be extracted from the resistor junction points, an example of these waveforms is shown in the next figure (Fig. 3). This is an oscillogram of a sine, cosine and inverted sine waveforms which can be identified by their maximum amplitude swing. These waveforms will be observed to pass through zero four times during one cycle of displacement. If two equal resistors connected between

sine and cosine, and two between cosine and inverted sine another two waveforms will be
generated at the resistor junctions which will be the average for the input waveforms. These
will cross zero (with reference to the sine) at 45°, 135°, 225°, and 315°. Similarly if we apply
adjacent waveforms across resistor pairs we will generate a further family of curves which
will further sub-divide the complete cycle. The oscillograph will be seen to be an approximate sub-division of the cycle into thirty-two parts at the zero crossing level. The inaccuracies present are due mainly to the sine nature of the applied initial waveforms. By using
suitable value resistors, however, in the curve family generator resistance chain the inaccuracies can be eliminated.

FIG. 3. Waveform oscillogram.

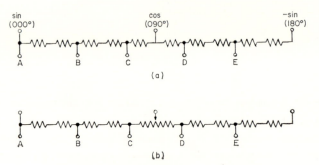

FIG. 4. Curve family generator (divide by 10).

If serrisoidal signals are obtained from the grating and applied across the network shown
in the next figure (Fig. 4) then we will achieve an accurate sub-division of ten as shown in
the next figure (Fig. 5). In practice perfect serrisoidal waveforms are difficult to achieve
due to the slight difference of the line widths in the grating but it will be seen that the accuracy of sub-division is dependent on the linearity of the waveform.

If we apply each waveform denoted A, B, C, D and E in the figure to a level detector
circuit which is adjusted to "switch" when the waveform passes through zero in either direction (this circuit is essentially a stable schmitt circuit with very small and controllable

hysteresis) and if the hysteresis is adjusted to be slightly more than the noise level present to prevent the device switching with noise, the resultant waveforms from the level detectors are shown in the next figure (Fig. 6) and they will be seen to present a shift-register type of code. This means that each discreet position can be identified by observing the condition of only two of the input waveforms; the decode of the ten parts of the example shown is

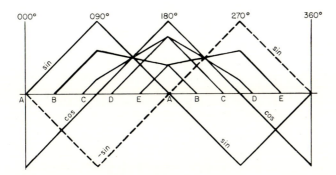

FIG. 5. Family of curves.

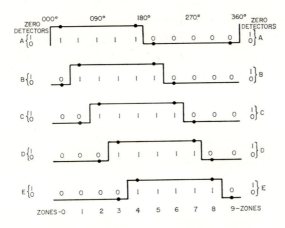

FIG. 6. Waveforms from level detectors.

zone 0 = A and not B, zone 1 = B and not C, zone 2 = C and not D, zone 3 = D and not E, zone 4 = E and A, zone 5 = not A and B, zone 6 = not B and C, zone 7 = not C and D, zone 8 = not D and E, zone 9 = not E and not A.

These positions are identified by blobs in the figure and will be seen to be unique for these particular positions or zones. Having found a method of identifying the waveforms it is only necessary to devise a logical switching decoder which will follow the above order. NOR logic elements have been used because of the powerful NOR decisions (an output for only one set of input conditions) and the built-in gain achieved. The logical diagram for the NOR decoder is shown in the next figure (Fig. 7). The Boolean algebraic terms are shown and comply with the above requirements.

By the means just described we can convert each cycle of information from the grating into ten or more unique sub-divisions by a simple process of amplification and logical selection of the grating signals. These sub-divisions can be related to Row B of the previous figure and it is now only necessary that each cycle of ten counts be identified by the next track of intelligence (which can be either from a coarser grating or a shaft encoder).

The next figure (Fig. 8) shows a grating/encoder system using a 100 line grating and a decimal encoder of three decades. One cycle from this grating will represent 0·01 in which,

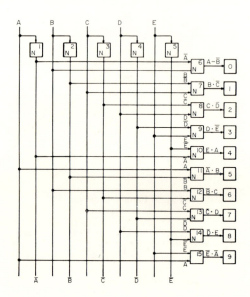

FIG. 7. Interpolator, decoder and decimal display.

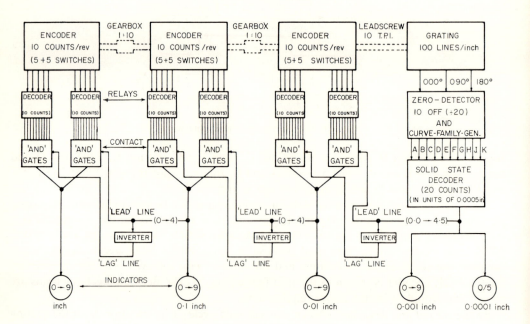

FIG. 8. Grating/encoder system.

for a ten turns per inch lead screw is one tenth of a revolution or 36° of rotation. It will be remembered from a previous slide that the point of switching would require to be accurate only to $\pm \frac{1}{4}$ of a cycle which, in this case is $\pm 9°$ of rotation. Providing the tolerance of all the possible effects on the lead screw is within this figure of $\pm 9°$ then such a system can be adopted. If not then the grating must be chosen with a line density which will provide adequate coverage for the encoder and lead screw tolerances (for instance, if both a 100 line and a 10 line per inch grating were used this would still provide the same resolution but give a total possible tolerance of $\pm 90°$ on the lead screw).

It will be observed from the figure that "and" gates have been inserted after the decoders of the shaft encoder. This is because it was felt that the time taken for relays to drop out and take up a new configuration at each switch point would be objectionable, as spurious numbers would appear at the indicator during this transition, otherwise the and gates could have been inserted between the encoder and the decoder which would have reduced the number of decoders necessary to one. The recent co-operation of a commercial digitizer manufacturer has resulted in the production of a digitizer suitable for this system, it is essentially a ten segment switch which is fitted with two brushes, a lead brush and a lag brush. The lead and lag interrogation lines are simply fed to the lead and lag brushes of the digitizers and the output for the digitizers are fed directly to the indicator lamps. This makes it possible to eliminate the decoders as shown in this figure. It will be observed in the figure that the 100 line per inch grating has been divided by 20 which yields a resolution of 5/10ths of a thousandth of an inch.

A decimal system has been used as it is suitable for manual operation but a binary system was also constructed and has obvious advantages for direct digital control operations. The grating chosen was 128 lines per inch, which when divided by 16 yielded a sub-division or "bit" of 1/2048 of an inch. A four turn per inch lead screw was used with a commercial binary encoder giving 32 counts per revolution (each count equals 1/128 in) and counting 32 revolutions. This gave a total stroke of 8 in and a binary output of 14 binary places. By interrogating the commercial shaft encoder in the manner previously described it is possible to extract the binary information direct without recourse to a decoder.

CONCLUSIONS

The measuring system just described lends optical grating measuring techniques the added convenience of absolute digital read out and makes fuller use of the inherent detail of positional information present in the grating. Normal manufacturing tolerances are well within those necessary for the construction of satisfactory transducers.

The system is flexible in that a grating or gratings can be chosen whose unit length is sufficient to provide for the possible tolerances involved in the lead screws, etc., affecting the accuracy of the shaft encoder switching. The grating can then be divided by any number up to a possible maximum of 30 or 40 so that the finest bit size will provide sufficient resolution and accuracy (the only theoretical limitation to the possible amount of sub-division is when the sub-division size becomes comparable with the hysteresis of the system which is dictated by the signal to noise ratio of the grating signals). As in all digital systems the inherent accuracy will be ± 1 bit.

Further work has been done on the system to enable the complete transducer to be constructed from one multi-track grating, the multi-track grating comprised a 1000 line per inch grating, a 100 line per inch grating, a 10 line per inch grating, a grating of 1 pattern per inch and a grating of 1 pattern per 10 in. The tracks on the grating are interrogated in a manner similar to that just described. This type of transducer, using gratings only, is more expensive as electronic interpreters such as those described for the grating in this paper are required, one for each track of the grating. The transducer is suitable for linear measurement on machines operated by linear devices, such as hydraulic rams and where no rotating members are present. A transducer can be constructed to suit any particular physical requirements, or the economics of a particular problem, by using either a multi-track grating only or a combination of shaft digitizers and one or more gratings. Rotary motions can also be

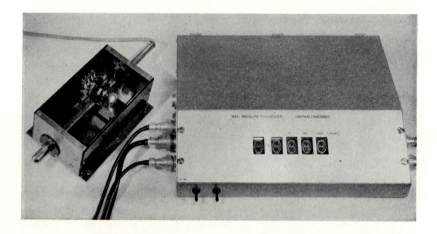

Fig. 9. Prototype linear transducer and interpolator.

Fig. 10.   NEL absolute measuring system applied to turning lathe.

measured and controlled when the system is used in conjunction with a multi-track radial grating, this has obvious advantages in the accurate indexing of machine tables and in the field of accurate rotary displacement measurement such as in radio telescopes, etc.

Figure 9 shows a prototype linear transducer and interpolator. Figure 10 shows the measuring system fitted to a turning lathe to measure the position of the tool-holder. Two gratings have been used: a 10 line per inch and a 100 line per inch which provide hundredths of inches, thousandths of inches and half-thousandths of inches on the indicator, while tenths of inches and inches are derived from two digitizers fitted to the lead screw of the tool-holder slide. Figure 11 shows in some greater detail the linear transducer and rotary digitizers fitted to the turning lathe.

FIG. 11. Grating transducer and commercial digitizers.

CRITICAL FEATURES OF THE SYSTEM

The critical features will be discussed in the logical sequence from the light source through to the indicator.

Lamp

The ideal light source would be an infinitely small spot, this would provide better collimation of light for a given set of conditions (i.e. lenses, etc.) but in practice the lamp filament is of a finite length and, aiming for a serrisoidal output waveform, this filament length is best held in the same plane as the lines of the grating.

Collimation

The collimation of light should be as good as possible so that an accurate replica of the grating structure impinges on the index and photocell, otherwise stray angular light destroys the structure and the resultant waveform at the photocells deteriorates from a serrisoid into "rounded" version which approaches a sinusoid.

Grating/Index

The grating and index (a section of the same grating) should have equal mark/space ratio with each line truly opaque and as "clean" as possible. This is particularly desirable in coarser gratings where less "averaging" is done as the photocell' 'observes" fewer lines of the grating.

Photocells

The silicone photo-voltaic type has been found best for this work. As the prime reference for the system depends on accurately defining the zero-points of the sine and cosine wave-forms from the grating it is important that these points are stable for different degrees of lamp brightness, grating density and temperature. The two photocells are illuminated from the 000° (sine) and 180° (− sine) indices. The zero level of the composite output wave-form occurs when both indices are passing half-light; in this condition the two photocells will deliver equal and opposite voltages which will cancel to zero. The cells must therefore be checked as a matched pair by being exposed to a common light source over a range of light brightness; if equality cannot be achieved (zero out under these conditions) then a matching resistor must be inserted and adjusted. This precaution will ensure stability when variations in light brightness or grating density are encountered. The matched pairs of photocells should be checked over a temperature range.

From Fig. 5 it will be seen that curves B and C are formed from curve A and the cosine, also that curves D and E are formed from the cosine and the negated, that is −sine. A fall in signal level will not alter the point at which any curve crosses zero, only the slope will alter. In practice some rounding of the wave-forms at the top and bottom is experienced; if this is severe enough to encroach to the point at which a generated "slave" waveform is crossing zero, then some definition will be lost due to hysteresis and variations of amplitude in the waveforms could introduce error in some of the "slave" waveforms. Thus the amount of accurate sub-division possible is largely dependent on the purity of the serrisoidal signal from the photocells.

Photocell Amplifier

The important factors for a suitable amplifier are: stability over an appropriate temperature range (as met in industrial environments), adequate gain with low impedance output, high stability of the "zero in, zero out" condition. There are several commercial solid state amplifiers which meet these conditions.

Curve Family Generator

This is shown in Fig. 4 and the main object is to generate a family of curves from the basic sine/cos curves which cross zero at the appropriate displacement point. Where input waveforms are good serrisoids and the sine/cos relationship is accurate, 4(a) will be sufficient (all resistors equal); where the sine/cos relationship is out of correct phase relationship 4(b) will permit a variable offset for the cosine inputs and will rectify this. When the input waveforms tend to be sinusoidal (a grating of 1000 or more lines per inch) then one must use suitable value (sin/cos) resistors.

Level Detector

The ideal level detector changes state as the input waveform goes through zero, is very stable and has very small and controllable hysteresis. The best circuit found and used in the device is that of ref. 1 and, although relatively expensive, is considered ideal. The permitted hysteresis is adjusted to be slightly greater than the noise level (about 30 mV) which, in practice, give a displacement hysteresis of less than 0·0001 in when a 100 line per inch grating is being divided.

FIG. 12. Finalized digital readout system (D.R.O.), using 3-track grating, fitted to lathe; resolution is 0·0001 in and stroke is 20 in diameter. (N.B. fittings on left is for in-process gauging and is not part of the absolute measuring system.)

Logic Circuits

Commercial NOR gates (Mullard Norbit) were used with success initially but, to reduce cost and size and to run the complete system from a ± 12 V supply, NOR gates are now constructed on plug-in cards from transistors and resistors. The only requirement is that they switch rapidly over a suitable temperature range. Emitter followers are used to light the indicator lamps and provide a register output for binary subtractors, etc.

REFERENCES

1.   B. MURARI, *A Transistorized Level Detector*. SGS Fairchild Report AR 113, p. 10, Figs. 3–7. Milan: Societá Generale Suniconduttori Fairchild, 1963.

NOTE

The N.E.L. absolute digital measuring system is now manufactured under the trade names of "TELETRAK" and "AUTO-SCALE" by Messrs. Whitwell Electronic Developments Ltd., Glasgow, and Moore Reed (Industrial) Ltd., Andover, respectively.

LOW-COST NUMERICAL CONTROL SYSTEM

A. Russell

Ministry of Technology National Engineering Laboratory,
Machinery Group, Machine Tools and Metrology Division

SUMMARY

This system is based on NEL absolute displacement transducers as the measuring elements. These transducers were the subject of the paper "A method of extracting absolute digital displacement information from optical gratings" presented to the Seventh International MTDR Conference.

Digital signals are used to give coarse positioning and accurate analogue signals for fine positioning and servo control. The stringency of the digital apparatus is relaxed by a factor of 100 by this method and one set of digital apparatus can give point-to-point control of many axes.

1. INTRODUCTION

Since the introduction of the NEL absolute displacement transducer, which was the subject of the paper "A method for extracting absolute digital displacement information from optical gratings" presented at the Seventh International MTDR Conference, it has been observed that accurate analogue control signals can be generated from the optical grating at very little additional cost. Experiments were carried out on a lathe slide-way, using a servo-amplifier and printed circuit motor, and the results were felt to justify the construction of an $x - y$ drilling machine under full tape control.

This feature is, at present, unique to the absolute transducers used but these are more expensive than incremental transducers; the cost of the additional equipment was therefore kept to a minimum.

2. THE TRANSDUCER

The NEL absolute displacement transducer is described elsewhere[1] and only the main features and method of forming the control signals will be given here.

The transducer can be purchased in various forms to suit particular applications and can have a resolution of 0·0001 inch in a stroke length of 100 in (1 part in 10^6). Indication is by in-line Nixie tubes and datum can be altered at will by switches on the indicator panel. Being absolute in nature the device can be switched off at any time without losing datum and is also immune to electrical interference. The basis of measurement is a linear optical grating.

The readout data are also available in electrical form as digital binary-coded-decimal signals which can be used for logging purposes or digital control.

As the machine member is traversed, two waveforms in quadrature are emitted from the grating reading-head; these will be sinusoidal from fine gratings and serrisoidal from coarse gratings, but, for convenience, serrisoids will be used in the illustrations. The quadrature signals are amplified by an amplifier circuit which also provides an inverted signal at its output. The four resultant signals can be referred to as sine (0°), $\overline{\text{sine}}$ (180°), cosine (90°) and $\overline{\text{cosine}}$ (270°).

By applying the four signals to a sine-cos potentiometer as shown in Fig. 1 ten wave-forms are made available, one of which will cross the zero-volts level each 36° of traverse. These are labelled 0°, 36°, 72°, etc., in Fig. 1 and their displacements are shown in Fig. 2.

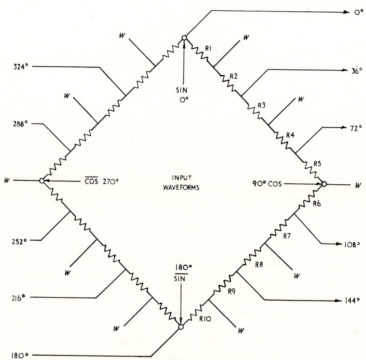

FIG. 1. Sin-cos potentiometer.

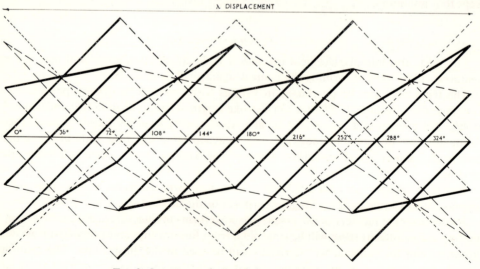

FIG. 2. Output waveforms of sin-cos potentiometer.

It will be appreciated that, as there is 90° displacement at the zero-volts level between the sin (0°) and cos (90°) waveforms, if 2/5 of the potential of these two waveforms is tapped off, this will give a new waveform which will traverse the zero-volts level at 90° × 2/5 = 36°. The amplitude of the "slave" waveforms at any angle θ will be: for the sin/cos "slaves" (sine waveform):

$$V = \cos \theta + \left\{ (\sin \theta - \cos \theta) \times \frac{R1 + R2}{R1 + R2 + R3 + R4 + R5} \right\} \tag{1}$$

and for the cos/$\overline{\sin}$ "slaves"

$$W = \cos \theta + \left\{ (\overline{\sin} \theta - \cos \theta) \times \frac{R9 + R10}{R6 + R7 + R8 + R9 + R10} \right\} \tag{2}$$

A suitable potential divider can be chosen which will produce any desired number of "slave" waveforms per cycle from the grating. The potential divider would consist of a chain of resistors which would yield, at the resistor junctions, the potential division (P), for sin/cos voltages.

$$P_{s/c} = \frac{R1}{R1 + R2} = \frac{\sin \theta}{\sin \theta - \cos \theta} \tag{3}$$

and for cos/$\overline{\sin}$ voltages

$$P_{c/s} = \frac{R4}{R4 + R5} = \frac{\cos \theta}{\cos \theta - \overline{\sin} \theta}. \tag{4}$$

For a decimal system the number of "slaves" required is ten or a multiple of ten.

The ten waveforms of Fig. 2 are converted into digital form at the zero-crossing level in such a way that the zones between waveforms can be displayed as the numerals $0 \rightarrow 9$ in the least significant stage of the readout; $0° \rightarrow 36° = 0$, $36° \rightarrow 72° = 1$, $72° \rightarrow 108° = 2$, etc.

It will be observed from Fig. 1 that a further similar set of waveforms will be available at the resistor-pair junctions. These are marked W and will be phase-displaced from the readout waveforms of Fig. 2 by 18°. These are the control waveforms and are used to servo-lock the system into the midpoint of the commanded least-significant digit of the input data.

3. DIGITAL COARSE CONTROL

Coarse control of the machine-tool axis is effected by the binary-coded-decimal (BCD) output from the axis readout. This is fed (Fig. 3) to a subtractor via (if appropriate) an x gate where it is subtracted from the position input data of the control tape. The BCD "difference" information contains the sign and magnitude of the error between the instantaneous position of the axis and the commanded position of the input data. This digital information is converted into analogue form and applied to the control amplifier; the motor then drives the axis to the commanded position at a speed limited by the speed information of the control tape. When the balance is reached the subtractor output will be zero and the motor stops.

The "staircase" coarse control waveform is shown in Fig. 4. The "steps" represent discrete least significant digits from the readout and are bounded by the zero-crossing points of the waveforms of Fig. 2. This does not constitute a very satisfactory servo-system as the control voltage is zero at the balance to cause the system to move out of balance. A one-

A. RUSSELL

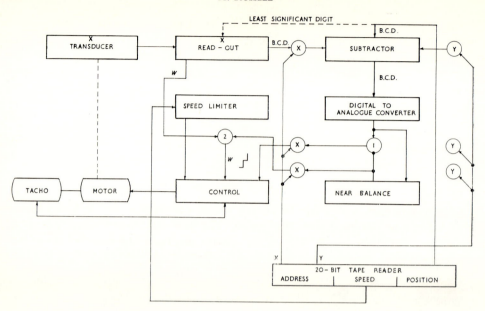

FIG. 3. Point-to-point servo-lock control system.

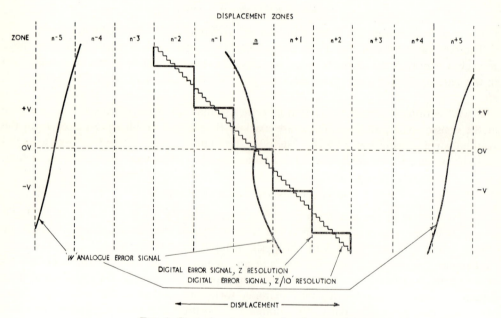

FIG. 4. Control system error signal comparison.

digit out of balance causes a step function to be applied to the system and this can drive the servo out of balance in the opposite direction. Reducing the gain of the system can help but results in a very "slack" servo.

4. ANALOGUE FINE CONTROL

Many methods have been devised in the past for providing fine control. These include "vibration" of the transducer by at least one least significant digit so that the balance point becomes a proportional signal based on the "dwell" times above and below balance.

A simpler method, which does not introduce moving parts, is to enlist the appropriate analogue waveform W as a fine control signal. This is shown as "analogue error signal" in Fig. 4 and is almost ideal for the purpose. It will be seen to be of correct polarity for control purposes to within ± 5 least-significant digits. Beyond this it would cause the system to "lock" to a position ± 10 units in error. The digital control is thus relaxed as it need only attain the balance point ± 5 units, where the analogue signal can take over control. In practice, where the least significant digit is one thousandth, the digital requirements are ± 5 thousandths and the analogue control signal has a repeatability better than one-tenth. Thus the digital requirements are relaxed by two orders of magnitude.

The method of using waveform W is shown in Fig. 3. The appropriate waveform is selected by the least significant digit of the input data and is fed from the readout to the 2 gate. When the axis is near balance an output is delivered from the near balance unit, this "inhibits" or closes the 1 gate which then prevents the digital control signal from reaching control; however, it also opens the 2 gate and the analogue signal takes over.

5. PRACTICAL CONSIDERATIONS

It will be realized that the machine-tool axis is now "locked" into position without the need for any mechanical braking, and is held by virtue of its own transducer's waveform. This means that the tape reader can be stepped to the next instruction and a further operation carried out while the previous axis will remain servo-controlled in position until again instructed to take up a new position. In this way one set of digital apparatus (subtractor, D/A converter, near balance unit and tape reader) can control any number of axes and cutting functions in a programmed sequence.

A cheap 20-bit tape reader is found useful for this application as a complete instruction of address (2 bits), speed (2 bits) and position (16 bits) can be contained in one row of punchings. The electronic storing of information is not then necessary.

The motor used is a printed circuit torque motor which has very low mechanical and electrical time constants, has good stalling torque properties and is cheap.

Where the system is used for drilling and boring operations some overshoot of position can be tolerated and a high traverse rate realized; however in applications that do not permit overshoot a tacho generator can be used as a velocity feed-back and overshoot can be prevented at the expense of slower traverse rates.

6. CONCLUSIONS

The main advantages of the system described are:
(a) control of many axes with one set of apparatus,
(b) digital control requirements are relaxed by two orders,
(c) axes are servo-controlled during cutting operation, and
(d) no expensive data storage necessary.
There are also the advantages of absolute measurements.

It is obvious that, where digital readout is not necessary as in high-speed plotting, etc., the complete control system could be achieved using only the grating analogue signals. Work is continuing on this at NEL.

ACKNOWLEDGEMENTS

This paper is published by permission of the Director of the National Engineering Laboratory, Ministry of Technology. It is Crown copyright and is reproduced by permission of the Controller of H.M. Stationery Office.

REFERENCE

1. A. RUSSELL, An Absolute Digital Measuring System using an Optical Grating and Shaft Encoder. *Instrum. Rev.*, **13** (5), 1966, 174–175.

A NEW GENERATION OF CONTINUOUS PATH N/C SYSTEMS

A. A. SHUMSHERUDDIN

Numerical Control Section, Mechanical Engineering Laboratory,
English Electric Co., Ltd., Whetstone, Leicester

NOMENCLATURE

X or ΔX	distance moved by X slide of machine tool
Y or ΔY	distance moved by Y slide of machine tool
Z or ΔZ	distance moved by Z slide of machine tool
L or ΔL	distance moved by cutter vectorially in space
V	instantaneous vector feed velocity
V_i	initial feed velocity
V_0	final feed velocity
a	vector acceleration
R	radius of curvature
S	deceleration distance
anc.func.	ancillary function
prep.func.	prep. function

Logic Symbols

$\&$	"and" gate
$\overline{\&}$	"nand" gate
S	set trigger
R	reset trigger
SS	set steering
RS	reset steering
1	output goes to 1 when set
0	output goes to 0 when set

INTRODUCTION

A few years ago, point-to-point N/C machines were used strictly for point-to-point type machining, such as drilling and tapping, and continuous path N/C was used for contouring only and had to be computer programmed.

Over the years, however, on the one hand point-to-point systems have become more complex. In order to reduce material handling and setting up time, in addition to drilling and tapping they are required to do straight line milling and sometimes planetary milling. Also, in order to improve programming efficiency, a certain number of point-to-point computer programmes have emerged. Continuous path systems, on the other hand, have tended to be simplified; this is because over 65% of machining operations do not require complicated contouring and hence contouring systems have had to be designed along more

1191

general lines, such that they can be manually programmed for simple operations such as drilling, straight line and circular milling, and computer programmed for contouring. Furthermore, owing to recent advances in electronic fabrication, the cost difference has somewhat narrowed. The difference in cost is determined by interfaces such as servos, measuring systems, paper tape readers, etc., rather than the electronics. As part of a general investigation into the applications of N/C systems, the Mechanical Engineering Laboratories of the English Electric Co., Ltd., at Whetstone, are examining second generation systems which enable simple machining operations such as point-to-point work, straight line and circular milling to be programmed manually but much more efficiently and cheaply than older systems, since the programmer does not have to do any of the tedious time-consuming calculations required by the older systems. At the same time, all the features of computer programming for more complicated shapes, multiaxes cutter movements and lower cost, are preserved.

GENERAL DESCRIPTION OF SYSTEM

Figure 1 shows the general scheme of the system. If the component to be manufactured is of a simple nature, such that the cutter path can be described by straight lines and circular arcs, then manual programming is used and the part programme, which is punched on paper tape using a flexowriter, is fed directly to the director, or interpolator, which controls the machine tool. If, on the other hand, the workpiece has an artistic shape or is complicated,

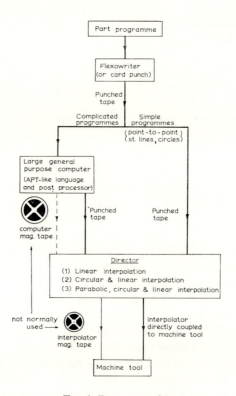

Fig. 1. Data processing.

then an APT-like language (APT, 2CL, EXAPT or any similar language) is used and the programme is run on a general-purpose computer which punches out the paper tape for feeding to the director.

The basic design of the director is such that it can perform linear and circular interpolation; however, it can easily be modified to linear only or even parabolic interpolation. In addition, it has built-in acceleration control, multi-axis capability, feed rate over-ride, ancillary function controls and provision is made for cutter radius compensation if desired.

Owing to the general design features of the director, several fall-out systems are possible. For example, if a vast amount of complicated contouring work is to be done, such as in aircraft application, then a magnetic tape link between the general-purpose computer and director may be used. Alternatively, if it is desired to share the director with up to four machine tools, then a magnetic tape link can also be used. On account of the built-in acceleration control, low performance electric servo-drives can be used, particularly for point-to-point and simple milling applications involving small to medium sized machine tools. Thus, very cheap systems are possible. However, hydraulic servo-drivers are recommended for larger machines or when faster traverse rates and good surface finish are

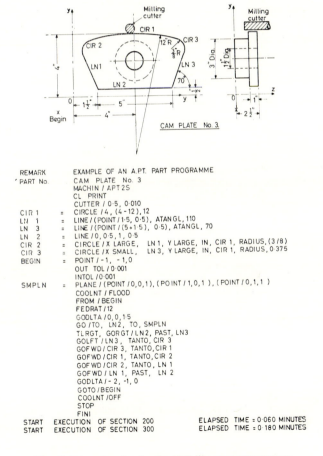

FIG. 2. Example of a simple APT-like part programme.

```
.....SECTN.3.....                                              CARD NO.010    TAPE NO.2
   CAM PLATE NO.3                                              CARD NO.130    TAPE NO.4
CUTTER/  0.5000                                                CARD NO.140    TAPE NO.6
COOLNT/ FLOOD
FROM/BEGIN                                                     CARD NO.150    TAPE NO.8
            X              Y              Z
         -1.0000000     -1.0000000      0.
DS IS       X              Y              Z                    CARD NO.160    TAPE NO.10
         -1.0000000     -1.0000000      0.5000000
DS IS/  L2  X              Y              Z                    CARD NO.170    TAPE NO.12
         -1.0000000      0.2500000      0.1000000
DS IS/  L2  X              Y              Z                    CARD NO.180    TAPE NO.14
          6.6750517      0.2500000      0.1000000
DS IS/  L3  X              Y              Z                    CARD NO.190    TAPE NO.16
          7.5800157      2.7363676      0.1000000
      C3(  0) = CIRCLE/   6.7578    3.0356   0.      0.6250     CARD NO.200    TAPE NO.18
DS IS/  C3                                                     CARD NO.200    TAPE NO.19
            X              Y              Z
          7.5901973      2.7643723      0.1000000
          7.6271917      2.9324686      0.1000000
          7.6305845      3.1045558      0.1000000
          7.6002438      3.2739833      0.1000000
          7.5373415      3.4342026      0.1000000
          7.4443080      3.5790211      0.1000000
          7.3247383      3.7028410      0.1000000
          7.1832532      3.8008764      0.1000000
          7.0253206      3.8693377      0.1000000
      C1(  0) = CIRCLE/   4.0000   -8.0000   0.      12.0000    CARD NO.210    TAPE NO.21
DS IS/  C1                                                     CARD NO.210    TAPE NO.22
            X              Y              Z
          6.8604754      3.9118601      0.1000000
          6.2206648      4.0475489      0.1000000
          5.5745242      4.1488978      0.1000000
          4.9238954      4.2156178      0.1000000
          4.2706329      4.2475188      0.1000000
          3.6165987      4.2445097      0.1000000
          2.9636572      4.2065992      0.1000000
          2.3136694      4.1338954      0.1000000
          1.6684880      4.0266054      0.1000000
          1.0300781      3.8845305      0.1000000
      C2(  0) = CIRCLE/   1.2422    3.0356   0.      0.6250     CARD NO.220    TAPE NO.24
DS IS/  C2                                                     CARD NO.220    TAPE NO.25
            X              Y              Z
          1.0008422      3.8772049      0.1000000
          0.8408965      3.8137377      0.1000000
          0.6964542      3.7202115      0.1000000
          0.5730953      3.6002394      0.1000000
          0.4755853      3.4584562      0.1000000
          0.4076910      3.3003389      0.1000000
          0.3720353      3.1319959      0.1000000
          0.3699956      2.9599304      0.1000000
          0.4199844      2.7363678      0.1000000   CARD NO.230    TAPE NO.27
DS IS/  L1  X              Y              Z          CARD NO.240    TAPE NO.29
          1.3249481      0.2500000      0.1000000
DS IS       X              Y              Z          CARD NO.250    TAPE NO.31
         -0.6750519     -0.7500000      0.1000000
DS IS/ BEGIN  X            Y              Z          CARD NO.260    TAPE NO.33
         -1.0000000     -1.0000000      0.           CARD NO.270    TAPE NO.35
                                                     CARD NO.280    TAPE NO.37
COOLNT/     OFF
   STOP
FINI

END OF SECTN3
```

FIG. 3. Computer print-out.

required. If only manual programming is necessary, then a slow speed mechanical tape reader can be used with further reduction in cost.

Figure 2 shows an example of a computer programme. It consists of three parts. The first part gives the title, part number and other general information. This is followed by a set of geometric definition statements which completely define the shape of the workpiece. Thus, the statement "LN2 LINE/0, 0·5, 1, 0·5" describes the side of the cam plate bounded by line "LN2" which is a straight line passing through the points $(0, 0·5)$ and $(1, 0·5)$ in the x–y plane. Similarly, the statement "CLR 1 = CIRCLE/4, (4–12), 12" describes the upper side of the cam bounded by a circle "CIR 1" with centre at $(4, -8)$ and 12 in radius. Having defined the geometric shape of the workpiece in this way, next the cutter must be directed to follow a specific path in order to machine the component and this is achieved by the motion command statements. For example, the statement "GOF D/CIR 1, TANTO CIR 2" implies that the cutter will move forward along circle "CIR 1" until the point of tangency between "CIR 1" and "CIR 2" is reached. The next statement "GOFWD/CIR 2, TANTO LN 1" causes it to move further along circle "CIR 2" to the point of tangency with "LN 1". In other words, the motion command statements are a form of back seat driving! The example in Fig. 2 is a relatively simple one. However, for instance, there are some 400 defini-

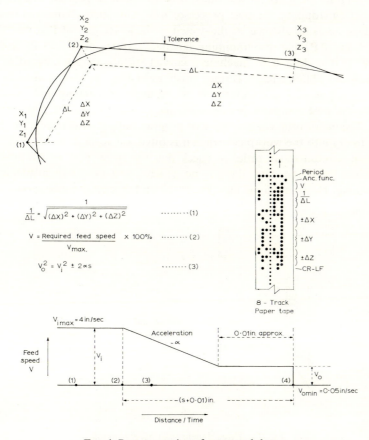

FIG. 4. Post processing of computed data.

tion statements in the APT language which will cope with most workpiece shapes, including three-dimensional shapes such as aerofoil forms.

When the programme is processed on a general-purpose computer, the computer works out a set of coordinates such that if these were joined by straight lines, the straight lines would not deviate from the desired cutter centre path by an amount greater than the tolerance specified by the programmer, as can be seen in Fig. 3. As expected, with this form of linear interpolation, a large number of points are required to fit the curved parts of the workpiece and very few are required for the straight edges. The print-out also indicates the parts of the workpiece for which the coordinates are calculated and the number of the punched card from which the data was obtained, in order to simplify the checking or altering of the programme.

Before the computed data can be used to control the machine tool, some further processing is necessary in order to transfer the data into a form which is acceptable to the particular control system in use and to punch it out on tape according to the required format. This is done by a special post processor programme which is run on the computer at the same time as the APT-like programme and, therefore, does not introduce an additional step. In this case, as can be seen from Fig. 4, what the post processor does is transform the coordinate information into incremental form and punch out a block of data for each straight line segment of the cutter path, as shown. Since it is necessary to prevent overshoot at sharp corners and during stopping, the post processor must anticipate changes in feed speed, V, and make adjustments in advance, as illustrated in the lower part of Fig. 4.

The present-day APT-like programming languages apply strictly to three-dimensional programming. If it is desired to machine a curved surface tangentially so as to remove metal efficiently and obtain a better surface finish, as in Fig. 5b, then multi-axis movement of the cutter is required. The computation of the angle of pitching and yawing must be carried out by the post processor programme, and this can make the post processor very complicated. The major problem in multi-axis programming is associated with the tendency of the cutter to dig into the workpiece when it is tilted, on account of which a fair amount of computation is required, in order to check that the cutter path is still held within the desired tolerance band. To date, there is no general post processor available for multi-axis programming. However, some have been written for special applications.

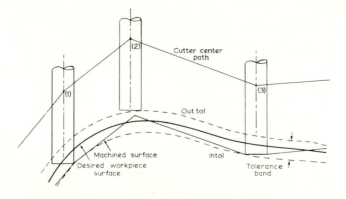

Fig. 5a. Three-dimensional programming.

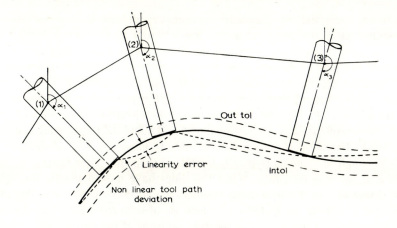

FIG. 5b. Multiaxis programming.

THE MACHINE TOOL DIRECTOR

The function of the director is to compute the cutter path from the input information on tape and accordingly control the slides of the machine tool with the aid of interfaces such as position sensors and servomechanisms.

One of two types of digital curve generators or interpolators are generally used in order

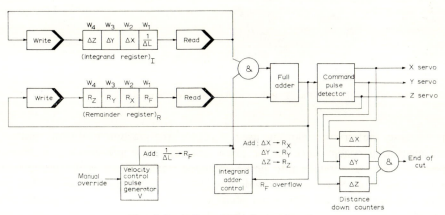

FIG. 6. Basic linear DDA interpolator.

to compute the cutter path. These fall into the category of Digital Differential Analysers (DDA) and Pulse Rate Multipliers (PRM) types. Figure 6 shows a block diagram of a simple linear DDA interpolator. It consists of two delay line registers of four word length, a full adder, down counters and some control circuitry. Let it be assumed that the data from the paper tape in Fig. 4, namely, $V, 1/\Delta L, \Delta X, \Delta Y, \Delta Z$ are read into the relevant word stores at the start of the cycle and the remainder registers are cleared. If now the velocity control pulse generator generates pulses at a rate proportional to the feed velocity V such that, every time a pulse appears, the words in the delay lines are circulated through the adder and only $1/\Delta L$ is added to R_F each time. Then, owing to the words in the remainder register being of finite length, R_F must overflow and the rate of overflow will be proportional to $V/\Delta L$. If it is also arranged that every time an overflow pulse from R_F is detected, $\Delta X, \Delta Y$ and ΔZ are added to R_x, R_y and R_z respectively, then overflows or command pulses from R_x, R_y, R_z must result at rates proportional to $\Delta X, \Delta Y$ and ΔZ respectively and the required feed velocity V. Therefore, if these pulses are used to move the X, Y, Z slides of a machine tool through small increments (typically 0·0001 in) then the cutter must generate the true vector path at the true feed rate. When the slides have moved exactly through the programmed distances, the distance down counters would become zero and this information could be used to initiate the transfer of a new set of data from tape into the registers, in order to commence the generation of the next segment of the cutter path.

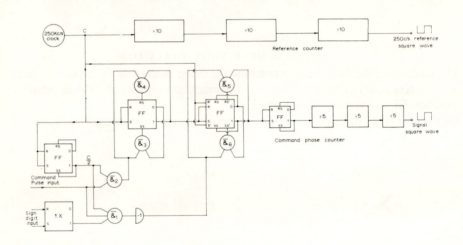

FIG. 7. Pulse to phase analogue converter.

In this case, the servomechanisms used to control the machine tool slide, and the associated measuring system work on the phase analogue principle, hence it is necessary to convert the command pulses into phase analogue signals which is illustrated in Fig. 7. The three decade reference counter divides the 250 kc/s clock frequency by 1000 to produce a 250 c/s square wave. The command phase counter also has the same length, however, some logic circuitry is added to it such that whenever a command pulse appears, an extra pulse is added or one is stopped, according to the sign digit, so as to advance or retard the signal square wave phase by 1/1000th of a cycle respectively. This would correspond to 0·0001 in if 360° change in phase corresponded to a slide movement of 0·1 in, which is determined by the position measuring system.

Figure 8 shows the phase analogue method of controlling an axis of the machine tool using an inductive, capacitive, or photo-electric table position measuring system or simply a resolver suitably geared to the leadscrew for less accurate work. A detailed description of this system was presented in ref. 1.

N/C contouring systems based on the above method of linear interpolation are suitable for application where only computer programming is used. Majority of applications, however, require manual programming for which both linear and circular cutter path generation is essential. Such numerical control systems based on a similar DDA design principle

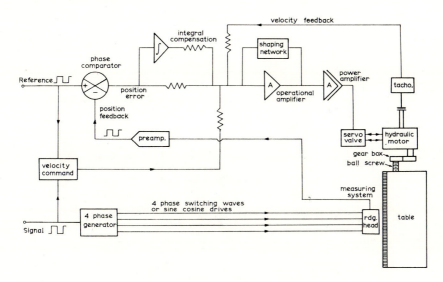

FIG. 8. One axis of a continuous path servo system.

have been described by Walker,[2] Henegar[3] and Mayorov.[4] Walker's N/C system utilizes magnetic tape input and requires the use of a general-purpose computer or a special computer and lacks the direct manual programming facility common to paper tape input systems. Although the N/C systems described by Henegar and Mayorov have the advantages of paper tape input and direct manual programming facilities, the programming of such systems are not straightforward and require the programmer to do a fair amount of time-consuming and tedious calculations specially for three-dimensional applications.

Figure 9 shows a block diagram of the combined circular and linear DDA interpolator which was designed to have much simpler manual programming facility and does not require the programmer to do any calculations such as working out the feed rate number (FRN) for instance; hence manual programming time is greatly reduced. The principle of operation of this interpolator is somewhat similar to the system of Fig. 6, in that during linear interpolation it operates in the same way. However, during circular interpolation, in the X–Y phase for example, as shown in the figure, the extra half adder/subtractor causes an increment to be added to ΔY every time an X-command pulse is generated and a decrement to be subtracted from ΔX every time a Y-command pulse is generated. Hence it is seen that if in the integrand register, ΔZ is set equal to zero and ΔX and ΔY are made equal to the initial radius vector components ΔY^1 and ΔX^1, so that also

$$\frac{1}{\Delta L} = \frac{1}{\sqrt{[(\Delta X^1)^2 + (\Delta Y^1)^2]}} = \frac{1}{R}$$

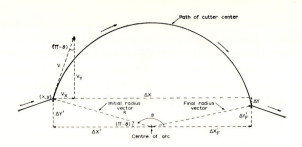

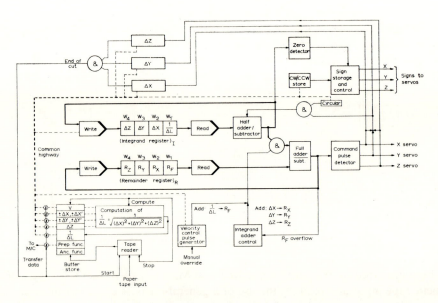

FIG. 9. Basic linear or circular DDA interpolator.

it follows that by similar triangles and the fact that the instantaneous overflow pulse rates reaching the Y and X axes are proportional to $\Delta X^1(V/R)$ and $\Delta Y^1(V/R)$ respectively by comparison with the linear cutter path generation case,

$$V_y = \frac{dy}{dt} = \Delta X^1 \frac{V}{R} = R \cos{(\pi - \theta)} . \frac{V}{R} \qquad (1)$$

and

$$V_x = \frac{dx}{dt} = \Delta Y^1 \frac{V}{R} = R \cos{(\pi - \theta)} . \frac{V}{R} \qquad (2)$$

where θ is the angle of sweep of the radius vector R. Hence by equations (1) and (2)

$$\frac{dy}{dx} = -\cot{\theta} \qquad (3)$$

The origin being chosen to be the arc centre, $\cot \theta = x/y$; so that eq. (3) becomes $dy/dx = x/y$, which, upon integration and substitution of the initial conditions $y = 0$, when $x = +R$ or $-R$ gives the equation of a circle $y^2 + x^2 = R^2$. Hence it is seen that by progressively varying the values of ΔX and ΔY in the integrand register by means of the half adder/subtractor, causes a continuous change in the overflow rate from R_x and R_y, which results in a progressive variation in slope such that a circular cutter path is generated at constant vector feed speed, since by eqs. (1) and (2) $V_y^2 + V_x^2 = V^2 \cos^2 \theta + V^2 \sin^2 \theta = V^2$.

The overflow pulse commands from the remainder register are converted into phase–analogue form in the same way as in Fig. 7, whether these pulses are added or subtracted is determined by the sign storage and control logic. All this does is select the signs in accordance with the signs of the radius vector components ΔX^1 and ΔY^1 and the preparation function programmed on the tape, so that the right quadrant and direction of rotation of the circular cutter path is generated. The preparation function is chosen according to whether a straight line, clockwise circular cutter path or counterclockwise circular cutter path is required. Where more than one quadrant of a circular path is to be generated the changes in sign at the $90°$ points (i.e. when $\cos (\pi - \theta)$ and $\sin (\pi - \theta)$ in eqs. (1) and (2) may change sign) is initiated by a zero detection circuit which is activated when the zero state of any of the words in the integrand register is reached. The end point of the circular cutter path is detected by the distance down counters which are initially set to the distances to be moved along the axes ΔX, ΔY and ΔZ, and therefore count down to zero when the end point is reached. While one segment of the cutter path is being generated, data from paper tape for the next segment is read into the buffer stores following which $1/\Delta L$ or $1/R$ is computed by circuits which are similar to those used in conventional digital computers, and are also stored in the buffer stores as indicated in the figure. Subsequently, when the end of cut signal is received from the distance down counters, the data from the buffer stores is transferred to the appropriate registers in order to commence the generation of the next segment of the cutter path.

The delay lines in the above interpolators can be of the magnetostrictive, piezoelectric, mercury column, scratch pad, or ferrite core type. These low level devices, however, are prone to electrical interferences which are common to machine shops, hence high level solid state shift registers had to be used. Furthermore, the electronic computing circuits, which are in this case similar to a general-purpose computer, are required to be very fast in operation and use up a lot of hardware, particularly the portion of the circuits used to calculate $1/\Delta L$ and this led to a high overall cost of the system. It was therefore decided, as a later development, to design and build another system which would combine the advantages of simpler programming with lower cost.

A much neater, more compact and economical interpolator design than the above mentioned digital differential analyser type can be achieved by use of the pulse rate multiplier (PRM) principle. Although, currently, about three or four different makes of N/C contouring systems are available on the market which utilizes this technique, only one of them have been described, by Evans and Kelling,[5] in the literature in some detail. Nevertheless, all these systems require time-consuming, complicated manual programming; for instance, the programmer has to calculate the so-called feed rate number

$$\text{FRN} = \frac{V}{\sqrt{[(\Delta X)^2 + (\Delta Y)^2 + (\Delta Z)^2]}}$$

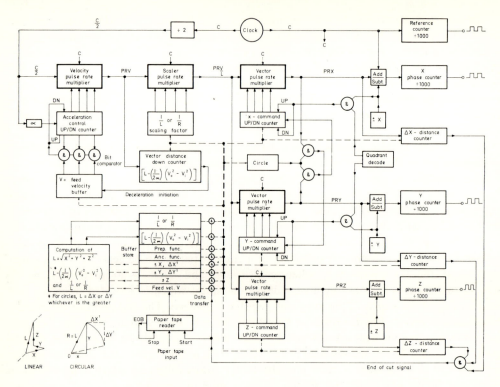

Fig. 10. Block diagram of pulse rate type interpolator.

Figure 10 shows the block diagram of the new generation of directors based on the PRM technique, which do not require such time-consuming and complicated manual programming and, at the same time, require much less hardware than the DDA type described above. In order to understand how this works, it is essential to know first what a pulse rate multiplier does.

Figure 11 shows one decade of a pulse rate multiplier. It consists of a four-bit decade counter, flip-flops FF1 to FF4, which counts in the 1-2-4-5 mode as the wave forms show, but this could be 1-2-4-2 or some other mode. The gates $\overline{\&}1$ to $\overline{\&}4$ are so connected that when the respective multiplier input bit, i.e. bistables BS1 to BS4, is in the 1 state the gates generate 1, 2, 4 or 8 pulses on the output line respectively for every 10 input pulses. Thus, as can be seen in the lower part of the figure, a multiplier input of 5 (0101 in BCD) produces exactly $10 \times (5/10) = 5$ pulses and an input of 9 (1001 in BCD) produces exactly $10 \times (9/10) = 9$ pulses for every 10 input pulses, or $50 \times (9/10) = 45$ pulses for every 50 input pulses on an average. In general, the following rules apply to pulse rate multipliers.

1. Any number of decades can be cascaded to extend the digital accuracy to any desired number of digits. Binary or BCD operations are possible.

2. Both input and output pulses can be irregular, hence output of one multiplier can be fed to the input of another.

3. The multiplier input can be changed suddenly in the time interval between successive input pulses. Alternatively, the input can be changed incrementally by, for example, connecting the multiplier input bistables so as to form a BCD up counter, down counter, or an

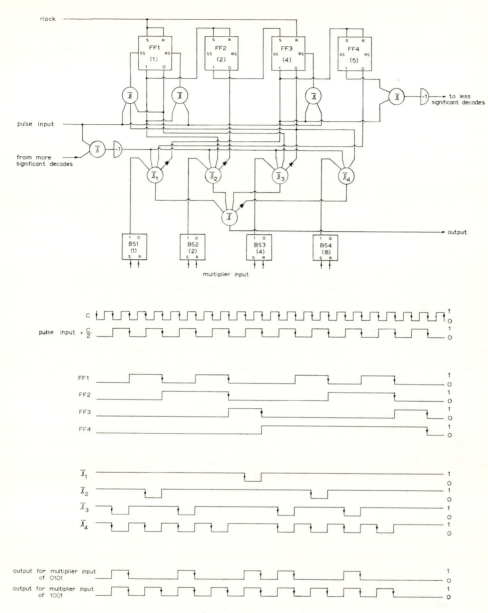

FIG. 11. Digital pulse rate multiplier.

up/down counter and causing it to count up or down, provided the counter is made to change state in synchronism with the input pulses, preferably in the time interval in between.

4. They have the advantage of being much faster than conventional digital multipliers, require less hardware, and, consequently, are cheaper than other types of digital multipliers.

In Fig. 10 the pulse rate multipliers are the same as described above, except they are shown in block form. A sub-division of the clock frequency $C/2$ is multiplied by the feed velocity

V so as to produce an output pulse rate PRV. The actual input to the "velocity pulse rate multiplier" is held in the up/down counter which tracks the input velocity V stored in the feed velocity buffer. Thus, if the value of V is changed suddenly, the bit comparator causes the up/down counter to count up or down as required at a clocking rate α, so that the rate of change of V is proportional to α, the required acceleration. Subsequently, for straight line cutter paths V is multiplied by the scaling factor $1/L$, and by the programmed X, Y and Z movements in the "vector pulse rate multipliers", which produce command pulse rates proportional to $V.X/L$, $V.Y/L$ and $V.Z/L$. Therefore, since X/L, Y/L, Z/L form the unit space vector, the cutter must follow the true vector path at the correct feed rate. Conversion from pulse to phase analogue command takes place in the same manner as described previously. The distances moved along the axes are totaled up by the distance counters, which provide the end-of-cut signal in order to initiate the start of the next segment of the cutter path when the required movements have taken place. For circular arc generation, the initial radius vectors are fed into the "command up/down counters" of the X and Y axes "vector pulse rate multipliers", which are cross-coupled to form the familiar sine–cosine generator. Thus, when say a clockwise circular arc is programmed on tape as shown in the bottom left hand of the figure, the appropriate preparation function on tape energizes the cross coupling gates such that whenever a Y-axis command pulse PRY appears the X-command up/down counter which is initially set to the value of Y is made to count down one. When an X-axis command pulse appears, the Y-command up/down counter, which is initially set to X, is made to count up one. Hence, by the same mathematical argument presented in the case of the DDA-type interpolator, this progressive incrementing of the Y-axis command and decrementing of the X-axis command results in a circular cutter path. The "up" or "down" mode of the command counters is selected by the quadrant decode logic according to the quadrant of circle programmed by the preparation function. The end point is detected by the distance counters in the same way as for the straight line generation case. In the case of circles the scaling factor is made equal to $1/R$. With this arrangement of pulse rate multipliers large amount of hardware economy is possible. For instance, one common set of counters and gates can be used for the X, Y and Z vector pulse rate multipliers and distance counters can be eliminated by suitable scaling. The interpolator can also be extended up to 5 axes.

When computer programming is used, the computation of the scaling factors $1/L$ or $1/R$ and deceleration points can be done by the post processor programme and fed in on tape. For manual programming, however, this computation is done within the director as indicated in the block diagram. Deceleration during stopping or at sharp corners is achieved by resetting the feed velocity buffer to some small value at a point $[L - (\frac{1}{2})(V_0^2 - V_i^2)]$ from the start of the segment. This latter point is detected by integrating the pulse rate V into a down counter which is initially set to $[L - (\frac{1}{2})(V_0^2 - V_i^2)]$.

Since errors in velocity and acceleration do not affect machining accuracy the computation of $1/L$ and $[L - (\frac{1}{2})(V_0^2 - V_i^2)]$ need only be approximate to say 1%. These computations can be performed very economically through the use of an iterative technique of using pulse rate multipliers to solve differential equations as follows.

Figure 12 shows a method of squaring. By setting up a distance X in the down counter A and counting down at the clock frequency C a number of pulses dx equal to the distance X are generated. These are fed to the pulse rate multiplier B in order to produce a pulse rate XDX which is subsequently doubled and integrated in the up counter to C to produce X^2. If this procedure is repeated with X, Y, Z in turn, the summation $(X^2 + Y^2 + Z^2) = L^2$

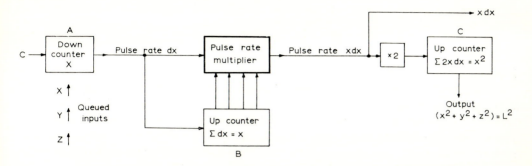

FIG. 12. Squaring and summing of numbers by iterative integration using pulse rate multipliers.

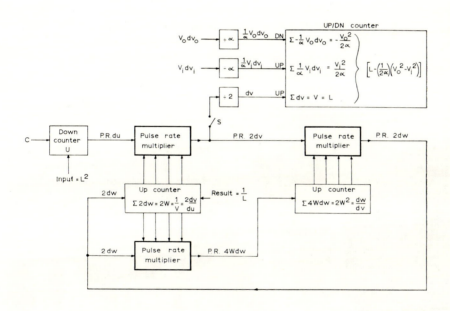

Let it be assumed that $L^2 = U$ ········(1)

Consider the differential equations: $\dfrac{dv}{du} = \dfrac{1}{2V}$, then $V^2 = U$ and $V = \sqrt{U} = L$ ········(2)

$\dfrac{dw}{dv} = 2W^2$, then $\dfrac{1}{W} = 2V$ and $2W = \dfrac{1}{V}$ ········(3)

By (1),(2) & (3) therefore $2W = \dfrac{1}{V} = \dfrac{2dv}{du} = \dfrac{1}{V} = \dfrac{1}{\sqrt{L^2}} = \dfrac{1}{L}$

FIG. 13. Computation of $1/L$ and $L - (\tfrac{1}{2}a)(V_0^2 - V_i^2)$.

results in the up counter C. A circuit for obtaining the square root of L^2 and the reciprocal $1/L$ by the simultaneous iterative solution of the differential equations

$$\frac{dv}{du} = \frac{1}{2v} \quad \text{and} \quad \frac{dw}{dv} = 2w^2$$

is shown in Fig. 13.

It consists of three pulse rate multipliers. The up counters form the multiplier inputs and they are connected up as shown. To start with, assume switch S is open. The number equal to $L^2 = U$ is preset into the down counter and by clocking it down to zero with a clocking rate C a stream of pulses each having a significance du (typically $(0{\cdot}0001)^2$ in^2) is produced such that $\sum du = L^2$. Bearing in mind the foregoing explanation of the principles of pulse rate multipliers and also the fact that if a pulse rate such as dV say is fed into an up counter which is initially set to zero, the counter would accumulate the pulses and therefore its contents would be $\sum dv = v$, it follows that at the end of the count down process, $1/L$ would result in the up counter as indicated, since the pulse rate multipliers are connected up to satisfy the equations as shown in Fig. 13. If at the start switch S was closed, the square root L becomes stored in the up/down counter. Therefore, if the pulse rates $V_0 dV_0$ and $V_i dV_i$, which can be generated by the previous circuit of Fig. 12 (instead of $x\,dx$) are also counted into it in the manner shown, the required result $[L - (\tfrac{1}{2}d)(V_0^2 - V_i^2)]$ is obtained. Serial division by the constants 2 and a are performed by counters having modulo 2 and a respectively.

A photograph of a machine tool and the director is shown in Fig. 14 for completeness.

Fig. 14. The machining centre.

The main feature of the N/C contouring system is that it does much more than existing systems and is, at the same time, much cheaper, which is evident from the above description. The machining centre is currently being used for advanced research in the numerical control field associated with three-dimensional and multiaxes programming, adaptive control and optimization of the metal cutting process, and for the solution of various problems such as are connected with cutter length and diameter compensation, manual overrides, table feed drives, surface finish, measuring systems and interpolation techniques.

ACKNOWLEDGEMENT

Permission to publish this paper by the Authorities of the Mechanical Engineering Laboratories, English Electric Co., Ltd., Whetstone, Leicester, is gratefully acknowledged.

REFERENCES

1. A. A. SHUMSHERUDDIN, Shaping the Response of an Electro-hydraulic Machine Tool Servomechanism. *Proc. 7th Int. Mach. Tool Des. & Res. Conf.*, p. 575. Pergamon Press, 1966.
2. D. F. WALKER, Data Processing for Numerical Control of Machine Tools. *J. Brit. I.R.E.* **26**, No. 6, 501 (1963).
3. H. B. HENEGAR, New Continuous Path System Uses DDA Interpolator. *Control Engineering*, 71 (Jan. 1961).
4. F. V. MAYOROV, *Electronic Digital Integrating Computers—DDA*. Iliffe, 1964.
5. J. T. EVANS and L. U. C. KELLING, Inside the Mark Century Numerical Controls. *Control Engineering*, 112 (May 1963).

IN-PROCESS CONTROL OF LATHES IMPROVES
ACCURACY AND PRODUCTIVITY

M. Bath and R. Sharp

Ministry of Technology, National Engineering Laboratory, Machinery Group,
Machine Tools and Metrology Division

1. INTRODUCTION

This paper deals with improvement of the dimensional accuracy of a lathe workpiece. The finished size is continuously monitored close to the cutting point during machining, and productivity of the machine is increased because the correct dimension is cut automatically and it is not necessary to stop the machine to check workpiece dimensions. Methods are discussed of reducing errors which arise from tool wear, machine misalignment, imperfect slideways, workpiece deflection and tool changing. These features are likely to be particularly valuable when finished size tolerances are critical.

The technique is applicable to the three principle lathe-cutting operations, namely, outside diameters, inside bores, and face cutting of shoulders and flanges. The hardware developed is readily adaptable to conventional machine tools for cutting cylindrical components. This is a restricted class of product, but constitutes a substantial porportion of total cut products.

The methods applicable to a centre lathe are described and the extent of the improvements obtained is shown. Further work indicates how the productivity and accuracy of a capstan lathe can be improved.

2. CONDITIONS WHERE IN-PROCESS CONTROL IS NEEDED

Conventional NC machine tools take their measurement reference from scales fitted to the machine slideways. This has disadvantages if dimensional tolerances are critical and if any of the following sources of error exist. In these conditions in-process control can contribute to accuracy.

2.1 Tool Wear

With some materials, tool wear during the production of a batch, or even during one cut, may be sufficient to cause errors in excess of the specified tolerance.

2.2 Machine out of Adjustment

Maladjustment of the machine, for example an offset tailstock which causes the production of a tapered component, or inaccurate slideway motion, i.e. the slideway may not be straight, which imposes machine form errors on the job, may contribute to loss of accuracy.

2.3 Workpiece Deflection

Certain classes of flexible workpieces, such as chuck-held unsupported cantilever components or thin flanges, may have errors introduced by deflection under cutting forces.

Such form errors are not even repeatable over a batch as cutting forces may vary as a function of tool wear. There are limitations to what can be done and other areas that require further study. Clearly a job section that is so flimsy that it deforms under its own weight presents problems for individual consideration.

2.4 *Replacement of Throw-away Type Tool Tips*

If a tool tip is changed the measurement datum will have to be re-established for each change.

3. A METHOD OF IN-PROCESS CONTROL FOR A LATHE

Automatic measurement and inspection has, in the past, generally relied on checking the job dimensions at one or more discrete spots; this check is usually made after the finishing cut is complete. An inspection of this type usually leads to an unacceptable percentage of workpieces scrapped because they are out-of-tolerance. The single point tool used for lathe work makes attractive the proposition of monitoring the workpiece dimension immediately it is cut, in a narrow region adjacent to the tool point. To illustrate the method, Fig. 1 shows a diagram for cutting an outside diameter. The system uses a caliper holding a pair of pneumatic proximity gauges.[1] One gauge is mounted on the tool-post carriage and detects the position of a ram carrying the toolholder. The other is supported on a pulley arrangement moving in conjunction with the first air gauge, as a caliper to span the diameter to be measured. A measuring scale head is carried by the pulley mechanism providing a digital display of the diameter spanned by the caliper.[2, 3] Any deviation of the workpiece diameter from specification results in a change in the sum of the outputs from the proximity gauges. These outputs are summed by an amplifier, the output of which drives an actuator to move the tool in the direction that will eliminate the error or reduce it to acceptable proportions. As the proximity gauges are linear over a very useful range, any movement of the tool away from the workpiece axis, caused by faulty slideway motion, will be corrected by the system. Design of the air gauges is outlined in the Appendix. Form and size

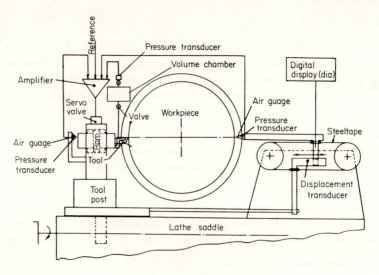

Fig. 1a. In-process control.

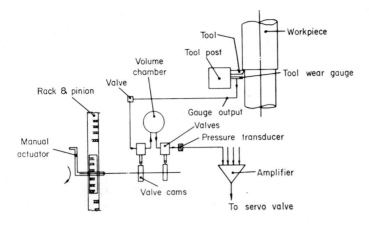

FIG. 1b. Tool setting and tool wear control system.

errors are thus fully accounted for, but only half of the error caused by tool post deflection, tool wear or inaccurate tool replacement is corrected. Attention to rigidity will eliminate the first while the second and third are discussed below.

A photograph of a Churchill–Redman centre lathe fitted with in-process control of diameter is shown in Fig. 2. It incorporates all of the features of the diagram of Fig. 1.

FIG. 2. Churchill–Redman centre lathe fitted with in-process control.

3.1 *Tool Wear*

A method of correcting for tool wear is to employ a third proximity gauge to monitor the distance between the tool post and workpiece surface. As tool wear occurs gradually with time, it is not necessary to place the third gauge very closely to the tool. In Figs. 1a and 1b, schematic elevation and plan, it is shown axially behind the cutting point. The correction for tool wear is fed into the summing amplifier by a simple and hold technique. A valve selecting the sampling time is operated at convenient points of the cutting cycle.

A view of early experimental equipment is shown in Fig. 3, where an 18-in long bar is being cut. The tool wear proximity gauge is seen following the tool.

3.2 *Tool Tip Replacement*

It is desirable in batch production to be able to replace the tool, or throw away type tool tip, without disturbing the measuring datum of the system. This can be accomplished by

FIG. 3. Early experimental equipment at N.E.L.

operating the tool wear correction valve each time the tool tip is changed. Providing that the new tool tip can be fitted within a few thousandths of an inch of its normal position, operation of this correction device will reinstate the cutting accuracy of the in-process control system.

4. OPERATION AND PERFORMANCE OF EXPERIMENTAL SYSTEM

The method of operation has been designed for manual setting up. Referring to Fig. 2, the cross-slide is moved until the diameter desired is shown on the digital readout. The cut is commenced and, as soon as the air gauge monitors the cut surface, automatic operation takes over and the finished size is governed by in-process control.

The ability to cut a long parallel bar is an elementary illustration of the performance to be expected. An 18-in long bar was cut on a lathe, the tail-stock deliberately off set to cut a 0·007-in taper. The profile of the bar is shown in Fig. 4. In this lathe there was also a lack

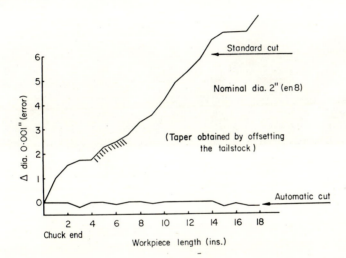

FIG. 4. Diagram for a workpiece turned between centres with and without in-process control.

of straightness of over 0·001 in owing to faulty saddle movement along the slideway. Controlled by the caliper, the automatic cut was parallel and straight to well within $\pm$ 0·0002 in. Thus higher precision can be obtained with less time and skill spent in adjusting the machine tailstock. The profile straightness of the job is also better than could be obtained from the standard machine.

An example of a workpiece that deflects is shown in Fig. 5. The bar is 0·5-in diameter En 8 steel with a 5-in length unsupported by a tailstock. Turned normally the profile has an error of 0·005 in along its length; this is reduced to 0·0003 in by automatic control as shown in Fig. 6. The error characteristics of a batch of fourteen such workpieces are shown in Fig. 7; the size drift is caused by tool deterioration. Using in-process control, fourteen similar bars were produced with an overall error within $\pm$ 0·0004 in. Although this operation is perhaps unlikely to be attempted in practice, it is typical of a class of workpieces that involve job deflection. Similar methods apply to the facing of thin flanges. The above results show the value of in-process control in maintaining dimensional accuracy throughout large batch production.

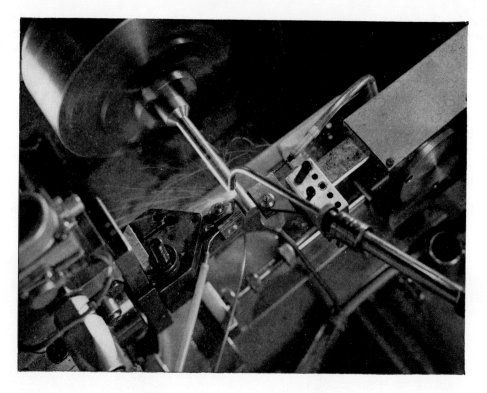

FIG. 5. Workpiece subject to deflection.

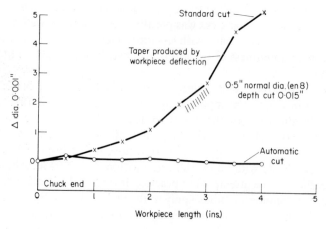

FIG. 6. Variation in diameter for a chuck-held workpiece with and without in-process control.

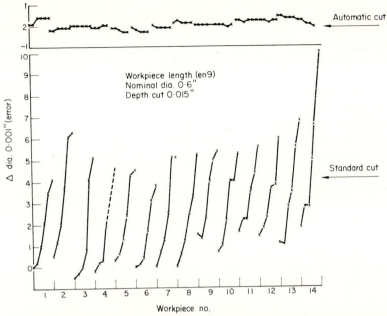

FIG. 7. Variation in diameter for 14 chuck-held workpiece with and without in-process control.

FIG. 8. Basic caliper applied to boring.

Internal bores are more difficult both to cut and to measure accurately than are outside diameters. Figure 8 shows how the basic caliper mechanism can be adapted to the boring operation. The experiment, the results of which are shown in Fig. 9, shows that a standard cut resulted in a taper of 0·002 in and that this was corrected by automatic operation to within 0·0003 in.

Results of using the tool wear device described in section 3.1 are shown in Fig. 10. The operation shows an error over five workpieces that is a combination of taper plus tool wear after installing a new tool tip. Under the same conditions, with automatic correction, the error was less than ± 0·0003 in.

In-process control can maintain accuracy when tool tip replacement is involved as shown in Fig. 11. This shows errors obtained in a batch of thirteen diameters cut with tool tip replacement between each operation. The solid line shows errors generated in a conventional system and the dashed line the errors when the tool wear datum correction was employed following each standard cut.

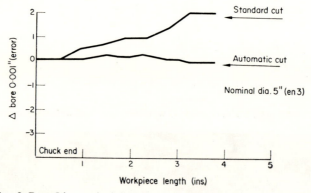

Fig. 9. Bore 5-in nominal diameter with and without in-process control.

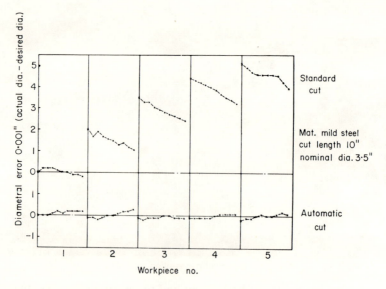

Fig. 10. Comparison of errors in components, turned between centres, with and without in-process control.

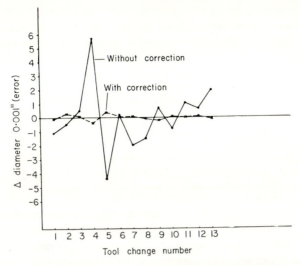

FIG. 11. Change workpiece diameter due to tool change with and without in-process control.

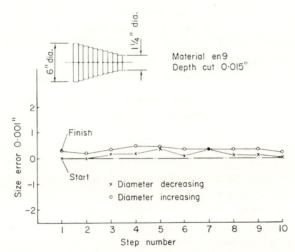

FIG. 12. Size errors in a ten-step workpiece using in-process control.

The accuracy with which the caliper system can be set by means of the built-in readout measuring transducer was also measured. Using the readout a workpiece was turned with ten arbitrary steps in diameter from $1\frac{1}{4}$-in to 6-in diameter. All operations were performed in sequence without stopping the machine; the sizes obtained were found to be within $\pm$ 0·00025 in, as shown in Fig. 12.

It is noteworthy that to produce this workpiece by conventional "stop the machine and manually check the size by micrometer" methods would have taken about two hours for the finishing cuts. The time to make the ten finishing cuts by in-process control was 15 min. In some instances, therefore, the accuracies being comparable, in-process control can show an increase in productivity by a factor of 8 over conventional methods of measurement and inspection.

The work of this paper applies to a machine tool where the workpiece revolves and the tool is quasi-stationary. On boring machines the workpiece is often stationary with a rotating tool. Such a machine tool offers the prospect of a useful outcome from future work on in-process control.

5. SAVINGS TO BE EXPECTED

To summarize.

5.1 Cost Reduction

In general a production cost reduction can be expected. For the "ten-step" test-piece described in section 4, the time to machine the finishing cuts was reduced from 2 hr to 15 min.

5.2 No Set-up Time Delay

On workpieces required urgently, perhaps in batches, this method can be put into operation without delay. Other methods, for instance copy turning, require substantial advance preparation in providing and fitting a template. The manually operated in-process control method may also compare favourably with tape controlled machines in the lead time required to set up a program for operation.

5.3 May Eliminate Some Grinding Operations

Unless the workpiece material is too hard to cut with a single point tool, the accuracy achievable by in-process control on a lathe may eliminate the necessity of finishing with a grinding operation, subsequent to rough machining.

5.4 Reduction of Scrap Rate

Automatic monitoring of size during machining, instead of inspection after a cut, should reduce the proportion of workpieces scrapped.

6. AUTOMATIC STATISTICAL CONTROL

For some classes of production a simplified form of in-process control may be adequate. Some reduction in cost of the system could be obtained in exchange for giving up some of the advantages of continuous control. For instance, if form errors can be overlooked and inspection is limited to size measurement at a single point, then a very simple system of automatic inspection is possible.

Figure 13 shows a mechanism employing a retractable caliper engaging the workpiece on the machine during the machining cycle. A fluidic computer is used in the feedback control to reset the tool post for size errors. This computer detects any pronounced size drift but does not respond to random errors. It is initially adjusted by setting in appropriate statistical parameters to optimize the dimensional accuracy. The operator has only to set the dimensional limits to suit a particular component, with reference to tables of data. A cheap and rugged form of stepping-motor controls the toolholder and completes a very effective form of statistical in-process control.

Figure 14 shows an error drift characteristic over some 40 components from a capstan lathe, with an overall error of 0·0035 in over this batch. The characteristic obtained with automatic correction reduced this error to less than $\pm 0·0005$ in.

Productivity is increased as no manual inspection and computation of data is required, and the machine can be corrected without stopping the machining cycle.

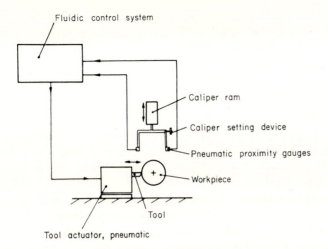

Fluidic control system

Caliper ram

Caliper setting device

Pneumatic proximity gauges

Workpiece

Tool

Tool actuator, pneumatic

FIG. 13. Automatic statistical control.

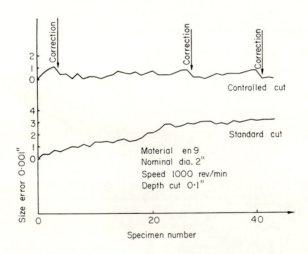

FIG. 14. Error comparison with automatic statistical control.

CONCLUSIONS

The improvements in productivity and dimensional accuracy to be obtained from a lathe by the use of in-process control have been examined. To illustrate the features of continuous in-process control, comparison has been made with a statistical sampling method of in-process control. Both methods are very relevant to the need for increasing the productivity of both manual and automatic lathes, at the same time reducing the demands on highly skilled operators.

ACKNOWLEDGEMENTS

This paper is published by permission of the Director of the National Engineering
Laboratory, Ministry of Technology. It is Crown copyright and is reproduced by permission
of the Controller of H.M. Stationery Office.

APPENDIX

N.E.L. Pneumatic Proximity Gauge

The proximity gauges used in the closed loop system are particularly significant and were
specially developed for the purpose. Alternative electrical, optical, or contacting gauges
were rejected as less suitable. The air gauge was found to be very practical and capable
of blowing away light swarf and coolant.

The design, which uses hypodermic tubing as the restrictor, minimizes the air volume
inside the gauge and allows a greater part of the supply pressure to be used. A typical
gauge is shown in Fig. 15. The performance characteristics, shown in Fig. 16, are out-
standing in possessing a useful linear range of measurement together with a frequency
response which can be as high as 100 c/s.

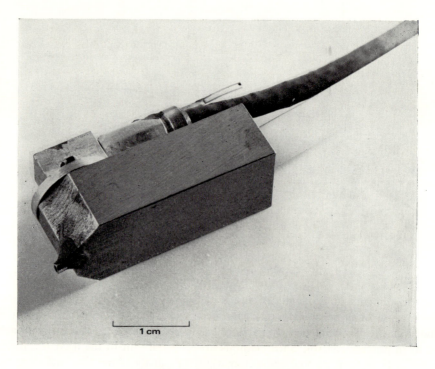

1 cm

FIG. 15. Typical in-process gauge.

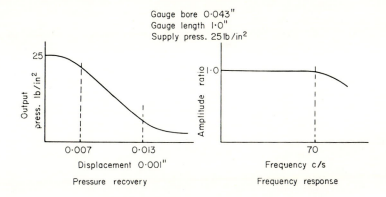

FIG. 16. Pneumatic proximity gauge characteristics.

REFERENCES

1. M. BATH and D. BROWN, A Pneumatic Proximity Gauge. NEL Report No. 322.
2. A. RUSSELL, An Absolute Digital Measuring System using Optical Grating and Shaft Encoder. NEL Report No. 233.
3. R. SHARP, R. MACLEAN and A. MCCLINTOCK, A Fluidic Absolute Measuring System. Second Cranfield Fluidics Conference, Cambridge 1967. Paper No. E3.

CONTRIBUTION TO THE DEVELOPMENT OF PNEUMATIC LOGIC ELEMENTS AND THEIR APPLICATION TO MACHINE TOOLS

Jiří Stivín

VÚOSO

DURING the past two years pneumatic logic elements have been in development in VÚOSO. A universal threshold logic element with 7 inputs and 2 outputs has been developed and tested. Various kinds of additional elements such as micro-switches, pneumo-hydraulic and pneumo-electric converters, tape reader, have also been developed and applied to the control of a grinding machine and of a single-spindle automatic lathe. The paper describes basic design and experience.

Until quite recently, machine tools were controlled exclusively by means of electric or electromechanic elements, even when the machines were equipped with hydraulic or pneumatic power means. The monopoly of electric controls is mainly due to the variety of well-developed components available and to the great experience gained in their use. But in spite of that, attempts have been made recently to apply pneumatic controls to machine tools. The following advantages are expected:

Increased reliability and service life.

Less expensive systems and maintenance.

Opportunity of employing useful pneumatic logic functions.

Some new types of pneumatic position transducers, pneumatic tape reading units and pneumatic display units seem to be very promising, too, and can play a special part in the economical evaluation of both systems.

In VÚOSO a system of pneumatic control for machine tools has been developed on the basis of an original concept. This concept has some advantages, especially in the possibility of designing complex logic circuits with a minimum number of elements.

The principle of this pneumatic control system is based on a universal multi-argument pneumatic logic element which is capable, in conformity with its interconnection, of performing practically all necessary logic functions including the threshold ones. It is diagrammatically illustrated in Fig. 1(A) and consists of two main parts.

The first part 0 (Fig. 1(A)) represents the output where the pressure signal is switched over from one output line to another. It has the form of two flat channels (Fig. 1(B)) with feeding points N_1, N_2 and output points F_1, F_2. The channels are actuated by a pin p (Fig. 1(A), (B)) which has two possible positions at the ends of a short vertical stroke. In the upper position it shuts the connection N_2-F_2 and simultaneously opens the point F_2 to the exhaust into the atmosphere.

Simultaneously the connection N_1-F_1 is opened, pressure air occurs at F_1 while the exhaust from F_1 is shut. In the lower position of the pin p the inverse stage occurs. The channels are made of special plastic (see Fig. 2).

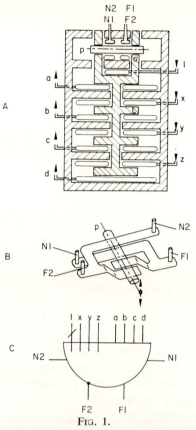

FIG. 1.

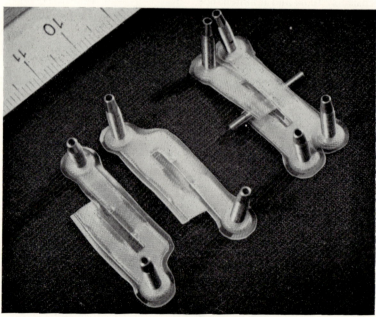

FIG. 2.

The second part of the element converts the input signals into corresponding positions of the pin p and controls in this way the stages of the output. It has the form of a vertically displaceable ladder l carrying at the top the pin p. The ladder is driven by means of eight plastic bags 1, x, y, z, a, b, c, d (see Figs. 1(A) and 3). The group of bags a, b, c, d elevates the ladder; group 1, x, y, z pushes it down. The bag 1 is permanently filled with air and has half the surface area of all other bags, the surfaces of which are equal in area.

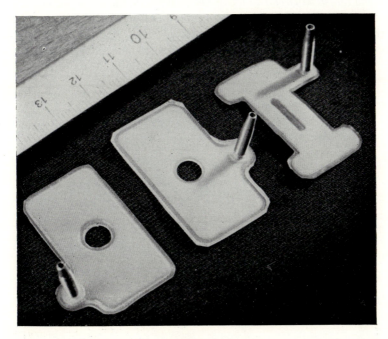

FIG. 3.

Let us denote by 0 the stage corresponding to the lower position of the ladder with the pin p. The upper position corresponds to a stage 1. Thus stage 0 signifies pressure on F_2, no pressure on F_1. Stage 1 means: no pressure on F_2, pressure on F_1. If no one of the x, y, z, a, b, c, d inputs receives pressure signal stage 0 occurs as result of pressure in the bag 1, which represents in this way a preloading spring. Otherwise the position of the ladder is the result of the sum of forces exerted by the other bags and depends in this way on the combination of pressure signals x, y, z, a, b, c, d.

The logic diagram corresponding to the described function is shown in Fig. 1(C).

As an example, the following basic functions are illustrated in Figs. 4, 5.

Four-input function "OR":
$$F_1 = a \vee b \vee c \vee d$$

Four-input function "NOR":
$$F_2 = \bar{a}.\bar{b}.\bar{c}.\bar{d}$$

Four-input function "AND":
$$F_1 = a.b.c.d$$

Four-input function "NAND":
$$F_2 = \bar{a} \vee \bar{b} \vee \bar{c} \vee \bar{d}$$

Also it is possible to create memory or flip-flop circuits by interconnecting any of the outputs F_1, F_2 with any of the inputs in a kind of feed-back line.

An important feature of this element is that it makes it possible to create the threshold functions. This simplifies the logic circuits and reduces the number of necessary elements, which is important not only from the point of view of cost, but improves also the reliability of the system.

We understand *threshold circuit* to mean one in which some of the inputs x, y, z are permanently connected with pressure and thus represent a threshold, the height of which depends on the number G_1 of inputs used in this way. The condition for changing the output state from 0 to 1 is that the number G_2 of pressure signals a, b, c, d be $G_2 > G_1$.

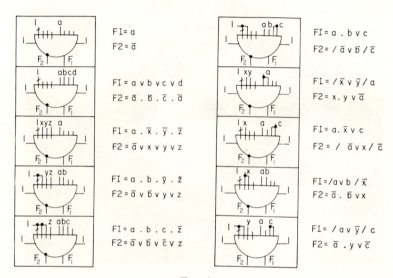

FIG. 4.

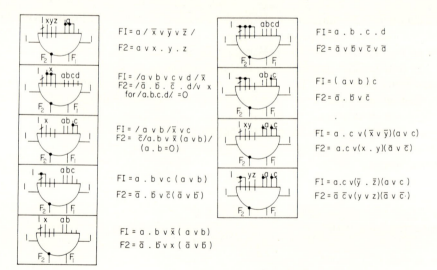

FIG. 5.

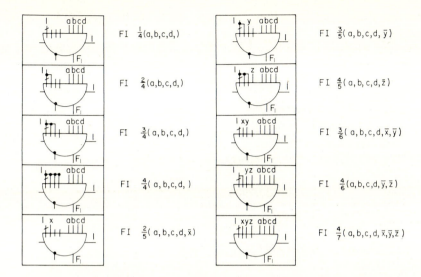

FIG. 6.

LOGIC   MEMORY   FUNCTIONS

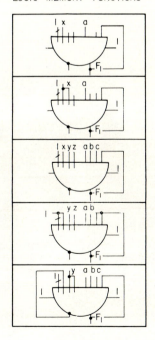

FIG. 7.

Figures 4–8 show other basic logic functions which can be obtained by means of the described element.

The described element is shown in Fig. 9. It can work with pressures from 0.5 up to 1.5 kp/cm² and supply 10 l./min at 1.0 kp/cm². The switching time is about 5 msec and the service life amounts to 10^8 switching operations. The permissible pressure tolerance of the

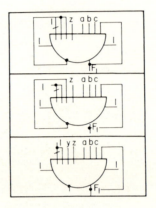

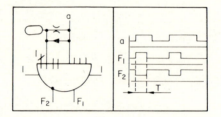

Fig. 8.

Fig. 9.

input signal is better than 1–20% or 0 + 20% on an arbitrary input for an arbitrary con-nection (also for 4-input function "AND").

A number of other complementary pneumatic elements have also been developed. These are switches and converters. Tests showed that it is not necessary to cater for instan-taneous mechanical switching as in the case of electric micro-switches because the signal emitted by the light switch is shaped to a rectangular signal by the logic element connected in feedback.

Because of the good results with plastic channels in the logic elements, these were simi-

FIG. 10.

FIG. 11.

larly used for microswitches. The channel possesses three inlets so that it can work in the positive as well as in the negative way. The throttling of the channel between the inlets is made by means of the rounded edges of a two-arm lever actuated mechanically by a spring system. A small dimension (see Fig. 10) is obtained in this way ($10 \times 25 \times 40$ mm), which is of great advantage. The described design was used for limit switches as well as for built-up rotary switches illustrated in Fig. 11.

For actuating linear operating movements, e.g. for driving hydraulic valves, corrugated plastic members were developed (Fig. 12) which were built into pressed sheet caps and dimensioned in such a way as to replace electromagnets of the conventional valves. A progressive

FIG. 12.

technology allows the corrugated plastic members to be made of arbitrary length so that they may be used instead of metal cylinders with pistons, which have bad efficiency and which are difficult to start after a longer time of rest. Figure 13 shows a comparison of electromagnets and the corresponding corrugated plastic members.

For electro-pneumatic and pneumo-electric converters, the combination of the current electric equipment with the pneumatic elements was used.

For the pneumo-electric converter, an electric microswitch is enclosed by a yoke that carries a pneumatic control element completed with a little spring in order to obtain the needed characteristic (Fig. 14).

For the electro-pneumatic converter an electric relay was used, and instead of contacts a plastic channel was employed (Fig. 15).

A very important element replacing the usual plug-boards is a pneumatic punched tape block-reading unit, shown in Fig. 16.

The movement of the tape is here derived from a permanently rotating roller driven by a small induction motor. In principle, the control of the reading unit consists on the one hand of a little roller pushing the tape against the rotating roller and, on the other hand, of a little brake block for stopping the tape. These two parts operate in an inverse tact and are controlled pneumatically. The control signal is given by the passage of the stopping hole on the tape between two nozzles pushed elastically against the tape. Otherwise the tape passes quite freely.

As soon as a stopping signal arrives the little roller effecting the movement of the tape is lifted and the tape is simultaneously braked. The tape stopped, two little rubber blocks with respective holes are pushed against it, and ensure in this way reading of the information without leakage of compressed air. After finishing the operation a signal for a further step

Fig. 13.

Fig. 14.

FIG. 15.

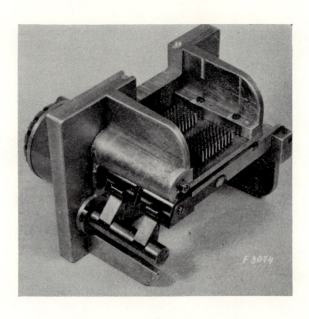

FIG. 16

is emitted and the whole process is repeated. Only two logic elements are needed for the control of the reading unit.

The reading unit described is convenient for parallel reading. For the control of double-carriage machine tools it is provided with two tapes, for each carriage separately. It is advantageous for simple programming. Velocity of the tape is 300 to 500 mm/sec. The over-run after braking is 0·5 mm approximately and is compensated by the position of the stopping nozzles. Each block possesses 60 informations and can be read 5 times per second.

For the present, two single-spindle double-carriage lathes, a cylindrical grinding machine and an eccentric press have been equipped with the described pneumatic control. These machines are tested in operation and the following present experience can be stated:

The switching time of the pneumatic logic element does not seem to be decisive for the function of the machine tools, provided that it does not exceed tens of milliseconds. It is the delay of the signal in the piping that is important. Therefore it is absolutely necessary to assemble the logic members, decisive for the switching velocity, in the immediate proximity of the source of the control signal (e.g. limit switch).

An important delay occurs if the piping and other larger spaces are evacuated through the logic element only. In these cases special pneumatically controlled members must be used which evacuate those spaces through a large passage. During the tests, switching velocities were obtained of such a value that the overruns for the pneumatically controlled hydraulic feeds were not greater than those of the electrohydraulic feeds.

In order to obtain short switching times, it is advantageous to work as much as possible with the signal 1 and to connect the lowest number of logic elements in a sequence.

Pneumatically controlled hydraulic valves operate softer than the electromagnetic ones and without harmful shocks. The definite prospects of the possibility of application of the described pneumatic controls for machine tools will surely become clearer after a longer period of tests of the prototypes of those machine tools.

COMPUTERIZED DETERMINATION OF OPTIMUM CUTTING CONDITIONS FOR A FIXED DEMAND

S. M. Wu

Department of Mechanical Engineering, Department of Statistics,
University of Wisconsin, Madison, Wisconsin 53706

and

L. H. Tee

Department of Mechanical Engineering, University of Wisconsin

SUMMARY

The optimum cutting condition to yield maximum profit lies in the interval between V_p and V_a when a fixed demand volume has to be fulfilled. V_a is the cutting speed at the maximum allowable feed rate to meet the demand; and V_p is the theoretical maximum profit cutting speed.

Computer is used to delineate the machining economics by examples. Profit region, profit responses, demand volume and production capacity are investigated. The effect of the variation of the seven parameters in the basic mathematical models are also evaluated.

INTRODUCTION

Determination of economic optimum machining conditions has long been based on the minimum cost or the maximum production rate criteria. Recently, a new concept has been presented by using maximum profit criterion. The analytical approach of the economic optimization has academic value but lack of practicability for industrial application. This paper aims at further delineation of machining economics analysis by incorporating all previous theories and simplifying the solution. Computer is employed to reduce the complicated analysis into a practical selection of optimum operating conditions.

To provide a background for this investigation purpose, the two conventional criteria and maximum profit criterion will be briefly reviewed. The basic mathematical models involved in machining economics are the formulation of total cost C_u and total time T_u. They are:

$$C_u = C_o t_m + t_m \left(C_o t_c + C_t \right)/T + C_o t_h$$

$$= \frac{C_o \pi DL}{12Vf} + \frac{\pi DL V^{1/n-1} f^{m/n-1}}{12k^{1/n}} \left(C_o t_c + C_t \right) + C_o t_h \tag{1}$$

and $\quad T_u = t_m + t_c t_m/T + t_h$

$$= \frac{\pi DL}{12Vf} + \frac{\pi DL V^{1/n-1} f^{m/n-1} t_c}{12k^{1/n}} + t_h \tag{2}$$

1235

where C_o = cost of operating time, $/min,
$\quad C_t$ = cost of tool, $/edge,
$\quad C_u$ = total cost per piece, $/pc,
$\quad D$ = diameter of workpiece, in,
$\quad f$ = feed rate, ipr,
$\quad k$ = constant,
$\quad L$ = axial length of cut, in,
$\quad m,n$ = exponential constants in generalized tool-life equation,
$\quad T$ = tool life, min/edge,
$\quad t_c$ = tool changing time, min/edge,
$\quad t_h$ = handling time, min/pc,
$\quad t_m$ = machining time per piece, min/pc,
$\quad T_u$ = total time per piece, min/pc,
$\quad V$ = cutting speed, sfpm.

The minimum cost cutting speed $V_{\min}$ is derived by setting $(dC_u/dV) = 0$ using eq. (1). Unit cost is then minimized by employing the maximum allowable feed.

$$V_{\min} = \frac{k}{f^m \left[(1/n - 1)(t_c + C_t/C_o)\right]^m} \tag{3}$$

Similarly, the equation for maximum production rate cutting speed $V_{\max}$ is derived by setting $(dT_u/dV) = 0$ using eq. (2).

$$V_{\max} = \frac{k}{f^m \left[(1/n - 1)t_c\right]^n} \tag{4}$$

Between these two conventional criteria of minimum cost and maximum production rate, it is known that there exists a range of cutting conditions from which an optimum condition may be selected to yield the maximum profit. Stemming from this fact, maximum profit criterion is thus brought up by Okushima and Hitomi,[4] Brown,[1] Armarego and Russell,[5] and Wu and Ermer.[6] Okushima and Hitomi employ a simple linear break-even chart for analyzing the maximum profit cutting speed V_p. Their determination of maximum profit cutting speed being based on particular n values. Brown, Armarego and Russell present an analytical evaluation of the mathematical models. Wu and Ermer employ the fundamental marginal principle of economics for analysis; and the concept of demand function correlating revenue has also been presented.

To simplify the analysis and make use of the computer, a fixed demand volume and constant revenue are assumed. Examples are used throughout this investigation to illustrate the practicability of the analysis. The approach being adopted is deterministic.

The profit concept is first investigated to obtain a clear picture regarding the profit region and profit responses. The relationship between demand volume, demand contour and production capacity is then analyzed. An optimum cutting speed interval V_p–V_d is proposed within the profit region.

The basic mathematical models involve seven parameters. Operating cost C_o and income I_u are investigated for the effects of their variations on the profit analysis. The remaining five parameters, namely n, C, C_o, t_h, t_c, are evaluated by a 2^{5-1} fractional factorial design.

Profit Region, Profit Contour, and Profit Response

The profit response of a machining operation can be expressed on two bases, one on a per unit item basis, the other on operating time basis. Minimum cost criterion claims a maximum profit return based on per unit item basis, i.e.

$$P = I_u - C_u \tag{5}$$

where

P = profit, \$/pc,
I_u = income per piece (excluding material cost), \$/pc,
C_u = total cost per piece, \$/pc.

When C_u is a minimum, P is a maximum. However, by using eq. (5), a maximum profit response can be obtained with an unjustifiable length of operating time which is evidently not an optimum. Therefore, an optimum solution should clearly involve a profit expressed on unit time basis rather than on unit item basis. The profit response equation being employed in this investigation is:[5]

$$P = \frac{I_u - C_u}{T_u} \text{ (\$/min)}$$

$$= \frac{I_u - \dfrac{\pi DLV^{1/n} f^{m/n} C_t}{12Vfk^{1/n}}}{\dfrac{\pi DL}{12Vf} + \dfrac{\pi DLV^{1/n-1} f^{m/n-1} t_c}{12k^{1/n}} + t_h} C_o \tag{6}$$

By assigning a zero profit for eq. (6), i.e. $P = 0$, or

$$\frac{I_u - \dfrac{\pi DLV^{1/n} f^{m/n} C_t}{12Vfk^{1/n}}}{\dfrac{\pi DL}{12Vf} + \dfrac{\pi DLV^{1/n-1} f^{m/n-1} t_c}{12k^{1/n}} + t_h} - C_o = 0 \tag{7}$$

boundaries of zero profit can be defined by specific values of feeds f and cutting speeds V which satisfy eq. (7). Among these specific values, there exists a lowest feed rate and a corresponding highest cutting speed which satisfy both eqs. (7) and (3).[5] For any feed above this lowest feed rate, two speeds satisfy eq. (7). Likewise, for any speeds lower than this highest speed, two feed rates may be found to satisfy eq. (7). In this manner, a region is thus bounded by the zero-profit boundaries. This region is termed "profit-region". Within this region, positive profit response (gain) will be realized, while beyond this region, negative profit response (loss) occurs. The profit responses discussed in this paper refer to positive profit responses.

From eq. (6), we can obtain a new equation by taking $(dP/dV) = 0$, i.e.

$$I_u - \left(\frac{V_p^{1/n} f^{m/n}}{k^{1/n}}\right)\left[(C_t t_h + I_u t_c)(1/n - 1) + \frac{\pi DLC_t}{12fV_p n}\right] = 0 \tag{8}$$

where V_p is the maximum profit cutting speed. Equation (8) represents the maximum profit response whose locus, when plotted on a f–V log–log coordinates, can be approximated by a straight line. An example is given to illustrate the "profit-region" and maximum profit locus, and to show how the conventional criteria are related with the profit region.

An Example

(a) *Profit Region.* A single pass, single point dry turning operation is performed assuming parameters known.[8] They are:

tool material $=$ HSS
$t_c = 1$ min/edge
$t_h = 1$ min/pc
$C_o = \$0.08$/min
$C_t = \$0.50$/tool
$D = 3$ in
$L = 6$ in
$d = 0.2$ in (depth of cut)

generalized tool-life equation is given as:

$$VT^{0.125}f^{0.67} = 6.5$$

and selling price is assumed to be a constant, i.e.

$I_u = \$1.00$/pc (excluding material cost).

Substituting these above values in eq. (7), zero-profit boundaries can be obtained as shown in Fig. 1 by varying feed f and speed V. The region bounded within the envelope of zero-profit boundaries is the "profit-region".

Using eq. (8), the maximum profit locus can be obtained which represents the maximum profit responses at different feeds and corresponding speeds V_p. From eqs. (3) and (4), we can superimpose on the profit region the minimum cost and maximum production rate loci. Notice that the minimum cost locus lies to the left of the maximum profit locus, while the maximum production rate locus lies to the right of the maximum profit locus. Also, the minimum cost locus intersects the maximum profit locus at a zero profit point.

(b) *Profit Contour.* Profit contours can be plotted on a f–V log–log coordinates using eq. (6) as shown in Fig. 2. The contours have a rising ridge characteristics. The direction of the ridge leads to conditions of higher profit responses, which explains why a maximum allowable feed rate is always preferred.

The profit contours are more widely separated at the left of the maximum profit locus than at the right. This indicates that any slight variation in cutting speed at the left of the maximum profit locus will cause smaller change in profit response than at the right. For example, when feed f equals 0.015 ipr, V_p equals 71 sfpm (profit response equals \$0.085909/min). An increase of speed V from 50 to 52 sfpm (left of V_p) will cause an increase of \$0.004295/min (0.056176 to 0.060471) of profit response, while an increase of speed from 90 to 92 sfpm (right of V_p) will result in a decrease of \$0.013497/min (0.034619 to 0.021122) of profit response—a three times variation of profit response for same unit change of speed.

The maximum profit locus is approximated by a straight line on a f–V log–log coordinates whereas the minimum cost and maximum production rate loci can be superimposed as straight lines shown on Fig. 2. The relationship of the three loci can be used for the purpose of the maximum profit approximation. For the example cited above, at any feed rate, the choice of the minimum cost speed V_{min} is preferred to the maximum production rate speed V_{max} not only in that V_{min} yields higher profit, but also, slight deviation from V_{min} will not cause a severe change in profit response.

(*c*) *Profit Response.* Profit response *P* versus cutting speed *V* at different feed rate *f* is plotted in Fig. 3. The maximum profit response increases with increasing feed, while maximum profit cutting speed V_p decreases accordingly. The range of cutting speeds which define the profit region also decreases with increasing feed.

With reference to the maximum profit response at different feeds, the right portion of profit response curve is steeper than the left portion. This indicates that a change of speed

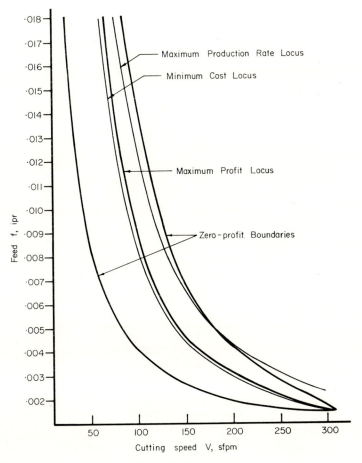

FIG. 1. Profit region.

which is higher than V_p will cause a greater rate of change of profit response than a change of speed which is lower than V_p. This phenomenon is analogous to that which was being discussed in "PROFIT CONTOUR"—that is, the profit contours are more widely separated at the left of the maximum profit locus than at the right.

Note that the profit region as shown in Fig. 1 is a combination of the cutting speed ranges of different feed rates on the zero profit ordinate of Fig. 3. If the locus of the maximum profit response defined for every feed rates in Fig. 3 is plotted, it is the maximum profit locus shown in Fig. 1. If we further imagine a *f–V* plane slicing through the ordinates of Fig. 3, then profit contours of Fig. 2 can be obtained.

(*d*) *Sensitivity*. Sensitivity of profit response increases with increasing feed which can be seen from the increasing steepness of the response curve accompanying increasing feed shown in Fig. 3. To facilitate further exploration of the sensitivity of profit responses, the profit response at $f = 0.018$ ipr is chosen as an example. In Fig. 4, an additional scale is given to show the relative profit response expressed in terms of percentage.

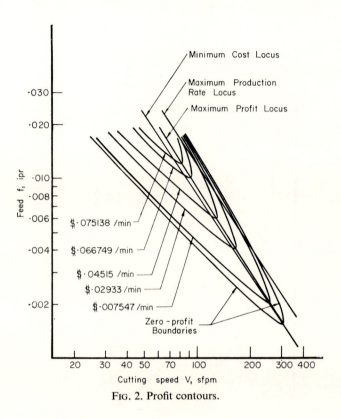

FIG. 2. Profit contours.

TABLE 1. THE RANGES OF CUTTING SPEEDS FOR AT LEAST 90% MAXIMUM PROFIT RESPONSE

Range (sfpm)	Cutting speed V (sfpm)	Feed f (ipr)
17	54–71	0.018
21	71–92	0.012
22	80–102	0.010
27	114–141	0.006
30	150–180	0.004

To obtain at least a 90% maximum profit response, the range of cutting speed is 17 sfpm as given on Fig. 4. For lower feed rates, the range of cutting speed to yield an at least 90% maximum profit response will gradually increase as shown in Table 1. This simple graph provides a quick method of detecting the sensitivity of the profit response at various feed rates.

The relative profit response for V_{min} and V_{max} can also be readily obtained. The minimum cost cutting speed yields an approximately 97% of the maximum profit, while maximum production rate cutting speed yields 73% of the maximum profit.

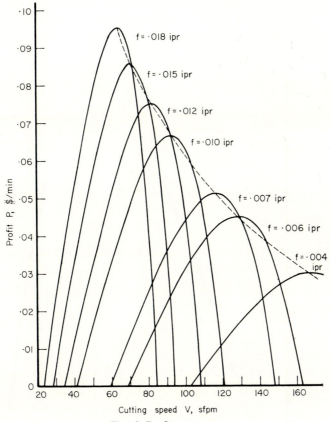

FIG. 3. Profit responses.

DEMAND

Wu and Ermer[6] assume a certain demand function from which corresponding total revenue is evaluated for maximum profit. To simplify the problem, the demand factor is treated as a constant in this paper. By so doing, demand volumes required for a specified period of time can then be calculated independently. For example, if the demand is 2000 pcs/month, assuming 20 working days per month, 8 working hours per day, the demand is therefore 0.208 pcs/min. If this demand volume is within feasible capacity of one machine tool, one machine tool will be employed, or it can be allocated to two machine tools each handling a capacity of 0.104 pcs/min. The constant demand volume W can be expressed in term of $1/T_u$ (pcs/min), i.e.

$$1/W = T_u = t_m + t_c(t_m/T) + t_h$$

$$= \frac{\pi DL}{12Vf} + \frac{\pi DLV^{1/n-1}f^{m/n-1}t_c}{12k^{1/n}} + t_h \tag{9}$$

Demand Volume, Demand Contour and Production Capacity

Demand volume can be plotted versus cutting speed V at feed rates from 0.006 to 0.020 ipr by substituting necessary data from the example into eq. (9) as shown in Fig. 5. When we speak of machining conditions satisfying a certain demand volume requirement, it is equivalent to saying that those machining conditions has a production capacity equal to the demand volume requirement. In this sense, Fig. 5 also shows the production capacities of various machining conditions.

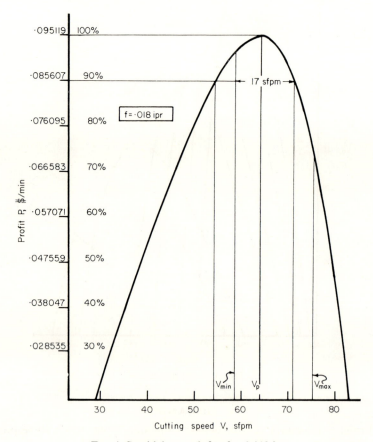

FIG. 4. Sensitivity graph for $f = 0.018$ ipr.

For every individual feed rate, there is a maximum demand volume. In other words, every feed rate has a maximum production capacity. If the locus of the maximum demand volume defined for different feed rates is plotted on a f–V coordinates, it is the same maximum production rate locus represented by eq. (4). Therefore, a feed of, say, 0.009 ipr cannot fulfill a demand volume of 0.1680 pcs/min because its maximum production capacity as shown in Fig. 5 is only 0.1663 pcs/min.

By extending a horizontal line from a specified demand volume, intersection points on the demand volume curves represent machining conditions of equal production rate. For example, points c ($f = 0.017$ ipr, $V = 62$ sfpm) and d ($f = 0.014$ ipr, $V = 79$ sfpm) have

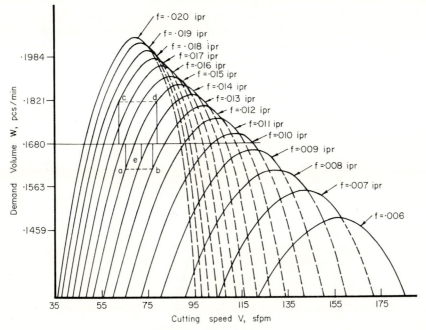

FIG. 5. Demand volume.

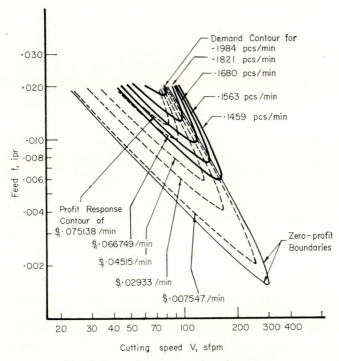

FIG. 6. Demand contours and profit contours.

different cutting conditions but have equal production rate of 0.1821 pcs/min. If we combine points *c d* and other points having same demand volume, a demand contour of 0.1821 pcs/min can be obtained. Five different demand contours are being plotted on a *f–V* log–log coordinates in Fig. 6 ranging from 0.1984 pcs/min to 0.1459 pcs/min.

The significance of demand contours being presented in Fig. 6 lies in the fact that they embody the three production situations being encountered on production line. For a

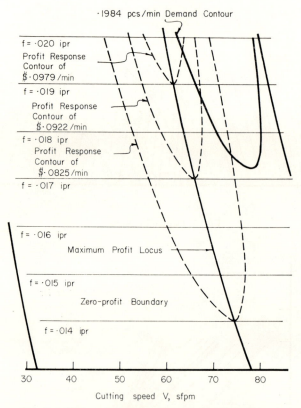

FIG. 7. 0.1984 pcs/min demand contour and profit contours.

specified demand volume requirement, its demand contour may be defined on the *f–V* plane, therefore, machining conditions chosen along the contour will just fully satisfy the demand volume requirement. Similarly, machining conditions within the envelope of the contour will result in over-production, while outside the contour there will be under-production.

The degree of either over or under-production of certain machining conditions can be detected by the distance differ from the required demand volume line. In Fig. 5, assume a required demand volume of 0.1680 pcs/min. Operating conditions on point *a* (*f* = 0.014 ipr, *V* = 65.5 sfpm) will produce the same degree of under-production as onpo int *b* (*f* = 0.012 ipr, *V* = 77 sfpm), whereas point *e* (*f* = 0.013 ipr, *V* = 72 sfpm) has lower degree of under-production.

Cutting Speed V_d to Fulfill the Demand

The profit response contours of Fig. 2 can be superimposed on the demand contours to obtain a cutting condition which will fulfill the demand and yield a certain profit. For example, from Fig. 5, point *a* has production capacity of 0.1613 pcs/min, which gives the profit response as 0.076 S/min shown in Fig. 6; point *b* has the same production capacity of 0.1613 pcs/min, but yields a profit response of $0.0732/min only.

To further delineate the importance of the demand volume contours, the demand contour of 0.1984 pcs/min is chosen from Fig. 6 and plotted in Fig. 7. It is obvious that for this example as shown in the graph, all feed rates below 0.01727 ipr cannot produce enough at any speed to meet the required demand volume of 0.1984 pcs/min. If the maximum allowable feed rate is 0.01727 ipr, only one speed—the vertex of the demand contour ($V = 78$ sfpm)—will just fully meet the demand volume of 0.1984 pcs/min, and this is a relatively rare case in actual operations.

A general case will be that of a maximum allowable feed rate which is above 0.01727 ipr. In this case, each feed rate has two cutting speeds of equal production capacity of 0.1984 pcs/min, e.g., at $f = 0.018$ ipr, $V = 70.5$ and 80 sfpm. It is seen that the lower speed always yields a higher profit response than the higher speed. For example, at $f = 0.018$ ipr, $V = 70.5$ sfpm, profit is $0.0877/min, while at $f = 0.018$ ipr, $V = 80$ sfpm, profit is only $0.0424/min.

For a specified maximum allowable feed rate, we shall denote the lower cutting speed on the demand contour by V_d. If the maximum allowable feed rate is 0.018 ipr, then V_d will be 70.5 sfpm for a fixed demand of 0.1984 pcs/min.

The Optimum Cutting Speed Interval $(V_p–V_d)$

Under the assumption of a fixed demand volume, it is clear that the maximum profit cutting speed V_p yields a maximum profit but it will not necessarily be able to fulfill the demand. In order to meet the demand, overtime expenses have to be paid, which will affect the operating cost and change the basis for comparison. On the other hand, if we operate at V_d, it will fulfill the demand but will not necessarily yield the maximum profit. Therefore a more realistic optimum cutting speed seems to be at an interval between $V_p–V_d$. This can be best explained by an example. The 0.1680 pcs/min demand contour from Fig. 6 is selected and drawn on cartesian coordinates in Fig. 8. Different values of V_{min}, V_p, V_d and V_{max} at different feed rates are shown in the figure. Table 2 tabulates the profit responses and production capacities of V_{min}, V_p, V_d and V_{max} for feed rates of 0.010 and 0.012 ipr.

At feed rate equals 0.010 ipr, profit responses for V_{min}, V_p, V_d and V_{max} are $0.065158, $0.066749, $0.061 and $0.041/min respectively. Both V_{min} and V_p have higher profit response than V_d, but operating at V_{min} leads to an under-production volume of 0.0158 pcs/min, while operating at V_p results in 0.008 pcs/min under-production volume. At V_{max}, the profit response is lower than V_d, and an over-production volume is piled up. A direct consequence could be the shortening of regular working hours. Both overtime operation and working hour shortening further lead to unforeseen economic factors which may be undesirable.

For overtime operation, operating cost C_o is consequently increased. Since profit response is expressed on unit time basis, such an increase in C_o and the extension of operating time will therefore vitiate the profit response calculated in the assumed regular working period. Similarly, for working hour shortening, the profit return in a shorter length

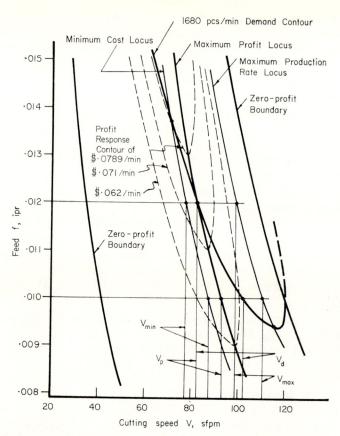

Fig. 8. Optimum cutting speed intervals ($V_p - V_d$) for 0.1680 pcs/min demand contour.

TABLE 2. PROFIT RESPONSE AND PRODUCTION CAPACITY OF V_{min}, V_p, V_d, V_{max} AT FEED RATES OF 0.010 AND 0.012 IPR

Feed (ipr)	Cutting speed (sfpm)		Profit ($/min)	Production capacity (pcs/min)	
0.010	V_{min}	87	0.065158	0.1522	(under production of 0.0158 pcs/min)
0.010	V_p	93	0.066749	0.1600	(under-production of 0.008 pcs/min)
0.010	V_d	102	0.061	0.1680	(fully satisfied)
0.010	V_{max}	111	0.041	0.1713	(over-production of 0.0033 pcs/min)
0.012	V_{min}	77	0.073211	0.1613	(under-production of 0.0067 pcs/min)
0.012	$V_p = V_d$	82	0.075138	0.1680	(fully satisfied)
0.012	V_{max}	98.7	0.049	0.1805	(over-production of 0.0125 pcs/min)

of operating time may not be a maximum in the long run. In other words, the $0.066749/min profit response at V_p may not be the true maximum any more. Therefore, between the speed interval of V_p and V_d, there should exist an optimum cutting speed which will yield the highest profit response, and at the same time fulfill the required demand volume with appropriate adjustment of operating time.

Note that at special cases, V_d can coincide with V_{min}, V_p or V_{max}. For example, when $f = 0.0138$ ipr, V_d equals V_{min} at 70 sfpm; when $f = 0.012$ ipr, V_d equals V_p at 82 sfpm; and for $f = 0.0094$ ipr, V_d equals V_{max} at 116 sfpm.

A Three-dimensional Model of Demand Volume and Profit Responses

A three-dimensional model for the example being used is schematically drawn in Fig. 9. The three axes are profit response P, cutting speed V and feed rate f. Profit responses of several feed rates and the demand contour of 0.1984 pcs/min which lies within the profit

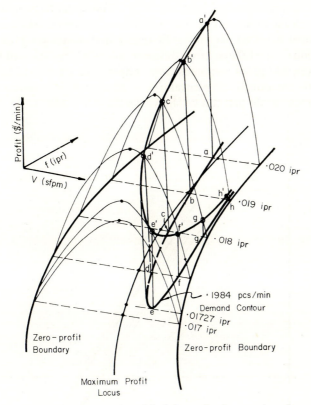

FIG. 9. Three-dimensional model of demand volume and profit responses.

region are shown. The maximum profit locus plotted on the f–V plane is the projection of the maximum profit responses of the different feed rates on the f–V plane.

The profit responses for the demand of 0.1984 pcs/min at a specific feed rate can be obtained in the model by projecting the specific points on the demand contour to the profit response curve. For example, at $f = 0.020$ ipr, speeds h and b both meet the demand volume and their corresponding profit responses are at the points h' and b'. The lengths of the projection, say bb' and hh' represent the magnitude of the profit response at respective conditions.

The projection length of aa' is greater than bb' and is gradually decreasing along the direction a–b–c up to h. Since b and h, c and g, d and f are pairs of different cutting conditions with equal production capacities and feed rates, it can be seen that the lower speeds (a, b, c, d in the model) yield higher profit than the higher speeds (f, g, h), which has been shown in Figs. 6 and 7.

Effect of the Variation of I_u and C_o on the Profit Region Analysis

The basic mathematical models involve seven parameters, namely, I_u, C_o, n, C, C_t, t_h, t_c. Among these seven parameters, it is obvious that income I_u and operating cost C_o are of paramount importance as shown in eq. (6). The effects of the variation of I_u and C_o on the profit analysis are therefore further investigated by examples; whereas the remaining five parameters are evaluated by a simple 2^{5-1} fractional factorial design.

(a) *Effect of I_u.* From the previous example, with other data being held constant, income I_u is varied to evaluate its effect on the profit region and maximum profit locus. I_u is originally assumed equal to \$1.00/pc, two different values of \$0.80/pc and \$1.10/pc are substituted into eq. (6), and profit responses from output data are plotted in Fig. 10. Since I_u takes on three different values, three envelopes of zero-profit boundaries and the three corresponding maximum profit loci can be shown on the f–V plane.

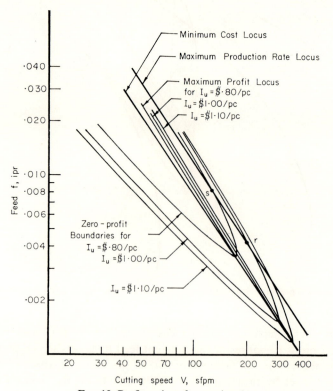

FIG. 10. Profit regions for varying I_u.

Minimum cost and maximum production rate loci are independent of I_u (refer to eqs. (3) and (4)), thereby remaining unaltered under variation of I_u. Both loci are superimposed on the figure as reference.

When I_u decreases from \$1.10/pc to \$0.80/pc, the profit region decreases. The envelope of zero-profit boundaries becomes smaller. This implies that the feasible choice of cutting conditions that are capable of yielding profit is also decreased. Correspondingly, as I_u is decreasing from \$1.10/pc to \$0.80/pc, the maximum profit loci are shifting toward the minimum cost locus. This implies that at relatively low income, minimum cost locus can serve as an approximation to the maximum profit locus for this example.

The variation of I_u has an interesting effect on the "limiting" feed rates for the maximum production rate locus. When I_u equals \$1.00/pc, maximum production rate locus yields negative profit (loss) at feeds lower than 0.0045 ipr (point r). When I_u equals \$0.80/pc, it will yield negative profit response at feeds lower than 0.008 ipr (point s). This "limiting" feed may continually rise as I_u is further decreased.

(b) *Effect of C_o.* Again from the previous example, with other data being held constant, operating cost C_o is varied. Aside from the original assumed value of \$0.08/min, two different C_o values of \$0.10/min and \$0.12/min are substituted into eq. (6). Three envelopes of zero-profit boundaries and three minimum cost loci at C_o equals 0.12, 0.10 and \$0.08/min are plotted in Fig. 11. Minimum cost locus shifts parallel to the right by an increment which conforms to the variation of C_o.

As C_o is increasing from \$0.08/min to \$0.12/min, the envelope of zero-profit boundaries becomes smaller. Profit region is decreasing, and profit response for a fixed cutting conditions within the profit region is also decreasing. Comparing the three intersection points

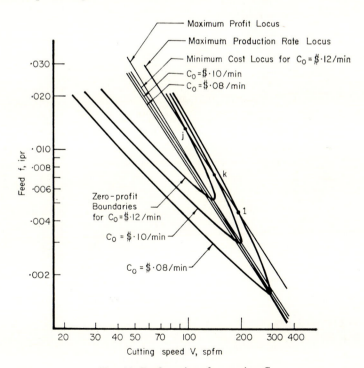

FIG. 11. Profit regions for varying C_o.

(j, k, l) of the maximum production rate locus with the three zero-profit boundaries for C_o, similar effect on the "limiting" feed rates can be observed as in the change of I_u.

The maximum production rate locus is unaffected by the variation of C_o as V_{max} is independent of_o C shown in eq. (4). The maximum profit locus is also unaffected by the variation of C_o, as V_p is independent of C_o shown in eq. (6). Note that for relatively high C_o value, the maximum profit locus can be approximated by the minimum cost locus.

Evaluation of Other Operating Parameters (i.e. n, C, C_t, t_h, t_c) by 2^{5-1} Fractional Factorial Design

A 2^{5-1} fractional factorial design is used to evaluate the effects of the variation of the five parameters, n, C, C_t, t_h, t_c. The responses selected for the evaluation are speed V_p and the corresponding maximum profit which can be readily calculated using eqs. (8) and (6). A deviation of $\pm 20\%$ from the original assumed values is chosen, the low level being -20% deviated, and high level $+20\%$ deviated. Maximum allowable feed, constant I_u and constant C_o are used. In this example, $I_u = \$1.00/\text{pc}$, $C_o = \$0.08/\text{min}$, and assume that the maximum allowable feed is 0.010 ipr.

The two level of the five variables and its transforming equations are as follows:

	low level	high level
n	0.10	0.15
C	112	172
C_t ($/tool)	0.4	0.6
t_h (min/pc)	0.8	1.2
t_c (min/edge)	0.8	1.2

$$x_1 = \frac{n - 0.125}{0.025}$$

$$x_2 = \frac{C - 142}{30}$$

$$x_3 = \frac{C_t - 0.5}{0.1}$$

$$x_4 = \frac{t_h - 1}{0.2}$$

$$x_5 = \frac{t_c - 1}{0.2}$$

Table 3 shows the design matrix for 16 trials. Maximum profit cutting speed V_p is first calculated by eq. (8), maximum profit response is then calculated by eq. (6). Take Trial 1 for example:

With $I_u = 1.00$
$\quad f = 0.010$
$\quad n = 0.1$
$\quad C = 112$ $(C = k/f^m)$
$\quad C_t = 0.4$
$\quad t_h = 0.8$
$\quad t_c = 1.2$

Substituting above data into eq. (8), we obtain V_p equals 78 sfpm. Then, substituting the above data together with V_p ($= 78$) and $C_o = 0.08$ into eq. (6), we obtain the maximum profit response as \$0.05478/min. The responses for the remaining 15 trials are calculated in the same manner and are tabulated in Table 3.

The main effects and two factor interaction effects on V_p and maximum profit response are calculated and shown in Table 4. The ANOVA for maximum profit response is given

TABLE 3. DESIGN MATRIX AND RESPONSES OF 2^{5-1} FRACTIONAL FACTORIAL DESIGN

Trial no.	Design matrix					V_p	Maximum profit (S/min)
	x_1	x_2	x_3	x_4	x_5 $(x_1x_2x_3x_4)$		
1	-1	-1	-1	-1	1	78	0.05478
2	1	-1	-1	-1	-1	69	0.03535
3	-1	1	-1	-1	-1	124	0.11843
4	1	1	-1	-1	1	110	0.08943
5	-1	-1	1	-1	-1	76	0.05178
6	1	-1	1	-1	1	65	0.02847
7	-1	1	1	-1	1	119	0.11131
8	1	1	1	-1	-1	106	0.08470
9	-1	-1	-1	1	-1	78	0.04846
10	1	-1	-1	1	1	68	0.02838
11	-1	1	-1	1	1	122	0.10110
12	1	1	-1	1	-1	111	0.08022
13	-1	-1	1	1	1	76	0.04402
14	1	-1	1	1	-1	65	0.02430
15	-1	1	1	1	-1	120	0.09823
16	1	1	1	1	1	104	0.07171

TABLE 4. EFFECTS OF VARIABLES

Variable		Main effect on V_p	Main effect on maximum profit
(n),	E_1	-11.875	-0.0231875
(C),	E_2	$42{\cdot}625$	0.054955
(C_t),	E_3	-3.625	-0.00519
(t_h),	E_4	-0.375	-0.009736
(t_c),	E_5	-0.875	-0.0015275

Variable		2 factor interaction effect on V_p	2 factor interaction effect on maximum profit
$(n - C)$,	E_{12}	-1.625	-0.0028137
$(n - C_t)$,	E_{13}	-0.875	-0.0008525
$(n - t_h)$,	E_{14}	-0.125	0.001425
$(n - t_c)$,	E_{15}	-0.125	-0.0000925
$(C - C_t)$,	E_{23}	-0.875	-0.0005225
$(C - t_h)$,	E_{24}	-0.125	-0.0033425
$(C - t_c)$,	E_{25}	-0.625	-0.00048
$(C_t - t_h)$,	E_{34}	0.125	-0.001585
$(C_t - t_c)$,	E_{35}	0.125	0.00059
$(t_h - t_c)$,	E_{45}	-0.125	0.0000775

in Table 5. Two factor interaction effects are considered insignificant and are used as residual. It can be seen that of the five factors under investigation, only n and C are significantly important as expected. It is interesting to see that a deviation of $\pm 20\%$ for C_t, t_h, and t_c has very insignificant effects on the profit response.

TABLE 5. ANALYSIS OF VARIANCE TABLE FOR MAXIMUM PROFIT RESPONSE

Source of variation	Sum of squares	Degree of freedom	Mean squares	Ratio
x_1	0.0021504	1	0.0021504	21.316
x_2	0.0120802	1	0.0120802	119.747
x_3	0.0001077	1	0.0001077	1.0676
x_4	0.0003791	1	0.0003791	3.7579
x_5	0.0000093	1	0.0000093	0.0921
$x_1 x_2$	$0.316676(10)^{-4}$	1		
$x_1 x_3$	$0.02907(10)^{-4}$	1		
$x_1 x_4$	$0.0812248(10)^{-4}$	1		
$x_1 x_5$	$0.00034225(10)^{-4}$	1		
$x_2 x_3$	$0.01092025(10)^{-4}$	1		
$x_2 x_4$	$0.446892(10)^{-4}$	1	sum $= 1.0088038(10)^{-4}$	
$x_2 x_5$	$0.009216(10)^{-4}$	1	(RESIDUAL)	
$x_3 x_4$	$0.1002984(10)^{-4}$	1		
$x_3 x_5$	$0.013925(10)^{-4}$	1		
$x_4 x_5$	$0.0002402(10)^{-4}$	1		

CONCLUSIONS

1. The application of computer for evaluation of economic optimum machining conditions shows a promising future in the respect that complicated problems can be readily delineated with simplicity and practicability.

2. The optimum cutting condition to yield maximum profit lies in the interval between V_p and V_d when a fixed demand volume has to be fulfilled. V_d is the lower cutting speed given the demand contour at the maximum allowable feed rate, and V_p is the theoretical maximum profit cutting speed.

3. The effect of the variation of income I_u and operating cost C_o on the profit analysis is investigated. Significant change of profit regions can be readily observed. The remaining five parameters (n, C, C_t, t_h, t_c) involved in the basic mathematical models are evaluated by a simple 2^{5-1} fractional factorial design. No significant changes are shown for parameters C_t, t_h, and t_c with a $\pm 20\%$ variation.

ACKNOWLEDGEMENTS

This research was supported by the National Science Foundation under a grant (NSF-GK-1695).

Use of the University of Wisconsin Computing Center was made possible through support, in part, from the National Science Foundation, other United States Government agencies and the Wisconsin Alumni Research Foundation (WARF) through the University of Wisconsin Research Committee.

REFERENCES

1. R. N. BROWN, On the Selection of Economical Machining Rates, *Int. Jnl. Prod. Res,*. **1**, No. 2, 1962, 1–22·
2. S. M. WU, Tool Life Testing by Response Surface Methodology, Part I and II. *ASME Trans.* Series B, May 1964, pp. 105–110, 110–116.

3. S. M. Wu, Analysis of Rail Steel Bar Welds by Two-level Factorial Design, *Welding Journal*, Research Supplement, April 1964, pp. 179S-183S.
4. K. Okushima and K. Hitomi, A Study of Economic Machining: An Analysis of the Maximum-profit Cutting Speed, *Int. Jnl. Prod. Res.*, **3**, No. 1, 1964, 73–78.
5. J. A. Armarego and Russell, Maximum Profit Rate as a Criterion for the Selection of Machining Conditions, *Int. Jnl. of Machine Tool Design & Research*, **6**, No. 1, 1966, 15–23.
6. S. M. Wu and D. S. Ermer, Maximum Profit as the Criterion in the Determination of the Optimum Cutting Conditions, *ASME Trans.* Series B, Nov. 1966, pp. 435–442.
7. J. A. Armarego and Russell, The Maximum Profit Rate Criterion Applied to Single Pass Shaping and Milling Process, *Int. Jnl. of Machine Tool Design & Research*, **7**, No. 2, 1967, 107–121.
8. *Manual of Cutting of Metals*—ASME, 1952.

THE APPLICATION OF NUMERICAL CONTROL TO CENTRE LATHE TURNING OPERATIONS—THE USE OF COMPUTING PROGRAMMING TO OBTAIN THE PARAMETERS FOR MACHINING

D. French, D. A. Milner

The University of Aston in Birmingham

and

W. J. Weston

English Electric Co. Ltd.

SUMMARY

The application of numerical control to the centre lathe turning of shafts covers a wide field of problems which has received little attention in Europe until the last three years. Analyses show that 70% of machining work requires some turning operation, but relatively little attempt has been made to apply numerical control to the centre lathe.

For the numerical control of machine tools it is necessary to determine:

(a) Tool motions.

(b) Cutting parameters.

(c) Tools required.

(d) Operations sequence.

This paper illustrates the use of a computer program to select the appropriate cutting parameters. A series of equations are presented and incorporated into a computer program to calculate the cutting parameters, using only the known facts of the machine tool, cutter and workpiece.

Throughout the paper the material considered is steel. PERA Report No. 142 states: "the high proportion (the figure mentioned is over 80%) of ferrous metals, particularly carbon steels, used in industry indicates that they should form the basis of any research and development concerned with machining".

NOMENCLATURE

HPCIN3	horsepower to remove one cubic inch of metal per minute
BH	Brinell hardness number of the workpiece
UTS	ultimate tensile strength of the workpiece
Cl	force constant for the workpiece material
F	factor connecting feed rate and the horsepower to remove one cubic inch of metal per minute
f	feed rate in inches per rev
OUTDIA	outside diameter of the workpiece
FINDIA	final diameter of the workpiece
HPG	gross horsepower of the machine tool
SWING	swing over the bed of the machine tool
EFF	mechanical efficiency of the machine tool
FDMIN	minimum feed rate obtainable on the machine tool, in/rev

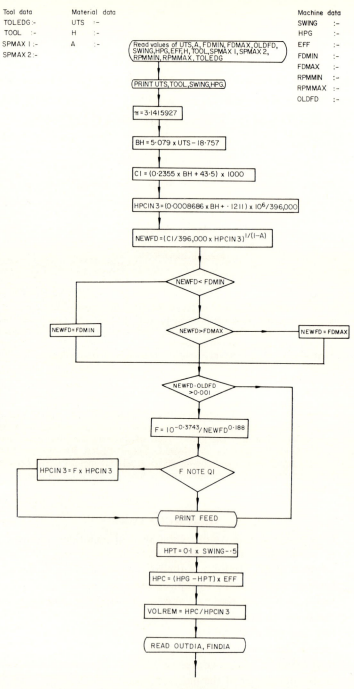

Fig. 1. Flow chart.

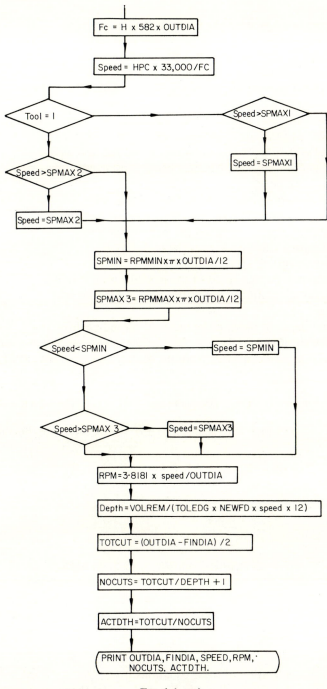

FIG. 1 (cont.)

FDMAX	maximum feed rate obtainable on the machine tool, in/rev
RPMMIN	minimum speed obtainable on the machine tool, rpm
RPMMAX	maximum speed on the machine tool, rpm
TOOL	TOOL = 1 represents a high speed steel tool
	TOOL = 2 represents a tungsten carbide tool
SPMAX1	maximum speed for machining with HSS tool, ft/min
SPMAX2	maximum speed for machining with a carbide tool, ft/min
TOLEDG	cutting tool edge, in
H	factor in Prof. Koenigsberger equation, $FC = H \times 582 \times D$
A	index of feed rate in equation $FC = Cl \times f^A \times t$
OLDFD	OLDFD = 0 for first iteration of feed rate calculation
N	number of outside diameters being considered
HPT	idle horsepower of the machine tool
HPC3	horsepower available at the tool cutting edge
NEWFD	feed rate calculated by the programme, in/rev
VOLREM	volume of metal removed per minute, in^3/min
FC	tangential cutting force on the tool, lbf
SPEED	surface cutting speed, ft/min
SPMIN	minimum cutting speed of the machine tool, ft/min
SPMAX3	maximum cutting speed of the machine tool, ft/min
DEPTH	depth of cut
TOTCUT	total depth of metal to be removed on radius, in
NOCUTS	number of cuts
RPM	speed in rpm converted from the surface cutting speed, SPEED

EQUATIONS USED FOR COMPUTING THE CUTTING PARAMETERS

$$HPCIN3 = (0{\cdot}0008686 \times BH + 0{\cdot}1211) \times 10^6/396,000 \qquad (1)$$

$$BH = 5{\cdot}079 \times UTS - 18{\cdot}757 \qquad (2)$$

For carbon steels

$$Cl = (0{\cdot}445 \times BH + 62) \times 1000 \qquad (3)$$

For free machining carbon steels

$$Cl = (0{\cdot}2355 \times BH + 43{\cdot}5) \times 1000 \qquad (4)$$

For Ni-Cr steels

$$Cl = (0{\cdot}333 \times BH + 62) \times 1000 \qquad (5)$$

$$F = 10^{-0{\cdot}3743}/f^{0{\cdot}188} \qquad (6)$$

(See Appendix for consideration of these equations.)

A block diagram, or flow chart, shown in Fig. 1 illustrates the inter-relationships between operations, tests or comparisons and input and output statements. The computer used was an ELLIOTT 803. A print of the ALGOL program with output is shown in Fig. 2.

```
PROGRAM TO DERIVE TECHNOLOGICAL DATA FOR TURNING OPERATIONS'
BEGIN REAL TOLEDG,UTS,H,A,SWING,HPG,HPT,HPC,BH,C1,EFF,FDMIN,FDMAX,
      HPCIN3,NEWFD,OLDFD,F,VOLREM,FC,SPEED,SPMIN,SPMAX3,PI,
      DEPTH,TOTCUT,ACTDTH,RPM'
  INTEGER TOOL,SPMAX1,SPMAX2,RPMMIN,RPMMAX,NOCUTS,I,N'
  READ N,UTS,A,FDMIN,OLDFD,SWING,HPG,EFF,H,TOOL,SPMAX1,
      SPMAX2,RPMMIN,RPMMAX,TOLEDG'
 BEGIN REAL ARRAY OUTDIA,FINDIA(1:N)'
  SWITCH S:=S1,S2,S3,S4,S5,S6,S7'
  PRINT ££L?TENSILE STRENGTH OF WORKPIECE= ?,SAMELINE,UTS,
      £TONS PER SQ INCH£L?CUTTER MATERIAL= ?,SAMELINE,
      TOOL,££L?LATHE CAPACITY= ?,SAMELINE,SWING,£INCHES
      £L?GROSS HORSE POWER OF LATHE= ?,SAMELINE, HPG'
  PI:=3·1415927'
  BH:=5·079*UTS×18·757'
  C1:=(·445*BH+62)*1000'
  HPCIN3:=(·0008686*BH+·1211)*10**6/396000'
  NEWFD:=(C1/(396000*HPCIN3))**(1/(1×A))'
  IF NEWFD LESS FDMIN THEN BEGIN NEWFD:=FDMIN'
                                          GOTO S1
                        END
  ELSE IF NEWFD GR FDMAX THEN BEGIN NEWFD:=FDMAX'
                                          GOTO S1
                               END
  ELSE GOTO S1'
  S1:IF NEWFD–OLDFD GR ·001 THEN BEGIN F:=10**(·3743)/NEWFD**·188'
                                          GOTO S2
                             END
     ELSE GOTO S3'
  S2:IF F NOTEQ 1 THEN BEGIN HPCIN3:=F*HPCIN3'
                                          GOTO S3
                        END
     ELSE GOTO S3'
  S3:PRINT ££L?FEED= ?,SAMELINE,NEWFD,£INCHES PER REV
           £L?OUTDIA ?,SAMELINE,££S6?FINDIA ?,
           ££S6?SPEED ?,££S7?RPM ?,££S9?NOCUTS ?,
           ££S2?ACTDTH ?'
  HPT:=·1*SWING×·5'
  HPC:=(HPG×HPT)*EFF'
  VOLREM:=HPC/HPCIN3'
  FOR 1:=1 STEP 1 UNTIL N
  DO BEGIN READ OUTDIA(I),FINDIA(I)'
           FC:=H*582*OUTDIA(I)'
           SPEED:=HPC*33000/FC'
           IF TOOL=1 THEN GOTO S4 ELSE GOTO S5'
     S4:IF SPEED GR SPMAX1 THEN BEGIN SPEED:=SPMAX1'
                                          GOTO S6
                                END'
     S5:IF SPEED GR SPMAX2 THEN BEGIN SPEED::=SPMAX2'
                                          GOTO S6
                                END'
     S6:SPMIN:=RPMMIN*PI*OUTDIA(I)/12'
        SPMAX3:=RPMMAX*PI*OUTDIA(I)/12'
        IF SPEED LESS SPMIN THEN BEGIN SPEED:=SPMIN'
                                          GOTO S7
                             END
        ELSE IF SPEED GR SPMAX3
                        THEN BEGIN SPEED:=SPMAX3'
                                          GOTO S7
                             END
```

FIG. 2. ALGOL program.

```
                ELSE GOTO S7'
            S7:RPM:=3·8181*SPEED/OUTDIA(I)'
                DEPTH:=VOLREM/(TOLEDG*NEWFD*SPEED*12)'
                TOTCUT:=(OUTDIA(I)×FINDIA(I))/2'
                NOCUTS:=TOTCUT/DEPTH+1'
                ACTDTH:=TOTCUT/NOCUTS'
                PRINT ££L2??,SAMELINE,OUTDIA(I),££S2??,
                    FINDIA(I),££S2??,SPEED,££S2??,RPM,
                    ££S2??,DIGITS(2),NOCUTS,££S5??,ACTDTH
            END
    END
END'
```

FIG. 2 (*cont.*)

ANALYSIS OF RESULTS

The program (Fig. 2) was drawn up to enable machine laboratory tests to be carried out using the values given below:

Material: EN 1A free machining steel.

Tensile strength for $2\frac{1}{2}$ to 4 in bar = 23 tonf/in².

Outside diameter of bar to be machined = 3·875 in.

Machine: Dean, Smith and Grace Centre Lathe.

Horsepower = 7·5.

Swing over the bed = 17 in.

Minimum rpm available = 9·8.

Minimum feed available = 0·0012 in/rev.

Maximum rpm available = 900.

Maximum feed available = 0·070 in/rev.

An efficiency of 0·8% was assumed.

Tool: $\frac{1}{2}$ in tungsten carbide tipped tool.

$Cl = (0·2355 \times BH + 43·5) \times 1000.$

The lathe was fitted with a Cranfield Force Dynamometer in order to compare the tangential cutting force with the value calculated by the program. The combination of a feed rate = 0·0035 in, a cutting speed = 363 rpm, and a depth of cut = 0·1875 in, gave the best results and were obtained with $H = 0·2$ and $A = 0·8$.

DATA USED FOR THE TURNING TESTS

Data which is basic to all twelve tapes:

$N = 1$, UTS = 23, FDMIN = 0·0012, FDMAX = 0·08, OLDFD = 0, SWING = 17, HPG = 7·5, EFF = 0·8, TOOL = 2, SPMAX1 = 200, SPMAX2 = 400, RPMMIN = 10, RPMMAX = 900, OUTDIA = 3·875, FINDIA = 3·5.

Table 1 illustrates differences between the data tapes.

This value is only reached in the practical tests with a cutting speed of 310 rpm and a feed rate in excess of 0·01 in/rev.

It should be noted that the feeds and speeds available on the lathe are not infinitely variable and so its becomes necessary to select the value which is closest to that provided by the computer program. This is a modification which could be incorporated in the program.

Having carried out the turning tests, a series of data tapes were compiled to run with the program. H was varied from 0·1 to 1 and this comprised one set of data tapes. A second set varied A from 0·6 to 0·95.

TABLE 1.

Data tape No.	A	H	TOLEDG
1	0·8	0·5	0·5
2	0·85	0·5	0·5
3	0·8	0·5	1·0
4	0·85	0·5	1·0
5	0·8	0·2	0·5
6	0·85	0·2	0·5
7	0·8	0·2	1·0
8	0·85	0·2	1·0
9	0·8	0·3	0·5
10	0·85	0·3	0·5
11	0·8	0·3	1·0
12	0·85	0·3	1·0

It should be noted that with $H = 0.2$
$FC = 0.2 \times 582 \times 3.875$
$FC \simeq 450$

SURVEY OF OTHER WORK WHICH COULD BE INCORPORATED IN THE PROGRAM FOR OPTIMIZATION OF THE PARAMETER

There are many other equations which could have been incorporated in the program for parameter selection. Having obtained a working program, this could then be used as a basic for further work employing these other equations. The running of the program using these alternative equations will enable the optimum to be obtained.

The two chief considerations in metal cutting become the need:

(a) To reduce the machining time to a minimum.
(b) To optimize tool life.

Other criterion may be:

(c) A combination of (a) and (b).
(d) Attainment of a certain surface finish.
(e) Attainment of a certain accuracy.
(f) A combination of (d) and (e).

Taylor's equation for tool life has been extended to include factors of feed rate and speed.

Taylor's equation $\quad V \times T^n = \text{constant}$ \hfill (7)

$$V \times T^n \times f^c \times t^d = \text{constant} \tag{8}$$

There is little published work on these aspects and the constants are known only for a few materials. Boston[5] has published values for the indices as follows: $n = 1/6$, $c = 0.77$, $d = 0.37$. In general, as feed rate or depth of cut is increased, the cutting speed must be decreased to keep the tool life constant but the volume of metal removed is increased if the depth is increased because of its low exponent.

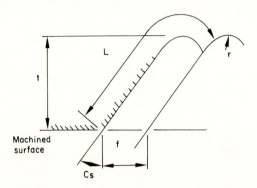

r = Tool nose radius
Cs = Plan approach angle

Fig. 3. Tool nomenclature.

The effect of nose radius and plan approach angle of the tool are not included in eq. (8). These two factors have, however, been incorporated in Woxen's equation, which is based on the concept of the chip equivalent, q.

$q = A_0/L$

where q = chip equivalent,
         A_0 = cross-sectional area of chip before removal,
         L = contact length of the tool edge.

A_0 is a measure of the quantity of heat liberated at a certain cutting speed and L is a measure of the quantity of heat carried away by the chip, the tool and the workpiece.

From Fig. 3,

$$q = f \times t/L \qquad (9)$$

$$q \simeq f \times t$$

$$\frac{(t - r)(1 - \sin Cs)}{\cos Cs} + \frac{(90 - Cs)\pi r}{180} \qquad (10)$$

The tool life equation may therefore be expressed in terms of cutting speed and chip equivalent.

$$V \times T^n \times q^m = \text{constant} \qquad (11)$$

From eq. (10)

$$q = C \times f$$

where C represents the other factors in eq. (10).

Therefore combining eq. (11) and the modified version of eq. (10)

$$V \times T^n \times (C \times f)^m = \text{constant} \qquad (12)$$

The total cost of machining[5] per piece may be broken down as follows:
 (a) Set-up and idle time costs — Ks
 (b) Machining costs — Km
 (c) Tool changing costs — Kc
   Grinding costs — Kg
   Depreciation — Kd

(a) $Ks = Oc \times ts$ (13)

where Oc = overheads and direct labour costs,
      ts = total set-up and idle time per piece.

(b) $Km = Oc \times L \times \pi \times D/f \times V$ (14)

(c) Kc. From eq. (12) it must be noted that the tool life at which the tool will be changed must be declared. This will be defined by the wear.
 There are several types of wear:
   Flank wear.
   Crater wear.
   A combination of flank and crater wear.
   Thermal cracking.
 Flank wear is taken as the deciding factor. This occurs on the side relief angle and the height of the wear band is usually used as the limiting criterion, the value selected being within the range 0·015 to 0·030 in.
 Let Tfw = tool life for a wear land of fw

 Therefore $V \times Tfw^n \times (C + f)^m = \lambda$ (15)

λ, n and m are parameters for given workpiece and tool materials. The wear and turning time relationship is linear between the initial fast wear rate at the commencement of cutting and the fast wear rate in the later stages of tool degeneration.
 From a plot of flank wear against turning time the following relationship may be obtained. $F = Ft + zt$ where Ff is the intercept on the wear axis and z is the slope both taken over the linear range

$$Kc = Oc \times tc \times L\pi D/(f \times V)(V/\lambda)^{1/n} \times (C \times f)^{m/n} \times (Ffw - Ff)/(F - Ff)$$ (16)

where tc = tool changing time.

 Kg. This can be expressed as the cost of a regrind multiplied by the number of regrinds per piece.[9] The number of regrinds per piece will be similar to that used in eq. (16). The cost may be broken down into three parts: a basic cost which is independent of the amount of grinding required plus the cost of grinding to remove the wear land plus an additional cost to allow a safety margin, as wear may have varied along the cutting edge.
   Basic cost = $Og1$
   Grinding cost = $Og2 \times F \times \sin c$
 where c = clearance angle.
   Safety cost = $Og3 \times s$
 where s = safety margin.

Therefore

$$Kg = (Og1 + Og2 \times F \times \sin c + Og3) \times \left[\frac{L\pi D}{fV} (V/\lambda)^{1/n} (Cf)^{m/n} \left(\frac{Ffw - Ff}{F - Ff} \right) \right] \quad (17)$$

Kd. The depreciation in value between successive regrinds is the difference between its initial and final values divided by the number of lives that may be expected to be obtained from it. If the total amount which can be removed in direction y is M (see Fig. 4), then the possible number of regrinds is $M/(F \times \sin c + s)$.

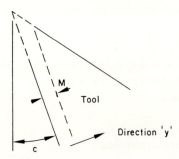

Fig. 4. Tool nomenclature.

The actual number is one plus this, as the tool is used once before any grinding is required. If Wa is the difference between initial and final values:

$$Kd = [Wa/(1 + M/F \sin c + s)] \left[\frac{L\pi D}{fv} (V/\lambda)^{1/n} (Cf)^{m/n} \left(\frac{Ffw - Ff}{F - Ff} \right) \right] \quad (18)$$

The total cost, Kt, will therefore be represented by the summation of eqs. (13), (14), (17) and (18).

$$Kt = Oc \times ts \times Oc \times \frac{L\pi D}{fV} + \frac{L\pi D}{fv} (V/\lambda)^{1/n} (Cf)^{m/n}$$

$$\times \left[\left(\frac{Ffw - Ff}{F - Ff} \right) (Oc \times tc + Og1 + Og3s + Og2F \sin c + Wa/(1 + M/F \sin c + s) \right]$$

$$(19)$$

If a given amount of flank wear is permitted and eq. (19) is differentiated with respect to (a) cutting speed and (b) feed rate, then the resulting expressions for minimum costs, obtained by equating the differential forms to zero, are mutually inconsistent. As long as $m < 1$, the most economical conditions will be obtained by machining at the largest possible feed rate consistent with machine power, workpiece rigidity, surface finish and dimensional accuracy. With this feed rate, $f1$, the most economical cutting speed, V_0 is given by:

$$V_0 = \left(\frac{Oc}{H(1/n - 1)} \right)^n (Cf1)^{-m} \lambda \quad (20)$$

where $H =$ the term within the square brackets in eq. (19) and C is a function of the depth to be cut (outside diameter—finish diameter), nose radius and the plan approach angle.

Over a practical range of values the effect of feed rate is negligible. The value for the nose radius is set as large as possible consistent with keeping chatter to a minimum, this enables a better surface finish to be obtained. The value of the plan approach angle is determined by the nature of the workpiece and the desired direction of chip flow.

This analysis has been made for brazed on tipped cutting tools but may be extended to cover throwaway tipped tools also. For such a tool

$$H = \left(\frac{Ffw - Ff}{F - Ff} \right) (Oc \times ti \times Wt/T) \tag{21}$$

where ti = indexing time + ($1/T$ of ts for one tip)

T = number of cutting edges

Wt = cost of one tip.

The value of the term $1/n - 1$ in eq. (20) has been determined for various cutting tool materials. For carbides = 3, HSS = 7, and ceramics = 1·6. Values of C, m and λ must be determined. Tables and monograms are in existence which obviate the need for actually using these values when solving eq. (20). For the application of this approach to the computer program it would be necessary for these monograms to be converted into equation form.

CONCLUSION

Machining Parameter Selection

Each set of parameters provided by the program, from the twelve data tapes, were tried out in the machine laboratory. All were found to be practicable, the best one being:

Cutting speed — 363 rpm

Feed rate — 0·0035 in/rev

Depth of cut — 0·1875 in

No. of cuts — 1

The feed rate was still low, but this value could be increased by reducing the value of the index A, e.g. for a workpiece with a UTS of 30 t/in² being machined on a lathe with a gross horsepower of 5 and a swing of 15 in, a feed rate of only 0·0035 in/rev is obtained for $A = 0.9$, but one of 0·080 in/rev for $A = 0.7$.

The computer program could then be used in conjunction with the APT (Automatic Programmed Tooling) language. The APT language analyses the workpiece geometry and the tool motions required for machining. The motion statements defined by APT could, perhaps, be superseded in the future by the use of light pen in-line with a computer.

APPENDIX

To Find the Horsepower to Remove one Cubic inch of Metal Per Minute

The horsepower required to remove one cubic inch of metal per minute is converted into several forms of energy:

(a) As shear energy on the shear plane.

(b) As friction energy on the tool face.

(c) As surface energy due to the formation of a new surface area in cutting.

(d) As momentum energy resulting from a momentum change associated with the metal as it crosses the shear plane.

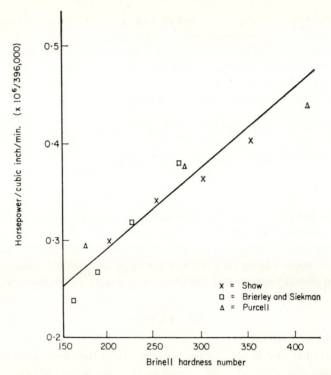

FIG. 5. Scatter diagram of the horsepower required to remove one cubic inch/min against Brinell hardness number.

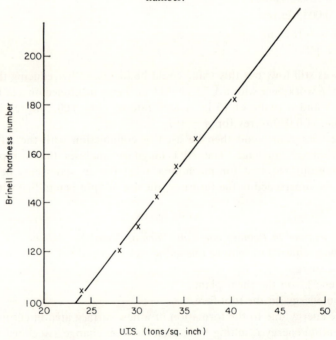

FIG. 6. Brinell hardness number against ultimate tensile strength.

Values of horsepower and Brinell hardness number are illustrated from three different sources[7, 9, 10] in Fig. 5. The results tabulated by Shaw[10] give a line which approximate to the mean of the three sets of data. By using linear regression theory the following equation was obtained:

$$HPCIN3 = (0{\cdot}0008686 \times BH + 0{\cdot}1211) \times 10^6/396{,}000 \tag{1}$$

It is also usual for the ultimate tensile strength (UTS) of the workpiece to be known rather than the Brinell hardness number (*BH*). From Fig. 6 the equation connecting these two variables is given by

$$BH = 5{\cdot}079 \times UTS - 18{\cdot}757 \tag{2}$$

To Determine the Feed Rate

The basic equation for deriving horsepower is:

$$HP = FORCE \times VELOCITY/33{,}000$$

where   FORCE is in lbf
and   VELOCITY is in ft/min.

The forces acting on the tool are shown diagrammatically in Fig. 7. The vertical force (*FC*) will be used in calculating the power required for cutting as it constitutes approximately 99% of the total power used.

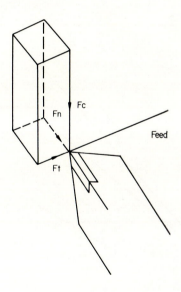

Fc = Vertical or tangential force
Ft = Feed force
Fn = Thrust force

FIG. 7. Forces acting on a tool.

Substituting in this equation the horsepower at the cutting edge (HPC), the vertical force (FC) and the cutting speed (VC)

$$HPC = FC \times VC/33{,}000$$

and
$$HPCIN = VC/33{,}000 \times t \times f \times VC \times 12 \qquad (7)$$

where $t =$ depth of cut in inches,
      $f =$ feed rate in inches per revolution and
HPCIN3 $=$ horsepower to remove 1 cubic inch of metal per minute,
whence

$$FC = f \times 396{,}000 \times HPCIN3 \times t \qquad (8)$$

The vertical force (FC) is relatively independent of cutting speed, except at low speeds where increase in the speed tends to produce a decrease in the cutting force. Its effect may be neglected on the reasonable assumption that for optimum conditions, whether minimum machining time or maximum tool life, the values will not lie in this range; however, the vertical force (FC) is related to the feed rate and depth of cut by a power law

$$FC \propto f^a \times t^b.$$

Hence
$$FC = Cl \times f^a \times t \qquad (9)$$

where Cl is a force constant for the material.

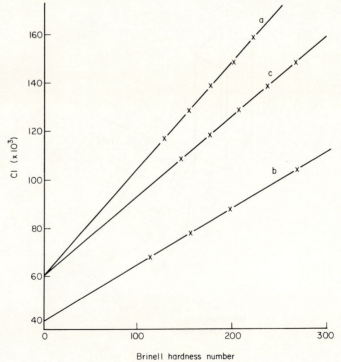

Fig. 8. Graph of Cl against BH.

From Fig. 8 the equation connecting Cl and BH is given by:

(a) For carbon steels   (3)

$$Cl = (0.445 \times BH + 62) \times 1000$$

(b) For free machining carbon steels

$$Cl = (0.2355 \times BH + 43.5) \times 1000 \qquad (4)$$

(c) For Ni-Cr steels

$$Cl = (0.333 \times BH + 62) \times 1000 \qquad (5)$$

Equating eqs. (8) and (9) and substituting $b = 1$ in eq. (9).

then $\qquad f \times 396{,}000 \times \text{HPCIN3} \times t = Cl \times f^a \times t$

whence $\qquad f^{1-a} + Cl/396{,}000 \times \text{HPCIN3}$

and $\qquad f = (Cl/396{,}000 \times \text{HPCIN3})^{1/(1-a)} \qquad (10)$

By taking logarithms of "factor F"$^{(10)}$ and feed rate (see Figs. 9 and 10) a linear relationship was found to exist and was represented by the equation:

$$\log F = -\,0.188 \times \log f - 0.3743$$

from which

$$F = 10^{-0.3743}/f^{0.188} \qquad (6)$$

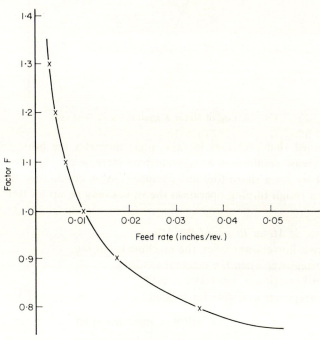

FIG. 9. Factor F against feed rate.

Having calculated the feed rate, the factor F is found. If $F \neq 1$, then the feed rate will affect the horsepower (HPCIN3) and a new value must be calculated.

$$\text{New HPCIN3} = F \times \text{old HPCIN3} \qquad (11)$$

From this new value of horsepower (HPCIN3), a new feed rate may be obtained. The values of feed rate and horsepower (HPCIN3) finally arrived at may then be used in the further calculations.

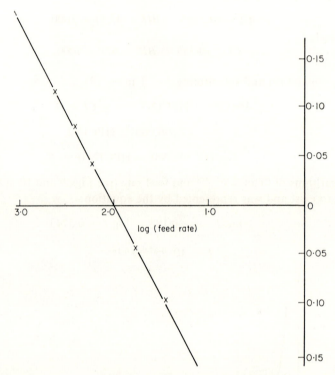

FIG. 10. Log of factor F against log of feed rate.

It should be noted that decreases in rake angle increases the horsepower (HPCIN3). (A one degree decrease resulting in an approximate increase of 2%.) The values of horsepower (HPCIN3) are for a sharp tool and average friction conditions. If the tool is about to be replaced in a rough turning operation the values may be up by 100%.

To Find the Volume of Metal Removed Per Minute

    HPG = gross horsepower when the machine is cutting.
    HPI = horsepower when the machine is idling.
    HPN = net horsepower, i.e. HPG − HPI.
    HPC = horsepower available for cutting,

$$\text{i.e. HPN} \times \text{mechanical eff.}$$

Therefore $$\text{HPC} = (\text{HPG} - \text{HPI}) \times \text{eff.} \tag{12}$$

From ref. 5,

    HPI $$= 0.1 \times \text{swing} - 0.5 \tag{13}$$

where swing = the swing of the lathe over the bed.

Then the volume of metal removed per minute is obtained by dividing the horsepower at the cutting edge by the horsepower required to cut one cubic inch of metal per minute, i.e.

$$VOLREM = HPC/HPCIN3 \tag{14}$$

To Determine Speed

This is obtained using $HPC = FC \times VC/33,000$

whence
$$VC = HPC \times 33,000/FC \tag{15}$$

For a well-designed lathe and a small overhang of the tool the limiting factor should only be the workpiece. For a workpiece of length L, and diameter, D (in inches) the limiting value of the vertical cutting force (FC) is given by:

$$FC = 2 \times \pi \times E \times D^4 \times s/3 \times L^3 \tag{16}$$

where $E =$ Young's modulus in psi

and $s =$ the permissible lateral deflection of the workpiece.

Koenigsberger has assessed s as 0·002 in and using a value of 1/6 for the D/L ratio and $E = 30 \times 10^6$

$$FC = 2 \times \pi \times 30 \times 10^6 \times L^3 \times D \times 0 \cdot 002/3 \times 6^3 \times L^3$$

$$FC = 582 \times D \tag{17}$$

However, it would be unwise to work on the maximum value of the vertical cutting force (FC) which this equation represents and so a factor H is introduced and the equation becomes

$$FC = H \times 582 \times D \tag{18}$$

An identifier H has been introduced instead of a numerical value so that several values may be used in the computer program data tapes and thus the effect of H on the machining parameters can be ascertained.

To Determine the Depth of Cut, Total Depth to be Cut, Number of Cuts and the Actual Depth

The depth is obtained from the equation:

$$t = VOLREM/f \times VC \times 12 \times TOLEDG \tag{19}$$

where TOLEDG is the length of tool edge, for example for a form tool.

Total depth to be cut = outside diameter − inside diameter/2   (20)

Number of cuts = total depth to be cut/t (21)

This must be rounded off to the nearest whole number above the value obtained. A new depth of cut, or actual depth, must then be calculated.

Actual depth = total depth to be cut/number of cuts.   (22)

REFERENCES
1. Pera Report No. 142. Industrial machining practice—a survey of machining operations.
2. Pera Report No. 103. Performance of ceramic and carbide tools when turning alloy steels.
3. Pera Report No. 61. Performance of ceramic and carbide tools when turning alloy steels.

4. Pera Report No. 163. Throwaway tip turning tools—an investigation of the relationships between machining variables and tool life.
5. R. C. Brewer and R. R. Rueda, A Simplified Approach to the Optimum Selection of Machining Parameters. *Engineers' Digest*, **24**, No. 9, Sept. 1963.
6. J. B. Pond, Programmer for Selecting Feeds and Speeds and Tool Geometry. *Cutting Tool Engineering*, 16 May 1966.
7. J. Purcell, The Basic Analysis of Machine Tool Selection. *Machinery*, 20 Nov. 1963.
8. E. J. Weller. Feeds and Speeds as Viewed by a Production Engineer. *Cutting Tool Engineering*, Jan.–Feb. 1965.
9. R. G. Brierley and H. J. Siekman, Machining Principles and Cost Control. McGraw-Hill.
10. M. C. Shaw, Metal Cutting Principles. The Technology Press. M.I.T.
11. Battelle Memorial Institute. An Evaluation of Present Understanding of Metal Cutting. A.S.T.E. publication.
12. Computer Decides Machining Conditions for Numerically Controlled Lathes. *Metalworking Production*, 13 October 1965.

STUDIES OF CUTTING TEMPERATURE CONTROL APPLIED TO A LATHE SPINDLE SPEED

R. A. BILLETT

School of Engineering, Bath University of Technology

INTRODUCTION

In the metal cutting process the tool life is an important factor in the economics of production. Previous research has indicated that, for a wide range of conditions during metal cutting, the tool life or tool wear rate is a function of the tool–work interface temperature. Cutting speed is one of the most familiar variables in the cutting process, it is often prescribed in an effort to obtain an economic or acceptable tool wear rate and it is established that, for given cutting conditions, the temperature is a function of cutting speed. The wear rate thus varies with the temperature which, in turn, varies with the independent variable of cutting speed.

The work presented in this paper is based on the concept of temperature as the independent variable in the cutting process with velocity and wear rate as dependent variables. The work–tool thermal e.m.f. is taken as a measure of temperature and is used as the controlled variable of a closed-loop control system so that the spindle speed of a lathe is continuously controlled to produce a pre-demanded interface thermal e.m.f.

It is suggested that the use of thermal e.m.f. feedback control of speed, and in some cases feed, will give improved control of tool wear rates compared with use of such concepts as constant cutting speed.

NOMENCLATURE

A, B, C, K	constants in tool wear equations
E	thermal e.m.f. of work–tool thermocouple
T	tool life, min
V	cutting speed, ft/min
W	flank wear land, in
d	depth of cut (chip width), in
f	feed per rev (uncut chip thickness), in
k, k'	constants of thermal e.m.f. equation
n, p	tool wear exponents
r	exponents of thermal e.m.f. equation
t	time
θ	temperature, °K

PREVIOUS WORK

There has been a considerable study of the thermal e.m.f. produced at the tool–chip interface, sometimes known as the Herbert–Gottwein thermocouple, which has been used to estimate the tool-temperature in many metal cutting research studies since the early work[1, 2] in the 1920's. Later studies have discussed various forms of tool wear and have indicated relationships between tool wear or wear rate and tool temperature, under various cutting conditions such as

$$dW/dt = Ae^{-K/\theta} \tag{1}$$

which was postulated by Takeyama[3] for the diffusion type of tool wear which occurs over a wide range of higher tool temperatures.

Other work[4] has suggested that the relation between cutting speed and temperature is of the form

$$\theta = BV^p \tag{2}$$

and this is supported approximately by Takeyama above certain cutting speeds. When used with the Taylor equation

$$VT^n = \text{constant} \tag{3}$$

this leads to

$$dW/dt = C\theta^{1/np} \tag{4}$$

Since $1/np$ is often in the range between 10 and 100, equation (4) indicates that large changes in wear rate can occur with fairly small temperature changes.

Some of the difficulties and inaccuracies associated with the toolwork thermocouple method when used to estimate temperature have been discussed in such work as that of Chao, Li and Trigger[5] and recently Braiden[6] presented studies related to the calibration of such thermocouples.

The thermal e.m.f. has been used in a control system for the adaptive control of a numerically controlled milling machine. This work, reported by Centner[7] in 1964, was carried out by the Bendix Corporation under a USAF contract. In this work, the thermal e.m.f. signal was used together with torque, vibration, feed and speed feedback signals in an optimizing, or adaptive, control system. The thermal e.m.f. was checked against a constraint level to prevent excessive temperature and was also used in the computing circuits to produce an estimate of tool wear rate based on previously obtained empirical data. The estimate of tool wear rate was then fed to the optimizing part of the adaptive controller. In this complex system there was no proposal or work reported which produced continuous control of the temperature by using the thermal e.m.f. as the controlled variable of a simple closed loop system. Results presented in the Bendix work indicated that tool wear rate was a function of metal removal rate as well as temperature.

During the course of the study to be discussed in this paper, some work has been carried out on a lathe with thermal e.m.f. control at R.A.E. Farnborough, some initial results from which have been published.[8] These relate to the use of a carbide tool at a constant feed rate and depth of cut and discuss the variation in cold junction temperature in the case of an inserted tool tip with no coolant.

It is relevant to mention the work of Trent, who has shown[9] that conditions of boundary lubrication do not exist at the tool–chip interface. For the thermocouple or Seebeck effect to be valid the equivalent of a welded junction is required, such as has been shown by Trent to exist at the tool tip.

CONCEPT AND OBJECTIVES

The concept of the use of thermal e.m.f. control of cutting speed is illustrated in Fig. 1. In the uncontrolled or open loop mode with constant cutting speed, the trend is for the thermal e.m.f. to increase with time as a result of tool wear. This causes an increase in the tool wear rate until eventual, and perhaps premature, breakdown of the tool occurs. In the controlled or closed loop mode the trend is for the velocity to decrease with time

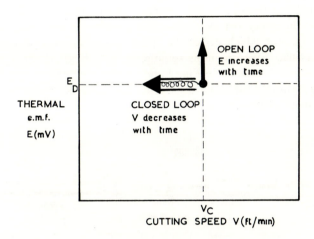

FIG. 1. Concept of closed loop thermal e.m.f. control.

so that the temperature and, as could be suggested from earlier work,[3] the tool wear rate remains constant and controlled. As the response of different control systems vary, the thermal e.m.f. can be considered to be distributed about the demanded value, the range of such deviation depending on the system used.

It was decided to use the thermal e.m.f. as the controlled variable rather than any other tool wear rate or temperature measurement. The use of thermal e.m.f. has the following advantages and disadvantages:

Advantages
1. Simple and cheap experimental set-up.
2. Should provide better means of tool wear rate control than methods in common use.
3. Provides a voltage which can be easily used in a control system.
4. Fast response time.

Disadvantages
1. Accuracy with which mean temperature and wear rates can be estimated is not known.
2. Affected by variation in tool and work materials, to a greater or lesser extent.

The aim of the study, of which the work presented forms a part, is to investigate and develop the use of machine tools fitted with temperature control as research aids and as production tools. The object of this paper is to present a progress report of some metal-cutting aspects of the use of a lathe fitted with thermal e.m.f. control rather than to examine in detail the methods of control which are subject to frequent modifications to suit various cutting test requirements. The basic control system uses established control techniques and equipment in the sub-systems with some adaptation for the particular duty.

EXPERIMENTAL EQUIPMENT

The basic control system consists of a sensor which provides the output signal, an error-detection sub-system and a speed variation sub-system. In this case the work–tool thermocouple is the sensor, the output of which is amplified, smoothed and compared with the demand signal which has been previously selected, usually by means of a manual potentiometer setting. The difference between these two signals, i.e. the error, is then amplified and fed to a servo-motor which drives the speed change control of a mechanical Kopp Variator which, in turn, drives the lathe spindle.

The arrangement used is shown in Fig. 2 and necessarily resulted in a longer time constant than ideally desirable for some tests. This has since been improved by the use of electronic motor speed control.

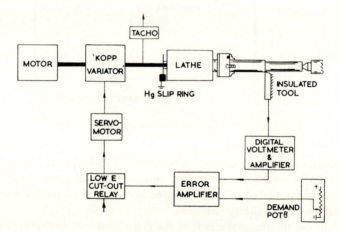

Fig. 2. Experimental arrangement.

Initial work was carried out with both the work and tool insulated and with a mercury slip-ring as the rotating electrical contact. It was later found that there was little change in the noise level if the slip-ring was grounded to earth; this was done since it eliminated the necessity to insulate the work at the chuck and tailstock. A coolant supply was provided for use as required.

The instrumentation available for short-term and long-term data recording including a digital milli-voltmeter, storage oscilloscope and x–y plotter.

The tools used for the majority of tests were 6:5:2 high-speed steel to SAE M2. In general, high-speed steel gives a lower thermal e.m.f. than carbides, but was more suitable for high wear rate tests with the limited power available. It will be appreciated that various compositions of high-speed steel give different temperature/thermal e.m.f. relationships and therefore accelerated tool wear or rapid breakdown will occur at different e.m.f. values for various compositions. For example, tests with a 5% Co high-speed steel gave values of 1·4–1·5 mV compared with 3·4–3·9 mV for rapid wear regions of the 6:5:2 steel, when cutting En 1a.

Most of the cutting tests were carried out on En 1a steel. Other materials were used and tests on En 8 and stainless steel are briefly reported. The specifications of En 1a and En 8 are:

	C	Si	Mn	S	P
En 1a	0·07–0·15	<0·1	0·8–1·2	0·2–0·3	<0·07
En 8	0·35–0·45	0·05–0·35	0·6–1·0	<0·06	<0·06

The tools used for the tests presented here were ground to the specification (0, 15, 10, 5, 15, 20, 0·015 in). Six inch lengths of $\frac{1}{2}$ in square material were used and there were no problems experienced with heating of the cold junction.

EXPERIMENTAL WORK

The initial experimental work of this study included both open loop and closed loop tests. Many open loop studies have been made by other investigators, some of whom have been mentioned earlier. The tests carried out in the present work were made in order to obtain information and experience which could be of use in the closed loop mode of operation and as a result may consider phenomena with a different emphasis to some of the earlier work.

Tests were carried out in order to determine the rapid breakdown or "burn-out" thermal e.m.f. of the tools when used in tests with increasing cutting speed, i.e. "accelerated wear tests". These were carried out either by facing at constant spindle speed or by parallel turning at increasing spindle speed. During these tests variations in thermal e.m.f. with cutting speed were observed for a range of conditions including variations in feed and depth of cut. Flank wear tests were carried out at constant spindle speed for comparison purposes and the rate of increase of thermal e.m.f. during such tests were obtained.

Open loop tests were carried out both with and without coolant in order to determine the reduction in thermal e.m.f. under given conditions when coolant was applied.

Closed loop tests were carried out, mainly with 4-in diameter bar, on facing and parallel turning operations at various values of demanded thermal e.m.f., feed and depth of cut, both with and without coolant. Flank wear studies were made during some of the parallel turning operations.

For the tests with constant thermal e.m.f., values of cutting speed were obtained which were associated with various feeds and depths of cut for given cutting conditions.

Where different workpiece materials were used, an attempt was made to determine relative machinability, from the aspect of tool wear, by testing to obtain a burn-out e.m.f. and by observing the variation in steady state spindle speed for the different materials for given cutting conditions and demanded thermal e.m.f. values. All results presented were obtained without the use of coolant, except for the particular results discussed separately.

RESULTS AND DISCUSSION

E–V Results and Accelerated Wear Tests

The general form of the thermal e.m.f./cutting speed plot shows a decreasing slope up to a relatively high value of E when there is an increase in slope, followed by rapid tool wear, failure or "burn-out". This is illustrated in Fig. 3(a). For tests with less severe cutting conditions the E–V relationship was found to be almost linear. Such a result is shown in Fig. 3(b) for the 6:5:2 tool and En 1a. In a number of cases changes in the chip form occur with increasing speed, these are often accompanied by changes in temperature as may be expected. This can be seen in Fig. 3(b).

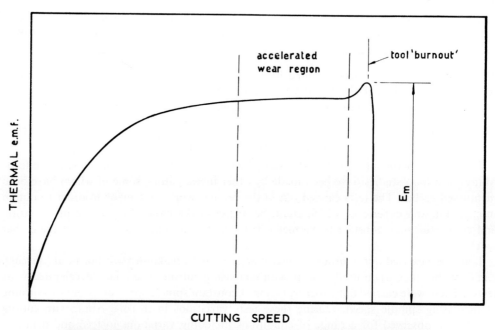

FIG. 3(a). General form of thermal e.m.f. (E)–cutting speed (V) relationship.

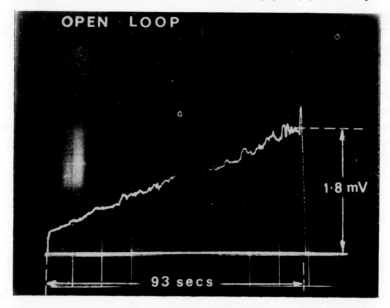

FIG 3(b). Facing with constant spindle speed and light cut–near linear relationship.

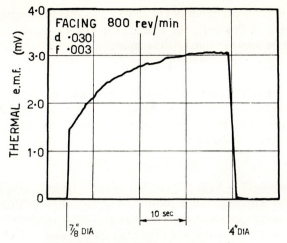

FIG. 3(c). Facing test showing general form.

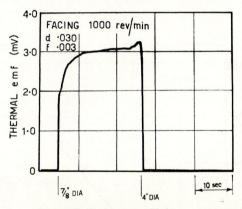

FIG. 3(d). Facing test with burn-out.

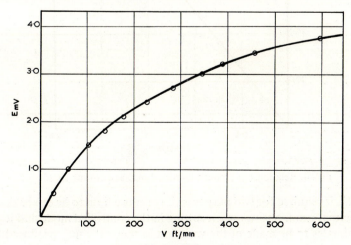

FIG. 3(e). Parallel turning test.

With an increase in the feed, depth of cut or speed so that the thermal e.m.f. was above the region 1·5–2·0 mV in this case, the relationship became nonlinear and over certain limited ranges could be approximated by the equation

$$E = kV^m \tag{5}$$

which for linear relationship between thermal e.m.f. and temperature is in agreement with equation (2). The exponent m varied, however, from values of 0·2 with $E > 2·5$ mV to 1·0 for lower values of E. The relationship was found to be better approximated in a number of cases by the relationship

$$E = E_m(1 - e^{-k'V}) \tag{6}$$

where E_m is the "burn-out" voltage. Figure 3(c) shows a typical curve from an open loop facing test and a curve with burn-out is shown in Fig. 3(d). Results from a parallel turning test with increasing speed are shown in Fig. 3(e) and these results are plotted in Fig. 4, using a value for E_m of 4·1 mV estimated from the experimental curve. The linearity of this graph indicates the goodness of fit of equation (6) for this case. Equation (6) can be used to obtain an estimate of E_m from data obtained at lower cutting speeds where the tool wear is not serious.

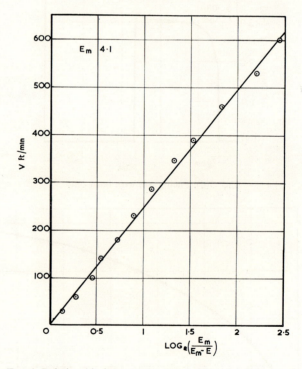

Fig. 4. Relationship between thermal e.m.f. and cutting speed.

The value of E_m is required for two purposes: it permits a limit to be established for values of demanded thermal e.m.f. settings when cutting in the controlled mode and it also enables approximate estimates to be made of the variation in thermal e.m.f. produced by different workpiece materials when machinability testing, by consideration of differences in their

E_m values. It was noted that during the tests at accelerated wear rates there was local welding or seizure on the flank wear land similar to the built-up edge which occurs on the rake face.

Comparision Between Open Loop and Closed Loop Facing

Many of the results discussed have been obtained by open loop facing tests, a comparison between this mode and the closed loop controlled mode is shown in Figs. 5(a) and 5(b). For similar facing cuts the controlled cut, with a demanded thermal e.m.f., E_D, of 1·6 mV,

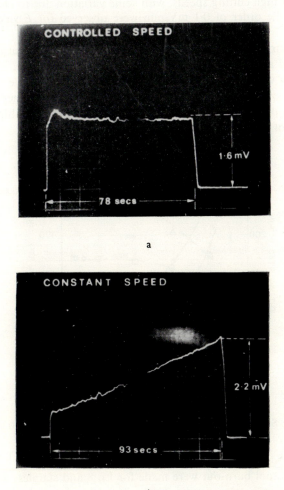

FIG. 5. Facing tests, d 0·020, f 0·002. (a) Constant spindle speed (V 500). (b) Thermal e.m.f. control (E_D 1·6).

resulted in a cutting time of 78 sec and a thermal e.m.f. of near to 1·6 mV throughout, with the exception of a small overshoot compared with the uncontrolled or constant spindle speed mode where the cutting time was 93 sec and a maximum thermal e.m.f. of 2·2 mV was reached.

The controlled mode gives an actual reduction of 16% in the cutting time and by using an approximate estimate for tool wear based on the early equations and shown in the Appendix, an estimated reduction in tool wear of some 29%. Further work on this aspect is in progress and results of a limited number of tool wear tests indicate various reductions in tool wear, some as great as 50% in addition to a reduction in cutting time.

Open Loop and Closed Loop Parallel Turning Tests

The open loop turning tests showed an increase in E as tool wear developed. This was more pronounced at high cutting speeds, with some variation due to changes in chip form. Values of dE/dt of the order of 0·1 mV/min were observed for E values around 3·0 mV (d 0·050, f 0·002).

The closed loop turning tests showed an overall decrease in the cutting speed with time as the tool wore and also showed a slight reduction in the wear rate.

Tests with various values of demanded thermal e.m.f. indicated the ease with which the variation in cutting speed with feed or depth of cut could be obtained. Typical curves of cutting speed, feed, depth of cut and metal removal rate relationships for a demanded thermal e.m.f. (E_D) of 1·0 mV are shown in Fig. 6.

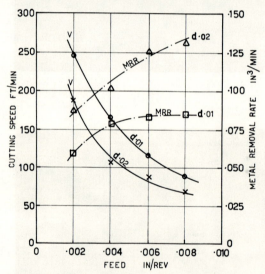

FIG. 6. Steady state cutting speeds for various conditions (E_D 1·0 mV).

Machinability Indication

Open loop tests to tool burn-out were made for En 8 and stainless steel. The curves of thermal e.m.f. for these metals were similar to those mentioned earlier. The value of E_m for En 8 with 6:5:2 was found to be 4·3 mV, almost the same as En 1a, whilst that for stainless steel was 7·4 mV.

Parallel turning tests were then made on each material with closed loop operation with a demanded e.m.f. of 62·5% of the E_m value. It was considered that this would enable a cut to be taken without undue tool wear over a short length. The feeds were varied and the steady state or mean cutting speed which resulted was noted for each case. The cutting speed for En 1a is taken as the reference and the speeds for the other materials are expressed as a ratio—Fig. 7 shows the relative machinability as indicated by such a test—the results are also tabulated in Table 1. It is relevant to point out that the material used for the total of nine results given in Table 1 was of the order of 2 in³, and it is considered that such a test could be of value in cases where relative machinability indication is required. In addition an adequate number of results for a statistical analysis of relative machinability can be easily and quickly obtained.

TABLE 1

Depth of cut 0·40	Material	En la	En 8		Stainless steel	
	E_D mV	2·5	2·65		4·6	
		V_1	V_2	V_2/V_1	V_3	V_3/V_1
Cutting speeds ft/min	$f = 0·001$	520	305	0·585	92	0·177
	$f = 0·002$	420	210	0·50	70	0·167
	$f = 0·004$	255	135	0·528	52	0·20

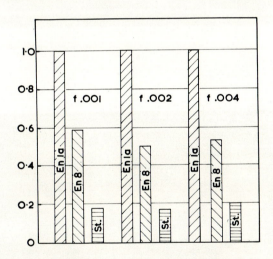

FIG. 7. Indicated relative machinability of En 8 and stainless steel, referred to En la, for d 0·050, 62·5% E_m and various feeds.

A greater accuracy in the determination of the required thermal e.m.f. setting to give the same interface temperature for various materials can be obtained by separate calibration of the workpiece materials with the tool steel. This is normally a lengthy process and, as has been stated, liable to some variation.

Effects of Coolant

It was found that variation in the thermal e.m.f. when coolant was applied gave an indication of the effectiveness of application. Figures 8a and 8b show the thermal e.m.f. and cutting speed variations with time which resulted from the application of coolant in the open loop and closed loop modes respectively. These results are from tests with $d = 0·050$

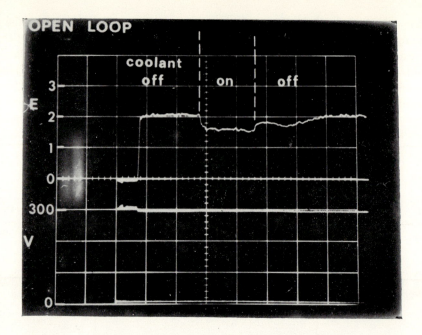

a

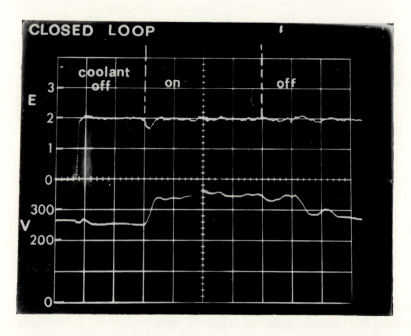

b

FIG. 8. Effect of coolant on thermal e.m.f. and cutting speed. d 0·050, f 0·003. (a) Uncontrolled, V 300. (b) Controlled, E_D 2·0.

and $f = 0.003$. In the open loop test it can be seen that the thermal e.m.f. falls from 2·0 mV to around 1·5 mV on application of the coolant. On cessation of the coolant, when any surplus was removed with a cloth, it is noted that the value of E only gradually returns to the initial value. This may be due to the lubricant effect of the small remaining coolant film which could reduce the area of chip–tool interface seizure.

In the controlled test—Fig. 8b—the application of coolant is seen to cause a decrease in thermal e.m.f. and an increase in cutting speed from 260 ft/min to around 340 ft/min. Cessation of coolant results in a gradual decrease of speed. It was observed that the point of application of the coolant was an important factor in obtaining maximum effectiveness and this was readily observed on the digital voltmeter for the uncontrolled cases.

Effects of Chip Form

Changes in the chip form were observed during a number of tests, particularly those over a wide cutting speed range, and there were also variations in the chip ductility. These changes were, in many cases, accompanied by a change in the thermal e.m.f. produced. The chips shown in Fig. 9 were produced on the same constant speed cut and showed a variation in thermal e.m.f., the close coiled chip giving a value of 2·9 mV and the open chip a value of 3·2 mV.

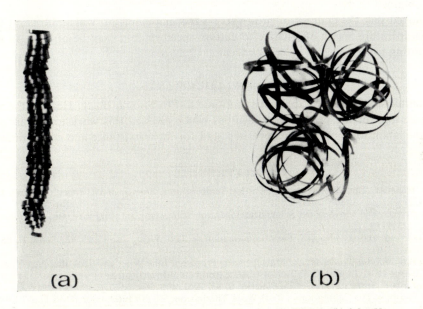

(a)   (b)

FIG. 9. Variation in chip form—open loop test. (a) 2·9 mV. (b) 3·2 mV.

APPLICATIONS AND CONCLUSIONS

It has been indicated that the concept of thermal e.m.f. feedback control of machine tools can have a number of applications which include the following:

1. As a research tool: Such a machine enables studies to be carried out with an improved control of variables and with the cutting speed as a dependent variable.
2. As a machinability test aid: The rapid evaluation of a cutting speed, for a given cut

and materials, enables comparative assessments to be easily made. A large number of tests can be carried out so that a reasonable statistical sample is obtained and, in addition a method of testing which is based on actual metal cutting is often to be preferred over those which relate to the various physical properties of a material.

3. As an integral part of the machine tool in production: such a system may be applied to various machine tools, e.g., centre lathes, with particular application to conditions of varying diameter cutting such as facing or copy turning. It has advantages over constant cutting speed arrangements in that correction is made for such changes as:

(a) Tool wear.

(b) Material inhomogeneity and metallurgical changes.

(c) Coolant variation or failure.

The control system can be easily adapted to vary feed if required.

In view of the relatively low cost of this type of system, it is possible that it could fill, to some extent, the gap between conventional machine tools and adaptive numerically controlled machine tools which have yet to be shown to be economically viable in the fully adaptive forms proposed.

The work presented in this paper is of the nature of a progress report; work is in progress to obtain further information and evidence relating to the indications observed in the results obtained to date.

It is hoped that the concept of prescribing a thermal e.m.f. or, with calibration curves, a tool temperature, as an alternative to cutting speed or feed may be of value in certain metal-cutting processes.

ACKNOWLEDGEMENTS

The author would like to express his gratitude to Professor J. Black, Head of the School of Engineering, for the provision of facilities which enabled this work to be carried out, and to the laboratory, office and workshop staff for their assistance and co-operation.

REFERENCES

1. E. G. HERBERT. The Measurement of Cutting Temperatures, *Proc. I. Mech. Engrs.* (*London*), **1**, 1926, 289–329.
2. K. GOTTWEIN. Die Messling der Schneidentemperatur beim Abdrehen von Flusseisen, *Maschinenbuu*, **4**, 1925, 1129–1135.
3. H. TAKEYAMA and R. MURATA. Basic Investigation of Tool Wear, *Trans. ASME*, Series B, **85**, 1963, 33–38.
4. J. DATSKO. *Material Properties and Manufacturing Processes.* John Wiley and Sons, Inc., New York, 1966.
5. B. T. CHAO, H. L. LI and K. J. TRIGGER. An Experimental Investigation of Temperature Distribution at Tool–Flank Surface, *Trans. ASME*, Series B, **83**, 1961, 496–504.
6. P. M. BRAIDEN. The Calibration of Tool/Work Thermocouples, 8th International Machine Tool Design and Research Conference, September 1967.
7. R. M. CENTNER. Development of Adaptive Control Techniques for Numerically Controlled Milling Machines, USAF Tech. Doc. Report No. ML-TDR-64-279, The Bendix Corp., Michigan, 1964.
8. N. F. SHILLAM and C. E. WILKINSON. Control of the Cutting Speed on a Machine Tool using the Work-piece/Tool Thermal Electric Voltage, RAE Tech Memo ADW17. RAE Farnborough, 1967.
9. E. M. TRENT. Conditions of Seizure at the Tool/Work Interface, Proc. Conf. on Machinability 1965, Iron and Steel Institute Report No. 94, 1967, pp. 11–18.

APPENDIX I

If the wear rate is given by

$$dW/dt = C\theta^{1/np}$$

total wear is thus

$$W = C \int_0^t \theta^{1/np} \, dt$$

Put
$$\frac{1}{np} = q$$

For constant θc,

$$W_c = k \, \theta_c{}^q \, t_c$$

In Fig. 5a, θ is an approximate linear function of t, i.e.,

$$\theta = \theta_1 + at$$

$$W_v = k \int_0^{t_v} (\theta_1 + at)^q \, dt$$

In this linear case $p = 1$; values of n are usually between 0·05 and 0·10 for H.S.S. Therefore assume a value for q of 15; it could be expected to be higher for $p < 1$. Earlier workers have estimated values for q of 20 and 23.

Using the data from Fig. 5, approximate temperatures are

θ_c 580°K

θ_1 420°K

a 2·58

and the time $t_c = 78$ sec

and $t_v = 93$ sec

The ratio of wear for the facing cut

$$\frac{W_v}{W_c} = \frac{93 \, (660^{16} - 420^{16})}{240 \times 16 \times 78 \times 580^{15}}$$

$$\frac{W_v}{W_c} = 1·42 \text{ or } W_c = 0·71 \, W_v$$

Thus the wear during the constant temperature cutting is only 71% of that for the case with variable temperature (constant spindle speed facing) and the indicated reduction in wear is thus 29%.

In this illustrative and simplest case initially T_a was 605 °R and 0.01 for H.S.S. Therefore assume a value for q of 15, it could be greater in the higher T_a of 25° Earlier research have estimated values for q of 20 and 25.

Using the data from Fig. 5, approximate temperatures are:

$$ T_a = 605 \ °R $$
$$ q_1 = 95 \ °R $$

and the time $t_a = 0 \ sec.$

In the table of seal for the loadng valve.

That the seal during the constant temperature portion is so small, and that the seal variable temperature portion. Single speed during the upstart is so small.

A THEORETICAL INVESTIGATION INTO THE CHARACTERISTICS OF A FEEDBACK CONTROLLED DAMPING UNIT

A. BOYLE and A. COWLEY

Machine Tool Engineering Division, Department of Mechanical Engineering, University of Manchester Institute of Science and Technology

NOTATION

A area of piston
B oil bulk modulus
C damping constant
D Ratio of system damping to critical damping
F exciting force
f natural frequency ratio (ω_n/Ω_n)
$G1$ position feedback transducer sensitivity
$G2$ acceleration feedback transducer sensitivity
g normalized exciting frequency (ω/Ω_n)
j operator
K stiffness of main system
K_1 gain relating servo valve input and spool displacement
K_2 gain relating spool displacement and flow rate
K' gain relating servo valve input and flow rate
k absorber spring stiffness
M mass of main system
m mass of absorber
P_1 cylinder pressure
P_s supply pressure
q_v valve flow rate
R resistance of hydraulic restrictor
s Laplace operator
V volume of oil at pressure P_1
X_{dyn} amplitude of harmonic structural response
X_{st} static structural displacement
Y auxiliary mass displacement
δ spool displacement
μ mass ratio (m/M)
Ω_n main system natural frequency
ω forcing frequency
ω_n auxiliary system natural frequency

INTRODUCTION

The vibrational behaviour of machine tools has been the subject of much investigation in recent years. Both the manufacturer and the user are fully aware of the desirability of high dynamic as well as static stiffness of the basic machine structure. To achieve the required stiffness characteristics due consideration must be given to the distribution of elasticity and mass throughout the machine structure. The use of model techniques[1] and, more recently, computer aided design[2,3] has done much to assist in this area. Unfortunately neither of these design aids permits a sufficiently accurate investigation to be conducted into the damping properties of alternative structural configurations. The contribution to the overall damping of the various structural resonances made by individual damping sources is of considerable interest.

The small inherent damping of machine tool structures arises from two basic sources viz.

1. Internal (material) damping.
2. Friction damping.

INTERNAL DAMPING

This source of damping is hysteretic in nature and depends for its existence upon a material's internal frictional losses which occur under cyclic loading.

Experimental investigations conducted into the damping properties of various materials have on the whole been restricted to simple constructional elements examined under controlled laboratory conditions. The problem of assessing the damping contribution made by each element of a machine tool structure is considerably more difficult.

There are some indications that material damping is often of little significance when compared with other damping sources. Eisele and Bauer[4] give the logarithmic damping decrements of cast iron and steel as 0·002 and 0·001 respectively. The higher damping capacity of cast iron is not of such great significance, however, because of the possibility of achieving higher values with steel by suitable designs. Peters[5] asserts that in the case of lathes material damping accounts for only a small proportion of the overall damping.

FRICTION DAMPING

In investigating the influence of damping and loading conditions upon the friction damping produced in fabricated beam elements Heiss[6] concluded, as did Kienzle[7] that a major contribution to the damping was provided by the presence of rubbing faces in the structure. Opitz and Bielefeld[8] when considering beams with Peters type ribbing showed conclusively that a fabricated steel beam could possess higher damping than a cast iron construction. The increased damping of the fabricated structure was attributed to the presence of an abrasive action created in the interrupted welded seams. Experiments conducted on individual elements and complete machines by Kettner and Kienzle,[9] Sommer,[10] Opitz,[11] Drumm[12] and Katzenschwanz[13] have further emphasized the importance of interface joints. Results have revealed that often damping due to friction can amount to ten times the value normally associated with material losses.

Whilst suitably designed elements and interface joints can considerably increase the damping inherent in a structure it remains desirable to increase this still further. For this reason a number of types of vibration absorber are finding an increasing number of applications in the machine tool field.

VIBRATION ABSORBERS

The so-called vibration absorber may take a number of forms as illustrated in Fig. 1. The undamped absorber as shown in Fig. 1a is, of course, an idealization. The absorber

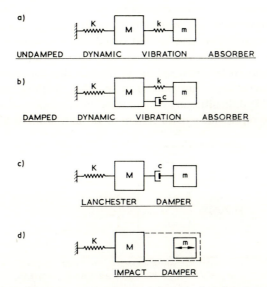

FIG. 1. Schematic arrangements of dynamic damping systems.

system comprising the spring k and auxiliary mass m is coupled to the main system (K and M). Its effect is to modify the response of the latter from that of a single resonance at a frequency of

$$\Omega_n = \sqrt{\left(\frac{K}{M}\right)}$$

to a pair of undamped resonances, one above and one below the main system resonant frequency. Such an absorber finds applications only where it is required to remove a system resonant frequency from the vicinity of a constant exciting force.

The damped absorber shown in Fig. 1b differs from its undamped counterpart inasmuch as it converts the system undamped resonance into a pair of damped resonances. This type of absorber, about which more will be said later, has found numerous applications in many branches of engineering.

A degenerate form of the damped dynamic absorber results when the spring connection to the auxiliary mass is removed. The resulting absorber system is known as the Lanchester damper (Fig. 1c). The free and forced vibration characteristics of such a damper have been presented by Hahn[14] and Brock[15] who give the conditions for optimum operation.

The best known application of the Lanchester damper to machine tools has been in connection with the suppression of vibrations in overhung boring bars.[14,16]

Yet another form of absorber which has been successfully applied in the machine tool field is the impact (or acceleration) damper depicted in Fig. 1d. This type of damper is essentially similar to the Lanchester system, the difference being in the manner by which energy is dissipated. In the case of the impact damper energy is dissipated as heat, due to the repeated collisions which take place between the unconstrained auxiliary mass and the main system mass.

The impact damper has been applied in a number of cases to machine tools, notably by Rzhkov[17] in the form of a lathe tool holder. A more recent account of the operation and application of the damper has been presented by Sadek and Mills.[18]

The theoretical behaviour of the vibration absorbers shown in Fig. 1 has been understood for some considerable time and the last two decades have seen the increasing use of such damping devices in machine tool applications. One limitation which is shared by all the passive arrangements so far discussed lies in the fact that for satisfactory operation the auxiliary system characteristics must be closely matched to those of the main system. In this respect considerable difficulties are encountered in machine tool applications since the machine tool structure represents a main system of variable stiffness and inertial properties.

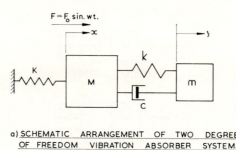

a) SCHEMATIC ARRANGEMENT OF TWO DEGREE OF FREEDOM VIBRATION ABSORBER SYSTEM.

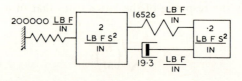

b) CALCULATED VALUES FOR OPTIMUM SYSTEM.

FIG. 2. A passive absorber system.

To illustrate the sensitivity of an absorber damped system to changes in the main system stiffness consider the general case of the damped absorber of Fig. 2. The equation governing the motion of the system when subjected to a harmonic force is derived in Appendix 1 and becomes:

$$\left|\frac{X_{\text{dyn}}}{X_{\text{st}}}\right| = \left\{\frac{K^2(k - m\omega^2)^2 + (KC\omega)^2}{[(K - M\omega^2)(k - m\omega^2) - mk\omega^2]^2 + C^2\omega^2[K - (M + m)\omega^2]^2}\right\}^{\frac{1}{2}}$$

This equation governs the response of the main system to harmonic loading. For a specific mass ratio (m/M) which should be as large as practicable the optimum values of the absorber spring stiffness and damping constant are approximated by the following expressions:[19]

(i)
$$k = \frac{\mu K}{(1 + \mu)^2}$$

and (ii)
$$C = \left[\frac{1 \cdot 5mk\mu}{(1 + \mu)^3}\right]^{\frac{1}{2}}$$

The general form of the response of the main mass to a harmonic force is shown in Fig. 3a for a mass ratio of 0·1 and for four values of damping, where the parameter D is the ratio of actual to critical damping constants. It can be shown that the two points P and Q at which the response curves for $D = 0$ and $D = \infty$ intersect are the intersection points for all

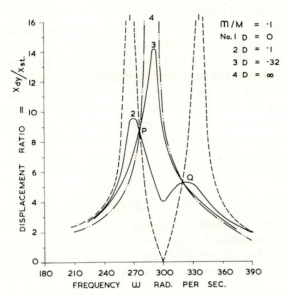

FIG. 3a. Frequency response curves, for a damped system, for changes in the absorber system damping.

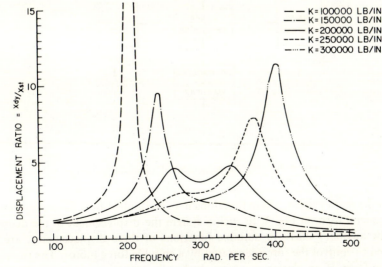

FIG. 3b. Frequency response curves, for a damped system, for changes in the main system spring stiffness.

intermediate values of damping. Condition (i) above ensures that these points have equal displacement ordinates whilst condition (ii) approximates to a value of C for which the tangents to the response curve at P and Q are horizontal (i.e. peaks occur at these points). Any departure from the optimum conditions will lead to one unnecessarily large resonant peak.

The numerical values for an optimized system are given in Fig. 2b. The resulting frequency response is shown by the unbroken curve of Fig. 3b. This figure illustrates the effect of changes in the main system stiffness both below and above that value at which optimization has been determined (i.e. $K = 2 \times 10^5$ lb in^{-1}). It is thus clearly undesirable to permit the stiffness of the system to vary without adjusting the absorber characteristics accordingly. Manual adjustment of the absorber spring and damper values may in certain circumstances provide an adequate solution but in general atuomatic compensation is desired.

A theoretical analysis of one type of self-optimizing absorber has been presented by Bonesho and Bollinger,[20] who demonstrated, with the aid of a simple electric network analogy, the idea of self optimization. Unfortunately the authors refrained from suggesting any form of practically suitable mechanism to provide the necessary variation in the absorber's physical characteristics.

ACTIVE DAMPERS

The basic requirement of a dynamic damping system is that it should be capable of imparting a force to the main system which resists the movement of the latter. The reaction of this force is supported by the inertial property of the auxiliary mass. The magnitude and direction of such a restoring force should be capable of control according to the forced

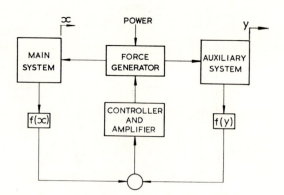

FIG. 4. Schematic arrangement of a control system.

deflection of the main system. These functions may be fulfilled by a control system depicted in its most general form in Fig. 4. In such a system the displacement of the main structure and/or of the auxiliary mass, or suitable functions of these displacements, are amplified and then used to control the force generated by a suitable force motor. The force generator may take the form of an electro-magnetic, hydraulic or pneumatic motor. In considering the suitability of these alternatives for a preliminary analysis a hydraulic system was selected since this appeared to possess the two essential features required for the damping of machine tool structures, viz.

(a) high force/size ratio

(b) adequate speed of response.

A SERVO-CONTROLLED HYDRAULIC DAMPER

The arrangement chosen for analysis is shown in Fig. 5. A differential area hydraulic ram links the main structure to an auxiliary mass (m). The fluid flow rate to the larger area side of the ram is controlled by a three-way electrohydraulic flow control valve whilst the smaller area side is connected to the source of constant supply pressure. For a 2:1 differential area ram the spool valve ensures that with the latter in its neutral position $P_1 = P_s/2$ and

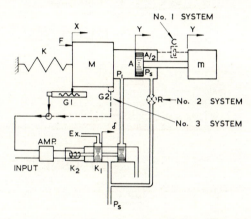

FIG. 5. Diagrammatic arrangement of a servo-controlled auxiliary mass damping system.

therefore the ram imparts zero force between the two masses. Any deflection of the structure (M) is detected by means of G_1, amplified and used to drive the spool in the required direction. For example, deflection of M to the right causes the spool to move to the right thus allowing fluid to flow from P_s to the large area ram chamber. The resulting increase in P_1 causes a force to act on M opposing the initial deflection.

The equations of motion of the basic system, excluding

(i) Viscous damper C,

(ii) Hydraulic resistance R,

(iii) Acceleration feedback loop G_2,

are developed in Appendix II (eq. 9) and the resulting characteristic equation is given by:

$$\frac{VMm}{A^2B}s^4 + \left(M + m + \frac{VmK}{A^2B}\right)s^2 + \frac{mK'G}{A}s + K = 0$$

The absence of the cubic term is an immediate indication of instability. This is a result of having neglected leakage flow across the ram and of assuming the valve flow rate to be proportional to the instantaneous spool deflection. It is well known that for such a system leakage has a stabilizing effect, as does the actual pressure dependent flow characteristic of a spool type valve (i.e. $q_v = \delta\sqrt{(P_s - P_1)}$). However, these and other simplifying assumptions outlined in Appendix II are justified at the preliminary stages of analysis. The unstable nature of the basic system may be avoided in a number of ways and three methods are adopted here.

(a) *System* 1

In the first system to be considered a viscous damper (C) is connected between the structure and the auxiliary mass. The equations describing the motion of the resulting system

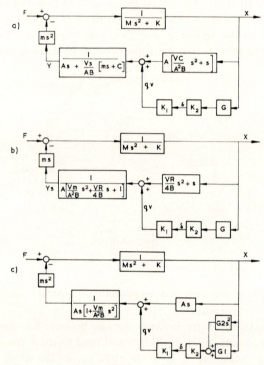

FIG. 6. Block diagrams for (a) No. 1 system; (b) No. 2 system; (c) No. 3 system.

are derived in Appendix II and a block diagram is shown in Fig. 6a. The characteristic equation of the system is given by:

$$\frac{MmV}{A^2B}s^4 + \frac{VC}{A^2B}(M+m)s^3 + \left[(M+m) + \frac{mKV}{A^2B}\right]s^2 + \left[\frac{KVC}{A^2B} + \frac{mK'G}{A}\right]s + K = 0$$

In comparing this equation with that characterizing the conventional damped absorber of Fig. 2a (eq. (5), Appendix I) it is evident that, in the absence of position feedback ($K'G = 0$), the hydraulic damping arrangement is equivalent to a passive absorber, the mechanical spring having been replaced by a compressible oil column of stiffness:

$$k = \frac{A^2B}{V}$$

The influence of the position feedback sensitivity ($K'G$) upon the response of the system to external loading may be deduced from the location of the roots of the characteristic equation. This will be seen to provide a useful means of optimizing the system parameters as well as ensuring stability of the closed loop system.

Consider a structure whose effective stiffness and mass are as given in Fig. 2b ($K = 2 \times 10^5$ lbf in^{-1}, $M = 2$ lbf sec ^{2}in^{-1}).

For a given mass ratio (m/M) it is required to examine the effect of both position feedback gain ($K'G$) and viscous damping coefficient (C) upon the response of the structure to an external force.

Figure 7 shows the loci of two of the roots[†] of the quartic characteristic equation for a mass ratio of 0·1. The full lines represent the loci corresponding to a constant value of damping coefficient (C) whilst the broken lines indicate constant values of feedback gain. The ranges of values of both C and $K'G$ have been constrained to contain the roots within the left half of the complex S plane, thus ensuring stability of the control system.

The significance of Fig. 7 may be understood by considering the reproduced region shown in Fig. 8. This figure shows the loci for values of C of 200 and 210 lbf sec in^{-1}, and for values of $K'G$ between 4000 and 6000 in^2 sec^{-1}. The figure indicates the location of two of the roots of the characteristic equation for $C = 210$ lbf sec in^{-1} and $K'G = 5500$ in^2 sec^{-1}.

The amplitude response of the structure to an external force may be deduced from Fig. 8 by constructing lines from each root to the origin. The length of such lines then gives the natural frequency of the modes and the modal damping ratios are given by the cosine of the appropriate angle β.

Figure 7 clearly indicates that low values of C together with high values of $K'G$ result in a lower natural frequency which approaches 300 rad/sec and a higher natural frequency of about 860 rad sec^{-1}. Upon increasing C and reducing $K'G$ a point is reached where the two

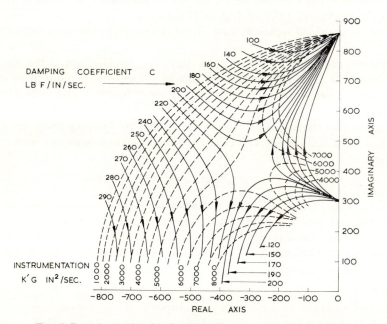

Fig. 7. Root locus plot showing lines of constant viscous damping.

† The remaining two roots are conjugates of those shown and have not been plotted.

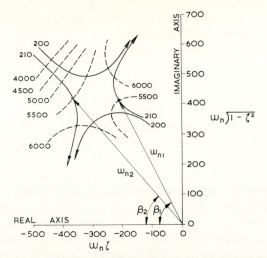

FIG. 8. Position of roots for $C = 210$ lbf sec in^{-1}, and $K'G = 5500$ in^2 sec^{-1}, giving natural
frequency of system (ω_n) and damping $(\cos \beta = \zeta)$.

roots coincide, thus giving only one natural frequency. The location of the two coincident
roots will be within the central region of Fig. 7 and will be obtained for:

$$C \simeq 205 \text{ lbf sec in}^{-1}$$

and $$K'G \simeq 5750 \text{ in}^2 \text{ sec}^{-1}$$

Any departure from this combination of the above parameters will lead to two separate
natural frequencies having different degrees of damping.

An alternative graphical representation of the system performance is indicated in Figs.
9 and 10.

Figure 9 shows the influence of $K'G$ upon the forced displacement of the structure for a
particular value of C. It is evident that the smaller values of $K'G$ result in an undesirably
high first resonant peak whilst excessively high values of feedback gain yield a large second
resonant peak.

If the magnitude of the largest peak is plotted against the damping coefficient C curves of
the form shown in Fig. 10 result. It is clear from these curves that an optimum condition is
achieved for:

$$C = 150 \text{ lbf sec in}^{-1}$$

and $$K'G = 8000 \text{ in}^2\text{sec}^{-1}$$

It is also interesting to note that with a value of C of 150 lbf sec in^{-1} the maximum
displacement ratio does not exceed its static value of unity for values of $K'G$ between 6000
and 12,000 in^2 sec^{-1} and therefore the performance of the damper is not critically dependent
upon this parameter.

(b) System 2

A similar end result can be achieved if the dashpot C of Fig. 5 is omitted and a hydraulic
resistance (R) inserted in the pressure supply line to the ram. The detailed analysis of such
a system is given in Appendix II and a block diagram is shown in Fig. 6b. The system's
characteristic equation becomes:

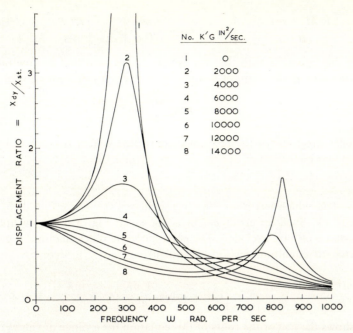

FIG. 9. Frequency response of system for $C = 150$ lbf sec in^{-1}.

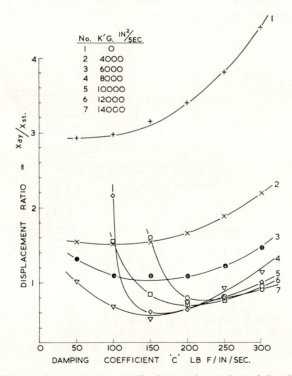

FIG. 10. Maximum resonant amplitude at various values of C and $K'G$.

$$\frac{MmV}{A^2B} s^4 + \frac{VR(M + m)}{4B} s^3 + \left[(M + m) + \frac{KmV}{A^2B}\right]s^2 + \left[\frac{KVR}{4B} + \frac{K'Gm}{A}\right]s + K = 0$$

Comparing this equation with that for system 1 it is seen that the hydraulic resistance acts in the same way as the dashpot C and the two systems become equivalent for:

$$R = \frac{4C}{A^2}$$

The root locus diagram for the resistance controlled system is given in Fig. 11 and the frequency response relationships appear in Figs. 12 and 13.

(c) *System* 3

A further system which has been examined achieves the desired result by means of acceleration feedback. Such an arrangement is shown in Fig. 5 if the dashpot (C) and the hydraulic resistance (R) are excluded. The corresponding block diagram is given in Fig. 6c.

The characteristic equation, derived in Appendix II (eq. 17), becomes:

$$\frac{MmV}{A^2B} s^4 + \frac{mK'G2}{A} s^3 + \left[(M + m) + \frac{VmK}{A^2B}\right]s^2 + \frac{mK'G1}{A} s + K = 0$$

Figure 14 shows the loci of the roots of this equation for a range of values of acceleration and position feedback sensitivities. Optimization may be facilitated with the aid of Fig. 14 by the procedure adopted for the previous systems.

A typical set of forced displacement responses is shown in Fig. 15, which indicates the presence of two resonant peaks.

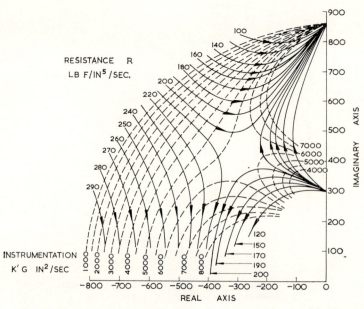

FIG. 11. Root locus plot showing lines of constant hydraulic resistance.

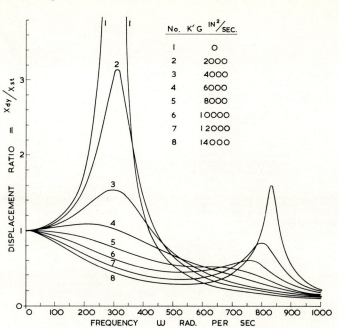

FIG. 12. Frequency response of system for $R = 150$ lbf/in^5/sec.

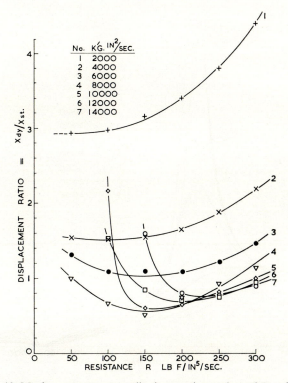

FIG. 13. Maximum resonant amplitude at various values of R and $K'G$.

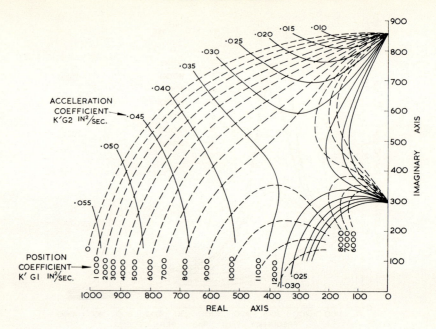

Fɪɢ. 14. Root locus plot showing lines of constant acceleration.

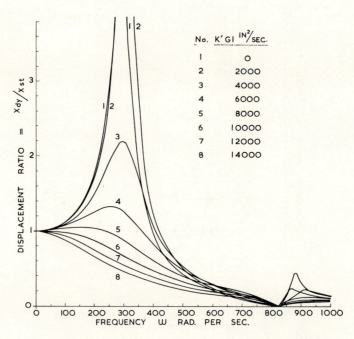

Fɪɢ. 15. Frequency response of system for $K'G2 = 0.015$ in² sec.

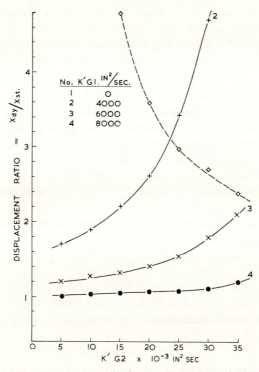

Fɪɢ. 16 Maximum resonant amplitude at various values of $K'G1$ and $K'G2$.

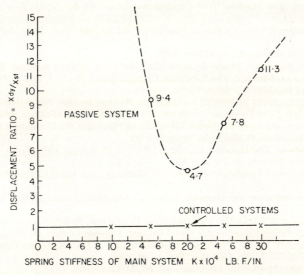

Fɪɢ. 17. Plots of the maximum displacement ratio produced in system by changes in the main system spring stiffness.

Figure 16 shows the variation in the magnitude of the larger resonant peaks with acceleration sensitivity. The figure indicates that high position feedback gain, combined with low acceleration feedback gain leads to the suppression of both resonant peaks. The lower limit of acceleration gain is dictated by stability considerations.

CONCLUDING REMARKS

The performance of the active damping systems considered may be measured by comparison with the characteristics of a conventional passive turned absorber. Figure 17 shows the maximum normalized peak amplitude for a passive absorber which has been designed to operate in conjunction with a structure of stiffness $K = 2 \times 10^5$ lbf in^{-1}. The deterioration in the performance of such a tuned absorber when the natural frequency of the structure changes as a result of variations in the structural stiffness is clearly apparent.

In the case of the controlled system, it has been shown that system parameters may be felected to suppress structural resonances. The influence of variations in structural stiffness upon the performance of such a controlled damper is indicated in Fig. 17.

The results presented have been obtained by considering simplified models of both the structure and the control systems.

The effects of system non-linearities and deficiencies in the response of the control system components remain to be investigated. However, this initial study has provided sufficiently encouraging results to justify the experimental investigation which is now under way.

REFERENCES

1. J. BIELEFELD, Model Experiments on Machine Tool Elements. *Industrie-Anzeiger* No. 80, October 57.
2. S. TAYLOR and S. A. TOBIAS, Lumped Constants Method for the Prediction of the Vibration Characteristics of Machine Tool Structures. *Proc. 5th International M.T.D.R. Conference*, Pergamon, 1964.
3. A. COWLEY and M. A. FAWCETT, Analysis of Machine Tool Structure by Computing Methods. *Proc. 8th M.T.D.R. Conference*, Pergamon, 1967.
4. BAUER and F. EISELE, Investigations into the Damping Behaviour of Different Coating Materials. 3 Fo. Ko. Ma. München.
5. J. PETERS, Damping in Machine Tool Construction. *Proc. 6th Int. Mech. Tool Des. Res. Conf.*, Manchester, 1965, p. 23.
6. A. HEISS, Vibration Characteristics of Machine Tool Frames. *Ver. Deut. Ing. Forschungshaft* 429, 1949/50, p. 1.
7. O. KIENZLE, German Patent No. 872895, 1953.
8. H. OPITZ and J. BIELEFELD, Model Tests on Machine Tools. Aachen Research Report No. 900, 1960.
9. H. KETTNER and O. KIENZLE, The Vibration Characteristics of a C.I. and a Steel Lathe Bed. *Werkstattstechnik und Werksleiter*, No. 9, 1939.
10. G. SOMMER, Development of Welded Steel Lathes. *Welding Journal*, No. 4, 1950.
11. H. OPITZ, The Static and Dynamic Characteristics of Machine Tool Frames. *Microtecnic*, No. 5, 1959.
12. H. DRUMM, Contributions to Frictional Damping, in Particular the Damping Effect of Welded Joints. Dissertation, Technische Hochschule, Munich, 1958.
13. N. KATZENSCHWANZ, The Dynamic Stability of Welded Structures with respect to the requirements of the Machine Tool Industry. 5. Fo. Ko. Ma. p. c15, 1961.
14. R. S. HAHN, Design of Lanchester Damper for Elimination of Metal Cutting Chatter. *Trans. ASME*, **73**, 1951, 331.
15. J. E. BROCK, A Note on the Damped Vibration Absorber. *Journ. Applied Mech. (Trans. ASME)* **68**, 1946, A-284.
16. Design and Testing of a Slug-Damped Overhung Boring Bar. PERA Report No. 96, December 1961.
17. D. I. RHZKOV, Vibration Damper for Metal Cutting. *Engineers Digest* **14**, 1953, 246.
18. M. M. SADEK and B. MILLS, The Application of the Impact Damper to the Control of Machine Tool Chatter. *Proc. 7th Int. Mech. Tool Des. Res., Conf.* 1966.
19. J. P. DEN HARTOG, *Mechanical Vibrations*. 4th ed. McGraw-Hill, 1956, p. 99.
20. J. A. BONESHO and J. G. BOLLINGER, Advances in Machine Tool Design and Research. *Proc. 7th Int. Mech. Tool Des. Res. Conf.* 1966, p. 243.

APPENDIX I

Equations of Motion of Classical Damped Vibration Absorber

The equations of force equilibrium at the main and auxiliary masses of Fig. 2a may be written

$$F(s) = Ms^2 X(s) + KX(s) + k[X(s) - Y(s)] + Cs[X(s) - Y(s)] \qquad (1)$$

$$0 = ms^2 Y(s) + Cs[Y(s) - X(s)] + k[Y(s) - X(s)] \qquad (2)$$

From (2)

$$Y(s) = \frac{(Cs + k)X(s)}{ms^2 + Cs + k} \qquad (3)$$

Substituting for $Y(s)$ from (3) into (1) gives

$$\frac{X(s)}{F(s)} = \frac{ms^2 + Cs + k}{Mms^4 + C(M + m)s^3 + [m(K + k) + Mk]s^2 + CKs + Kk} \qquad (4)$$

The system characteristic equation is obtained by equating the forcing function $F(s)$ to zero. Therefore,

$$\frac{Mm}{k} s^4 + \frac{C}{k}(M + m)s^3 + \left[\frac{mK}{k} + M + m)\right] s^2 + \frac{CK}{k} s + K = 0 \qquad (5)$$

For a harmonic forcing function, $F(s) = F_0 \sin \omega t$, the harmonic response of the main mass becomes

$$X_{\text{dyn}} = \frac{(-m\omega^2 + jC\omega + k)}{Mm\omega^4 - jC(M + m)\omega^3 - (Km + mk + Mk)\omega^2 + jCK\omega + Kk}$$

The static displacement is given by

$$X_{\text{st}} = \frac{F_0}{K}$$

and the displacement ratio (modulus) becomes

$$\left|\frac{X_{\text{dyn}}}{X_{\text{st}}}\right| = \left\{\frac{K^2(k - m\omega^2)^2 + K^2 C^2 \omega^2}{[(K - M\omega^2)(k - m\omega^2) - mk\omega^2]^2 + C^2\omega^2(K - M\omega^2 - m\omega^2)^2}\right\}^{\frac{1}{2}} \qquad (6)$$

This expression is most frequently quoted in terms of the following parameters:

μ mass ratio, m/M

Ω_n main system natural frequency, $\sqrt{\left(\dfrac{K}{M}\right)}$

ω_n auxiliary system natural frequency, $\sqrt{\left(\dfrac{k}{m}\right)}$

f frequency ratio, ω_n/Ω_n

g normalized exciting frequency, ω/Ω_n

D ratio of system damping to critical damping. C/C

Equation (6) then becomes

$$\left|\frac{X_{\text{dyn}}}{X_{\text{st}}}\right| = \left\{\frac{(f^2 - g^2)^2 + (2Dg)^2}{[(1 - g^2)(f^2 - g^2) - \mu g^2 f^2]^2 + (2Dg)^2(1 - g^2 - \mu g^2)^2}\right\}^{\frac{1}{2}} \tag{7}$$

APPENDIX II

Derivation of Characteristic Equation and Amplitude Ratio for Active Damping Systems

Consider the system shown in Fig. 5. Firstly, in the absence of acceleration feedback $(G2 = 0)$, hydraulic resistance $(R = 0)$ and damping $(C = 0)$, the system equations of motion will be formed.

The displacement of the structure (M) is detected by means of a suitable transducer G. This signal is used to drive the spool of an electrohydraulic flow control valve.

It is assumed that the spool responds instantaneously to any input command signals. The relationship between spool displacement (δ) and structural displacement (X) becomes:

$$\delta = GK_2 . X$$

In considering the flow characteristics of the servo valve a further simplifying assumption has been adopted and it is assumed that the change in flow rate is related to spool displacement by:

$$q_v = K_1 \delta$$

Therefore

$$q_v = K'G . X \text{ where } K' = K_1 . K_2 \tag{1}$$

Force equilibrium at the mass M gives

$$F(s) = (Ms^2 + K)X(s) + P_1 A - \frac{P_s A}{2} \tag{2}$$

and at the mass m

$$0 = ms^2 Y(s) + \frac{P_s A}{2} - P_1 A \tag{3}$$

Substituting from (3) into (2) gives

$$F(s) = (Ms^2 + K)X(s) + ms^2 Y(s) \tag{4}$$

The oil flow rate from the servo valve is comprised of two components (leakage flow is neglected in this analysis). Firstly the compressible flow is given by:

$$\frac{V}{B} s P_1$$

and secondly the flow required to displace the piston is

$$As[Y(s) - X(s)]$$

Therefore

$$q_v = As[Y(s) - X(s)] + \frac{V}{B} s P_1 \tag{5}$$

The compressibility term can be related to the auxiliary system (m) displacement by

$$\left(P_1 - \frac{P_s}{2}\right)A = ms^2 Y(s)$$

Differentiating this expression w.r.t. time yields

$$sP_1 = \frac{ms^3}{A} Y(s) \tag{6}$$

Substituting (6) into (5) gives

$$q_v + AsX(s) = As\,Y(s)\left(1 + \frac{Vm}{A^2B}s^2\right) \tag{7}$$

and substituting for q_v from (1) gives

$$s\,Y(s) = \frac{(K'G + As)X(s)}{A\left(1 + \dfrac{Vm}{A^2B}s^2\right)}$$

The force/displacement properties of the structure may now be obtained by substituting the above expression into eq. (4). Therefore

$$F(s) = \left[Ms^2 + K + \frac{(K'G + As)ms}{A\left(1 + \dfrac{Vm}{A^2B}s^2\right)}\right] X(s)$$

The structural receptance thus becomes

$$\frac{X(s)}{F(s)} = \frac{1 + \dfrac{Vm}{A^2B}s^2}{(Ms^2 + K)\left(1 + \dfrac{Vm}{A^2B}s^2\right) + ms\left(s + \dfrac{K'G}{A}\right)} \tag{8}$$

The characteristic equation of the system is obtained by equating $F(s)$ to zero and becomes

$$\frac{VMm}{A^2B}s^4 + \left(M + m + \frac{VmK}{A^2B}\right)s^2 + \frac{mK'G}{A}s + K = 0 \tag{9}$$

System 1

The effect of incorporating a viscous damper (C) between the main and auxiliary mass is to modify eqs. (2) and (3), which become

$$F(s) = (Ms^2 + Cs + K)X(s) - Cs\,Y(s) + P_1A - \frac{P_sA}{2} \tag{10}$$

and

$$0 = (ms^2 + Cs)\,Y(s) - CsX(s) - P_1A + \frac{P_sA}{2} \tag{11}$$

The resulting structural receptance then becomes

$$\frac{X(s)}{F(s)} = \frac{1 + \dfrac{V(ms^2 + Cs)}{A^2B}}{(Ms^2 + K)\left[1 + \dfrac{V}{A^2B}(ms^2 + Cs)\right] + ms\left[\dfrac{K'G}{A} + s\left(1 + \dfrac{VC}{A^2B}s\right)\right]} \tag{12}$$

The characteristic equation is therefore given by

$$\frac{MmV}{A^2B} s^4 + \frac{VC}{A^2B}(M + m)s^3 + \left[(M + m) + \frac{mKV}{A^2B}\right] s^2 + \left[\frac{mK'G}{A} + \frac{KVC}{A^2B}\right] s + K = 0 \quad (13)$$

System 2

In considering the behaviour of system 2, incorporating a hydraulic restrictor R in the pressure supply line to the cylinder, eqs. (2) and (3) describe force equilibrium if P_s is replaced by P_2 where

$$P_2 = P_s + R.q_v$$

The effect of such a modification results in a receptance of

$$\frac{X(s)}{F(s)} = \frac{\dfrac{mV}{A^2B} s^2 + \dfrac{VR}{4B} s + 1}{(Ms^2 + K)\left[\dfrac{mV}{A^2B} s^2 + \dfrac{VR}{4B} s + 1\right] + ms\left[\dfrac{K'G}{A} s + \dfrac{VR}{4B} s^2\right]} \quad (14)$$

The characteristic equation becomes

$$\frac{MmV}{A^2B} s^4 + \frac{VR(M + m)}{4B} s^3 + \left[(M + m) + \frac{KmV}{A^2B}\right] s^2 + \left[\frac{KVR}{4B} + \frac{K'Gm}{A}\right] s + K = 0 \quad (15)$$

System 3

In the case of the acceleration feedback system eqs. (2) and (3) are valid but the valve flow rate now becomes

$$q_v = K'G1 X(s) + K'G2s^2 X(s)$$

where $K'G1$ represents the gain of the position feedback loop (originally $K'G$) and $K'G2$ is the acceleration feedback gain constant.

The resulting receptance becomes

$$\frac{X(s)}{F(s)} = \frac{1 + \dfrac{Vm}{A^2B} s^2}{(Ms^2 + K)\left(1 + \dfrac{Vm}{A^2B} s^2\right) + ms\left[s + \dfrac{K'G1 + K'G2}{A} s^2\right]} \quad (16)$$

and the characteristic equation is given by

$$\frac{MmV}{A^2B} s^4 + \frac{mK'G2}{A} s^3 + \left[(M + m) + \frac{VmK}{A^2B}\right] s^2 + \frac{mK'G1}{A} s + K = 0 \quad (17)$$

THE DESIGN OF ACTIVE DAMPING FOR ELECTROHYDRAULIC CYLINDER FEED DRIVES

R. BELL and A. DE PENNINGTON

Machine Tool Engineering Division, Department of Mechanical Engineering,
University of Manchester Institute of Science and Technology

SUMMARY

The mechanical damping of feed drives employing anti-friction bearings is minimal. The absence of effective damping imposes a severe restriction on the use of the full bandwidth of the system. This limitation has a detrimental influence on the design of the control system. This paper gives a design procedure for providing adequate damping at the load by employing acceleration or load pressure feedback for electrohydraulic cylinder feed drives.

NOMENCLATURE

A	effective area of cylinder
B	bulk modulus of the oil
K	minor loop gain constant for generalized analysis
K_A	drive amplifier gain constant
K_{cq}	normalized minor loop gain constant for the coincident complex pole condition
K_p	position loop gain constant
K_v	linearized, unloaded, flow gain of the servovalve
K_a	minor loop feedback gain constant for negative acceleration feedback
K_{ta}	minor loop feedback gain constant for negative transient acceleration feedback
K_1	gain constant relating load velocity to the error signal
K_2	minor loop normalized feedback gain constant
K_3	variable gain constant for the position loop
M	load mass
s	Laplace operator
s_1	time scaled operator
T	transient feedback "differentator" time constant
T_{AV}	the effective time constant for the servovalve and drive amplifier
T_D	transient feedback lagging time constant
T_M	time constant of the torque motor coil
y	output displacement
α	normalized value of T_{AV}
β	normalized value of T_D
ω_l	angular natural frequency of the load
ω_v	angular natural frequency of the servovalve
ξ_{cq}	damping factor of the coincident pole quadratics
ξ_l	damping factor of the load resonance
ξ_v	damping factor of the servovalve resonance

INTRODUCTION

The need to introduce damping into machine tool feed drives is often felt but the solutions that are available are not free from disadvantages. The most obvious solution is to introduce slideway friction by the use of plain slideways. The damping action of plain slideways is, however, a function of the table velocity[1] and additional damping is only introduced at low feed speeds and is dependent on the open loop drive being completely stable (i.e. no stick-slip phenomena). It is found that slideway friction is helpful in certain positioning systems; the ill considered use of this approach can lead to a loss of accuracy or the presence of non-linear oscillations. The use of viscous damping is generally out of the question as the magnitude of the damping constant required is not compatible with the power rating of the feed drive.

The use of a hydraulic cylinder in the feed drive introduces the possibility of adding damping by the use of viscous leakage across the load ports of the servo valve. This technique is inexpensive and effective as it is applied across the oil column which is the most compliant sector of the drive. There are two objections to the use of viscous leakage damping; the first is the inefficiency of the drive and the resultant reduction in loop gain, the second is that it enhances threshold non-linearities. In most cases the leakage flow required is completely excessive but in some systems the technique provides an adequate design solution.

The advantages of antifriction bearings often make their use essential; under these conditions the maximum attainable steady state and dynamic accuracy is influenced by the degree of damping at the load. The use of feedback signals to modify the drive dynamics has been advocated in the past.[2-4] The principal application of these techniques has been in the field of flight control devices. This feedback can either be implemented by hydro-mechanical networks in the servovalve or with the aid of transducers. One limitation to the former is that the majority of electrohydraulic control valves that are commercially available for machine tool systems do not permit the introduction of additional fluid feedback paths. The use of dynamic pressure feedback[4] in the servovalve is possible but the additional cost of tuning the mechanical networks to the system natural frequency requires careful consideration.

The application of transducer feedback has the considerable advantage of maximum flexibility. This paper discusses the effectiveness of negative acceleration feedback and negative transient acceleration feedback to provide adequate damping to feed drives that consist of a four-way, electrohydraulic flow control valve, an equal area cylinder and a table on anti-friction guideways. The generalized analysis is presented in sufficient detail to allow any system to be designed with the aid of this analysis. An example of the application of these techniques and a discussion of a number of practical points is also included.

THE MATHEMATICAL MODEL FOR THE ELECTROHYDRAULIC DRIVE

The mathematical model employed in this paper is linear. The absence of non-linear friction at the load confines the source of non-linearities to the servo-valve and the cylinder seal friction. These non-linearities make it impossible to apply the following analysis to either signals that cause the valve flow to saturate or to the threshold region of the valve. These two factors do not place any practical restriction on the application of this analysis to the design of machine tool feed drives.

It has been shown[5] that the linearized response of the load displacement, $y(s)$, to an error signal $\epsilon(s)$ can be written in the form

$$\frac{y(s)}{\epsilon(s)} = \frac{K_1}{s(1 + sT_M)\,[(s^2/\omega_v{}^2) + 2\xi_v\,(s/\omega_v) + 1]\,[(s^2/\omega_l{}^2) + 2\xi_l\,(s/\omega_l) + 1]} \tag{1}$$

where K_1 is the product of a number of system parameters and $= K_A K_v/A$
the load natural frequency $\omega_l \doteqdot \sqrt{2\beta A^2/VM}$
the damping factor of the load ξ_l in this class of drive is $\gg 0.1$.

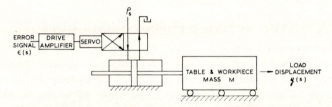

FIG. 1. Schematic diagram of an electrohydraulic cylinder drive.

The other two terms of the transfer function are due to the servovalve. The time constant, T_M of the servovalve torque motor coil is a significant factor. The quadratic lag due to the servovalve is normally of secondary significance, as the natural frequency, ω_V, of a two stage electrohydraulic valve is typically in excess of 1000 rad/sec. Consider a representative example of a cylinder drive with a load natural frequency of 25 Hz and if it is assumed that antifriction slideways are employed, then the load damping factor would be typically of the order of 0.08. The servovalve has a natural frequency of 190 Hz and a damping factor of 0.7. The break frequency of the torque motor coil ($1/2\pi T_M$) is 30 Hz. The open loop poles of this drive are plotted in Fig. 2. This diagram shows the dominant influence of the lightly damped load poles and the relative insignificance of the servovalve quadratic lag.

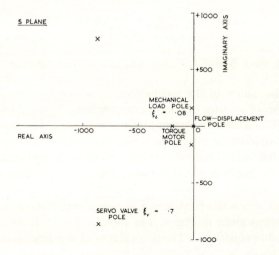

FIG. 2. The open loop pole plot, for a typical electrohydraulic cylinder drive.

In general it has been found possible to reduce the complexity of eq. (1) to the form given below:

$$\frac{y(s)}{\epsilon(s)} = \frac{K_1}{s(1 + sT_{AV})\left[(s^2/\omega_l^2) + 2\xi_l(s/\omega_l) + 1\right]} \tag{2}$$

where the time constant T_{AV}, is introduced. This time constant is then equivalent to the combined influence of the two transfer functions that have been omitted. The evaluation of the time constant, T_{AV}, is discussed later in the paper. The transfer function given in eq. (2) will now be taken to be an adequate representation of the relationship between $y(s)$ and $\epsilon(s)$.†

THE ACTIVE DAMPING TECHNIQUES

It has been shown earlier[5] that either negative acceleration or negative transient acceleration feedback can be used to introduce active damping into lightly damped cylinder servo-mechanisms. The use of velocity signals is to be avoided as they do not improve the dynamics of this type of servomechanism. The acceleration signal can be obtained by using an accelerometer or a differential pressure transducer across the load ports of the servovalve. It should be noted that the pressure transducer must not be allowed to inject a d.c. component into the system as this will reduce the static stiffness of the system.

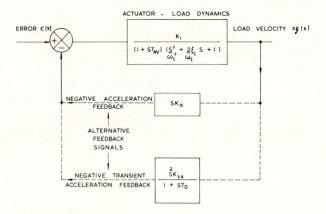

FIG. 3. Block diagram for active damping based on acceleration signal.

The analysis of the action of the acceleration signal on the damping of the system is presented in the form of a series of root locus diagrams. This approach has the merit of presenting the performance of the system under a wide range of parameter variations in a single diagram. It is desirable to confine all further theoretical work to the transfer function relating the error signal, $\epsilon(s)$ to the load velocity, $sy(s)$, as the pole due to the flow-displacement integration at the origin does not influence the damping action which is being examined.

The block diagram which has been found to give a realiable guide to the study of this form of compensation is given in Fig. 3. The valve and load dynamics are quoted in the simplified form stated in equation 2. The two options of negative acceleration and negative

† This simplification may not be adequate to cover the performance of single stage servovalves.[6]

transient acceleration feedback are included. The term K_a is the gain constant of the negative acceleration feedback path and is the product of the gain constants of the accelerometer, its amplifier and any other circuit elements before the summing point at the drive amplifier. The term K_{ta} is the gain constant of the appropriate signal path when negative transient acceleration feedback is employed. The difference between K_a and K_{ta} depends on the manner in which the transient lead term, $sT/(1 + sT_D)$, is incorporated into the system.

The damping of the mechanical load, i.e. the resonant single degree of freedom system formed by the effective stiffness of the cylinder oil column and the total load mass, can be directly assessed by studying the roots of the characteristic equation for the block diagram shown in Fig. 3.

The transfer function of a system with negative acceleration feedback is

$$\frac{sy(s)}{\epsilon(s)} = \frac{K_1}{\{s^3 . (T_{AV}/\omega_l^2) + s^2 [(1/\omega_l^2) + (2\xi_l T_{AV}/\omega_l)] + s[T_{AV} + (2\xi_l/\omega_l) + K_1 K_a] + 1\}} \tag{3}$$

The damping action of negative acceleration feedback can be estimated by examining the damping factor of the revised quadratic lag obtained from the characteristic equation of the transfer function of eq. (3), and then by computing the transient response of the load velocity.

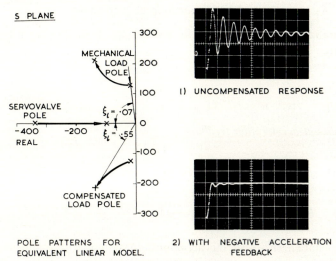

FIG. 4. An example of the use of negative acceleration feedback, to damp the output velocity/error response, of a typical electrohydraulic cylinder drive.

An example of the use of this technique is shown in Fig. 4. This cylinder servomechanism drove a load of 373 lb, had a load natural frequency of 22 Hz and a damping factor of approximately 0·08. The uncompensated, i.e. the open loop, poles are plotted and the transient response of load velocity to an error step input equivalent to 70% of full valve opening is shown. The velocity waveform shows the lack of damping in the roller bearing slideways. The lower transient photograph shows the damping obtained by the use of negative acceleration feedback. The considerable increase in damping is immediately apparent but as the characteristic equation is of third order the damped response does not follow the familiar single degree of freedom response. The root locus plot gives a more detailed appreciation of the action of the feedback signal. The re-location of the poles shows that the revised quadratic lag poles are better damped, ξ_l has been increased from 0·08 to 0·55, and the natural frequency of these poles is now 38 Hz. This trend is also shown in the lower test photograph. The modification of the pole due to the servovalve which is moved nearer the origin should be noted. This pole plays a visible role in determining the system response and a comparison of the photographs of the transients reveals that the leading edge of the compensated system is slower than the uncompensated response.

The example quoted above demonstrates the effectiveness of the technique but it is desirable to simplify the computation of the added damping. The problem of computation can be overcome by generalizing the analysis and producing a number of root locus diagrams which can then be used to design the majority of systems that might benefit from the application of either of these methods of compensation without anything more than a few brief calculations. It is necessary to introduce a number of parameters on which the generalized root locus diagrams can be based.[5] These are

$$a = T_{AV}.\omega_l, \qquad \beta = T_D.\omega_l \quad \text{and} \quad s_1 \equiv s/\omega_l$$

also $K_2 = $ feedback gain constant $= \omega_l K_a$ (negative acceleration feedback)
             $= \omega_l^2 K_{ta}$ (negative transient acceleration feedback)

and $K = K_1 K_2 = $ normalized minor loop gain constant.

The two generalized transfer functions for the actuator-load system with active damping can then be written[5] in the form:

(1) *Negative acceleration feedback*

$$\frac{s_1 y (s_1)}{\epsilon(s_1)} = \frac{K_1/\omega_l}{[s_1^3 . a + s_1^2 . (1 + 2\xi_l a) + s_1 n + t_1 . (a + \alpha\xi_l + K) + 1]} \tag{4}$$

(2) *Negative transient acceleration feedback*

$$\frac{s_1 y (s_1)}{\epsilon(s_1)} =$$

$$\frac{K_1 (1 + s_1\beta)/\omega_l}{[s_1^4.a\beta + s_1^3 (a + \beta + 2\xi_l a\beta) + s_1^2 (1 + 2\xi_l (a + \beta) + \alpha\beta + K) + s_1(a + \beta + 2\xi_l) + 1]} \tag{5}$$

The dynamic characteristics of the actuator-load system with compensating feedback can now be studied in a completely general manner. The uncompensated mechanical load damping factor ξ_l must now be decided. Experience has shown that a representative value for ξ_l is 0·08. The values of ξ_l for particular cases may lie in the range 0·1–0·05 but this variation is of little significance.

The root locus diagram for the minor loop relating output velocity to the error signal shows that with zero loop gain there are two complex mechanical load poles, a negative real pole due to the servovalve (at $-1/a$) and a zero at the origin due to the acceleration sensor. At a particular value of minor loop gain constant there are a modified pair of complex load poles and a negative real pole.

A computed family of root locus curves for the case of negative acceleration feedback is given in Fig. 5. The diagram is in two sections: the upper section shows the influence of α and K on one of the complex conjugate pairs of load poles. The factor $\alpha(= \omega_l T_{AV})$ gives the measure of the interaction between the servovalve lag and the load natural frequency; the factor K defines the gain constant for the minor loop. This figure shows that for values of α up to 0·5 considerable damping can be introduced into the dynamic response of the load. For α values greater than 0·5 the improvement is slight and in the limit the only influence is that of increasing the natural frequency but with little increase in damping. The second sector of the diagram shows the influence of α and K on the positioning of the negative real pole. In all cases the influence of K is to move the pole towards the origin. This effect is not very important for small values of α. However, in the range $\alpha = 1$–5 this pole dominates the transient response and the combined effect is not completely satisfactory.

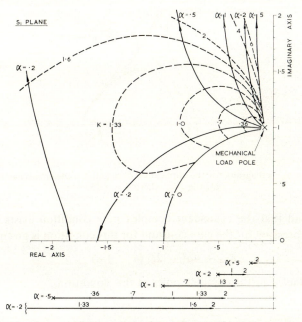

FIG. 5. Root loci for the normalized analysis of the influence of negative acceleration feedback.

The study of the use of negative transient acceleration feedback is a little more involved as the characteristic equation of the transfer function in eq. (5) is a quartic equation and the factors α, β and K have to be included. The zero gain pattern for this minor loop consists of (i) the mechanical load poles, (ii) servovalve pole (at $-1/\alpha$), (iii) a pole due to the lag in the feedback path (at $-1/\beta$) and (iv) a double zero at the origin. The closed loop transfer function consists of four poles (that can be in complex pairs) and one zero due to the feedback lag (i.e. at $-1/\beta$).

A typical case is shown in Fig. 6 for the case where $\alpha = 1·25$; the variables in this diagram are K and β. The single condition which is considered to be the optimum is for the curve where $\alpha\beta = 1$, here when $K = 0·89$ the four poles form two identical complex conjugate pairs, i.e. their damping factors and natural frequencies are the same. This may not be the optimum choice for certain applications. Consider the pole positions when $\alpha\beta = 1$

and $K > 0.89$, both pairs of poles exhibit a reduction in damping factor and the natural frequency of one pair is increased and the natural frequency of the pole pair moving towards the origin is decreased. In the context of this paper the best condition is defined as the coincident pole condition and the damping factor of these quadratics at this point, ξ_{cq}, is an important parameter.

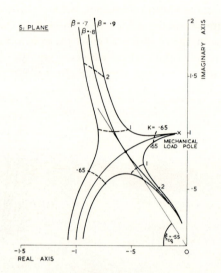

FIG. 6. The root locus diagram, for the application of negative transient acceleration feedback, for the case where $a = 1.25$.

It has been found that the coincident complex pole condition exists in the range $a = 0.285–3.51$. The expression for the gain constant for this condition is given by the expression

$$K_{cq} = 4a^2 - 2\xi_l (a + 1/a) \tag{6}$$

and the damping factor of the complex poles at this condition,

$$\xi_{cq} = \cos (-a) \tag{7}$$

where a is the real part of the coincident pole position. The damping action of negative transient acceleration feedback in the range of values, $a = 0.285–3.51$, can be described in one diagram if the important sector of the locus where $a\beta = 1$ is plotted, i.e. the part of the loci where $K \geqslant K_{cq}$. These curves are drawn in Fig. 7; two important points are shown in this diagram. The more important point is that the natural frequency of the load poles is not reduced by the introduction of optimum damping. It is also shown that the minimum value of ξ_{cq} ($= 0.54$) occurs for the case where $a = 1.0$. The values of K_{cq} and ξ_{cq} for the critical loci in range $a = 0.285–3.51$ are given in Table 1.

It must be stressed that this "optimum" condition may not be considered to be the only effective choice of compensation. The effect of increasing the loop gain, K, is to cause the two pairs of poles to separate, one pair move towards the origin, i.e. its natural frequency decreases, the other pair move away from the origin. The damping of both pairs of poles is not radically modified and therefore it can be concluded that a design based on the coincident pole condition will not be critically dependent on gain.

The coincident pole criterion is only applicable to the range of values $a = 0.285–3.51$. However systems are designed that involve a values outside this range and therefore the design problem for these conditions must be considered. The range where $a < 0.285$ need not be considered as it has been shown[5] that for small a values it is better to use negative acceleration feedback. An example of a high a value is shown in Fig. 8, the ability to introduce load damping is illustrated but the resultant system is relatively slow.

TABLE 1. COINCIDENT POLE DAMPING FACTORS

a value	Coincident quadratic damping factor, ξ_{cq}	Loop gain constant K_{cq}
1·0	0·54	0·845
0·5, 2·0	0·67	1·365
0·4, 2·5	0·77	1·876
0·33, 3·03	0·88	2·561
0·285, 3·509	0·99	3·301

The discussion of the application of negative transient acceleration feedback has been restricted so far to the study of the modification of the damping of the load poles. Attention must now be drawn to the important influence of the closed loop zero of the minor loop due to the lag in the feedback network (T_D, β in the generalized form). When $1 < a < 3.5$ the value of β is < 1 and the zero is sufficiently distant from the origin to exert little influence on the overshoot of the sub-system transient response. When $0.285 < a < 1.0$, the zero due to the lag approaches sufficiently near to the origin to act as a "dynamic magnifier" of the overshoot of the transient response due to the location of the poles. It is concluded, therefore, that negative acceleration feedback is the better technique in the range $0 < a < 1.0$ and that negative transient acceleration feedback should be used in the range $1 < a$. However, if the design of the position loop permits the introduction of a pole

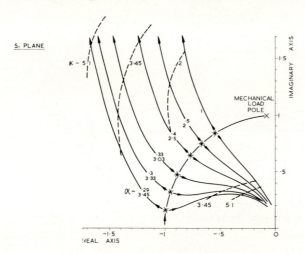

FIG. 7. Coincident complex pole loci, for the dynamics of the normalized minor loop, with negative transient acceleration feedback.

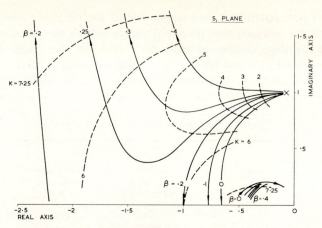

FIG. 8. The root locus diagram, for the application of negative transient acceleration feedback, for the case where $a = 5$.

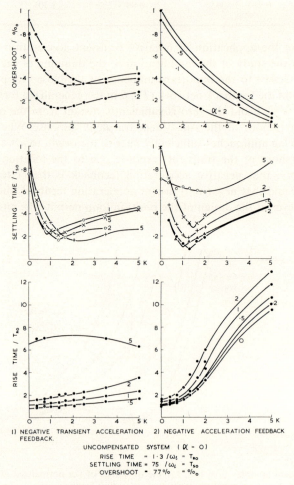

1) NEGATIVE TRANSIENT ACCELERATION   2) NEGATIVE ACCELERATION FEEDBACK
FEEDBACK.

UNCOMPENSATED SYSTEM ($a = 0$)
RISE TIME   = $1 \cdot 3 / \omega_t = T_{RO}$
SETTLING TIME = $75 / \omega_t = T_{SO}$
OVERSHOOT   = $77\% = \%_o$

FIG. 9. Data on the transient response of the linear model, for active damping.

(i.e. a first order lag) to cancel out the zero due to β (i.e. T_D) then the use of negative transient acceleration feedback can be extended to the lower values of α (i.e. to systems where the effective servovalve break frequency $(1/2\pi\,T_{AV})$ is much greater than the drive natural frequency.

The transient response of the linear model with either form of compensation has been computed and the relevant data is given in Fig. 9 in order to show the relative merits of the techniques. The rise time, overshoot and settling time have been normalized to the base of the response of the uncompensated actuator-load system when $\xi_l = 0.08$ and $T_{AV} = 0$. It is possible to design the compensation of the load with the aid of these curves on the basis of its transient response.

An example of these techniques will now be discussed in order to illustrate the suggested design procedure. The system consists of a four way servovalve, (flow rating, 3·65 gpm at a supply pressure of 1000 lbf/in²) which controls a symmetrical cylinder with an effective area of 5·3 in². The load natural frequency of the drive is 27 Hz. The servovalve developed full flow for an error signal of 4 volts. The results of frequency response tests on the servomechanism indicated that T_{AV} (the 45° lag frequency for the valve and drive amplifier response) is $\simeq$ 2·95 msec, the α value for this system is therefore 0·5 and the gain constant $K_1 = 0.66$ in/sec/volt. The load mass is supported on recirculating ball guideways and the load damping is found to be of the order of 0·08.

This information is all that is required to calculate quickly the damping that can be added and the minor loop parameters. As the α value is 0·5 negative acceleration feedback is expected to be the more effective technique. The root loci for $\alpha = 0.5$ given in Fig. 5 indicate that a high value of K will give a very poor response due to the real pole. The selection of $K = 1$ appears to be a good choice as the damping factor of the load poles is increased from 0·08 to 0·42. This choice can be checked by consulting Fig. 9.

The parameters for use of negative transient acceleration feedback are decided by (i) using the coincident pole condition $\alpha\beta = 1$, i.e. $\beta = 2$, and (ii) taking the value of $K = K_{cq} = 1.365$ (from Table 1). Under these conditions the damping of the load poles is increased from 0·08 to 0·67. The influence of the zero is assessed by consulting Fig. 9. The system constants must now be realized from the normalized values.

Case 1. Negative acceleration feedback

$$K = K_1\,K_2 = \omega_l\,K_a\,K_1 = 1.0$$

and
$$K_a = \frac{1.0}{0.66 \times 2\pi \times 27} = 0.0089 \text{ volts/in/sec}^2.$$

Case 2. Negative transient acceleration feedback

$$K = K_1\,K_2 = \omega_l^2\,K_1\,K_{ta} = 1.365$$

and
$$K_{ta} = \frac{1.365}{0.66 \times (2\pi\,.\,27)^2} = 7.2 \times 10^{-5}$$

$$T_D = \frac{\beta}{\omega_l}$$

$\therefore$
$$T_D = \frac{2.0}{0.5/T_{AV}} = 4\,T_{AV} = 11.8 \text{ msec.}$$

1320

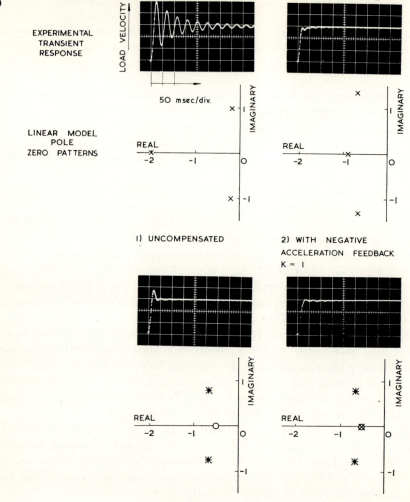

EXPERIMENTAL
TRANSIENT
RESPONSE

LOAD VELOCITY

50 msec/div.

LINEAR MODEL
POLE
ZERO PATTERNS

1) UNCOMPENSATED

2) WITH NEGATIVE
ACCELERATION FEEDBACK
K = 1

3) WITH NEGATIVE TRANSIENT
ACCELERATION FEEDBACK
K = 1·37

4) WITH NEGATIVE TRANSIENT
ACCELERATION FEEDBACK AND
ZERO CANCELLATION K = 1·37.

FIG. 10. Experimental results obtained with a system where $\alpha = 0·5$.

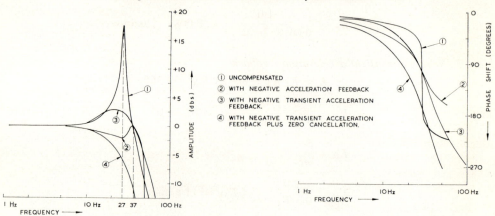

① UNCOMPENSATED
② WITH NEGATIVE ACCELERATION FEEDBACK
③ WITH NEGATIVE TRANSIENT ACCELERATION
FEEDBACK.
④ WITH NEGATIVE TRANSIENT ACCELERATION
FEEDBACK PLUS ZERO CANCELLATION.

FIG. 11. Steady state frequency response curves for system $[sy(s)/\epsilon(s)]$ where $\alpha = 0·5$.

The parameters calculated above were set up in the experimental system; the results obtained from the transient testing of the drive are shown in Fig. 10. Each of these tests was carried out with an input step signal equivalent to 70% of the rated valve flow. In each case the compensation achieved is very effective. The pole–zero pattern of the approximate linear model is included for each case to illustrate the good correlation between the model and the experimental system.

The fourth trace in Fig. 10 is included to show the influence of using zero cancellation. This would be achieved by inserting in the position loop, prior to the minor loop, a first order lag with a time constant of 11·8 msec (i.e. $= T_D$). The waveform for this case is the most damped and shows the merit of lag cancellation where the maximum load pole damping is required without the reduction of the load natural frequency in the intermediate values of α (0·3 → 1·0).

The transient responses in Fig. 10 show that the damping techniques are effective, but it is also desirable to study the equivalent set of steady state frequency responses which are given in Fig. 11. The uncompensated response peaks with a resonant magnification of 17·5 db. In this case negative acceleration is seen to be the most effective technique as the peaking is almost eliminated and the frequency of maximum amplitude has been increased from 27 Hz to 34 Hz a ratio 1·26 (the predicted increase from Fig. 5 is 1·4). The phase gradient at this frequency is reduced and this is a useful factor when the design of the position loop is considered.

PROBLEMS INVOLVED IN THE APPLICATION OF ACTIVE DAMPING TO MACHINE TOOL FEED DRIVES

The technical merit of employing acceleration signals has been explained but a certain number of important practical points must be discussed before a full appreciation of their application can be achieved. It is imperative that the maximum amplitude of the error signal should be restricted to avoid saturation of the servovalve, if consistent damping is to be achieved in the presence of large error signals. The negative feedback signal is ineffective if the servovalve spool is displaced to its extreme of stroke. The drive signal must therefore be limited to a level that just produces the maximum unloaded flow. This involves a simple addition to the circuitry and places no limitation on the electrohydraulic actuator.

The choice of the transducer to provide the damping signal is primarily an economic problem. The acceleration signal sensitivity required for low natural frequencies can present difficulties but additional signal amplification is often an inexpensive addition to the system design. It has been shown earlier[5] that pressure transducers can be employed as an alternative to an accelerometer. The pressure signal must always be decoupled in order to conserve the static stiffness of the servomechanism. The use of pressure transducers with high gain for low natural frequencies can present problems due to the influence of the seal friction component of the pressure signal. The influence of severe load disturbances can cause saturation of the accelerometer signal. Under this condition charge amplifiers employing an electrometer valve input stage possess a very poor rate of recovery and therefore the transducer sensitivity must be selected with care. The current development of FET input stages would appear to eliminate this problem and perhaps cheapen the transducer amplifier.

The evaluation of the effective servovalve time constant, T_{AV}, can only be done experimentally. If there is no interaction between the drive amplifier and the torque motor coil then the coil time constant quoted in the valve manufacturers' data will provide a rough guide. It is necessary to frequency response test the servomechanism in order to determine

T_{AV}. The time constant is calculated by identifying T_{AV} with the frequency of 45° phase lag for the unloaded valve. The frequency response curve must have the contribution of the load quadratic removed prior to this calculation. This procedure is readily justified if a number of drives are to be employed with the same type of servovalve because of the resultant simplification in the design of the servomechanism.

The determination of the α factor infers, in part, the selection of a servovalve. The curves in Fig. 9 show that a wide range of α values can be tolerated. It is desirable to make $1/2\pi T_{AV} > \omega_l$, i.e. $\alpha < 1$, but values greater than unity can be used. The particular case of $\alpha = 1$ is best avoided. The α value is not a fixed parameter; a certain degree of flexibility in system design can be introduced by modifying T_{AV}. This can be inexpensively achieved if servovalves which require low drive current (≈ 10 mA) are used. An example of this approach using current forcing is shown in Fig. 12. This is a servovalve which is widely used in machine tool control systems. The results shown in this figure are for maximum current swing. It is quite practicable to consider reducing the α value by a factor of three and it is therefore a valuable design parameter.

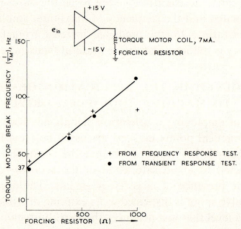

FIG. 12. The influence of current forcing on the frequency response of the torque motor coil of an electrohydraulic servovalve.

THE ROLE OF THE MINOR LOOP IN A POSITION CONTROL SYSTEM

So far the minor loop caused by using an acceleration feedback signal has been discussed in detail; it is now necessary to discuss its role in a position control system. The design of a feed drive servomechanism is, as yet, difficult to specify clearly and there are several contrasting opinions on this point. It is possible to show the value of active damping by a simulated example based on the experimental frequency response data given in Fig. 11. The block diagram for a position control system based on this data is drawn in Fig. 13. The evaluation of the control system consists of (i) designing the position loop parameters with the aid of Bode and Nicols diagrams and (ii) practical adjustment of the parameters for a desired response (this latter stage was simulated).

The first stage in this procedure is summarized in Fig. 14. These three frequency response curves were calculated for the arbitrary choice of a maximum peaking of 2 db for three possible conditions: (i) no compensation employed—system adjustment by gain constant K_3, (ii) the use of negative acceleration feedback as calculated earlier and adjustment of K_3 and (iii) the use of negative acceleration feedback, phase advance compensation and adjust-

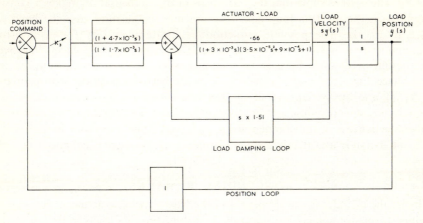

FIG. 13. Block diagram of a possible feed drive design.

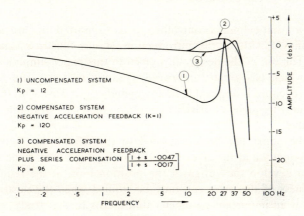

FIG. 14. The frequency response design of a position control system, based on the experimental results given in Fig. 11.

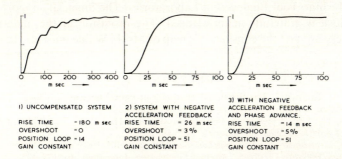

FIG. 15. Simulated transient responses, for a position control system based on the experimental data given in Fig. 11.

ment of K_3. The designs show that the introduction of compensation makes a considerable improvement in the position loop gain constant and in the third case the band width is increased to 37 Hz. The possible reduction in constant velocity following error by a factor of almost 10:1 and a 35% increase in band width are very significant improvements.

These design predictions have to be modified when checked under transient response conditions (see Fig. 15). A choice has to be made between position loop gain constant, K_p $(= K_1 K_3)$, and the system overshoot. The first waveform shows the limitations of the basic system, i.e. slow response and lightly damped load. The other two waveforms demonstrate the effectiveness of the compensation, and even if the overshoot is restricted to less than 5% the dynamic and steady state performance of the drive is enhanced.

CONCLUDING REMARKS

The hydraulic cylinder is the simplest and the least expensive actuator that can be used in a high performance numerically controlled machine tool. Its use is restricted by the compliance of the oil column, and in the extreme the servomechanism specification and the economics of drive design clash. The consideration of the use of active damping on electrohydraulic cylinder drives can extend the usefulness of this class of drive beyond the currently accepted limits.

It has been shown that cylinders controlling loads supported on antifriction bearings can be adequately damped without the introduction of non-linearities and the increase of threshold errors. The use of the design aids given in this paper simplifies the design of the servomechanism which has greatly improved dynamic response and better steady state behaviour. This procedure is applicable to any cylinder feed drive with the usually accepted range of load natural frequencies driven by any commercially available two-stage electrohydraulic servovalve.

The scope of the paper has been restricted to the relationship between the dynamic response of load velocity and the error signal. However, the data made available makes it readily possible to develop the design of the position loop to the required specification. The dynamic stiffness of the drive can only be studied when the position loop has been designed, but this is an important parameter that merits further study.

ACKNOWLEDGEMENTS

The authors wish to thank Professor F. Koenigsberger for being able to carry out this work in the Machine Tool Engineering Laboratories. The financial support of the Science Research Council is gratefully acknowledged. The provision of the material for Fig. 12 by Mr. S. K. Cessford and the considerable help of Mr. P. Walde must also be acknowledged.

REFERENCES

1. R. BELL and M. BURDEKIN, The Frictional Damping of Plain Slideways for Small Fluctuations of the Velocity of Sliding. *Proc. 8th M.T.D.R. Conference*, Manchester, 1967.
2. T. R. WELCH, The Use of Derivative Pressure Feedback in High Performance Hydraulic Servomechanisms. *Trans. A.S.M.E., J. Eng. for Industry*, Feb. 1962, p. 8.
3. A. C. MORSE, *Electrohydraulic Servomechanisms*. McGraw-Hill, 1963, pp. 143–169.
4. ANON. Dynamic Pressure Feedback. *Aircraft Engineering*, June 1960, pp. 171–176.
5. R. BELL and A. DE PENNINGTON, The Active Compensation of Lightly Damped Electrohydraulic Cylinder Drives using Derivative Signals. To be published.
6. Moog Industrial Division Catalogue 740.